The Handbook of Geographic Information Science

Blackwell Companions to Geography

Blackwell Companions to Geography is a blue-chip, comprehensive series covering each major subdiscipline of human geography in detail. Edited and contributed to by the disciplines' leading authorities each book provides the most up to date and authoritative syntheses available in its field. The overviews provided in each Companion will be an indispensable introduction to the field for students of all levels, while the cutting-edge, critical direction will engage students, teachers, and practitioners alike.

Published

The Handbook of Geographic Information Science

Edited by

John P. Wilson

and

A. Stewart Fotheringham

Blackwell
Publishing

BLACKWELL PUBLISHING
350 Main Street, Malden, MA 02148-5020, USA
9600 Garsington Road, Oxford OX4 2DQ, UK
550 Swanston Street, Carlton, Victoria 3053, Australia

First published 2008 by Blackwell Publishing Ltd

2   2008

Library of Congress Cataloging-in-Publication Data

The handbook of geographic information science / edited by John P. Wilson and A. Stewart
Fotheringham.
     p. cm. — (Blackwell companions to geography)
   Includes bibliographical references and index.
   ISBN 978-1-4051-0795-2 (hardback : alk. paper) — ISBN 978-1-4051-0796-9 (pbk. : alk. paper)
1. Geography—Data processing.   2. Geographic information systems.   I. Wilson, John P. (John
Peter), 1955– II. Fotheringham, A. Stewart.

G70.2.H356 2008
910.285—dc22

                                                                        2007008297

A catalogue record for this title is available from the British Library.

Set in 10/12pt Sabon
by Graphicraft Limited, Hong Kong
Printed and bound in Singapore
by Utopia Press Pte Ltd

The publisher's policy is to use permanent paper from mills that operate a sustainable forestry
policy, and which has been manufactured from pulp processed using acid-free and elementary
chlorine-free practices. Furthermore, the publisher ensures that the text paper and cover board
used have met acceptable environmental accreditation standards.

For further information on
Blackwell Publishing, visit our website:
www.blackwellpublishing.com

Contents

Figures

Contributors

Jochen Albrecht
Department of Geography, Hunter College, City University of
New York, 695 Park Avenue, New York, NY 10021, USA.
E-mail: jochen@geo.hunter.cuny.edu

Michael Batty
Centre for Advanced Spatial Analysis, University College London, 1–19
Torrington Place, London WC1E 6BT, UK. E-mail: mbatty@geog.ucl.ac.uk

Allan J. Brimicombe
Centre for Geo-Information Studies, University of East London,
Longbridge Road, Dagenham, Essex RM8 2AS, UK.
E-mail: a.j.brimicombe@uel.ac.uk

James D. Brown
Institute for Biodiversity and Ecosystem Dynamics, Universiteit van
Amsterdam, Nieuwe Achtergracht 166, 1018 WV Amsterdam, The Netherlands.
E-mail: brown@science.uva.nl

Chris Brunsdon
Department of Geography, University of Leicester, University Road, Leicester
LE1 7RH, UK. E-mail: cb179@lei.ac.uk

William E. Cartwright
School of Mathematical and Geospatial Sciences, RMIT University, Melbourne,
Victoria 3001, Australia. E-mail: w.cartwright@rmit.edu.au

Martin E. Charlton
National Centre for Geocomputation, National University of Ireland, Maynooth,
Ireland. E-mail: martin.charlton@may.ie

George C. H. Cho
Division of Health, Design and Science, University of Canberra, Canberra, ACT 2601, Australia. E-mail: george.cho@canberra.edu.au

David J. Cowen
Department of Geography, University of South Carolina, Columbia, SC 29208, USA. E-mail: cowend@gwm.sc.edu

Peter H. Dana
PO Box 1297, Georgetown, TX 78627, USA. E-mail: pdana@pdana.com

Yongxin Deng
Department of Geography, Western Illinois University, Macomb, IL 61455-1390, USA. E-mail: Y-Deng2@wiu.edu

Sara I. Fabrikant
Department of Geography, University of Zurich, Winterthurestrasse 190, CH-8057 Zurich, Switzerland. E-mail: sara@geo.unizh.ch

Peter F. Fisher
School of Informatics, City University, Northampton Square, London, EC1V 0HB, UK. E-mail: pff1@soi.city.ac.uk

Frederico Fonseca
School of Information Sciences and Technology, Pennsylvania State University, University Park, PA 16802-6823, USA. E-mail: fredfonseca@ist.psu.edu

A. Stewart Fotheringham
National Centre for Geocomputation, National University of Ireland, Maynooth, Ireland. E-mail: stewart.fotheringham@may.ie

Mark Gahegan
GeoVISTA Center, Department of Geography, Pennsylvania State University, University Park, PA 16802, USA. E-mail: mng1@psu.edu

John C. Gallant
CSIRO Land and Water, GPO Box 1666, Canberra ACT 2601, Australia. E-mail: john.gallant@csiro.au

Michael F. Goodchild
National Center for Geographic Information and Analysis, Department of Geography, University of California, Santa Barbara, CA 93106-4060, USA. E-mail: good@geog.ucsb.edu

Trevor M. Harris
Department of Geology and Geography, West Virginia University, Morgantown, WV 26506, USA. E-mail: trevor.harris@mail.wvu.edu

Gerard B.M. Heuvelink
Laboratory of Soil Science and Geology, Wageningen University and
Research Centre, PO Box 37, 6700 AA, Wageningen, The Netherlands.
E-mail: gerard.heuvelink@wur.nl

Michael F. Hutchinson
Centre for Resource and Environmental Studies, Australian National University,
Canberra ACT 0200, Australia. E-mail: hutch@cres.anu.edu.au

Geoffrey M. Jacquez
BioMedware, Inc., 515 North State Street, Ann Arbor, MI 48104-1236, USA.
E-mail: jacquez@biomedware.com

Piotr Jankowski
Department of Geography, San Diego State University, 5500 Campanile Drive,
San Diego, CA 92182-4493, USA. E-mail: piotr@typhoon.sdsu.edu

Christopher B. Jones
Department of Computer Science, Cardiff University, Newport Road, PO
Box 916, Cardiff, Wales CF24 3XF, UK. E-mail: c.b.jones@cs.cardiff.ac.uk

Joseph J. Kerski
Environmental Systems Research Institute, Inc., 1 International Court,
Broomfield, CO 80021-3200, USA. E-mail: jkerski@esri.com

Craig A. Knoblock
Information Sciences Institute, University of Southern California, Marina del
Rey, CA 90292, USA. E-mail: knoblock.isi.edu

Brian G. Lees
School of Physical, Environmental and Mathematical Sciences, University of
New South Wales at the Australian Defence Force Academy, Canberra,
ACT 2600, Australia. E-mail: b.lees@adfa.edu.au

William A. Mackaness
School of Geosciences, University of Edinburgh, Drummond Street, Edinburgh,
Scotland, EH8 9XP, UK. E-mail: william.mackaness@ed.ac.uk

David J. Martin
School of Geography, University of Southampton, Highfield, Southampton
SO17 1BJ, UK. E-mail: d.j.martin@soton.ac.uk

Harvey J. Miller
Department of Geography, University of Utah, Salt Lake City, UT 84112-9155,
USA. E-mail: harvey.miller@geog.utah.edu

Ashley Morris
School of Computer Science, Telecommunications, and Information
Systems, DePaul University, Chicago, IL 60604, USA. E-mail:
amorris@cti.depaul.edu

Timothy L. Nyerges
Department of Geography, University of Washington, Seattle, WA 98195-3550,
USA. E-mail: nyerges@u.washington.edu

Ross S. Purves
Department of Geography, University of Zurich – Irchel, Winterthurerstr. 190,
CH-8057 Zurich, Switzerland. E-mail: rsp@geo.unizh.ch

Andreas Reuter
European Media Laboratory, Schloss-Wolfsbrunnenweg 33, 69118 Heidelberg,
Germany. E-mail: andreas.reuter@eml.org

Vincent B. Robinson
Department of Geography, University of Toronto, Mississauga, Ontario L5L
1C6, Canada. E-mail: vbr@eratos.erin.utoronto.ca

Cyrus Shahabi
Integrated Media Systems Center, Viterbi School of Engineering, University
of Southern California, Los Angeles, CA 90089-2561, USA. E-mail:
cshahabi@pollux.usc.edu

Shashi Shekhar
Department of Computer Science and Engineering, University of Minnesota,
Minneapolis, MN 55455, USA. E-mail: shekhar@.cs.umn.edu

André Skupin
Department of Geography, San Diego State University, San Diego,
CA 92182-4493, USA. E-mail: skupin@mail.sdsu.edu

Nicholas J. Tate
Department of Geography, University of Leicester, Leicester LE1 7RH, UK.
E-mail: n.tate@le.ac.uk

David L. Tulloch
Center for Remote Sensing and Spatial Analysis, Rutgers, State University
of New Jersey, New Brunswick, NJ 08904, USA. E-mail:
dtulloch@crssa.rutgers.edu

Ranga Raju Vatsavai
Department of Computer Science and Engineering, University of Minnesota,
Minneapolis, MN 55455, USA. E-mail: vatsavai@cs.umn.edu

Daniel Weiner
Department of Geology and Geography, West Virginia University, Morgantown, WV 26506, USA. E-mail: daniel.weiner@mail.wvu.edu

John P. Wilson
Department of Geography, University of Southern California, Los Angeles, CA 90089-0255, USA. E-mail: jpwilson@usc.edu

May Yuan
Department of Geography, University of Oklahoma, Norman, OK 73019-1007, USA. E-mail: myuan@ou.edu

A-Xing Zhu
Department of Geography, University of Wisconsin, Madison, WI 53706-1404, USA. E-mail: azhu@facstaff.wisc.edu

Alexander Zipf
Department of Geoinformatics and Surveying, University of Applied Sciences FH Mainz, 561156 Mainz, Germany. E-mail: zipf@geoinform.fh-mainz.de

Geographic Information Science: An Introduction

A. Stewart Fotheringham and John P. Wilson

GIS, the acronym for Geographic Information Systems, has been around since the 1980s. Although one can impute a number of characteristics from the use of this acronym, at the heart of the term "systems" lies a computer software package for storing, displaying, and analyzing spatial data. Consequently, the use of the term GIS implies an **object** or tool which one can use for exploring and analyzing data that are recorded for specific locations in geographical space (see Cowen [1988] for an early article articulating this type of definition and Foresman [1998] for a rich and varied account of the history of Geographic Information Systems). Conversely, Geographic Information Science or GI Science, or more simply GISc, represents a much broader framework or *modus operandi* for analyzing spatial data. The term GI Science emphasizes more the **methodology** behind the analysis of spatial data (see Burrough [1986] for what was perhaps the first GIS text to promote such a framework and Chrisman [1999] for an article advocating an extended definition of GIS along these same lines). Indeed, one could define GI Science as: *any aspect of the capture, storage, integration, management, retrieval, display, analysis, and modeling of spatial data.* Synonyms of GI Science include Geocomputation, GeoInformatics, and GeoProcessing.[1]

The Breadth of GI Science

Under this definition, GI Science is clearly an extremely broad subject and captures any aspect connected with the process of obtaining information from spatial data. A feeling for this breadth can be seen in Figure 0.1 which describes a schematic of some of the elements that make up GI Science. At the top level, GI Science is concerned with the collection or capture of spatial data by such methods as satellite remotely sensed images, GPS, surveys of people and/or land, Light Detection And Ranging (LiDAR), aerial photographs, and spatially encoded digital video.[2] The key element here is to capture not only attribute information but also accurate information on the location of each measurement of that attribute. For instance, we might

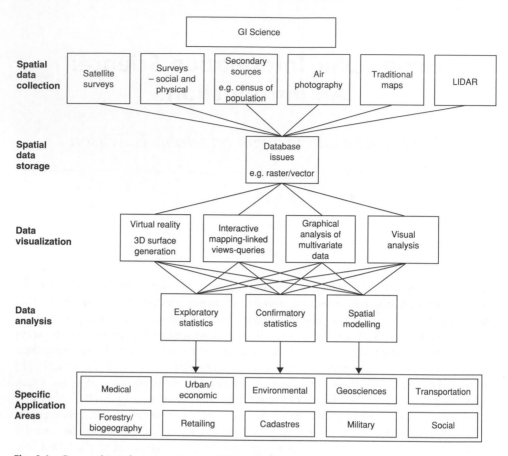

Fig. 0.1 Geographic Information Science (GISc): An Overview

ask people some information on themselves during a survey but we would also like to record some aspect of the location of that individual – this might be the location at which the survey took place, or the person's usual residence or their workplace or some other location. Similarly, if we measure some attribute such as the elevation above sea level and/or the precipitation at a set of points, we also need to know the locations of these points, otherwise the elevation and precipitation measurements are useless (see Corbett and Carter [1996] and Custer, Farnes, Wilson, and Snyder [1996] for two examples of what can be accomplished by combining locational information with elevations and precipitation measurements).

Once spatial data have been captured, they need to be stored and transmitted. This can create challenges as some spatial data sets can be extremely large. A census of population in the USA would contain, for example, records on almost 300 million people with the locational data on each person typically being his or her residence (hence the well-known problem of census data being a snapshot of where people are at midnight rather than during the day). In some countries, a decennial census has been replaced with a more continuous monitoring of the population in the form of a register which can be updated more regularly. Satellite imagery of the Earth's surface can generate terabytes of data and the move towards global data sets can lead to even larger data sets. The data storage and transmission demands of spatially encoded digital video have already been referred to in endnote 2 and pose challenges to current systems. Consequently, spatial data sets can be extremely large and finding ways to store, process, and transmit such large data sets efficiently is a major challenge in GI Science.

The next two levels of operations in the schematic in Figure 0.1 refer to the process of transforming data into information. We are currently living in a data-rich world that is getting richer by the day. In many operations, large volumes of spatial data are being collected and a major challenge in GI Science is to turn these data into useful information. Consider, for example, the following sources of spatial data (which are but a small sample from the complete set of sources):

- Censuses of population which typically occur every five or 10 years and which typically record information on each individual in each household and on the household itself;
- Customer databases held by retail-related companies which hold information on individuals submitted in various application forms or warranty cards;
- Traffic flow monitoring along streets or at intersections;
- LiDAR – low pass fly-overs by plane generating large volumes of detailed data on terrain features or urban areas;
- Digital Elevation Models (DEMs) captured via satellites or the US space shuttle which can be at a global scale;
- Health records either on the location of patients with particular diseases, used to study possible geographic influences on etiology or to assess the level of demand for various services in particular hospitals;
- Satellite remotely-sensed images or aerial photographs used to track land use change over time or to study the spatial impacts of various natural disasters or for various military uses such as tracking missiles or identifying targets;
- Satellite GPS used increasingly for general data capture of vehicles and individuals. This makes possible vehicle tracking, in-car navigation systems, precision agriculture, animal tracking, monitoring of individuals, and general data capture on the location of objects via GPS receivers. It is now possible to contemplate, as the UK is doing, tracking the movement of all vehicles and charging for per mileage road use instead of a flat road tax. Similarly, it is now possible via GPS to monitor a child's movements via a GPS watch linked to a central monitoring system that parents can access remotely via the World Wide Web (see http://www.wherifywireless.com for additional details). The linking of GPS to mobile phones will allow the tracking of friends so that one can query the location of a registered friend at any moment. The use of mobile phones to locate individuals by triangulation from mobile relay stations is already standard police practice in the case of missing persons. Some of these uses of course immediately raise important ethical and legal questions which need to be resolved. Just how much spatial information on ourselves are we prepared to have captured and stored?

Most organizations simply do not have the resources (measured in terms of personnel, knowledge, and/or software) to be able to make full use of all the data they routinely gather. There is a growing need for techniques that allow users to make sense out of their spatial data sets. This mirrors the general transformation of society from one dominated by the industrial revolution with its origins in the eighteenth and nineteenth centuries, to one dominated by the information revolution with its origins in the computer age of the late twentieth century. Two main sets of techniques exist to turn data into information: visualization and statistical/ mathematical modeling.

There is a vast array of techniques that have been developed for visualizing spatial data and it remains a very intense and fruitful area of research (see Dykes, MacEachren, and Kraak [2005] for one such treatment). Spatial data lend themselves to visualization because the data are geocoded and can therefore be represented easily on maps and map-like objects. Simply mapping spatial data can shed so much more light on what is being studied than if the data are presented in tabular form.[3] However, maps can also deceive (see Monmonier [1991] for a popular treatment of this topic) and there are many GI Science issues that need to be considered if spatial data are to be displayed to provide reasonably accurate information content. The development of algorithms for continuous cartograms, software to create pseudo-3D virtual reality environments, and hardware that allows digital video to be linked to a GPS is providing us with the means towards much more sophisticated visualization of spatial data than traditional 2D maps and there are great advances in this area yet to come. Because spatial data contain attribute and locational information, the data can be shown together as on a simple map of the distribution of an attribute or they can be displayed separately in different windows. The use of multiple windows for displaying spatial data is now commonplace and it can sometimes provide useful information to display a map in one window and a non-spatial display of the data (such as a histogram or scatterplot for example) in another window and to provide a link between the two (see GeoDA™ [Anselin, Syabri, and Kho 2006] and STARS [Rey and Janikas 2006] for examples of such systems). In this way, only data highlighted on the non-spatial display need to be mapped to show the spatial distribution of extreme values, for example. Alternatively, all the data can be displayed on a map and only the data points selected on the map need to be highlighted in the non-spatial display. Finally, many spatial data sets are multivariate and it is a major challenge to try to represent such complex data in one display.

The other set of methods to turn spatial data into information are those involving statistical analysis and mathematical modeling. Statistical analysis was traditionally dominated by what is known as "confirmatory" analysis in which a major objective was to examine hypotheses about relationships that were already formed. The typical approach to confirmatory analysis would be to develop a hypothesis about a relationship from experience or the existing literature and to use statistical techniques to examine whether the data support this hypothesis or not. Confirmatory statistical analysis generally depends on assessing the probability or likelihood that a relationship or pattern could have arisen by chance. If this probability or likelihood is very low, then other causes may be sought. The assessment of the role of chance necessitates the calculation of the uncertainty of the results found in a set of data (if we had a different data set, would the results perhaps be substantially different or pretty much the same?). In classical statistical methods, this calculation typically assumes that the data values are independent of each other. A major problem arises in the use of this assumption in spatial data analysis because spatial data are typically not independent of each other. Consequently, specialized statistical techniques have been developed specifically for use with spatial data (see Bailey and Gattrell [1995] for an informative and accessible summary of some of these techniques) and a great deal more research is needed in this area.

More recently, and probably related to the recent explosion of data availability, "exploratory" statistical techniques have increased in popularity. With these, the emphasis is more on developing hypotheses from the data rather than on testing hypotheses. That is, the data are manipulated in various ways, often resulting in a visualization of the data, so that possible relationships between variables may be revealed or exceptions to general trends can be displayed to highlight an area or areas where relationships appear to be substantially different from those in the remainder of the study region. A whole set of localized statistical techniques has been developed to examine such issues (for example, Fotheringham, Brunsdon, and Charlton 2000).

Finally, spatial modeling involves specifying relationships in a mathematical model that can be used for prediction or to answer various "what if" questions. Classical spatial models include those for modeling the movements of people, goods, or information over space and the runoff of rainwater over a landscape. There is a fuzzy boundary between what might be termed a mathematical model and what might be termed a statistical model. Quite often, models are hybrids where a formulation might be developed mathematically but the model is calibrated statistically. Where models are calibrated statistically from spatial data, one important issue is that it is seldom clear that all, or even most, relationships are stationary over space, usually an assumption made in the application of various modeling techniques. For instance, the application of traditional regression modeling to spatial data assumes that the relationships depicted by the regression model are stationary over space. Hence, the output from a regression model is a single parameter estimate for each relationship in the model. However, it is quite possible that some or all of the relationships in the model vary substantially over space. That is, the same stimulus may not provoke the same response in all parts of the study region for various contextual, administrative or political reasons – people in different areas, for example, might well behave differently. Consequently, specialized statistical techniques such as Geographically Weighted Regression (GWR) have been developed recently to allow for spatially varying relationships to be modeled and displayed (Fotheringham, Brunsdon, and Charlton 2002).

The final layer of Figure 0.1 represents some of the application areas of geocomputation which gives an indication of why it is such an important and rapidly growing area of study. Spatial data can be found in most areas of study and include many different types of data, such as:

- Geodetic – coordinate reference systems for locating objects in space;
- Elevation – recording heights of objects above mean sea level;
- Bathymetric – recording the depth of water bodies;
- Orthoimagery – georeferenced images of the earth's surface;
- Hydrography – data on streams, rivers and other water bodies;

- Transportation networks – roads, railways, and canals;
- Communication networks – the transmission of ideas and data across space;
- Cadastral – precise positioning of property boundaries;
- Utilities – the locations of pipes, wires and access points;
- Boundaries – electoral, administrative, school and health districts;
- Medical – the location of incidents of disease and patients with respect to the location of services;
- Crime – the location of police incidents;
- Environmental – habitats, pollution, and the impacts of natural disasters;
- Urban – the location of areas of high priority for social and economic intervention;
- Planning – the spatial impacts of locational decisions;
- Retailing – the location of consumers with respect to the location of services;
- Biogeography – the location of one species with respect to the location of one or more others.

We now turn to a brief discussion of the topics covered in this book. Given the enormous breadth of GI Science, it is clear that not everything can be included in this volume. However, in order to be as comprehensive as possible, we have tried to solicit contributions which have a fairly general application as opposed to being strictly about the use of GI Science in one particular field.

What Follows Next!

The remainder of this book is organized under six headings. We start each section with a brief description of the chapters that follow and the chapters, themselves, offer stand-alone treatments that can be read in any order the reader chooses. Each chapter includes links to other chapters and key references in case the reader wants to follow up specific themes in more detail.

The first group of six chapters looks at some of the recent trends and issues concerned with geographic data acquisition and distribution. Separate chapters describe how the production and distribution of geographic data has changed since the mid-1970s, the principal sources of social data for GIS, remote sensing sources and data, the possibilities of using spatial metaphors to represent data that may not be inherently spatial for knowledge discovery in massive, complex, multi-disciplinary databases, the myriad sources of uncertainty in GIS, and the assessment of spatial data quality.

The second section of the book explores some of the important and enduring database issues and trends. Separate chapters describe relational, object-oriented and object-relational database management systems, the generation of regular grid digital elevation models from a variety of data sources, the importance of time and some of the conceptual advances that are needed to add time to GIS databases, and new opportunities for the extraction and integration of geospatial and related online data sources.

The third section of the book consists of seven chapters that examine some of the recent accomplishments and outstanding challenges concerned with the visualization of spatial data. Separate chapters describe the role of cartography and interactive multimedia map production, the role of generalization and scale in a digital world, the opportunities to display and analyze a variety of geographical phenomena as surfaces, fuzzy classification and mapping in GIS, predictive rule-based mapping, multivariate visualization, and the ways in which digital representations of two-dimemsional space can be enriched and augmented through interactivity with users in the third dimension and beyond.

The fourth section of the book contains three chapters looking at the increasingly important task of knowledge elicitation. These chapters examine the role of inference and the difficulties of applying these ideas to spatial processes along with the process of geographic knowledge discovery (GKD) and one of its central components, geographic data mining, and the prospects for building the geospatial semantic web.

The next group of four chapters examines spatial analysis. The links between quantitative analysis and GIS, spatial cluster analysis, terrain analysis, and dynamic GIS are discussed here.

The six chapters of Part VI examine a series of broader issues that influence the development, conduct, and impacts of geographic information technologies. Separate chapters examine institutional GIS and GI partnering, public participation GIS, GIS and participatory decision-making, several participatory mapping projects from Central America to illustrate the dynamic interplay between conceptions of people and place and the methods used to survey them, the relationship between GIS, personal privacy, and the law across a variety of jurisdictions, and the major developments and opportunities for educating oneself in GIS.

Finally, Part VII examines future trends and challenges. Separate chapters examine the role of the World Wide Web in moving GIS out from their organization- and project-based roles to meet people's personal needs for geographic information, the emergence of location-based services (LBS) as an important new application of GIS, and two views of the challenges and issues that are likely to guide GI Science research for the next decade or more.

Closing Comments

This handbook seeks to identify and describe some of the ways in which the rapidly increasing volumes of geographic information might be turned into useful information. The brief introductions to topics offered in the previous section give some clues as to what we think is important here – the rapid growth in the number and variety of geographic data sets, finding new ways to store, process, and transmit these data sets, new forms of visualization and statistical/mathematical modeling, etc. To the extent that this book has helped to clarify the current state of knowledge and indicate profitable avenues for future research, it will have helped to educate and inform the next generation of geographic information scientists and practitioners. This generation will need to be more nimble than its predecessors given the rapid rate of technological change (innovation) and the tremendous growth of geographic information science, geographic information systems, and geographic information services that is anticipated in the years ahead. With this in mind, we hope the reader will tackle the remainder of the book with an opportunistic and forward-looking view of the world around them.

ENDNOTES

1   In some circumstances, Geomatics also is used synonymously with GI Science as in Geomatics for Informed Decisions (GEOIDE), the Canadian Network Centre of Excellence headquartered at Laval University (www.geoide.ulaval.ca). However, in other circumstances, such as in the naming of academic departments in the UK, the term Geomatics has been used to "re-brand" Departments of Surveying where its scope and purpose are much more restricted.

2   While we are used to seeing orthophotographs, photographs with associated files giving information on the location of each pixel so that operations can be carried out on the spatial relationships within the photograph, spatially encoded digital video allows the user to perform spatial queries and spatial analysis on video images. As one can imagine, the volumes of such data that need to be stored and transmitted create special challenges.

3   See Ian McHarg's 1969 book entitled "Design with Nature" for an influential book that documented how maps could be overlaid and used to evaluate the social and environmental costs of land use change. This book has been reprinted many times and still serves as an important text in many landscape architecture courses and programs.

REFERENCES

Anselin, L., Syabri, I., and Kho, Y. 2006. GeoDa: An Introduction to Spatial Data Analysis. *Geographical Analysis* 38: 5–22.

Bailey, T. C. and Gattrell, A. C. 1995. *Interactive Spatial Data Analysis.* New York: John Wiley and Sons.

Burrough, P. A. 1986. *Principles of Geographical Information Systems for Land Resources Assessment.* New York: Oxford University Press.

Chrisman, N. R. 1999. What does "GIS" mean? *Transactions in GIS* 3: 175–86.

Corbett, J. D. and Carter, S. E. 1996. Using GIS to enhance agricultural planning: The example of inter-seasonal rainfall variability in Zimbabwe. *Transactions in GIS* 1: 207–18.

Cowen, D. J. 1988. GIS versus CAD versus DBMS: What are the differences? *Photogrammetric Engineering and Remote Sensing* 54: 1551–4.

Custer, S. G., Farnes, P., Wilson, J. P., and Snyder, R. D. 1996. A comparison of hand- and spline-drawn precipitation maps for mountainous Montana. *Water Resources Bulletin* 32: 393–405.

Dykes, J. A., MacEachren, A. M., and Kraak, M. J. (eds). 2005. *Exploring Geovisualization.* Amsterdam: Elsevier.

Foresman, T. W. (ed.). 1998. *History of Geographic Information Systems: Perspectives from the Pioneers.* Englewood Cliffs, NJ: Prentice Hall.

Fotheringham, A. S., Brunsdon, C., and Charlton, M. E. 2000. *Quantitative Geography: Perspectives on Spatial Data Analysis.* Thousand Oaks, CA: Sage Publishers.

Fotheringham, A. S., Brunsdon, C., and Charlton, M. E. 2002. *Geographically Weighted Regression: The Analysis of Spatially Varying Relationships.* Chichester: John Wiley and Sons.

McHarg, I. L. [1969] 1994. *Design with Nature* (25th anniversary edn). New York: John Wiley and Sons.

Monmonier, M. 1991. *How to Lie with Maps.* Chicago: Chicago University Press.

Rey, S. J. and Janikas, M. V. 2006. STARS: Space-time analysis of regional systems. *Geographical Analysis* 38: 67–86.

Part I   Data Issues

This first group of chapters looks at some of the recent trends and issues concerned with geographic data acquisition and distribution. The first of these chapters, by David J. Cowen, describes how the production and distribution of geographic data has changed in the past three decades. This chapter walks the reader through the overlapping worlds of the National Spatial Data Infrastructure, Federal Geographic Data Committee (FGDC), framework data, metadata, standards, FGDC clearinghouses, Geospatial One-Stop, and the National Map, and thereby offers a summary of recent developments in the United States where publicly funded geographic data sets have been distributed at little or no cost to potential users for many years.

The second chapter in this group, by David J. Martin, reviews the principal sources of social data for Geographic Information Systems (GIS). The examples demonstrate how conventional administrative, survey, and census-based data sources are becoming increasingly integrated as national statistical organizations look towards data collection strategies that combine elements of each. This integration will probably produce higher spatial and temporal resolution and cross-scale data sets in future years that will, in turn, require entirely new geocomputational tools for effective visualization and/or analysis.

The third of the chapters, by Brian G. Lees, describes the remote sensing sources and data that are commonly used as inputs to GIS. The opportunities for extracting and updating spatial and attribute information in geographic databases from remotely sensed data are examined in some detail, and special attention is paid to the role of error within remote sensing and how insights about the behavior of spatial data in GIS and spatial statistics are feeding back into remote sensing and driving innovation in these rapidly evolving fields.

In the fourth chapter André Skupin and Sara I. Fabrikant explore the possibilities of using spatial metaphors to represent data that may not be inherently spatial for knowledge discovery in massive, complex, multi-disciplinary databases. This area of research is termed "spatialization" and the chapter discusses what kinds of data can be used for spatialization and how spatialization can be achieved. The authors conclude their chapter by noting that spatialization is a new and exciting area in which GI Science is challenged to address important cognitive and computational issues when dealing with both geographic and non-geographic data.

Ashley Morris starts out the fifth chapter by noting that uncertainty permeates every aspect of spatial data (including the assimilation and storage of geospatial features, operations on those features, and the representation of the results of these operations) and goes on to explain why fuzzy object-oriented databases provide a viable and attractive option for modeling uncertainty in spatial databases. These databases provide membership functions that aid in the storage and representation of objects with uncertain boundaries and the focus on features means that they are able to store and represent both vector- and raster-based objects. These features coupled with the use of multiple alpha-cuts provide extensible systems that can support objects with either crisp or ill-defined boundaries at any desired level of detail. The chapter concludes by noting that the storage, manipulation, and representation of objects with uncertain boundaries is likely to become more important as users become more sophisticated.

The final chapter in this group of six, by James D. Brown and Gerald B. M. Heuvelink, focuses on geographic data, it complements the previous chapter and deals with the assessment of spatial data quality. This information is essential if we are to manage social and environmental systems effectively and more generally, for encouraging responsible use of spatial data where knowledge is limited and priorities are varied. This chapter offers an overview of data quality and measures of data quality, the sources of uncertainty in spatial data, and some probabilistic methods for quantifying the uncertainties in spatial attributes. The conclusion notes several challenges that must be overcome if we are to estimate and use information on spatial data quality more effectively in the future.

Chapter 1

The Availability of Geographic Data: The Current Technical and Institutional Environment

David J. Cowen

The need for digital geographic data dates from the earliest computer-based applications in mapping, statistical analysis, and Geographic Information Systems (GIS) in the 1960s. Simply stated, without data there are no GIS applications. The evolution of today's robust GIS market can be directly linked to the availability of high quality data. The fact that the current US$5 billion market for geospatial data and services is expected to expand six fold in the next couple of years suggests that the production and distribution of GIS data to an eager user community has dramatically changed (Gewin 2004). This chapter presents an overview of the technical and institutional environment in which GIS data are made available.

The issues relating to the availability of geographic data have changed significantly over the past three decades. For GIS pioneers the question of data availability was a major concern. This usually meant that users had to wait for large mapping organizations to make the transition from the production of paper maps to the generation of digital representations of features on those maps. Of particular note are the efforts of the US Geological Survey to convert their topographic maps into digital line graphs which created the first nationwide geographic data base (Anderson, Marx, and Keffer 1985, USGS 1989). As these subsequently evolved into the Census TIGER line files (Broome and Godwin 2003) and derivative products, the GIS user community in the USA was provided with several nationwide data sources that were freely distributed. The use of these publicly funded data sources was fostered by a liberal federal data dissemination policy that encouraged federal data creators to "throw the data over the fence" (NRC 1990) and let an eager GIS community determine the best way to incorporate them into their applications. These federal data sources have been supplemented by a robust commercial sector. For example, Geographic Data Technology (http://www.geographic.com/home/index.cfm), a pioneer in commercial GIS data production, recently sold its geographic data assets for more than US$100 million. The combination of public and private data providers has fostered the rapid expansion of the GIS and location-based services market. The fundamental questions regarding the availability of geographic data today often focus on discovery, choices, and the legal and financial environment in which the data exist.

In 2006, a discussion of the availability of geographic data could simply review a series of web sites that provide access to spatial data. In fact, a novice user is quite likely to be overwhelmed by following a Google search for GIS data (June 3, 2006) that yields about 64,700,000 links. Among these links one will find an increasing number of GIS data portals that are maintained by libraries such as the one at Stanford University (http://www-sul.stanford.edu/depts/gis/web.html). A little probing into these sites will reveal the confusing and overlapping worlds of the National Spatial Data Infrastructure, Geospatial One-Stop, the Federal Geographic Data Committee, the National Map, the National Atlas, the Geography Network, the Map Store, the Map Machine, the Demographic Data Viewer, the Open Geospatial Consortium and many other data related sites. Ideally, the Electronic Government initiative to create Geospatial One-Stop (http://www.geo-one-stop.gov/) should satisfy the need for a common search engine for existing GIS data. At the same time hundreds of commercial sites are using address matching capabilities and geographic data visualizations that are accessed by millions of users every day. Unfortunately, a serious user who wants to acquire rather than simply view data soon discovers that the availability of geographic data is couched in a complex milieu of financial, institutional, legal, technical, and even security issues.

General Market Model

Acquiring geographic data involves a successful market transaction that links the producers and consumers. Lopez (1996) provides a useful conceptualization of the geographic data marketplace that exists to handle these transactions (Figure 1.1). On the supply side there is a robust commercial sector that is supplemented by a public spatial data infrastructure. These firms operate in an environment that is much different from the one that existed in the early stages of GIS. Instead of building geographic data by converting existing maps they now rely on high resolution data sources to identify features. Generally, the providers of geographic data include commercial companies that acquire geographic data by direct surveying, recording measurements from airborne- and satellite-based platforms, interpreting and analyzing the raw data through the use of GIS, photogrammetric or image processing techniques.

While these suppliers may provide data directly to users they are often assisted by value-added intermediaries. The value-added intermediaries are analogous to the retail sector. These companies take raw or native data generated by the suppliers and enhance it or provide time and space utility to it. That means that an end user may obtain the data in a more convenient or technically acceptable manner. For example, commercial street centerline vendors may improve accuracy and add additional attributes to data and provide the data on a DVD for a vehicle navigation system. Many federal agencies have established web-based marketplaces where users can acquire data in a variety of formats and prices.

A successful marketplace is one in which transactions provide revenue to the providers and satisfaction to the consumer. It is useful to conceptualize the following steps in such transactions:

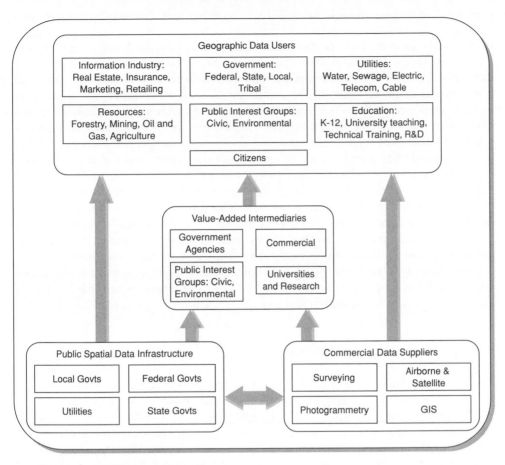

Fig. 1.1 Geographic data marketplace
http://books.nap.edu/books/0309092671/html/42.html

1   Awareness of the need for the data
2   Discovery of the source of the data
3   Understanding of the characteristics and quality of the data
4   Agreement on the price
5   Determination of a means of payment
6   Agreement on the restrictions on use
7   Acquisition of the data
8   Importing of the data into an application

There are obstacles associated with each step in this process. Many users are unaware of the type of data that they may need for a project. For example, the real estate developer may not know that the local county maintains a complete GIS-based multi-purpose cadastre that will provide him or her with valuation, taxation, and zoning information. In order to be effective, suppliers need to advertise the existence of data holdings. Government organizations do not usually have much experience in marketing, therefore there is often a disconnect between potential producers and consumers. Fortunately, an increasing number of organizations are creating metadata that advertises the contents of their data holdings and Internet-based search engines are discovering and serving these "electronic card catalogs" to the public. Many data providers also allow users to preview the data by viewing and querying it through a web-based mapping system. The standardized spatial metadata is a great boon to users who can quickly determine whether the data satisfies their intended needs. Users often find that free and easily accessed data are not of the appropriate scale or resolution or do not include the necessary attributes or reflect the current situation on the ground. The rapid expansion of the GIS data market is based on the realization the relatively small scale databases (1:100,000 or 1:24,000) created by Federal agencies are not appropriate for large scale applications in an urban setting. While the data available from commercial or non-federal public agencies are much more likely to meet user needs they are provided within a set of complex legal and financial covenants. These financial and legal issues often stop the transaction. The methods and media used to physically transfer data from a producer to a consumer have evolved along with the general information technology. Today, these methods include everything from simple download via an FTP site to ordering an external hard drive loaded with gigabytes of digital imagery (USGS, EROS Data Center 2005).

Institutional and Legal Environment

A sign that the GIS marketplace has matured significantly is the recent publication of a NRC report on licensing geographic data and services. The report provides a useful definition of geographic data: "location-based data or facts that result from observation or measurement, or are acquired by standard mechanical, electronic, optical or other sensors" (NRC 2004, p. 24). Surveyor's coordinates, parcel corners, and unprocessed data captured by a sensor platform all fall into the category of raw data. In a legal context courts have decided that such native data are facts and cannot be copyrighted. These unprocessed records may be subject to public disclosure under most Freedom of Information Acts (FOIA). While native geographic data have value to GIS professionals the value-added or preprocessed data is much more of a consumer good. For example, fully attributed georeferenced records in a multipurpose cadastre are extremely valuable products that can be marketed. These derived products can be thought of as "information" rather than "data" and may be protected by a copyright. The NRC (2004, p. 107) panel provides a clear statement on copyright: "Although geographic data equivalent to facts will not be protected by copyright, compilation of geographic data such as databases and data sets, as well as maps and other geographic works that incorporate creative expression may have copyright protection." According to the panel there are two dominant business models that govern the ownership of geographic data (NRC 2004, p. 63). In one case, all rights are sold to the purchaser but the vendor retains the right to use the work. In the other case, the rights are retained by the vendor, but customers are allowed to use the data under a license. In either case it is important to understand the characteristics of digital geographic data. As with digital music it is intangible and can be obtained and used concurrently by many consumers.

Since the government is such an important player throughout the "geographic value chain" the concept of public domain is important. Again the NRC (2004, p. 26) definition is useful: "Public domain information – information that is not protected by patent, copyright, or any other legal right and is accessible to the public without contractual restrictions on redistribution or use." It must be noted that there is a major difference between the way that the federal government and other levels of the public sector operate in the USA in this regard. The federal policy is based on the premise that data derive their value from use and it wishes to actively foster a robust market of secondary and tertiary users. Therefore, the federal model can be summarized as: "although exceptions exist, the general framework provided by the U.S. laws supports the current federal information data policy, which may be summarized as a strong Freedom of Information Act, no government copyright, fees limited to recouping the cost of dissemination and no restrictions on reuse" (NRC 2004, p. 81). This policy is most clearly articulated in OMB Circular A-130 which states that federal agencies will:

1. Avoid establishing, or permitting others to establish on their behalf, exclusive, restricted, or other distribution arrangements that interfere with the availability of information dissemination products on a timely and equitable basis.
2. Avoid establishing restrictions or regulations, including the charging of fees or royalties, on the reuse, resale, or redissemination of Federal information dissemination products by the public.
3. Set user charges for information dissemination products at a level sufficient to recover the cost of dissemination but no higher. They must exclude from calculation of the charges costs associated with original collection and processing of the information (U.S. Office of Management and Budget 2004).

The US federal policy is in stark contrast to much of the rest of the world. For example, in the UK the Ordnance Survey protects its mapping products (and data) by the 1988 Crown Copyright, Designs and Patents Act: "The Ordnance Survey operates under a carefully controlled Licensed Use Schedule and pricing policy that prohibits any use of Ordnance Survey Data which is not expressly addressed in this Licensed Use Schedule under the definition of 'Standard Licensed use' or which is not otherwise expressly permitted by Ordnance Survey is prohibited" (Her Majesty's Stationary Office 1988). Within the USA, the issues regarding ownership and restrictions on use of geographic data at the state and local government levels are extremely fragmented. Faced with limited budgets they have developed a variety of institutional frameworks to facilitate acquisition and sharing of data. These include: (1) ad hoc collaboration; (2) organized collaboration; (3) umbrella organizations; (4) contract work; and (5) agency assessments (NRC 2004, p. 84).

The divergent philosophies of local governments have resulted in a wide range of policies governing the distribution of their data assets. For example, Richland County, South Carolina operates under an ordinance (072-00HR) that states that "geographic information systems (GIS) data elements are distributed to customers in exchange for a data licensing and/or maintenance fee. Upon receipt of GIS data, customers must enter into a nontransferable data license agreement with the County" (Richland County, South Carolina 2004). At the other extreme, Delaware

County, Ohio provides a simple web interface for free downloads without any form of identification or registration.

Unfortunately, local government licensing programs often inhibit the development of the type of federal and local government partnerships that are needed to support programs such as the Census Bureau's TIGER modernization program and the USGS National Map. Consequently, federal taxpayers often have to foot the bill to create a poorer quality version of street centerline files than the ones that already exist in many local governments (Cowen and Craig 2004).

The role of government in the creation and distribution of geographic data is pervasive (Figure 1.2). The NRC (2004) panel provided a useful diagram of the logical downstream and upstream flows of geographic data to and from government as it moves from a data source to secondary and tertiary users. Whereas federal mapping organizations once maintained major in-house mapping capabilities they now rely heavily on contractual arrangements for the outsourcing of the acquisition of native data from the private sector.

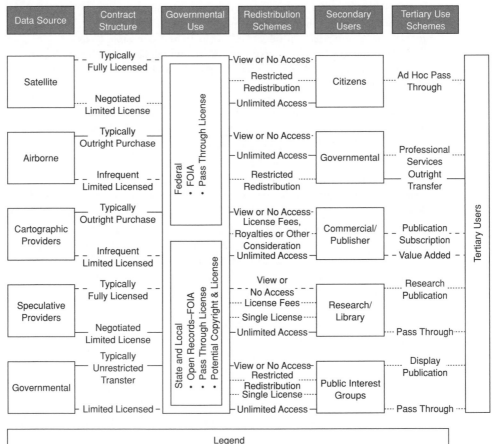

Fig. 1.2 Data flow to and from government
http://print.nap.edu/pdf/0309092671/pdf_image/60.pdf

In this context, a natural tension exists between the firms that produce the data and the public agencies that are the customers. The commercial producer would like to control the use and distribution of data, while under OMB A-130 guidelines the federal government must attempt to obtain outright licenses on data. This arrangement enables agencies to modify the data and place it into the public domain. From the commercial supplier's perspective this type of one-off arrangement limits the number of potential customers and thereby increases the unit cost to the agency. Even though it may purchase the data from a commercial provider the public agency becomes the rightful custodian of the data. As such, it is beneficial to be the sole source for the data. As its custodian the agency can ensure that it is properly maintained and provides proper quality control programs. For example, there needs to be only one official set of Census Bureau TIGER line files for Census processing and reapportionment.

The federal government also encourages reuse of the data and cannot be compensated for more than the marginal cost of distribution. The Internet, inexpensive mass storage, and high speed bandwidth has dramatically changed the data distribution process. Today, instead of responding to specific requests many agencies place their data holdings on an FTP site and enable users to select and download the data.

Standards

Whether we are talking about light bulbs or video tape, a user must be able to acquire a product with the assurance that it is going to work properly. The history of the last two decades of GIS is closely tied to the evolution of spatial data standards. Some of the earliest efforts focused on the need to establish standards that would enable data to be transferred between proprietary software systems. In fact, between 1980 and 1992 the federal government under the direction of USGS worked with the user community to develop the Spatial Data Transfer Standard (SDTS). This comprehensive effort could be viewed as the first serious effort to make geospatial data "more available."

The SDTS provides a neutral vehicle for the exchange of spatial data between different computing platforms. It provides a detailed description of the logical specifications, conceptual model, and spatial object types. The standard recognizes and defines the variety of formats that are involved with spatial data transfer (Figure 1.3). It also includes components of a data quality report and the layout of all needed information for spatial data transfer (Figure 1.4). It contains a catalog of spatial features and associated attributes for common spatial feature terms to ensure greater compatibility in data transfer.

From a policy viewpoint the current version of the SDTS standard (ANSI NCIT 320-1998) is mandatory for federal agencies and an increasing number of organizations distribute their data in this format. The SDTS website maintained by the USGS (http://mcmcweb.er.usgs.gov/sdts/) lists the current status of SDTS implementation and private sector involvement. Furthermore, there is a library of C language functions available for DOS or Unix operating systems which read and write the format used by SDTS. It should be noted, however, that SDTS is an exchange format that must be converted into an operational format supported by the GIS software.

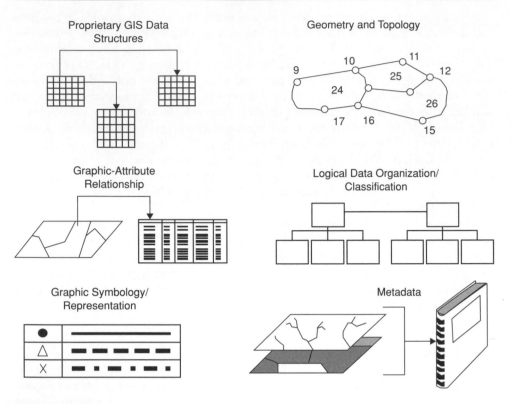

Fig. 1.3 Fundamental issues in spatial data transfer
http://mcmcweb.er.usgs.gov/sdts/training.html

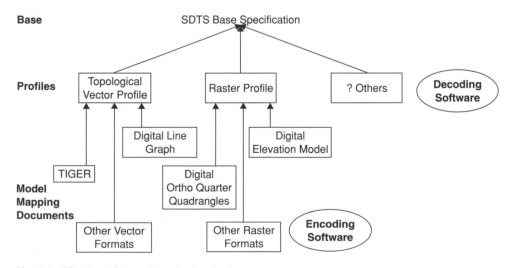

Fig. 1.4 The Spatial Data Transfer Standard
http://mcmcweb.er.usgs.gov/sdts/training.html

Open Geospatial Consortium

In some ways the SDTS effort evolved into the formal establishment of the more encompassing Open Geospatial Consortium (OGC) that includes 312 member organizations. The OGC has the mission: "to create open and extensible software application programming interfaces for GI Systems and other mainstream technologies." Through its efforts a high level of interoperability now exists between data providers and users. These open system specification efforts are particularly important within a web-based environment where users wish to develop applications that incorporate a variety of data that may be held in widely dispersed sites. Such applications may be simple real-time, location-based services or complex decision support systems used by mangers in times of emergency. The OGC has pushed the commercial side of the GIS business to accept open standards that have greatly improved interoperability.

The Federal Geographic Data Committee (FGDC) and National Spatial Data Infrastructure (NSDI)

While the OGC has focused on technical issues required to make data interchangeable between systems other efforts have addressed the institutional obstacles relating to GIS data. Any user soon discovers that there is a wealth of data available and that multiple representations of the same themes often exist. For example, almost every community has at least four versions of street centerlines (Census TIGER, USGS digital line graphs, and two commercial vendors.) These multiple representations vary in accuracy, currency, attributes, price, and scale. Unfortunately, until recently the uninformed user had little assistance in identifying the fitness for use of a particular geospatial data set. Since the problem of duplicative and overlapping data creation efforts was particularly acute within the federal government there has been a concerted effort to improve the way federal agencies operate. Most of these efforts can be encapsulated under the concept of the National Spatial Data Infrastructure (NSDI) which has been closely related to the creation of the National Research Council Mapping Science Committee in 1989. The committee was established to provide: "independent advice to society and to government at all levels on scientific, technical, and policy matters relating to spatial data." The NRC Mapping Sciences Committee defined the NSDI very broadly in a 1993 report as "the materials, technology, and people necessary to acquire, process, store, and distribute such information to meet a wide variety of needs. The committee described the components of the NSDI to be users, policies and procedures, institutional support, people, geographic information and the materials and technology" (NRC 1993).

The NSDI became part of the official federal lexicon in 1994 when President Clinton signed Executive Order 12906 "Coordinating Geographic Data Acquisition and Access: The National Spatial Data Infrastructure." That executive order further defined the NSDI as follows: "NSDI means the technology, policies, standards, and human resources necessary to acquire, process, store, distribute, and improve utilization of geospatial data (Office of the Federal Register 1994).

A critical component of the federal role has been a serious effort to minimize the barriers that inhibit access to federal data and to reduce redundant data collection efforts. One of the most significant steps in the process occurred in 1990 when the Office of Management and Budget issued Circular A-16 that provided "direction for federal agencies that produce, maintain or use spatial data either directly or indirectly in the fulfillment of their mission." OMB also used A-16 to establish the Federal Geographic Data Committee (FGDC). In effect, the FGDC was established to oversee the development of the NSDI.

One of the first tasks of the FGDC was to establish categories of geospatial data and to develop communities of interest for different thematic types of geospatial data. Its efforts resulted in the following subcommittees that represent the taxonomy of geospatial data: (1) base cartographic; (2) cadastral; (3) cultural and demographic; (4) geodetic; (5) geologic; (6) ground transportation; (7) international boundaries; (8) soils; (9) vegetation; (10) spatial water; (11) wetlands; (12) marine and coastal spatial data; and (13) spatial climate. These subcommittees have been overseeing the development of standards that facilitate the exchange of data and provide the community of GIS data users with a clear definition of terms and specifications. Table 1.1 lists the current standards that have been finalized by the FGDC, and

Table 1.1   Final Stage – FGDC Endorsed Standards (January 24, 2004)
http://www.fgdc.gov/standards/status/textstatus.html

Content Standard for Digital Geospatial Metadata (version 2.0), FGDC-STD-001-1998
Content Standard for Digital Geospatial Metadata, Part 1: Biological Data Profile,
    FGDC-STD-001.1-1999
Metadata Profile for Shoreline Data, FGDC-STD-001.2-2001
Spatial Data Transfer Standard (SDTS), FGDC-STD-002
Spatial Data Transfer Standard (SDTS), Part 5: Raster Profile and Extensions,
    FGDC-STD-002.5
Spatial Data Transfer Standard (SDTS), Part 6: Point Profile, FGDC-STD-002.6
SDTS Part 7: Computer-Aided Design and Drafting (CADD) Profile, FGDC-STD-002.7-2000
Cadastral Data Content Standard, FGDC-STD-003
Classification of Wetlands and Deepwater Habitats of the United States, FGDC-STD-004
Vegetation Classification Standard, FGDC-STD-005
Soil Geographic Data Standard, FGDC-STD-006
Geospatial Positioning Accuracy Standard, Part 1, Reporting Methodology,
    FGDC-STD-007.1-1998
Geospatial Positioning Accuracy Standard, Part 2, Geodetic Control Networks,
    FGDC-STD-007.2-1998
Geospatial Positioning Accuracy Standard, Part 3, National Standard for Spatial Data
    Accuracy, FGDC-STD-007.3-1998
Geospatial Positioning Accuracy Standard, Part 4: Architecture, Engineering Construction
    and Facilities Management, FGDC-STD-007.4-2002
Content Standard for Digital Orthoimagery, FGDC-STD-008-1999
Content Standard for Remote Sensing Swath Data, FGDC-STD-009-1999
Utilities Data Content Standard, FGDC-STD-010-2000
US National Grid, FGDC-STD-011-2001
Content Standard for Digital Geospatial Metadata: Extensions for Remote Sensing
    Metadata, FGDC-STD-012-2002

the extent of these standards efforts points to the complexity of the issues surrounding geospatial data.

Framework data

The FGDC also recognized the need for a common geographic base to create and maintain additional thematic layers. The FGDC foundation or backbone consists of the following layers: (1) geodetic control; (2) orthoimagery; (3) elevation; (4) transportation; (5) hydrography; (6) governmental units; and (7) cadastral information. Ideally, framework data provides a common base at a sufficiently high level of resolution and accuracy that any thematic layer based on the framework should be permanently maintained with a high degree of confidence. It should be noted that there is considerable debate about how to establish and fund partnerships to develop the framework layers. As noted previously, the issues surrounding federal and local government cooperation can be very perplexing.

Metadata

Arguably, the most successful program of the FGDC has been the promulgation of The Content Standard for Digital Geospatial Metadata. The need for a metadata standard was included in the 1994 executive order and became a cornerstone of federal policy over the past decade. To the geospatial data community metadata is analogous to a library's card catalog. The benefits of metadata are listed by FGDC (2005) as

1. Organize and maintain an organization's internal investment in spatial data.
2. Provide information about an organization's data holdings to data catalogues, clearinghouses, and brokerages.
3. Provide information to process and interpret data received through a transfer from an external source.

The detailed specification for creation of metadata is spelled out in great technical detail in several FGDC standards (Figure 1.5). The major elements of metadata are: (1) identification information; (2) data quality information; (3) spatial data organization information; (4) spatial reference information; (5) entity and attribute information; (6) distribution information; (7) multi-use sections; and (8) extensibility.

Metadata provides a user with the necessary information to make an informed decision about whether an available data set is appropriate for use in an application. Producers and consumers of geographic data have greatly benefited from this "truth in advertising" approach and software tools to create and maintain metadata as XML files are now commonplace.

FGDC security concerns

Access to spatial data is also impacted by maters of national security. In fact, after the events of September 11, 2001 many of the sources of spatial data that had been freely distributed were quickly retracted:

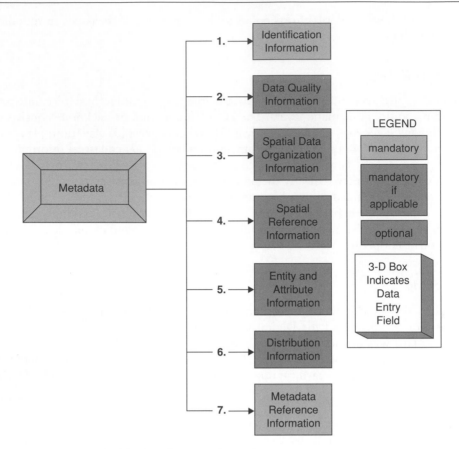

Fig. 1.5 Content standard for digital geospatial metadata
http://biology.usgs.gov/fgdc.metadata/version2/

After September 11, individual federal organizations withdrew some of their geospatial information that had been previously available to the public via agency websites and printed documents. These initial decisions were made under conditions of time pressures and without much top-level guidance. However, even under the best circumstances, several factors complicate the decisionmaker's task of determining which information sources have significant homeland security implications and, if so, whether some type of restrictions on public access are necessary (Baker, Lachman, Frelinger, et al. 2004).

When the GIS data resources were analyzed, the Rand group found that only six percent of the 629 federal data sets were judged to be potentially useful to attackers and was both useful and unique. These and other actions prompted the FGDC to establish The Homeland Security Working Group. Security, foreign policy, law enforcement, and privacy issues represent major challenges to policy makers considering geographic data access issues. There are difficult decisions relating to the proper balance between potentially harmful or intrusive uses and legitimate uses (see Chapter 29 by Cho, this volume, for a more detailed discussion of GIS, personal privacy, and legal issues). Therefore, the FGDC Homeland Security Working Group

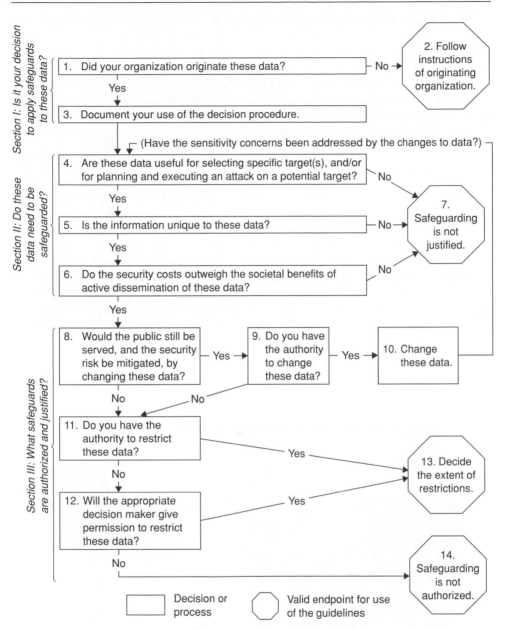

Fig. 1.6 FGDC decision tree for providing appropriate access to geospatial data in response to security concerns
http://www.fgdc.gov/fgdc/homeland/revised_access_guidelines.pdf

(2004) developed a set of guidelines for providing appropriate access to geospatial data in response to security concerns (Figure 1.6). This flowchart suggests that "blanket restrictions and classification on national security or law enforcement grounds are inadvisable except in unambiguous cases."

Under the Critical Infrastructure Information Act of 2002 the Department of Homeland Security has developed its own program to assemble important data from utility companies and local governments. This Protected Critical Infrastructure Information (PCII) Program solicits potentially sensitive information from private and other sources and if the information qualifies for protection it restricts the distribution. The PCII Program, creates a new framework which enables members of the private sector to, for the first time, voluntarily submit sensitive information regarding the nation's critical infrastructure.

FGDC clearinghouses

Organizations around the world have established their own methods for distributing geospatial data. In fact, a web search for "GIS clearinghouse" performed by the author on June 3, 2006 generated a list of more that 1,500,000 links. Most of these sites do not conform to a standard set of guidelines and are not linked into a comprehensive network. In contrast to these *ad hoc* sites, the FGDC has fostered the creation of official Geospatial Data Clearinghouses that follow a set of rigorous standards. The success of these official clearinghouses is directly linked to the widespread acceptance of the FGDC metadata standards and their adoption by public agencies (Figure 1.7). The key to the success of these clearinghouses is the ability to discover and acquire geospatial data by harvesting metadata. Fortunately, the library community developed an ANSI/ISO protocol for network-based search and retrieval of such metadata. At the current time this standard is

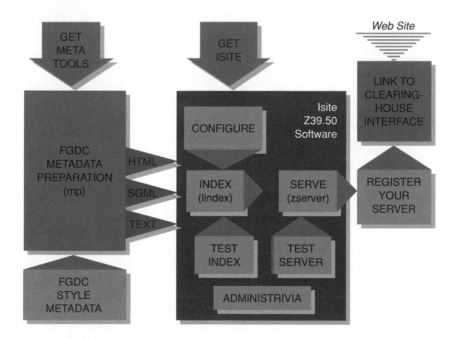

Fig. 1.7 FGDC steps to create a clearinghouse
http://www.fgdc.gov/clearinghouse/tutorials/imagetour.html

Table 1.2   NSDI Geospatial Data Clearinghouse Topics

Administrative and Political Boundaries
Agriculture and Farming
Atmospheric and Climatic Data
Base Maps, Scanned Maps, and Charts
Biologic and Ecologic Information
Business and Economic Information
Cadastral and Legal Land Descriptions
Earth Surface Characteristics and Land Cover
Elevation and Derived Products
Environmental Monitoring and Modeling
Facilities, Buildings, and Structures
Geodetic Networks and Control Points
Geologic and Geophysical Information
Human Health and Disease
Imagery and Aerial Photographs
Inland Water Resources and Characteristics
Ocean and Estuarine Resources and Characteristics
Society, Cultural, and Demographic Information
Tourism and Recreation
Transportation Networks and Models
Utility Distribution Networks

the Z39.50 protocol. By adopting this standard it was possible for the FGDC to establish geospatial data clearinghouses that could be assembled under a loosely structured federation (Figure 1.7). It should be noted that while organizations must provide metadata that accurately describes the data this does not ensure that the actual data is current or accurate.

Throughout the world there are more than 250 clearinghouse nodes which advertise that they have geographic data that they are willing to share. These nodes can be assessed through six FGDC clearinghouse gateways as follows: (1) Alaska Geographic Data Committee; (2) EROS Data Center; (3) FGDC; (4) NOAA Coastal Services Center; (5) Natural Resources Conservation Service (NRCS); and (6) ESRI. Each of the gateways provides a link to determine the status of the network of clearinghouses. They also utilize a standard NSDI search wizard to "smart select" servers and data. The first step in this search process involves the selection of a topic of interest from a list of themes (Table 2.2).

Geospatial One-Stop (http://www.geo-one-stop.gov/)

As noted above, in an ideal world there would be a one stop shop for geographic data. This is exactly the goal of Geospatial One-Stop (GOS) which is one of the President's Electronic Government (E-GOV) Initiatives in the USA. In the broadest sense GOS has been established as a web-based government gateway to the geographic data

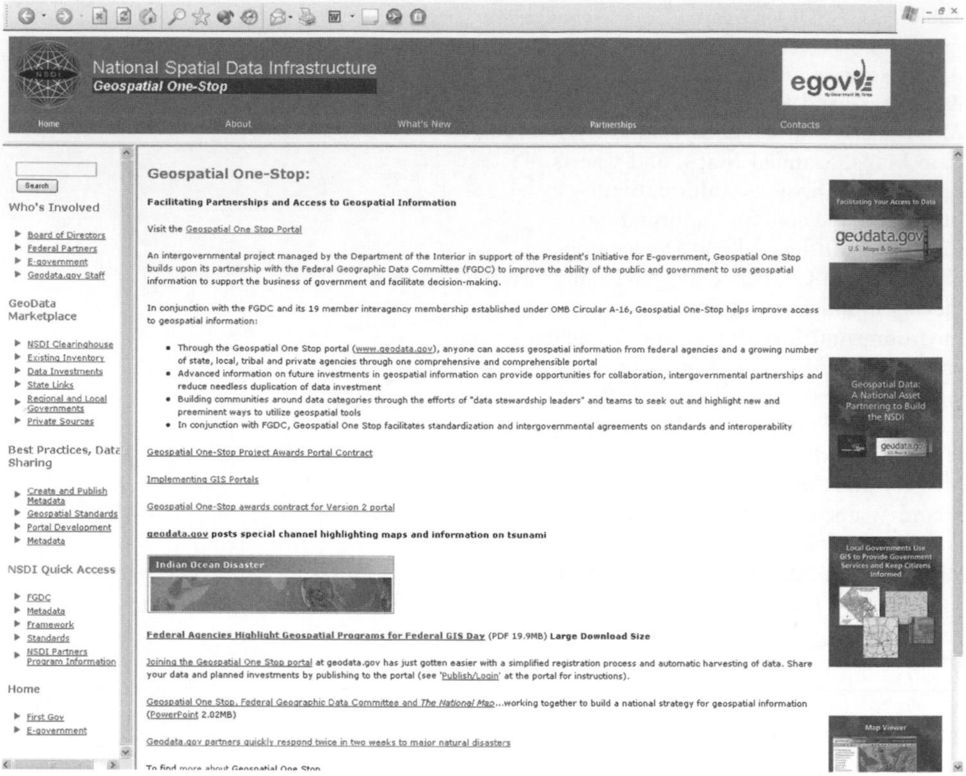

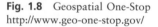

Fig. 1.8 Geospatial One-Stop
http://www.geo-one-stop.gov/

marketplace. GOS is a voluntary system that has been established to encourage organizations to publish geographic content such as maps, data and geographic activities, or events.

GOS provides direct links to a wide range of public sector clearinghouses (Figure 1.8). There are links to every state level clearinghouse, and private companies may also advertise their sites. An important function of GOS is to serve as a site for information on future investments in geospatial information. Posting details on this site about future data capture investments is designed to provide opportunities for collaboration such as intergovernmental partnerships.

While GOS represents a major step forward in providing a uniform starting point for locating geographic data it does not guarantee a consistent result and may not be the best way to locate desired data. For example, someone looking for wetlands data will be rewarded with a robust site for accessing the National Wetlands Inventory maintained by the US Fish and Wildlife Service (Figure 1.9). Other searches through the GOS portal will not result in such positive outcomes. For example, by following a link to Cultural, Society and Demographic data and the subcategory Law Enforcement one gets directed to the general FBI web site with no obvious links to any GIS data or mapping sites.

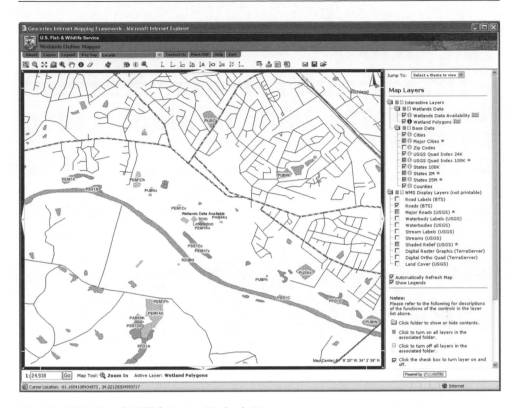

Fig. 1.9 US Fish and Wildlife Service Wetlands Mapper
http://www.fws.gov/data/IMADS/index.htm

The National Map (http://www.nationalmap.usgs.gov)

The USGS initiative to create a National Map is intended to complement the FGDC and Geospatial One-Stop (Figure 1.10). In fact, these three entities are organized within the National Geospatial Programs Office (NGPO) under the leadership of an Associate Director for Geospatial Information (ADGI) and Chief Information Officer.

The vision for the National Map is to create nationally consistent layers of high resolution digital orthoimagery, elevation and bathymetry, hydrography, transportation, structures, boundaries of government features, geographic names, and land cover. More importantly, the USGS envisions that all of these data themes will form seamless resources that are complete, consistent, integrated, and current. Technically, The National Map is a distributed series of web servers that cooperate "to coordinate and negotiate access to their data, develop protocols for data integration, develop data maintenance processes, and define data requirements." For example, North Carolina has established a state-level program (NC One Map; see http://www.nconemap.com/ for additional details) that serves as an umbrella for state, county, and local government partnerships. Therefore, as one interactively

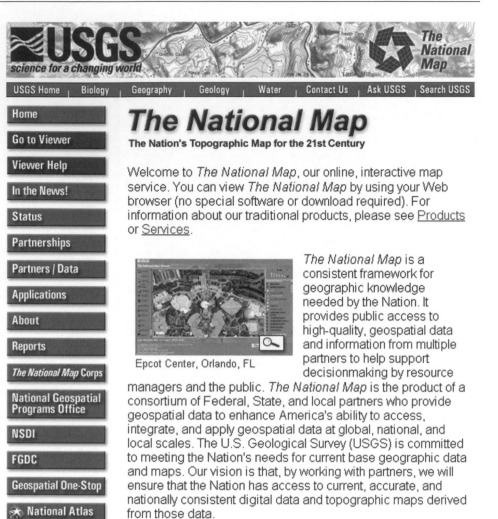

Fig. 1.10 The National Map
http://nationalmap.usgs.gov/

pans and zooms across North Carolina, the user is informed about several state and local governments that have agreed to share their data.

The National Map provides extensive web-based mapping and data discovery tools. One can view the data (quilt patches) that have been contributed by local partners. In some cases this includes high resolution orthophotography, building footprints, and even parcel boundaries. An important aspect of The National Map Viewer is that it supports download of some vector features and raster images.

The National Map also provides a link to the Seamless Data Distribution System (SDDS; Figure 1.11) that is maintained by the EROS Data Center to provide an online map interface to view the data sets listed in Table 1.3 that are available for download or media delivery.

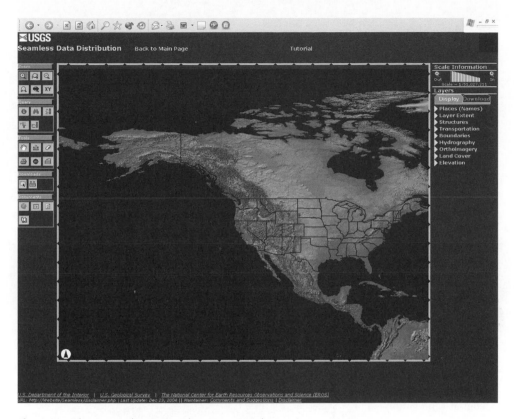

Fig. 1.11   The National Map Seamless Data Distribution System Viewer
http://seamless.usgs.gov/website/seamless/viewer.php

Table 1.3   USGS Seamless Data Distribution System

National Elevation Data set (NED) 1 Arc Second (~30 m resolution)
National Elevation Data set (NED) 1/3 Arc Second (~10 m resolution)
National Elevation Data set (NED) 1/9 Arc Second (~3 m resolution)
National Landcover Characterization Data set (NLCD)
Shuttle Radar Topography Mission (SRTM) 1 Arc Second (~30 m resolution) US
   Elevation Data set
Shuttle Radar Topography Mission (SRTM) 3 Arc Second (~90 m resolution) Global
   Elevation Data set
High Resolution Orthoimagery
1 meter Orthoimagery (limited areas)
Moderate Resolution Imaging Spectroradiometer (MODIS) Direct Broadcast US
   Normalized Difference Vegetation Index (NDVI) 7-day Composites
Bureau of Transportation Statistics (BTS) Roads Vector Data

Table 1.4 USGS Earth Explorer Data Sources

Satellite Imagery
  Advanced Very High Resolution Radiometer
  Declassified Satellite Imagery – 1 (1996)
  Declassified Satellite Imagery – 2 (2002)
  EO-1 Advanced Land Imager
  EO-1 Hyperion
  ETM+ (Landsat 7, June 1999–May 2003)
  ETM+ SLC-off (Landsat 7, July 2003–present)
  Landsat Orthorectified TM Mosaics
  MSS (Landsat 1–5, July 1972–October 1992)
  SPOT (Search Only)
  TM (Landsat 4–5, July 1982–present)

Aerial Photography
  Digital Orthophoto Quadrangles
  Digital Orthophoto Quadrangles – County
  National Aerial Photography Program (1987–present)
  National High Altitude Photography (1980–9)
  Space Acquired Photography
  Survey Photography
  USGS High Resolution Photography

Digital Line Graphs
  Digital Line Graph – 1:100,000 scale
  Digital Line Graph – Large Scale

Elevation
  Digital Elevation Model – 15 Minute
  Digital Elevation Model – 30 Minute
  Digital Elevation Model – 7.5 Minute
  National Elevation Data set (Predefined Areas)

Maps (Related Links)
  Digital Raster Graphics
  National Atlas of the United States

There is a wide range of options for acquiring data from the SDDS. These options range from free downloads of small amounts of data to the purchase of a 250 GB hard drive loaded with data for US$950 (Table 1.4).

The National Atlas

In 1997 the USGS provided some of the first web-based mapping functions with the National Atlas (Figure 1.12). For maps and data at a scale of about 1:2,000,000

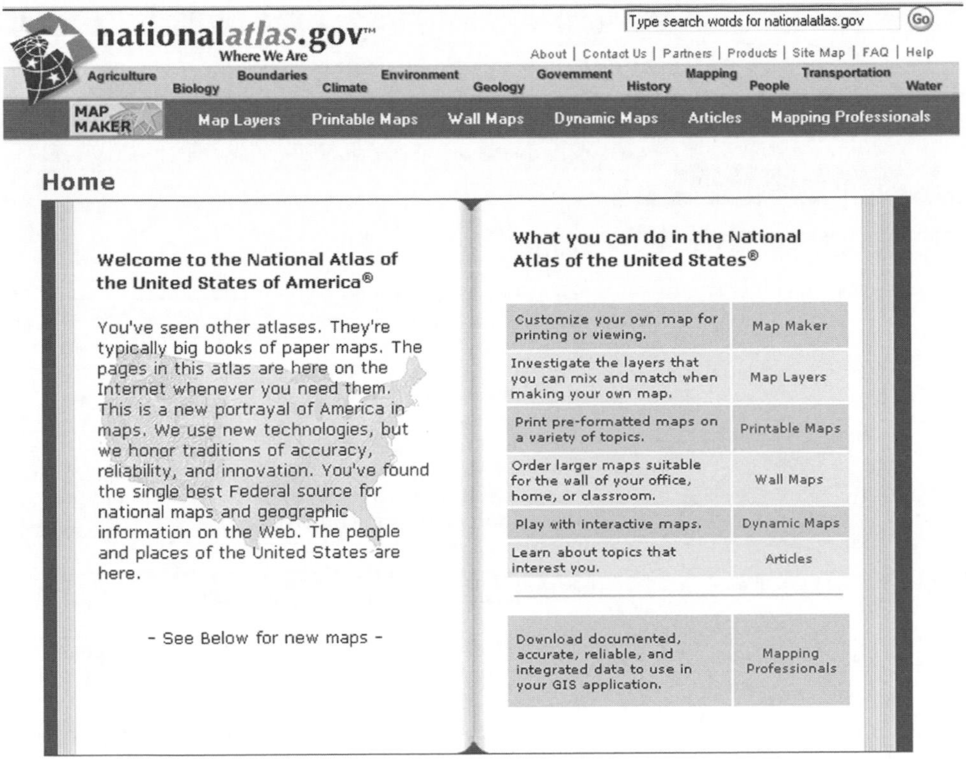

Fig. 1.12 The National Atlas
http://www.nationalatlas.gov/

the National Atlas provides an extremely useful way to discover, view, and even download data. The National Atlas is a useful national and regional atlas that has evolved into a source for acquiring spatial data.

ESRI

It would be remiss to discuss the current status of geographic data availability without including Environmental Systems Research Institute, Inc. (ESRI), an important worldwide commercial supplier of GIS software and services. ESRI serves as one of the six NSDI Gateways to Official NSDI Data Clearinghouses. It also operates the Geography Network (http://www.geographynetwork.com/) that is a NSDI clearinghouse node. In a dynamic web environment ESRI provides several ways to access the Geography Network (Figure 1.13). They provide the following categories of data access: (1) downloadable data; (2) dynamic data and maps; (3) offline data; and (4) clearinghouses. ESRI and other companies are providing a new form of web services to users. These services do not require the user to acquire specialized application software.

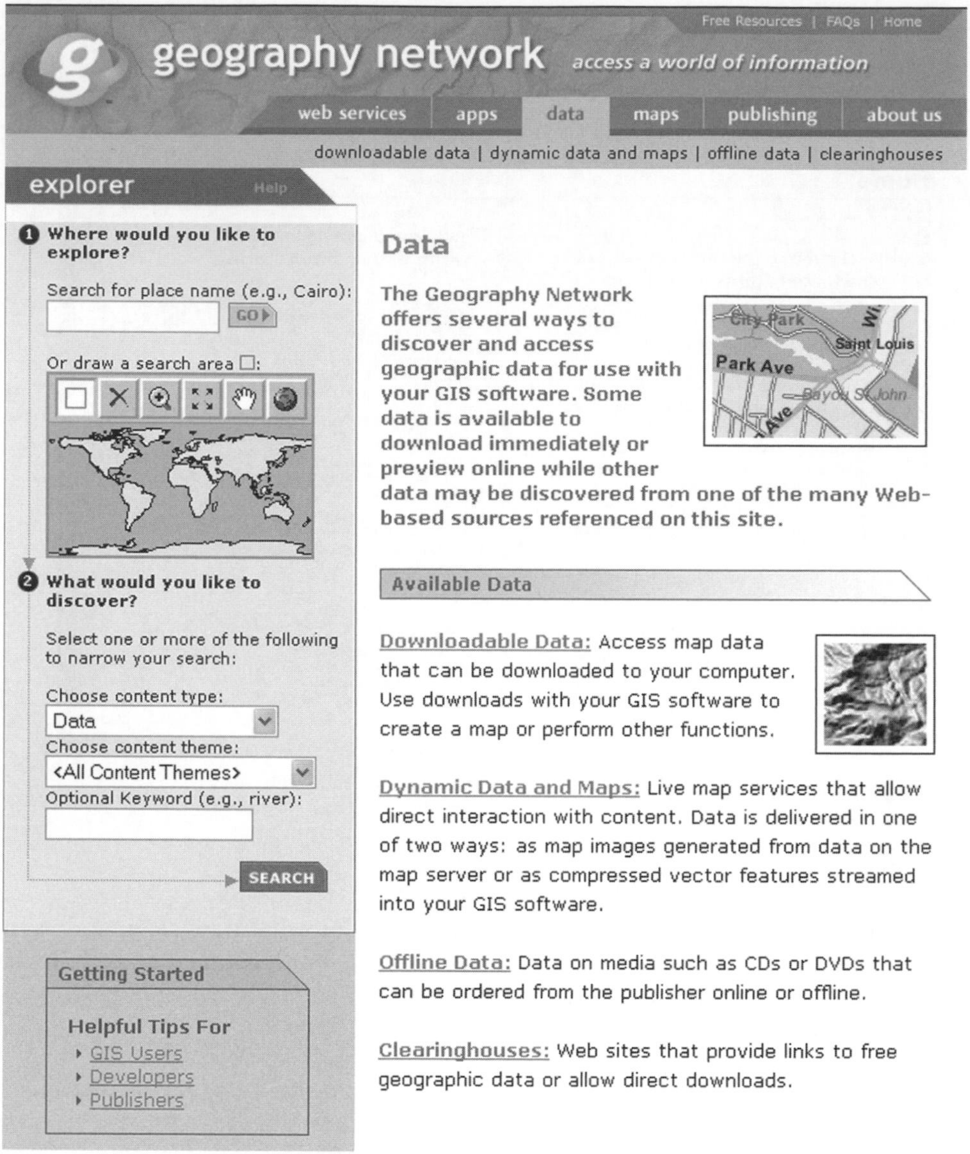

Fig. 1.13 The Geography Network
http://www.geographynetwork.com/

Other Initiatives

An additional aspect of geographic data availability involves a group of organizations that advocates various positions regarding the creation and distribution of geographic data. The data producer side is represented by the Management Association for Private Photogrammetric Surveyors (MAPPS). MAPPS is an association of firms in the surveying, spatial data, and geographic information systems fields. The objective of MAPPS is to "promote the business interests of the profession. Whether it is fighting unfair competition by government, universities or non-profit entities, or promoting qualifications-based selection, MAPPS enhances the ability of its member firms to participate in our great free enterprise system: the business of MAPPS is the business of maps" (see http://www.mapps.org/ for additional details).

An interesting organization on the consumer side of the business is the Open Data Consortium (ODC) which advocates the following principles:

1   Public information is a necessary component of the democratic process and open government;
2   The value of geospatial data is realized through its usage;
3   Widespread distribution and use of public geodata benefits the data steward's entire jurisdiction;
4   Public agencies increasingly store data electronically, and such digital data constitutes the public record;
5   In their roles as data custodians, public agencies have a responsibility to make data available both for citizen access, and to reduce duplication of effort among public agencies;
6   Public agencies need funding to develop, maintain, and distribute their data.

The fact that organizations such as MAPPS and ODC exist suggests that the issues surrounding the availability of geographic data are quite contentious and will continue to be in the future.

CONCLUSIONS

Over the past three decades an extremely robust industry has emerged that relies on a steady supply of geographic data that serve a growing number of public and private organizations that require accurate, current, and reliable information about the conditions of features on the Earth's surface. The expanding demand for geographic data will be fueled by location-based services that are ported to an ever increasing set of spatially aware devices (see Chapter 32 by Brimicombe, this volume, for additional details). The utilization of these data resources has the potential to improve the level of decision making and planning throughout society. They should enable companies in the private sector to be more efficient and responsive. The proper use of the data in the public sector will make government more accountable and more equitable in the way it protects the health and safety of its citizens. While we have witnessed exciting technical developments in the way we gather, use, and distribute geographic data we face many obstacles that often inhibit and frustrate the user community. There are several efforts such as Geospatial One-Stop and The National Map that are addressing the problems with locating, acquiring, and using geographic data.

REFERENCES

Anderson, K., Marx, R., and Keffer, G. 1985. A prospective case for a national land data system: Ten years later. In *Proceedings of the Fifty-first ASP – ACSM Annual Meeting*, Washington, DC, USA.

Baker, J., Lachman, B., Frelinger D., O'Connell, K., Hou, A., Tseng, M., Orletsky, D., and Yost, C. 2004. *Mapping the Risks: Assessing the Homeland Security Implications of Publicly Available Geospatial Information.* Santa Monica, CA: Rand National Defense Research Institute (available at http://www.rand.org/pubs/monographs/2004/RAND_MG142.pdf).

Broome, F. R. and Godwin, L. S. 2003. Partnering for the people: Improving the U.S. Census Bureau's MAF/TIGER Database. *Photogrammetric Engineering and Remote Sensing* 69: 1119–16.

Cowen, D. J. and Craig, W. 2004. A retrospective look at the need for a multipurpose cadastre. *Surveying and Land Information Science* 63: 205–14.

FGDC. 2005. Metadata. WWW Document, http://www.fgdc.gov/metadata/metadata.html

FGDC Homeland Security Working Group. 2004. FGDC guidelines for providing appropriate access to geospatial data in response to security concerns. WWW document, http://www.fgdc.gov/fgdc/homeland/revised_access_guidelines.pdf.

Gewin, V. 2004. Mapping opportunities. *Nature* 427: 376–7 (available at http://www.aag.org/nature.pdf).

Her Majesty's Stationary Office. 1988. Copyright, Designs and Patents Act 1988. WWW document, http://www.hmso.gov.uk/acts/acts1988/Ukpga_19880048_en_1.htm.

Lopez, X. R. 1996. Stimulating GIS innovation through the dissemination of geographic information. *Journal of the Urban and Regional Information Systems Association* 8(3): 24–36.

NRC (National Research Council). 1990. *Spatial Data Needs: The Future of the National Mapping Program.* Washington, DC: National Academy Press.

NRC. 1993. *Toward a Coordinated Spatial Data Infrastructure for the Nation.* Washington, DC: National Academy Press (available at http://books.nap.edu/openbook/0309048990/html/index.html).

NRC. 2003. *Weaving a National Map: Review of the U.S. Geological Survey Concept of the National Map.* Washington, DC: National Academy Press (available at http://www.nap.edu/books/0309087473/html).

NRC. 2004. *Licensing Geographic Data and Services.* Washington, DC: National Academy Press.

Office of the Federal Register 1994. Executive Order 12906. *Federal Register* 59(71): 17671–4.

Richland County, South Carolina 2005. Richland County Geographic Information Systems (RC GEO). WWW document, http://www.richlandmaps.com/.

USGS. 1989. *Digital Line Graphs from 1:100,000-scale Maps: Data Users Guide 2*: Reston, VA: US Geological Survey.

USGS EROS Data Center. 2005. Seamless Data Distribution System. WWW document, http://seamless.usgs.gov/.

US Office of Management and Budget. 2004. Circular A-130. WWW document, http://www.whitehouse.gov/omb/circulars/a130/a130.html.

Chapter 2

Social Data

David J. Martin

In this chapter we shall be concerned with data that relate to human populations and their activities. We shall explore the sources and spatial characteristics of social data with a view to their implications for Geographic Information Systems (GIS) analysis, but will not examine analytical techniques in detail as these will be covered in later chapters. There is continuing growth in the availability and sophistication of social data sources globally, although change in the key phenomena of interest is more subtle: issues such as social exclusion and accessibility remain as important in the early 2000s as they were to nineteenth-century commentators, although their manifestation is continually evolving (Dorling, Mitchell, Shaw, Orford, and Davey-Smith 2000, Thurstain-Goodwin 2003). Methods for geographic referencing and the value of linking data from different sources continue to have enormous importance in social GIS applications at a time when there is evidence of a retreat from conventional population censuses towards alternative mechanisms of social data gathering. Data collection environments, ethical considerations, and data protection measures are growing in significance, with new challenges emerging over detailed personal data and the principles governing their use.

The key sources for social data are administrative records, censuses, and surveys. A further dimension to social data is added by remote sensing, although the latter does not form a major emphasis in this chapter. Increasing resolution in geographic referencing is leading to more detailed attention on the ontologies of relevance to social data, specifically to more careful definition of households and residents and to continued interest in concepts of "neighborhood" (Kearns and Parkinson 2001). In part, this change of focus is due to the wealth of potential data. Household definition issues are more important when we can realistically access geographically referenced individual records than when the only available social data are aggregated across many hundreds of dwelling units. GIS has also become important to the actual collection and management of social data in addition to the traditional role, whereby GIS contributed to their visualization and analysis. The 2000/1 round of censuses has been significantly influenced by the use of GIS technology in national statistical organizations, and plans for future social data collection are being strongly influenced by geographic referencing strategies and data management considerations.

In the remainder of this chapter we shall visit each of these issues in more detail, beginning with the fundamental objects of interest in social GIS. We shall then move on to consider the principal social data sources, before dealing more specifically with representational considerations for the use of social data in GIS. These representational issues are also addressed in Martin (1999). Our discussion is intended to be international in application and examples have been drawn from a variety of countries, although this is necessarily an illustrative rather than exhaustive review.

Social Data Entities

The smallest social unit in GIS applications has usually been conceived as the individual person. Further, some definition of household or family unit is also elemental for many purposes, as is a definition of the dwelling unit. Dwelling units are frequently the smallest entity to which a geographic reference can be attached. Individuals and households are generally indirectly georeferenced, for example by linkage to a dwelling unit address or code. Events such as hospital admissions or unemployment episodes fall below the level of the individual person, and may be georeferenced either by linkage to a specific person or via some other known location: only recently has sufficient precision been available for such events to be treated individually in some GIS applications. Above the person and household scales a series of imposed geographic units may be of interest either for practical or analytical purposes and these could all be interpreted to a greater or lesser degree as measures of something called "neighborhood". A neighborhood in these terms may represent a very small aggregation of the elemental units – perhaps to a street block face or unit postcode – ranging up to a distinct settlement or identifiable administrative or social subdivision within a city.

The increasing resolution of available geographic referencing systems has brought definitional issues much more to the fore, and this is particularly relevant with the growing use of administrative sources for social GIS data. The 2001 UK censuses, for example, have been significantly challenged by the changing nature of households and dwellings. Conventional census definitions are increasingly unable to accommodate complex arrangements in which professionals dwell at a city apartment during the week and a family home outside the city at the weekend, or with children who spend different parts of the week with different parents, each of whom may be resident in households with other children from different partnerships. By contrast, similar basic dilemmas are observable in the 2001 South African census in which a single rural residential compound may contain several dwellings, each occupied by different branches of the same family yet considered by the inhabitants to be a single "household." In the face of such complex social realities, simple conventional classifications of households are inadequate.

For many years, social data have only been made available to most GIS users as aggregate counts for areas – typically census areas – and there has been relatively little concern with the detailed definition of elemental units such as persons and households from a geographic referencing perspective. Of more concern have been the ways in which the fixed reporting units could be related to the higher order entities such as neighborhoods and suburbs. The lowest available levels of geographic referencing vary widely between countries, but the general trend is towards increasing spatial resolution. Areally aggregated social data are particularly prone to ecological fallacy and modifiable areal unit problems that are very relevant to the social GIS user (Openshaw 1984, Fotheringham and Wong 1991). This long-recognized group of issues presents obstacles to statistical inference from areally aggregated data, which are sensitive both to the scale and specific aggregation designs applied, and in which individual-level associations may be obscured or even reversed by aggregation.

These problems arise primarily from the fact that the areal units over which social data are aggregated are usually defined in terms which are only weakly related to socio-economic characteristics. Thus a street block face or political division have meaningful interpretations for the purposes of organizing mail delivery or local elections, but there is no necessary reason why their boundaries should be coincident with the social transitions that might be recognized by residents. The extent to which meaningful social neighborhoods are indeed recognizable is subject to intense debate. Galster (2001) presents a review of neighborhood definitions noting several separate traditions in the academic literature according to whether the concern is with purely ecological perspectives or with more socially-oriented definitions. All of these suffer from weaknesses which make them problematic from GIS and policy perspectives, and almost all conceptualizations of neighborhood acknowledge that the phenomenon is recognizable at several spatial scales (Kearns and Parkinson 2001).

A further layer of complexity is added when the temporal dimension is to be taken into account. This operates at two scales: the long-term over which change occurs in the economic and social characteristics of neighborhoods, and real-time in which people travel through the settlement structure. Neighborhood-level change is often obscured by changes in reporting geographies, whereby administrative processes dictate redrawing of boundaries, making impossible direct comparison of the same small area between two consecutive censuses. Traditional data sources have generally captured night-time populations via residential locations. Increasingly, we are also interested in population flows and in daytime locations or even 24-hour population dynamics. For the purposes of cellular telephony, location-based services, emergency planning, and sophisticated retail modeling it is not sufficient to understand only the distribution of the population when they are mostly asleep at home at 2 a.m., but also necessary to be able to capture commuting and leisure flows which reveal the locations of office workers at lunchtime and sports fans traveling to a major weekend event. At present, our understanding of such 24-hour geographies is still formative, but already detailed data are being captured which will help to model such phenomena.

One of the greatest challenges to socio-economic GIS is finding appropriate data and representational models to match such imprecise concepts as household and neighborhood. It is important that this is not seen as a search only for an "optimal" data structure or zoning solution, but that we develop GIS models which more realistically capture the perceptual spatial knowledge on which most individual decision-making is based. In the following section we shall consider the available sources of social data, recognizing that none of these capture the full range of diversity and complexity encountered in social reality, and that value judgments are required at every stage of GIS representation and modeling.

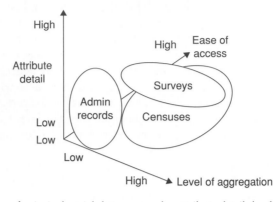

Fig. 2.1 Comparison of principal social data sources by attribute detail, level of aggregation and ease of access

Social Data Sources

We may usefully begin our consideration of social data sources by considering the three dimensions of spatial aggregation, attribute detail, and ease of access. Within this three-dimensional space we will place our three major sources of social data, namely administrative records, censuses, and surveys, as illustrated in Figure 2.1. Ease of access is not just the result of arbitrary procedural arrangements, but a key consideration in social data. This relates to local interpretation both of data protection measures and ethical issues associated with personal data – topics which are addressed more fully in Chapter 29 of this volume by Cho.

   Administrative data sources display the lowest geographic aggregation, but very low accessibility and only moderate attribute detail. Census data are more readily accessible, and offer more attribute detail in many respects, but cover a broader range of aggregation scales, often to quite coarse geographies. Survey data also range in aggregation scale, frequently available only for very large geographic units, but provide high levels of attribute detail and greater ease of access than most administrative sources. Survey and census data are usually subject to some minimum population threshold requiring aggregation before publication. The three principal sources of social GIS data thus occupy this space, differing in detail between specific data sources and national contexts, but generally adhering to the characteristics outlined here. Webber and Longley (2003) note that while government has tended to place reliance on censuses and information abstracted from administrative sources in its modeling of population need, commercial organizations are often forced to rely on the findings of syndicated market research surveys and to use geodemographic classifications as the bridge between survey results and denominator populations, indicating the importance of the accessibility axis in determining actual data use. It will become apparent that as national statistical organizations respond to contemporary data collection challenges, the traditional boundaries between censuses, surveys and administrative sources are becoming increasingly blurred.

Administrative records

Administrative records refer to data which are collected as a result of routine organizational activities. In some countries, particularly in Scandinavia, there are explicit population registers which aim to capture all members of the population. In the Netherlands, the integration of administrative data sources has replaced conventional census taking, with the census data now being derived from matched administrative records (van der Laan 2001). In many other national settings where explicit population registers do not exist there are nevertheless multiple population listings associated with government functions such as electoral registration, health care delivery, and personal taxation. These lists generally contain information not only on the registered population, but provide limited attribute information including basic personal or property characteristics and some information on service usage. Frequently such registers do not actually contain any specific denominator and records relate to service delivery events. For example, UK Hospital Episode Statistics (HES) record instances of patients in hospital. Thus a single patient, or even a single illness, may be represented by several distinct entries in the HES system. Vital events registration systems also fall into this category, recording instances of births and deaths but not any record of the base population. In each of these cases it is usually some level of the residential postal address which is used to georeference the event being recorded. Administrative sources may thus provide us with either denominator or event data. The quality of address or property listings again varies widely between nations, with the nature of property taxation having a major impact on the existence and quality of cadastral maps and records. Such lists may also be maintained by utility companies or postal services, and by commercial organizations in the form of customer records, frequently providing large sample sizes and very rich attribute information. Some real-time administrative data with high geographic resolution are being captured, for example by cellular telephone operators although to date access to such information has been highly restricted.

Coverage issues are paramount when considering the potential of administrative data sources for social GIS applications: very few listings short of a full population register or comprehensive cadastre actually capture the entire population or all dwelling units. Lists maintained for almost any non-statistical purpose are liable to systematic omissions and biases: electoral registers capture only those entitled to vote, thus omitting children, non-native residents and all those who choose not to register; postal service address lists capture only those addresses which receive post, thus omitting industrial and other premises which may still be locations of employment or other activities. The National Health Service Register, probably the UK's best approximation to a population register, is based on patient registration with a primary care doctor and in most areas represents an over-count of the population as there are no automatic mechanisms for removing patients who move away from an area but fail to inform their doctor (Haynes, Lovett, Bentham, Brainard, and Gale 1995). Understanding the utility of such administrative data sources requires a very careful consideration of the basic social data entities noted above in order to determine whether the definitions and assumptions implicit in the data source match those required for the GIS application. These considerations are additional to specific issues of geographic referencing, which are addressed below.

Administrative records that contain information about individual members of the population are also subject to various levels of data protection that may prevent their use outside the organization by which they were originally created. In general terms, data protection legislation seeks to preserve the anonymity of individuals and may further require that personal data are used only for purposes made known to the individuals at the time of data collection. A particular example of the challenges for GIS use arising from this situation relates to the use of medical information for statistical analysis and health care planning. The handling of such data is generally governed by nationally-specific guidelines such as the UK's Caldicott principles (Table 2.1), which grew out of a review of the ways in which patient information is used. These principles recognize that while there is a need to protect the confidentiality of individual patients, such information also underpins analyses that are essential to the effective organization of a health care system. Such guidelines typically place restrictions on the data owner and may involve the appointment of

Table 2.1 Summary of UK Caldicott principles for use of patient-identifiable data (adapted from http://www.learnonline.nhs.uk/info_gov/Caldicott.asp)

Principle	Summary	Explanation
1	Justify purpose for use	Every proposed data use should be clearly defined and scrutinized, and continuing use reviewed, by an appropriate data guardian.
2	Individually identifiable data should be used only where absolutely necessary	Individually identifiable information items should only be included where they are essential for the specific work purpose. The need for such data should be considered at each stage of the work.
3	Individually identifiable data should be kept to a minimum	Where use of individually identifiable information is considered to be essential the need for each individual data item should be considered and justified so that the minimum of identifiable information is made available.
4	Access to individually identifiable information should be on a strict need-to-know basis	Only those individuals who need access to individually identifiable information should have access to it, only have access to the items that they require.
5	Everyone with access to patient-identifiable information should be aware of their responsibilities	All those handling individually identifiable information should be made fully aware of their responsibilities and obligations to respect confidentiality.
6	Understand and comply with the law	All use of individually identifiable information must be lawful. Someone in each organization should be responsible for ensuring that the organization complies with legal requirements.

a review panel or data guardian to examine applications for data use. Administrative information collected by commercial organizations is generally additionally protected for reasons of competitive advantage and may thus be unobtainable outside the collecting organization or may be traded at a market price. Increasingly, administrative records provide new sources for aggregate statistics, such as the UK's new Neighbourhood Statistics system which is based on the aggregation of records in government administrative systems so as to provide small area data series that would not be available from other sources. Integration with census data is provided by the use of a common geographic framework (Office for National Statistics 2003).

Censuses

A frequently cited characteristic of censuses is that they provide the most spatially detailed geographic social data with the highest coverage of the population. It is therefore unsurprising that census data form a major component of most social GIS implementations. Censuses attempt to collect a broad range of social data on the entire population of a country, although this is becoming harder to achieve due to increasing non-compliance with census-taking. Census data are typically available for a hierarchy of geographic units ranging from neighborhood to national scales, and in many countries digital boundary data are available to support GIS analysis of census results. A comprehensive review of the UK Census data system leading up to the 2001 census is provided by Rees, Martin, and Williamson (2002). Census data are usually subject to specific confidentiality legislation which requires data protection measures be applied – such as data suppression for areas that fall below some population threshold size, and randomization or rounding of counts for small areas. Data for small areas are also subject to randomization and rounding so as to protect the identity of individuals with distinctive characteristics. A comparison of the attribute coverage of four national censuses from the 2000/1 round is given in Table 2.2. In each of these countries the internal structure of households is captured through a grid of questions about relationships between household members, although these differ in detail. More important differences reflect differing social needs, such as the interest in orphaning and childhood survival in South Africa, and the political acceptability of questions, such as the UK's lack of an income question, the topic being believed to be too sensitive to broach and that to do so will risk endangering responses to other questions. Most of these census questions tell us little about consumption habits and lifestyle, for which we must rely on administrative and particularly survey data sources.

Censuses are also notable for the range of output data products that they can provide. These include not only the most obvious areally aggregated data and associated geographic boundary frameworks that are widely used for thematic mapping and GIS, but also microdata samples (Public Use Microdata Samples in Canada and the USA; Household Sample Files in Australia; Samples of Anonymised Records [SARs] in the UK) and longitudinal data sets such as the Office for National Statistics Longitudinal Study in England and Wales. The latter two data types provide extremely rich attribute detail but at the cost of high levels of geographic aggregation to the extent that these data sets rarely appear in GIS applications concerned with small area analysis: in Figure 2.1 they represent that part of the

Table 2.2 International comparison of 2000/1 census topic coverage

	England and Wales 2001	USA 2000 short (S)/ long (L) form	Australia 2001	South Africa 2001
Accommodation type	H1, H2	L34, L35		H-23
Accommodation size	H3	L37, L38	46	H-24
Water/toilet/washing facilities	H4	L39		H-26, H-27
Lowest floor level	H5			
Heating (availability, fuel)	H6	L42		H-28
Kitchen facilities		L40		
Motor vehicles	H7	L43	45	
Telephone (availability)		L41		H-29
Household goods (PC, radio, TV, refrigerator)				H-29
Computer/Internet use			20, 21	
Mode of refuse disposal				H-30
Accommodation tenure business use, landlord	H8, H9	S2, L33, L44	47, 49	H-25
Household costs, rent		L45-L50, L52, L53	48	
Property value		L51		
Relationships within household	Grid	Grid	Grid	Grid
Sex	2	S5, L3	3	P-02
Age/date of birth	3	S6, L4, L18	4	P-03
Marital status	4	L7	6	P-05
Visitors/absent residents	Tables 1, 2		7, 44	P-11
Country/place of birth (of parents)	7	L12	11 (13,14)	P-09
Ethnic group/race	8	S7, S8, L5, L6	17, 18	P-06
Religion	10 optional		19 optional	P-08
Language at home/ English ability		L11	15, 16	P-07
Citizenship		L13	10	P-10
When arrived in country		L14	12	
General state of health	11			
Caring for others	12	L19		
Long term illness/disability	13	L16, L17		P-13
Address one year ago	14		8	
Address five years ago		L15	9	P-12
Whether parents alive				P-14, P-15
When moved to present address		L36		P-12
Live births/surviving children				P-20
Whether in education	5	L8	22, 23	P-16
Recent deaths in household				H-31
Term-time address	6			
Qualifications/completed education	16, 17	L9	25–30	P-17
Military service		L20		
Current economic activity/ employment history/availability for work	18–23	L21, L25, L26, L30	32, 42, 43	P-18
Employment status/job title/ work done	25, 27–9	L28, L29	33–5	P-19
Income		L31, L32	31	P-22
Employer, business	26, 33	L22, L27	36–9	P-19
Mode of travel to work	34	L23, L24	41	P-21
Hours worked	35		40	P-19

Values in cells are question numbers on census forms

census domain that is closest to large-scale survey data. Census interaction data describe flows of individuals between home and work addresses or between present and past addresses, offering a snapshot of residential migration and are sometimes published at sufficiently detailed geographic resolution to be a practical data set for use in GIS. However the two-dimensional nature of these interaction data present numerous additional data handling challenges (Stillwell and Duke-Williams 2003) not well handled by conventional GIS.

In the face of growing enumeration and coverage challenges at the start of the 2000s, several countries are moving away from traditional censuses as the primary source of official population data collection and looking towards increased use of administrative records and large surveys, with an associated explicit move toward modeled rather than fully enumerated social data, a theme reviewed more fully in Martin (2006).

Surveys

Large-scale social surveys supplement the information collected by censuses and administrative sources. Surveys give the opportunity to gather much more attribute detail, perhaps relating to a specific area of socio-economic activity, such as the UK's Labour Force Survey (LFS), or to fill gaps in other official statistical systems, for example in order to capture information on international travelers, such as the UK's International Passenger Survey (IPS). In the commercial world, market research surveys aim to capture household consumption and lifestyle character-istics and may often achieve large sample sizes, albeit with acknowledged biases in terms of the types of respondents from which information can be obtained. From the GIS perspective, these data sets share some of the main disadvantages of the census micro data sets, in that their relatively small sample sizes and sparse geographic coverage lead to data aggregation over large geographic regions. Consequently these data sources are rarely able to directly provide geographically detailed inputs to GIS applications.

The principal data-collection strategies of some major countries for the early years of the 2000s may now be best described as surveys rather than censuses: France is moving to a rolling census model, whereby each small area is surveyed once every five years (Durr and Dumais 2000) and the USA has adopted a continuous popula-tion survey known as the American Community Survey (Alexander 2000) to replace the information previously gathered by the "long form" census data (Table 2.2) on a rolling basis and to be used alongside a census short form in 2010.

Remotely sensed data

Mesev (2003a) makes it clear that sensing cities remotely is a difficult endeavor. Many of the socio-economic variations in which the social GIS users are likely to be most interested are not directly observable – whether from the ground or sky. Nevertheless, there is a distinct role for remote sensing in providing additional information to that captured by our three ground-based groups of data sources, particularly in rela-tion to urban land use and the estimation of population data for regions in which conventional measurement is impractical. The attribute information which can be captured by this means are generally limited to those aspects which are reflected by settlement extent, land use, and energy usage (see Chapter 3 by Lees in this volume for additional details on the role of remote sensing in delineating these features). Nevertheless these may be valuable clues to updating ground information in areas not covered by ground data collection, particularly where rapid development is taking place. Interested readers will find a contemporary review of techniques in Mesev (2003b), including methodologies for the complementary use of census and remotely sensed data.

Social Data Representation in GIS

Representation here refers to technical, as opposed to ethical, considerations following on from our discussion of social data sources. Methods relevant for the analysis of social data in GIS will be found in many chapters of this book, but all are affected by the initial data acquisition and representational model. The way in which we choose to represent social data in GIS is of particular importance because we are so often dealing with secondary data sets and our representational decisions are therefore overlaid onto a series of decisions already made by the data collecting organizations. An important first step is to recognize that social data in GIS share many general characteristics with all secondary social data sets, particularly in terms of the uncertainties associated with pre-publication collapsing of classifications, imputation of missing values, random modification, and rounding. It is wise to treat all such data as estimates rather than counts and to pay particular attention to uncertainties that may display spatial bias. Census under-enumeration, for example, is generally related to specific population sub-groups: both the original error and subsequent correction methods will therefore contain distinct spatial patterning, invisible in the headline published counts.

It is conventional to consider spatial objects as point, line, area, and surface types. All four representational models are applicable to social data although in conventional applications point and area types predominate. Data collected from individuals and households are generally conceptualized as point phenomena, although confidentiality considerations may necessitate areal aggregation prior to data publication. Information on transportation flows such as roads or services within a public transportation system are generally conceptualized as lines and captured as such. Little information is generally available about the socio-economic characteristics of routes and flows, although online public transportation timetables are beginning to provide route attributes that are suitable for spatial analysis and accessibility modeling (see Martin et al. (2002) and Okabe, Okunuki, and Shiode (2006) for additional details). Surprisingly few social phenomena are genuinely areal in nature: most area-based data are the result of aggregation. However, land ownership, political representation, and policy zoning are all examples of genuinely areal social phenomena for which exact boundaries may be identified.

Except in the very rare situations where the individual reveals his/her own spatial coordinates directly, for example by use of location-based services or a cellular telephone, we are forced to make decisions about how best to establish an appropriate geographical object to which to relate data about them. If we have administrative or survey information about an individual person, we must determine whether it relates most appropriately to their home or work address or to their presence at some other location. We must then attempt to match the description of this location to an object for which we have a known or measurable geographical location. Administrative and survey data sources most frequently provide either a full or partial postal address, perhaps making use of a national postal code system such as the UK's postcode (Raper, Rhind, and Shepherd 1992) or US zipcode: these codes provide relatively rapid geographic referencing at the neighborhood level as they are straightforward to match against national directories of codes and locations. However, the apparently simple task of textual address matching is error-prone and gives rise to many ambiguities. In many national contexts, there are no definitive address lists against which to compare recorded addresses and approximate matches may have to be made against street segments or even quite large neighborhoods: the maintenance of up-to-date address listings is impossible without extensive inter-agency collaboration. Both the UK and USA are engaged in projects to enhance existing address listings. In the USA, this involves the modernization of the Master Address File/TIGER lists used for the 2000 census (Vitrano, Pennington, and Treat 2004). Meanwhile the UK is struggling to achieve a national solution, most recently through the National Spatial Address Infrastructure (NSAI) initiative (Office of the Deputy Prime Minister 2005). These major infrastructural data sets are difficult to complete due to differences in the objects of interest of the various organizations and continual change in actual addresses. While social data may be directly associated with point locations by any of these methods, the techniques of point pattern analysis may be most applicable, although it is important always to consider the spatial distribution of the relevant denominator populations.

The 2001 census in England and Wales made use of GIS not just for enumeration district design, but also for the creation of an entirely separate output geography, designed for the specific needs of census data publication (Openshaw and Rao 1995, Martin 1998 2000). This output geography took particular advantage of the small size of the UK's unit postcodes, allowing output areas to be assembled by direct aggregation of postcodes. Many national statistical organizations could in theory aggregate from address-referenced census and administrative data to produce output data for any desired geographical units but the associated problems of disclosure control and data access are complex. Duke-Williams and Rees (1998) describe the differencing problem whereby a GIS user may intersect official counts for two very slightly different areas thereby obtaining counts for an intersection polygon that falls below acceptable confidentiality thresholds. In England and Wales the response has been to plan all new data publication for exact aggregations of the 2001 census output areas. Whatever the mechanisms for geography definition and aggregation, areally aggregated data remain the primary form of social data for use in GIS, although there are many circumstances in which complete area boundaries are unavailable and such data can only be georeferenced via centroids or other summary point locations associated with area identifiers.

In cases in which social data are required for areal units other than those for which they have been published, some form of areal interpolation is required. This is another enduring GIS challenge (Flowerdew and Green 1992). Crude area-based interpolation in which population-related counts are transferred between zones in proportion to the areas of intersecting polygons is unsatisfactory except where population is uniformly distributed within polygons and the use of lookup tables or ancillary distributional data is required. Goodchild, Anselin, and Deichmann (1993) note that this task of areal interpolation is actually one of estimating the characteristics of an underlying continuously varying surface and indeed many social phenomena which are expressed as densities per unit area or rates relative to denominator populations may helpfully be treated as surfaces (see Chapter 13 by Tate, Fisher and Martin in this volume for additional information about surfaces). However, no such phenomena are directly measurable in surface form, and surfaces are more usefully considered as a representational or analytical device which overcomes many of the difficulties associated with conventional area-based methods (Martin 1996, Lloyd, Haklay, Thurstain-Goodwin, and Tobon 2003). Further representational options are introduced when our interest is with areal units which may be modeled but not directly measured, such as the effective catchment area of a retail outlet or health center: such spatial objects may be constructed in many ways and are again a candidate for surface rather than area-based modeling.

CONCLUSIONS: IMPLICATIONS FOR SPATIAL ANALYSIS

In this chapter we have briefly reviewed the principal contemporary sources of social data for GIS. The central message of the chapter is that there are multiple approaches to the acquisition and representation of almost all social phenomena, but none of these are neutral in their impacts on subsequent GIS use. The data available in any national setting are a complex product of social priorities, organizational culture, and historical precedent. Good practice with social GIS data involves a clear ontology of the phenomena being modeled, a full understanding of data acquisition and georeferencing, and careful attention to the choice of representational models. Much analysis of social data in GIS tends to be focused on the manipulation of attributes without reference to geography, but, as all social GIS representations are to some extent spatial models, the spatial aspects cannot be ignored: they are rarely random!

Conventional administrative, survey, and census-based data sources are becoming increasingly integrated as national statistical organizations look to data collection strategies which combine elements of each. Geographic referencing is continually improving but high precision is often offset by data protection measures that result in spatially inappropriate aggregation. Exciting new challenges will be posed by moves towards higher spatial and temporal resolution and cross-scale data integration that require entirely new geocomputational tools such as those advocated by Gahegan in Chapter 16 of this volume.

ACKNOWLEDGEMENTS

The author is supported by Economic and Social Research Council Award RES-507-34-5001 as Coordinator of the ESRC/JISC 2001 UK Census of Population Programme.

REFERENCES

Alexander, C. H. 2000. The American Community Survey and the 2010 US Census. Paper presented at *INSEE-Eurostat Seminar on Censuses after 2001* (available at http://www.insee.fr/en/nom_def_met/colloques/insee_eurostat/pdf/alexander.pdf).

Dorling, D., Mitchell, R., Shaw, M., Orford, S., and Davey-Smith, G. 2000. The ghost of Christmas past: Health effects of poverty in London in 1896 and 1991. *British Medical Journal* 321: 1547–5.

Duke-Williams, O. and Rees, P. 1998. Can census offices publish statistics for more than one small area geography? An analysis of the differencing problem in statistical disclosure. *International Journal of Geographical Information Science* 12: 579–605.

Durr, J.-M. and Dumais, J. 2002. Redesign of the French census of population. *Survey Methodology* 28: 43–9.

Flowerdew, R. and Green, M. 1992. Developments in areal interpolation methods and GIS. *Annals of Regional Science* 26: 67–77.

Fotheringham, A. S. and Wong, D. W.-S. 1991. The modifiable areal unit problem in multivariate statistical analysis. *Environment and Planning A* 23: 1025–44.

Galster, G. 2001. On the nature of neighbourhood. *Urban Studies* 38: 2111–24.

Goodchild, M. F., Anselin, L., and Deichmann, U. 1993. A framework for the areal interpolation of socioeconomic data. *Environment and Planning A* 25: 383–97.

Haynes, R. M., Lovett, A. A., Bentham, G., Brainard, J. S., and Gale, S. H. 1995. Comparison of ward population estimates from FHSA patient registers with the 1991 census. *Environment and Planning A* 27: 1849–58.

Kearns, A. and Parkinson, M. 2001. The significance of neighbourhood. *Urban Studies* 38: 2103–10.

Lloyd, D., Haklay, M., Thurstain-Goodwin, M., and Tobon, C. 2003. Visualising spatial structure in urban data. In P. A. Longley and M. Batty (eds) *Advanced Spatial Analysis: The CASA Book of GIS*. Redlands, CA: ESRI Press, pp. 266–88.

Martin, D. J. 1996. An assessment of surface and zonal models of population. *International Journal of Geographical Information Systems* 10: 973–89.

Martin, D. J. 1998. Optimizing census geography: The separation of collection and output geographies. *International Journal of Geographical Information Science* 12: 673–85.

Martin, D. J. 1999. Spatial representation: The social scientist's perspective. In P. Longley, M. F. Goodchild, D. J. Maguire, and D. W. Rhind (eds) *Geographical Information Systems: Principles, Techniques, Applications and Management* (2nd edn). Chichester: John Wiley and Sons, pp. 71–80.

Martin, D. J. 2000. Towards the geographies of the 2001 UK Census of Population. *Transactions of the Institute of British Geographers* NS 25: 321–32.

Martin, D. J. 2006. Last of the censuses? The future of small area population data. *Transactions of the Institute of British Geographers* NS 31: 6–18.

Martin, D. J., Wrigley, H., Barnett, S., and Roderick, P. 2002. Increasing the sophistication of access measurement in a rural healthcare study. *Health and Place* 8: 3–13.

Mesev, V. 2003a. Remotely sensed cities: An introduction. In V. Mesev (ed.) *Remotely Sensed Cities*. London: Routledge, pp. 1–19.

Mesev, V. (ed.). 2003b. *Remotely Sensed Cities*. London: Routledge.

Office for National Statistics. 2003. *Proposals for an Integrated Population Statistics System*. London, Office for National Statistics Discussion Paper (available at http://www.statistics.gov.uk/downloads/theme_population/ipss.pdf).

Office of the Deputy Prime Minister. 2005. Towards the National Spatial Address Infrastructure: Outline Prospectus. WWW document, http://www.odpm.gov.uk/pub/578/TowardstheNationalSpatialAddressInfrastructurePDF235Kb_id1144578.pdf).

Okabe, A., Okunuki, K.-I., and Shiode, S. 2006. The SANET toolbox: New methods for network spatial analysis. *Transactions in GIS* 10: 535–50.

Openshaw, S. 1984. *The Modifiable Areal Unit Problem*. Norwich, GeoBooks Concepts and Techniques in Modern Geography No. 38.

Openshaw, S. and Rao, L. 1995. Algorithms for reengineering 1991 census geography. *Environment and Planning A* 27: 425–46.

Raper, J. F., Rhind, D. W., and Shepherd, J. W. 1992. *Postcodes: The New Geography*. Harlow: Longman.

Rees, P., Martin, D., and Williamson, P. (eds). *The Census Data System*. Chichester: John Wiley and Sons.

Stillwell, J. and Duke-Williams, O. 2003. A new web-based interface to British census of population origin: Destination statistics. *Environment and Planning A* 35: 113–32.

Thurstain-Goodwin, M. 2003. Data surfaces for a new policy geography. In P. A. Longley and M. Batty (eds) *Advanced Spatial Analysis: The CASA Book of GIS*. Redlands, CA: ESRI Press, pp. 145–70.

van der Lann, P. 2001. The 2001 census in the Netherlands: Integration of registers and surveys. Paper presented at *INSEE-Eurostat Seminar on Censuses after 2001* (available at http://www.insee.fr/en/nom_def_met/colloques/insee_eurostat/pdf/laan.pdf).

Vitrano, F. A., Pennington, A., and Treat, J. B. 2004. Census 2000 testing, experimentation, and evaluation program topic report 8: Address list development in Census 2000. Washington, DC: US Census Bureau.

Webber, R. and Longley, P. A. 2003. Geodemographic analysis of similarity and proximity: Their roles in the understanding of the geography of need. In P. A. Longley and M. Batty (eds) *Advanced Spatial Analysis: The CASA Book of GIS*. Redlands, CA: ESRI Press, pp. 233–66.

Chapter 3

Remote Sensing

Brian G. Lees

Remotely sensed data is an important source of information for many spatial decision support systems. As the information age has progressed and, almost concurrently, problems of over-population, food security, pollution, global warming, and national security have become ever more urgent, decision makers at all levels of government and industry are increasingly required to make decisions over shorter time cycles. These time cycles are now so short as to preclude extensive surveys to provide a solid information base on which to base considered planning. It would be hard to envisage, in this century, project funding being allocated to the 25- to 50-year-long national soil surveys of the last century. Rather, it is becoming common practice to use available archived data updated, where possible, with recent remotely sensed data or other reconnaissance tools. Remote sensing has become an indispensable tool for keeping geographic databases current and ready for rapid application.

Remote sensing, as a field, covers everything from digital scanning and optochemical photography from satellites and aircraft, laser and radar profiling, to echo sounding from ships. It can be classified as active or passive remote sensing depending on whether the system is sensing the reflections of its own transmissions or not. Passive systems sense reflections and re-radiations of the Sun's energy, both in the visible and middle infra-red wavelengths, and emitted radiation in the middle and thermal infra-red wavelengths. Active systems sense reflected radiation in the microwave, laser light, and sound bandwidths.

There is a considerable literature on how the Earth surface interacts with electromagnetic radiation in the microwave range. Some is reflected, some is absorbed, and some is transmitted. Colwell's (1983) huge two volume *Manual of Remote Sensing*, now out of print but widely available in libraries, gives an excellent overview of the detailed characteristics of this interaction. Later volumes in the series cover specialties such as radar remote sensing (Henderson and Lewis 1998), earth science applications (Rencz and Ryerson 1999), and natural resource management (Ustin 2004), keeping the series current and making it a first point of call for those seeking detailed information.

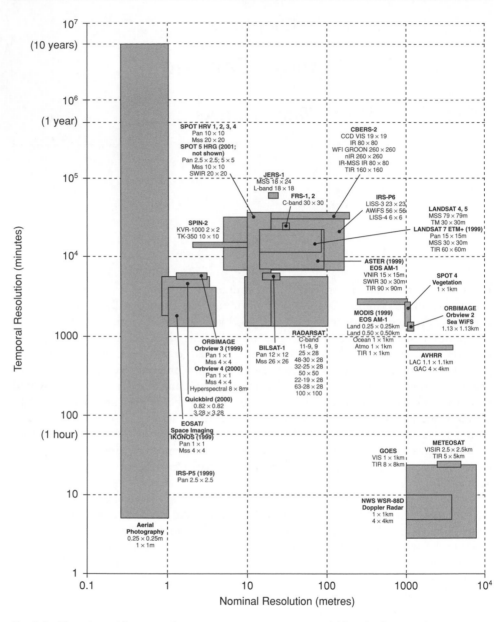

Fig. 3.1 There is a wide range of remote sensing instruments available. The diagram shows some typical platforms and instrument packages with a range of spatial resolutions (the wider boxes) and revisit schedules

This diagram is modified after Longley et al. (2001) which itself is a modification of a diagram in Jensen and Cowen (1999). The figure is in no sense comprehensive; an up-to-date reference list of remote sensing satellites and their instrument packages can be found at http://www.tbs-satellite.com/tse/online/mis_teledetection_res.html

There is a now a wide choice of systems with a range of spatial and spectral resolutions (Figure 3.1). Within this chapter we will concentrate on only those remote sensing sources and data which are commonly used as inputs to Geographic Information Systems (GIS), or have the potential to be so.

Remotely Sensed Imagery and GIS

Until the end of the Cold War and the declassification of military standard satellite remote sensing systems, most spatial information used in GIS was derived from cartographic and survey products. Most of these, in turn, were derived from aerial photography. It is only since 2001 that the satellite imagery available for civil purposes has been of a fine enough resolution to be of more direct use in GIS.

Remote sensing typically produces two kinds of information. It has the potential to record high-quality, relatively accurate, spatial information and poorer quality attribute information. This potential is limited by the spatial and spectral resolution of the particular instrument. Newcomers to remote sensing are often confused by the considerable "overselling" of the quality of attribute information being produced.

Many remote sensing platforms carry instruments with different characteristics matching an instrument with a high spatial and low spectral resolution with an instrument that has a moderate spatial and high spectral resolution. Specialized instruments that have both high spatial and high spectral resolutions tend to be used over limited areas only and are often carried on aircraft or very low orbit spacecraft.

Remotely sensed image data and cartographic data are two related, but different, forms of diagram in which there is a spatial treatment of spatial data. Both have a planar metric combined with planar topology. A photograph, however, is "realistic" while a map is symbolic (Lemon and Pratt 1997). The relationship of both the image and graphical diagrammatic representations to the original data is a homomorphism. Homomorphism is a many-to-one mapping in effect representing a pattern in the domain of the mapping for a simpler pattern in its range. Homomorphisms are important in establishing whether one system is a model of another and which properties of the original system the model retains. For each system one can construct a lattice of homomorphic simplifications. The inverse of homomorphism is not a mapping (Krippendorff 1986).

The translation between the "realistic" homomorphism of imagery (land cover) and the symbolic homomorphism of mapping (land use) is difficult to automate and even the most advanced systems still currently require the supervision and intervention of expert analysts. This is an area of active research. Brown and Duh (2004) discuss a model for translating from land use to land cover. This seems to be a possible achievement in the future, but a fully automated model for the more necessary translation from land cover to land use is hard to imagine. Partially automated systems for this are discussed below.

The following sections discuss the methods for extracting symbolic spatial data from imagery which are currently available.

Extracting Spatial Information from Remotely Sensed Data

Removing distortion

A considerable amount of processing is needed to derive high-quality spatial information from remotely sensed data. Geometric distortion must be removed and the data needs to be registered to some coordinate system. In addition, the relevant spatial information needs to be separated from the irrelevant information. These are not trivial tasks.

All remote sensing is subject to geometric distortion. Optochemical systems, even under perfect conditions, produce data where the scale increases away from the ground point closest to the camera. In addition, platform acceleration or deceleration, pitch, roll, and yaw all cause geometric distortion. These effects are not restricted to aircraft. Remote sensing satellites and ships both travel across the surface of dynamic fluids. In the former case, it is the upper atmosphere, and in the latter case, the sea. Like aircraft, they accelerate, decelerate, climb, descend, pitch, roll, and yaw. To avoid these problems, aircraft involved in remote sensing are given strict limits for each of these movements and imagery outside these limits is usually not accepted. It is not possible to do this for satellite remote sensing or ship-borne sonar scanning and, to harness the spatial information contained in these data, the errors and distortions which result from these displacements must be corrected. Geometric correction of images can be performed at several levels of complexity and accuracy.

Normally, these corrections are dealt with by the data distribution agency. The more corrections performed by the data provider, of course, the higher will be the cost of the imagery. Data providers also carry out some radiometric corrections as a matter of course. These corrections usually involve the removal of distortions due to sensor–solar geometry interactions, such as cross track brightening, solar glint, and hot spots. Increasingly, image data can be purchased "ready to go" with the geometric corrections already done, the data registered to some coordinate system, and the image digital numbers converted to physical units for modeling, inversion, or multi-date work.

For those who wish process their own raw remote sensing data, there is a wide range of processing systems available, a developed literature, and a whole remote sensing discipline.

Updating spatial information in geographic databases

There are some important issues when using remotely sensed data to update geographic databases. One of the fundamental geographic databases is the cadastre. This is a vector data set which records land title, prior interest in land, and other legal requirements regarding land ownership. It is usually derived from a primary, text-based, spatial database. If there is disagreement between the two then, in many legislatures, the text-based spatial database is given precedence. Digital databases derived from the cadastre tend to be comparatively inaccurate spatially, although they tend to be complete. Updating the digital cadastral database using remotely sensed data can keep it complete, but the updates cannot easily be transmitted back to the text-based spatial database because of their low spatial accuracy.

Geographic databases are usually based around the entity data model and comprise points, lines, and areas. Until the advent of high spatial resolution space imagery, the process of updating geographic databases involved the laborious comparison of the digital database with current orthophotos. With increased availability of high spatial resolution digital data this process can be partly automated. The process of updating the spatial information in a geographic database is comprised of three steps (Walter and Fritsch 2000).

First, changes between the original database and the more up-to-date imagery need to be identified. Second, attributes need to be corrected. In an urban area these might be street names or ownership details, extracted from other databases. In non-urban areas, attributes are usually related to land use or land cover. These can be derived from remotely sensed data using techniques discussed later in this chapter. Finally, the changes need to be stored in the geographic database and checked for quality.

The first stage is the most labor intensive and is still one in which expert supervision is necessary. At this stage, these two data sets have different data models and structures leading to a fundamental mismatch which must be overcome. This is because they are different homomorphs of spatial reality. The crudest approach is to classify the imagery and convert it from a field data model to an entity data model. However, statistical clustering and classification tools do not produce the cultural entities contained in geographic databases, rather they produce spectral entities. Some form of linking of spectral entities to produce cultural entities is necessary. In the past, this was one of the most time-consuming parts of processing remotely sensed data as it needed to be accomplished manually assessing the color, shape, relative size, texture, pattern, and context of the image. Multi-resolution segmentation allows a hierarchical network of the image objects to be constructed which, under supervision, greatly increases the speed at which this can be accomplished. Coupled with object relationship modeling, it becomes a powerful tool. It is worthwhile briefly describing how this works.

Conventional clustering or classification algorithms tend to rely only on the topological relationships between pixels in spectral space. Multi-resolution segmentation and object relationship modeling use both the topological relationship between pixels in spectral space and their topological relationship in geographic space. Pixels that are adjacent in geographic space and also proximate in spectral space are combined into image objects. Image objects which are adjacent in geographic space and proximate in spectral space can be further combined to give a hierarchical net of image objects with the user deciding at which level the hierarchy best represents a cultural landscape. Some software implementations allow image objects which are adjacent in geographic space and remote in spectral space to be combined on the basis of rules. For example, "sunlit roof" and "shadowed roof" are both elements of the cultural object "roof" and can be adjacent in geographic space but remote in spectral space.

Multi-scale or multi-resolution segmentation and object relationship modeling is becoming a mature methodology (Baatz and Schape 2000, Blaschke, Lang, Lorup, Strobl, and Zeil 2000, Hofmann and Reinhardt 2000, Manakos, Schneider, and

Ammer 2000, Blaschke and Hay 2001, Burnett and Blaschke 2003). Software which enables the user to formulate concepts and knowledge about how spectral entities can be linked to give cultural entities and enable automated, rule-based linking is now available (for example, eCognition, www.definiens-imaging.com).

Despite the significant improvements in processing methodology, it is important to recognize that difference between an older geographic database and a more recent image does not, in itself, imply change. An automated, computer-based process which achieves a result as good as a human operator in interpreting and identifying cultural entities is not yet a reality (Hofmann and Reinhardt 2000) and a difference between their results is expected. Nevertheless, these tools greatly accelerate the transformation of image data.

This approach produces areas and area boundaries. However, it is not an effective method of extracting the other components of an entity data structure, points and lines.

Extracting points and lines

The extraction of points and lines from remotely sensed data requires a different processing approach to the extraction of area features. There are a number of strategies which are being used, with varying degrees of success, to extract line features from remotely sensed data. The particular strategy used, and the results obtained, depend very much upon the type of imagery being processed.

One strategy looks at the digital numbers of a particular type of pixellated line feature. These would usually cluster about some mean value. While a spectral signature is usually not enough to enable the identification of the particular line feature completely, it can be used to refine exclusion criteria. Looking at the spectral characteristics of a neighborhood of pixels is another strategy that is sometimes used. Following these initial transformations, areas with similar length and width dimensions can be excluded if one is looking for line features. The problem here is that broken sections of continuous line may be filtered out. Filters that search for patterns which are weakly linked can also be used to extract line features.

To do better than this, the complexity of the approach increases rapidly. Hinz and Baumgartner (2003) describe an automatic extraction of urban road networks. Their system compiles knowledge about radiometric, geometric, and topological characteristics of roads in the form of a hierarchical semantic net. Pixels in the initial image which form a bright blob, or compact bright region, are interpreted as compact concrete or asphalt regions which, in the real world, may represent a simple junction. Pixels in the image which form an elongated bright region are interpreted as elongated flat concrete or asphalt regions which, in the real world, may represent a section of road pavement, and so on. This road model is extended using knowledge about context.

The context model uses relationships which may exist locally to reinforce the extraction of road features. For example, in denser urban areas, the road tends to parallel the front face of buildings. In higher resolution imagery, the presence of vehicles on roads or trees overshadowing roads, can occlude large parts of the road making it difficult to extract focal features. By identifying vehicles and trees, and tree shadows, as things which commonly occlude roads then the context model can link the line feature through the occlusion.

Once the road segments have been extracted, the network can be constructed geometrically using connection hypotheses. Put simply, most roads join up. Although Hinz and Baumgartner (2003) have demonstrated automatic extraction of urban road networks from remotely sensed imagery, their results emphasize that low error rates are only possible due to the expertise of the system developers in setting the parameters correctly. This sort of system is still at the stage of fundamental research and the reality is that extraction of point and line features from remotely sensed imagery still requires considerable human intervention. Choi and Usery (2004) describe a similar approach.

Extracting Attribute Information

The most common use of remotely sensed data in GIS, has been the generation of attribute information on land cover. There are thousands of papers discussing the processing of remotely sensed data to produce land cover information. We are now at the stage where the promise of remote sensing, at least in this area, is becoming a reality. Users can make the choice to process data themselves or download pre-processed products.

For an example of available pre-processed products, the University of Maryland Global Land Cover Facility lists a range of fine, moderate and course resolution products as does the NASA MODIS Land Discipline website. The latter offers:

Radiation budget variables
- Surface reflectance products
- LST and emissivity
- Snow and ice cover
- BRDF and albedo

Ecosystem variables
- Vegetation indices
- LAI and FPAR
- Vegetation production, NPP
- Evapotranspiration and surface resistance

Land cover characteristics
- Fire and thermal anomalies
- Land cover
- Vegetative cover conversion
- Vegetation continuous fields

There is extensive documentation to support each of these standard products. The land cover data set is supported by an algorithm theoretical basis document (Strahler, Muchoney, Borak, et al. 1999). This outlines the background and historical perspective to the algorithm development, the algorithm structure, and the mathematical description of the algorithm. It also covers testing and validation of the algorithm and discusses sources of error and uncertainty. For those wishing to incorporate standard products into their geographic databases, these highly processed products are invaluable. However, it is almost certain that one of the attribute data layers required is not available off-the-shelf and some processing will be required!

For those whose experience with remotely sensed data is limited, it is worth reviewing some important aspects of processing remotely sensed data to produce attribute information.

Accuracy

In much of the older remote sensing literature there is an unfortunate overstating of the accuracy of products. Without citing culprits, it was not unusual to find claims of better than 90 percent accuracy for land cover classifications. This, of course, is the result of a failure to track errors properly along with a misunderstanding of the statistical basis for classification and the impact on accuracy of mismatched data structures. Some early papers on this topic are still important.

Given the simplest case, in which we have two land cover types such as land and sea, a significant proportion of our pixels will be boundary pixels. These pixels cannot be accurately classified as either land or sea and are truly a mixture. The proportion of pixels which are truly mixels depends upon the shape of the land cover units and the size of the pixels (the spatial resolution of the sensor). This proportion is often very much higher than one might expect. In a mapping exercise for a coastal bay where the ratio of the average land cover unit area to the pixel area was 1 to 16, Jupp, Adomeit, Austin, Furlonger, and Mayo (1982) found that 54.9 percent of the total cells were boundary pixels (mixels).

Allocating mixels to one or other of the land cover classes results in errors of omission and commission; A or B type errors (Crapper 1980). Markham and Townshend (1981) and Townshend and Justice (1981) have shown that, over a limited range, classification accuracies can improve as the resolution becomes coarser. They also observed that, for a given scene, there may be an optimum resolution above and below which classification accuracies will decline. This optimum resolution will depend upon the within-theme spectral variation. Thinking about this error generation in GIS terms, one can see that is the natural result of representing entities using a grid data structure.

The problem is sometimes compounded when the land cover is better represented as a field, rather than as entities. This usually occurs when the land cover is determined by environmental, or edaphic, processes rather than cultural activities, yet the same sort of classification process is used. The classification of a constantly varying (in space) land cover into land cover types generates quite serious errors of omission and commission (Lees 1996b, Mather 1999). In this case, the error results from the use of an inappropriate ecological model which maps directly onto our cartographic model.

Improving the quality of attribute information

A strong theme of research in the field of remote sensing has been to improve the quality of the attribute information available from remotely-sensed image data. Since the mid-1990s techniques have been developed to enhance the attribute information available by combining the remotely-sensed image data with "ancillary" environmental and terrain data. Because the data distribution information required for the diverse data sets is not always available, and the time involved in set-up, parametric classifiers have become less favored than the non-parametric alternatives in this sort of synthesis (Benediktsson, Swain, and Ersoy 1990). It is common to find papers which suggest that parametric classifiers are unsuitable for classifying this sort of mixed data but Benediktsson, Swain, and Ersoy (1990) demonstrate how it can be accomplished. They also demonstrate that this approach is extremely time-consuming compared to the non-parametric alternatives.

Numerous studies of classification using decision trees (Lees and Ritman 1991), neural networks of various kinds (Benediktsson, Swain, and Ersoy 1990; Fitzgerald and Lees 1995, 1996), genetic algorithms and even cellular automata have been published to examine these alternatives (Lees 1996a). However, the most common form of delivery, and use, of vegetation information to spatial decision support systems remains the mapped thematic, choropleth, form. Lees (1996b) argued that the pre-processing of data to suit this data structure perpetuates the use of an inappropriate data model. Even following digital image analysis, results are conventionally presented using one of the least useful data types, categorical, in a form which mimics the entity data structure, although the data structure is usually still pixels. In many natural environments, a continuum of change is being represented as a series of overlapping gaussians. This leads inevitably to the generation of errors of omission and commission.

Lees (1996b) proposed a new method of representing vegetation data which avoided the errors of omission and commission that are generated using conventional techniques. This required the distribution of each species to be recorded as a field and stored separately in some form of a database. Unfortunately, a field-worker wishing for information about a point would have to scroll through, in the case of the Kioloa data set, the distributions of 41 tree species, 94 shrubs, and 108 understorey species. The increase in accuracy gained is seriously offset by the indigestible nature of the output. Lees and Allison (1999) discussed the practicalities of these approaches without offering a solution. It remains clear that there is a requirement for a single representation of vegetation distribution using the field data model which gives a synthetic view of the forest while avoiding the errors generated by classification.

With only few exceptions, most classifications of remotely sensed data are non-spatial analyses of spatial data. The data is analyzed and clustered using its topological relationships in spectral space. The spatial content of the image data usually remains un-investigated. Those studies that have examined the explicitly spatial analysis of this sort of spatial data have looked at strategies such as spatial signal detection, local spatial autocorrelation, and local regression.

Where local spatial statistical measures have been used in environmental remote sensing studies, most have used the Moran or Geary statistics (for example, Pearson 2002). Relatively few have used the Getis–Ord statistic, although Wulder and Boots (1998) have reviewed its use with remotely sensed data. More recently, Holden, Derksen, and LeDrew (2000) used this method to evaluate coral ecosystem health by identifying spatial autocorrelation patterns in multi-temporal SPOT images.

The Getis–Ord statistic (Getis and Ord 1992, Ord and Getis 1995), with its added information about brightness values, has shown itself to be particularly useful in dealing with remotely sensed data. This simple local spatial statistic has achieved results comparable to some of the more complex classifications of remotely sensed data published in the past which have attempted to include spatial context (Fitzgerald and Lees 1995) with far less effort.

CONCLUSIONS

The use of expert systems to automate much of the supervised processing of remotely sensed data into GIS data has increased greatly in the literature. In part this has been driven by the increased availability of high spatial resolution remotely sensed imagery and in part by increasing demand for such algorithms. Increasingly, insights into the behavior of spatial data from GIS activities are feeding back into remote sensing and driving innovation. Some of these innovations are an increased aware-ness of error, both spatial and attribute, within remote sensing and an increased interest in spatial statistics.

REFERENCES

Benediktsson, J. A., Swain, P. H., and Ersoy, O. K. 1990. Neural network approaches versus statistical methods in classification of multisource remote sensing data. *IEEE Transactions on Geoscience and Remote Sensing* 28: 540–51.

Baatz, M. and Schape, A. 2000. Multiresolution segmentation: An optimization approach for high quality multi-scale image segmentation. In J. Strobl, T. Blaschke, and G. Griesebner (eds) *Angewandte Geographische Informationsverarbeitung XII, Beiträge zum AGIT-Symposium*. Karlsruhe: Herbert Wichmann Verlag, pp. 12–23.

Blaschke, T., Lang, S., Lorup, E., Strobl, J., and Zeil, P. 2000. Object-oriented image pro-cessing in an integrated GIS/remote sensing environment and perspectives for environmental applications. In A. Cremers and K. Greve (eds) *Environmental Information for Planning: Politics and the Public*, vol. 2. Marburg: Metropolis Verlag: 555–70.

Blaschke, T. and Hay, G. J. 2001. Object-oriented image analysis and scale-space: Theory and methods for modeling and evaluating multiscale landscape structures. *International Archives of Photogrammetry and Remote Sensing* 34(4): 22–9.

Brown, D. G. and Duh, J. D. 2004. Spatial simulation for translating from land use to land cover. *International Journal for Geographical Information Science* 18: 35–60.

Burnett, C. and Blaschke, T. 2003. A multi-scale segmentation/object relationship modelling methodology for landscape analysis. *Ecological Modelling* 168: 233–49.

Choi, J. and Usery, E. L. 2004. System integration of GIS and a rule-based expert system for urban mapping. *Photogrammetric Engineering and Remote Sensing* 70: 217–24.

Colwell, R. N. 1983. *Manual of Remote Sensing*, vols 1 and 2. Falls Church, VA: American Society of Photogrammetry and Remote Sensing.

Crapper, P. F. 1980. Errors incurred in estimating an area of uniform land cover using Landsat. *Photogrammetric Engineering and Remote Sensing* 46: 1295–1301.

Fitzgerald, R. W. and Lees, B. G. 1995. Spatial context and scale relationships in raster data for thematic mapping in natural systems. In T. C. Waugh and R. G. Healey (eds) *Advances in GIS Research*. London: Taylor and Francis, pp. 462–75.

Fitzgerald, R. W. and Lees, B. G. 1996. Temporal context in floristic classification. *Computers and Geosciences* 22: 981–94.

Getis, A. and Ord, J. K. 1992. The analysis of spatial association by use of distance statistics. *Geographical Analysis* 24: 188–205.

Henderson, F. M. and Lewis, A. J. 1998. *Manual of Remote Sensing: Volume 2, Principles and Applications of Imaging Radar*. New York: John Wiley and Sons.

Hinz, S. and Baumgartner, A. 2003. Automatic extraction of urban road networks from multi-view aerial imagery. *ISPRS Journal of Photogrammetry and Remote Sensing* 58: 83–98.

Hofmann, P. and Reinhardt, W. 2000. The extraction of GIS features from high resolution imagery using advanced methods based on additional contextual information: First experiences. *ISPRS Journal of Photogrammetry and Remote Sensing* 33: 51–8.

Holden, H., Derksen, C., and LeDrew, E. 2000. Coral reef ecosystem change detection based on spatial autocorrelation of multispectral satellite data. WWW document, http://www.gisdevelopment.net/aars/acrs/2000/ts3/cost006pf.htm.

Jensen, J. R. and Cowen, D. J. 1999. Remote sensing of urban/suburban infrastructure and socio-economic attributes. *Photogrammetric Engineering and Remote Sensing* 65: 611–22.

Jupp, D. L., Adomeit, E. M., Austin, M. P., Furlonger, P., and Mayo, K. K. 1982. The separability of land cover classes on the south coast of N.S.W. In *Proceedings of LANDSAT '79 Conference*, Sydney, Australia.

Krippendorff, K. 1986. A dictionary of cybernetics. Unpublished report.

Lees, B. G. 1996a. Inductive modelling in the spatial domain. *Computers and Geosciences* 22: 955–7.

Lees, B. G. 1996b. Improving the spatial extension of point data by changing the data model. In M. F. Goodchild (ed.) *Proceedings of the Third International Conference on Integrating GIS and Environmental Modeling*. Santa Fe, NM, USA. Santa Barbara, CA: University of California, National Center for Geographic Information and Analysis: CD-ROM.

Lees, B. G. and Allison, B. 1999. A comparison between the accuracy of traditional forest mapping and more advanced, database oriented techniques. In W. Shi, M. F. Goodchild, and Fisher, P. (eds) *Proceedings of the International Symposium on Spatial Data Quality*. Hong Kong: Hong Kong Polytechnic University, pp. 442–8.

Lees, B. G. and Ritman, K. 1991. A decision tree and rule induction approach to the integration of remotely sensed and GIS data in the mapping of vegetation in disturbed or hilly environments. *Environmental Management* 15: 823–31.

Lemon, O. and Pratt, I. 1997. Spatial logic and the complexity of diagrammatic reasoning, *Machine Graphics and Vision* 6: 89–108.

Longley, P. A., Goodchild, M. F., Maguire, D. J., and Rhind, D. W. 2001. *Geographic Information Systems and Science*. Chichester: John Wiley and Sons.

Manakos, I., Schneider, T., and Ammer, U. 2000. *A Comparison between the ISODATA and eCognition Classification Methods on the Basis of Field Data*. ISPRS, Vol. XXXIII, Supplement CD. Amsterdam: ISPRS.

Mather, P. 1999. Land cover classification revisited. In P. M. Atkinson and N. J. Tate (eds) *Advances in Remote Sensing and GIS Analysis*. Chichester: John Wiley and Sons, pp. 7–16.

Markham, B. L. and Townshend, J. R. G. 1981. Land classification accuracy as a function of sensor spatial resolution. In *Proceedings of the Fifteenth International Symposium on Remote Sensing of the Environment*. Ann Arbor, MI, USA, pp. 1075–90.

Ord, J. K. and Getis, A. 1995. Local spatial autocorrelation statistics: distributional issues and an application. *Geographical Analysis* 27: 286–305.

Pearson, D. M. 2002. The application of local measures of spatial autocorrelation for describing pattern in north Australian landscapes. *Journal of Environmental Management* 64: 85–95.

Rencz, A. N. and Ryerson, R. A. 1999. *Manual of Remote Sensing: Volume 3, Remote Sensing for the Earth Sciences*. New York: John Wiley and Sons.

Strahler, A., Muchoney, D., Borak, J., Friedl, M., Gopal, S., Lamblin, E., and Moody, A. 1999. MODIS Land Cover and Land Cover Change: Algorithm Theoretical Basis. WWW document, http://modis-land.gsfc.nasa.gov/.

Townshend, J. R. G. and Justice, C. O. 1981. Information extraction from remotely sensed data: A user view. *International Journal of Remote Sensing* 2: 313–29.

Ustin, S. 2004. *Manual of Remote Sensing: Volume 4, Remote Sensing for Natural Resource Management and Environmental Monitoring*. New York: John Wiley and Sons.

Walter, V. and Fritsch, D. 2000. Automated revision of GIS databases. In K.-J. Li, K. Makki, N. Pissinou, and S. Ravada (eds) *Proceedings of the Eighth ACM Symposium on Advances in Geographic Information Systems*, Washington, DC, USA, pp. 129–34.

Wulder, M. and Boots, B. 1998. Local spatial autocorrelation characteristics of remotely sensed imagery assessed with the Getis statistic. *International Journal of Remote Sensing* 19: 2223–31.

Chapter 4

Spatialization

André Skupin and Sara I. Fabrikant

Researchers engaged in geographic information science are generally concerned with conceptualizing, analyzing, modeling, and depicting geographic phenomena and processes in relation to geographic space. GI scientists consider spatial concepts, such as a phenomenon's absolute location on the Earth's surface, its distance to other phenomena, the scale at which it operates and therefore should be represented and studied, and the structure and shape of emerging spatial patterns. Geographic location is indeed a core concept and research focus of GI Science, and this is well reflected throughout the many chapters of this volume. In recent years, however, it has become apparent that the methods and approaches geographers have been using for hundreds of years to model and visualize geographic phenomena could be applied to the representation of any object, phenomenon, or process exhibiting spatial characteristics and spatial behavior in intangible or abstract worlds (Couclelis 1998). This applies, for example, to the Internet, in which text, images, and even voice messages exist in a framework called cyberspace. Other examples include medical records that have body space as a frame of reference, or molecular data structures that build up the human genome. These abstract information worlds are contained in massive databases, where billions of records need to be stored, managed, and analyzed. Core geographic concepts such as location, distance, pattern, or scale have gained importance as vehicles to understand and analyze the hard-to-grasp and volatile content of rapidly accumulating databases, from real-time stock market transactions to global telecommunication flows. This chapter is devoted to the use of spatial metaphors to represent data that may not be inherently spatial for knowledge discovery in massive, complex, and multi-dimensional databases. It discusses concepts and methods that are collectively referred to as *spatialization*.

What Is Spatialization?

In very general terms, spatialization can refer to the use of spatial metaphors to make sense of an abstract concept. Such spatialization is frequently used in everyday language (Lakoff and Johnson 1980). For example, the phrase "Life is a Journey" facilitates the understanding of an abstract concept ("human existence") by mapping from a non-spatial linguistic source domain ("life") to a tangible target domain ("journey") that one may have actually experienced in the real world. The desktop metaphor used in human–computer interfaces is another example for a spatial metaphor.

The role of spatial metaphors, including geographic metaphors, is also central to the more narrow definition of spatialization developed in the GI Science literature since the 19990s (Kuhn and Blumenthal 1996, Skupin and Buttenfield 1997, Skupin, Fabrikant, and Couclelis 2002), which is the basis for this chapter. Spatialization is here defined as the systematic transformation of high-dimensional data sets into lower-dimensional, spatial representations for facilitating data exploration and knowledge construction (after Skupin, Fabrikant, and Couclelis 2002).

The rising interest in spatialization is related to the increasing difficulty of organizing and using large, complex data repositories generated in all parts of society. Spatialization corresponds to a new, visual paradigm for constructing knowledge from such data. In the geographic domain, interest in spatialization stems largely from the growing availability of multi-dimensional attribute data originating from such sources as multi-temporal population counts, hyperspectral imagery, and sensor networks. New forms of data, still largely untapped by geographic analysis include vast collections of text, multimedia, and hypermedia documents, including billions of Web pages. A number of examples are discussed in this chapter highlighting the role of spatialization in this context.

The focus on spatial metaphors hints at a fundamental relationship between spatialization efforts and GI Science, with relevance beyond the geographic domain. Many spatio-temporal techniques developed and applied in GI Science are applicable in spatialization, and the ontological, especially cognitive, foundations underlying the conceptualization and representation of space can inform spatialization research. That is particularly true for a group of spatializations collectively referred to as "map-like" (Skupin 2002b), which are discussed and illustrated in some detail later in this chapter.

Spatializations are typically part of systems involving people exploring highly interactive data displays with sophisticated information technology. Most current spatialization research is directed at defining and refining various parameters of such interactive systems. However, the result of a spatialization procedure could also be a static hardcopy map that engages the viewer(s) in a discussion of depicted relationships, and triggers new insights (Skupin 2004). For example, one could visualize all the scientific papers written by GI scientists in 2006 in the form of a map printed on a large poster and use this to inspect the structure of the discipline at that moment in time. This can then encourage and inform the discourse on the state and future of the discipline much as a neighborhood map facilitates discussion on zoning ordinance changes during a city-planning forum.

Who Is Working On Spatialization?

The main challenge faced by anyone embarking on the creation of spatializations is that insights and techniques from numerous, and often disparate, disciplines need to be considered. *Visualization* research is very interdisciplinary and conducted by a heterogeneous group of loosely connected academic fields. *Scientific visualization* (McCormick, Defanti, and Brown 1987) and *information visualization* (Card, Mackinlay, and Shneiderman 1999) are two strands of particular interest for this discussion, both drawing heavily on computer science. The former is concerned with the representation of phenomena with physically extended dimensions (for example, width, length, height), usually in three dimensions. Typical application examples are found in such domains as geology (rock formations), climatology (hurricanes), and chemistry (molecular structures). Scientific visualization has obvious linkages with *geographic visualization* (see Chapters 11 and 16 of this volume, by Cartwright and Gahegan respectively, for two treatments of this topic) whenever the focus is on depicting phenomena and processes that are referenced to the Earth's surface. In contrast, information visualization is concerned with data that do not have inherent spatial dimensions. Examples include bibliometric data, video collections, monetary transaction flows, or the content and link structure of Web pages. Most information visualizations are in essence spatialization displays. Spatialization is thus best interpreted in the context of information visualization, which is quickly maturing into a distinct discipline, including dedicated conferences, scientific journals, textbooks, and academic degree programs.

Within GI Science, interest in spatialization tends to grow out of the *geographic visualization* community, which in turn mostly consists of classically trained cartographers. It is not surprising then that GIScientists involved in spatialization research draw inspiration from traditional cartographic principles and methods (Skupin 2000). On the other hand, ongoing developments in geographic visualization have also led to interactive, dynamic approaches that go beyond the static, 2D map (see Chapter 17 by Batty, in this volume, for some additional discussion and examples of this type) and within which spatialization tools can be integrated.

Data mining and knowledge discovery share many of the computational techniques employed in spatialization (see Chapter 19 by Miller, this volume, for some additional discussion of geographic data mining and knowledge discovery), for example artificial neural networks. Many preprocessing steps are similar, such as the transformation of source data into a multidimensional, quantitative form (Fabrikant 2001), even if these data sources are non-numeric.

Ultimately, spatialization is driven by the need to overcome the limited capacity of the human cognitive system to make sense of a highly complex, multidimensional world. That is why *psychology* and especially *cognitive science* have become influential disciplines in this research area. In this context it should be pointed out that while this chapter focuses on visual depictions, spatializations could include multimodal representations involving other senses such as sound, touch, smell, etc. In fact, the term spatialization first became known in the context of methods for producing 3D sound and detecting 3D spatial relationships from sound.

Computer science is still the dominant academic home to most spatialization efforts and has led the development of fundamental principles and novel techniques, especially in the human–computer interaction (HCI) field (Card, Mackinlay, and Shneiderman 1999). Few areas of scientific work have devoted as much effort to spatialization as information and library science, particularly when it comes to the analysis of text and hypermedia documents (Börner, Chen, and Boyack 2002, Chen 2003).

What Kinds Of Data Can Be Used For Spatialization?

Spatialization methodologies can be applied to many different types of data. One possible division of these would focus on the degree to which they are structured, leading to a distinction between structured, semi-structured, and unstructured data (Skupin and Fabrikant 2003). This is useful in terms of highlighting basic data transformation difficulties often encountered in spatialization. For example, unstructured text data may lack a clear indication of where one data item ends and another begins and can have dimensions numbering in the hundreds or thousands, as contrasted with multidimensional data typically used in geospatial analysis, where one rarely encounters more than a few dozen dimensions. However, given the focus of this volume on GI Science, this chapter considers two broad data categories. First, we discuss geographically referenced data, which are of obvious relevance to GI scientists. Then, much attention is given to data that are not referenced to geographic space or even related to geographic phenomena.

Geospatially referenced data

Why would one want to apply spatialization to *geographically referenced data* if cartographic depictions have proven useful for over 5,000 years and continue to be at the heart of current geovisualization research? Consider one very common example, the geographic visualization of demographic change. One almost always finds either juxtaposed maps of individual time slices or change condensed into composite variables (for example, relative percentage of growth). This may be sufficient for the visual detection of *change as such*, but does not easily support detection of *temporal patterns of change*. While location is what vision experts and cognitive psychologists call "pre-attentive" (MacEachren 1995, Ware 2000), this is basically taken out of play when geographically fixed objects, such as counties, are visualized in geographic space in this manner. Spatialization can eliminate that constraint by creating a new, low-dimensional representation from high-dimensional attributes. For example, one could take multi-temporal, multi-dimensional, demographic data for counties, map each county as a point and, with defined temporal intervals, link those points to form trajectories through attribute space (Skupin and Hagelman 2005). Thus, change becomes visualized more explicitly (Figure 4.1). One can then proceed to look for visual manifestations of common verbal descriptions of demographic change, such as "parallel" or "diverging" development (Figure 4.2). Traditional cartographic visualization in geographic space may also fail to reveal patterns and relationships that do not conform to basic assumptions about geographic space, such as those expressed by Tobler's First Law of Geography (Tobler 1970). With spatialization one can take geographic location out (or control for it) while focusing on patterns formed in n-dimensional attribute space.

In practice, spatializations derived from geographically referenced data will tend to be used not in isolation but in conjunction with more traditional geographic depictions. Due to their predominantly two-dimensional form, geometric data structures and formats used in GI Systems (GIS) are applicable to spatializations. They can be displayed and interacted with in commercial off-the-shelf GIS. Most examples

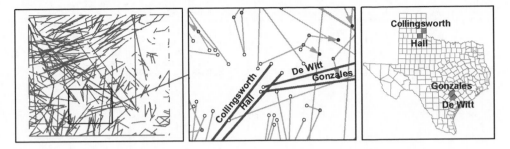

Fig. 4.1 Census-based visualization of trajectories of Texas counties based on data from 1980, 1990, and 2000 US population census
From Skupin and Hagelman 2005

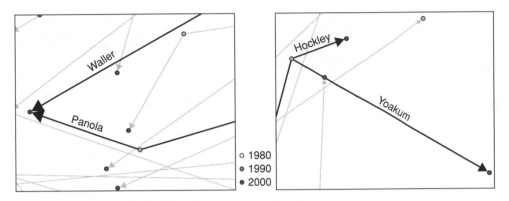

Fig. 4.2 Cases of convergence and divergence in a spatialization of Texas county trajectories
From Skupin and Hagelman 2005

shown in this chapter were in fact created in ArcGIS (Environmental Systems Research Institute, Redlands, California). Spatializations can also be juxtaposed to geographic maps, linked via common feature identifiers, and explored in tandem.

Many types of geographic data are suitable for spatialization. Population census data, for example, have traditionally been subjected to a number of multivariate statistics and visualization techniques, sometimes combined to support exploratory data analysis. Scatter plots and parallel coordinate plots (PCP) are established visual tools in the analytical arsenal. The spatialization methods discussed here do not replace these, but add an alternative view of multivariate data. In this context, it helps to consider how coordinate axes in visualizations are derived. In the case of the popular scatter plot method, each axis is unequivocally associated with an input variable. This is only feasible for a very limited number of variables, even when scatter plots are arranged into matrix form (Figure 4.3). Principal coordinate plots likewise exhibit clear association between axes and variables.

Contrast this with map-like spatializations, in which the relationship between input variables and display coordinates is far less obvious. Some even refer to the resulting axes as "meaningless" (Shneiderman, Feldman D, Rose A, and Grau 2000) and questions like "What do the axes mean?" are frequently encountered. They are difficult

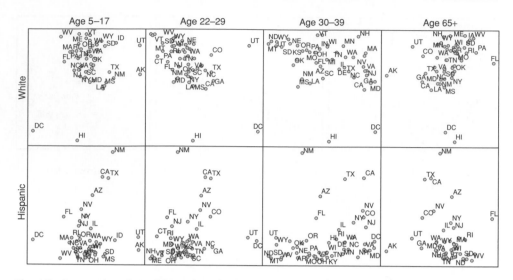

Fig. 4.3 Scatter plots derived from demographic data for US states

to answer, since in such techniques as multidimensional scaling or self-organizing maps *all* input variables become associated with *all* output axes. This allows a holistic view of relationships between observations (Figure 4.4). Figure 4.4 was derived by training an artificial neural network, specifically a self-organizing or Kohonen map (Kohonen 1995), with 32 input variables. Overall similarity of states becomes expressed visually through 2D point visualization. In addition, some of the input variables are shown as component planes in the trained Kohonen map to allow an investigation of relationships between variables.

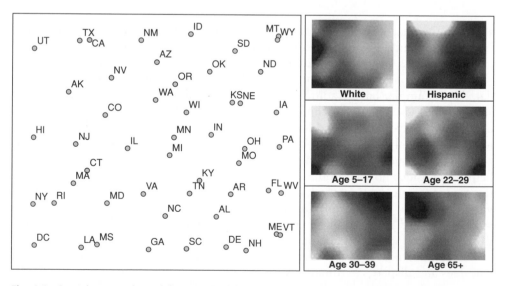

Fig. 4.4 Spatializations derived from 32 demographic variables using the self-organizing map method. Higher values in six (out of 32) component planes expressed as lighter shading

Data without geographic coordinate reference

Some of the most exciting and evocative developments in the visualization field in recent years have been efforts to apply spatial metaphors to non-geographic data or, more specifically, data that are not explicitly linked to physical space. Due to significant differences in how such data are stored, processed, and ultimately visualized, this section discusses a number of data types separately.

There are two broad categories of source data. One involves sources that already contain *explicit* links between data items, which in their entirety can be conceptualized as a graph structure. The goal of spatialization for this category is to convey such structures in an efficient manner in the display space. Hierarchical tree structures are especially common. A prime example is the directory structure of computer operating systems, like Windows or UNIX. Tree structures are also encountered in less expected places. For example, the Yahoo search engine organizes Web pages in a hierarchical tree of topics. The stock market can also be conceptualized as a tree, with market sectors and sub-sectors forming branch nodes and individual stocks as leaf nodes. Apart from such tree structures, data items could also be linked more freely to form a general network structure. The hypermedia structure of the World Wide Web is a good example, with Web pages as nodes and hot links between them. Scientific publications can also be conceptualized as forming a network structure, with individual publications as nodes and citations as explicit links between, generally pointing to the past. The exception might be preprints as they do not exist yet in their defining form. To illustrate this, we collected a few citation links from the *International Journal of Geographical Information Science* (*IJGIS*), starting with a 2003 paper by Stephan Winter and Silvia Nittel entitled "Formal information modelling for standardisation in the spatial domain." The result is an origin–destination table of "who is citing whom" (Table 4.1). Later in this chapter, a visualization computed from this citation link structure is shown.

The second major group of non-georeferenced source data treats items as autonomous units that have no explicit connections among each other. Spatialization of such data relies on uncovering *implicit* relationships based on quantifiable notions of distance or similarity. This requires first a chunking or segmentation of individual data items into smaller units, followed by a computation of high-dimensional relationships. For example, the spatialization of text documents may involve breaking up each document into individual words. The following computations are then based on finding implicit connections between documents based on shared terms (Skupin and Buttenfield 1996). Similarly, images could be spatialized on the basis of image segmentation (Zhu, Ramsey, and Chen 2000). Other examples for spatializations involving disjoint items have included human subject test data derived from user tracking and elicitation experiments (Mark, Skupin, and Smith 2001).

How Does Spatialization Work?

The types of data to which spatialization can be applied are so heterogeneous that there really is no single method. As was stressed earlier, spatialization tends to draw on many different disciplines and integrating these influences can be challenging. For

Table 4.1 Origin–destination table for a citation network formed by papers within the *International Journal of Geographical Information Science*

Link	From Author	From Title	To Author	To Title
1	Su et al. (1997)	Algebraic models for the aggregation . . .	Abler (1987)	The National Science Foundation . . .
2	Su et al. (1997)	Algebraic models for the aggregation . . .	Rhind (1988)	A GIS research agenda
3	Su et al. (1997)	Algebraic models for the aggregation . . .	Brassel and Weibel (1988)	A review and conceptual framework . . .
4	Sester (2000)	Knowledge acquisition for the . . .	Su et al. (1997)	Algebraic models for the aggregation . . .
5	Lin (1998)	Many sorted algebraic data models . . .	Su et al. (1997)	Algebraic models for the aggregation . . .
6	Lin (1998)	Many sorted algebraic data models . . .	Kosters et al. (1997)	GIS-application development with . . .
7	Winter and Nittel (2003)	Formal information modeling . . .	Lin (1998)	Many sorted algebraic data models . . .
8	Lin (1998)	Many sorted algebraic data models . . .	Takeyama and Couclelis (1997)	Map dynamics: integrating cellular . . .
9	Clarke and Gaydos (1998)	Loose-coupling a cellular automaton . . .	Takeyama and Couclelis (1997)	Map dynamics: integrating cellular . . .
10	Shi and Pang (2000)	Development of Voronoi-based . . .	Okabe et al. (1994)	Nearest neighborhood operations . . .
11	Shi and Pang (2000)	Development of Voronoi-based . . .	Takeyama and Couclelis (1997)	Map dynamics: integrating cellular . . .
12	De Vasconcelos et al. (2002)	A working prototype of a dynamic . . .	Takeyama and Couclelis (1997)	Map dynamics: integrating cellular . . .
13	Cova and Goodchild (2002)	Extending geographical representation . . .	Takeyama and Couclelis (1997)	Map dynamics: integrating cellular . . .
14	Wu (2002)	Calibration of stochastic cellular . . .	Takeyama and Couclelis (1997)	Map dynamics: integrating cellular . . .
15	Wu and Webster (2000)	Simulating artificial cities in a GIS . . .	Takeyama and Couclelis (1997)	Map dynamics: integrating cellular . . .
16	Wu and Webster (2000)	Simulating artificial cities in a GIS . . .	Batty and Xie (1994)	Modelling inside GIS: Part I. model . . .
17	Wu and Webster (2000)	Simulating artificial cities in a GIS . . .	Peuquet and Duan (1995)	An event-based spatial temporal . . .
18	Wu and Webster (2000)	Simulating artificial cities in a GIS . . .	Burrough and Frank (1995)	Concepts and paradigms in spatial . . .
19	Wu and Webster (2000)	Simulating artificial cities in a GIS . . .	Clarke and Gaydos (1998)	Loose-coupling a cellular automaton . . .

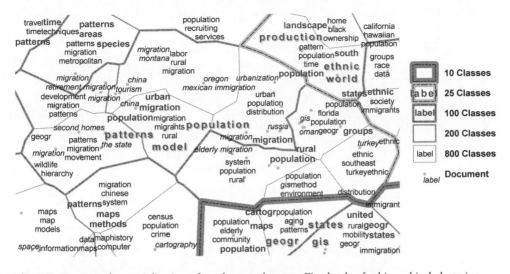

Fig. 4.5 Portion of a spatialization of conference abstracts. Five levels of a hierarchical clustering solution are shown simultaneously
From Skupin 2004

an example, consider the task of creating a map-like visualization of the thousands of abstracts that are presented at the annual meeting of the Association of American Geographers (AAG). This is an example of a *knowledge domain visualization* and would be useful in the exploration of major disciplinary structures and relationships in the geographic knowledge domain (Figure 4.5). Figure 4.6 shows the broad outline of a possible methodology for creating such a visualization. In the process, it also serves to illustrate the range of involved disciplines and influences, which include

- Information science and library science for creation of a term-document matrix, similar to most text retrieval systems and Web search engines (Widdows 2004);
- Computer science for the artificial neural network method used here (Kohonen 1995);
- GIS for storage and transformation of spatialized geometry and associated attributes;
- Cartography for scale dependence, symbolization and other design decisions.

Preprocessing

At the core of most spatialization procedures are techniques for dimensionality reduction and spatial layout. These tend to be highly computational, with very specific requirements for how data need to be structured and stored. Preprocessing of source data aims to provide this. In the case of well-structured, numerical data stored in standard database formats, preprocessing is fairly straightforward. For example, for single-year census data it will often involve only a few processing steps that can easily be accomplished using spreadsheet software, such as computation of z-scores, log transformations, or scaling of observations to fit into a 0–1 range.

The data to which spatialization is to be applied are, however, often not in a form that is amenable to immediate computation. In that case, much effort may

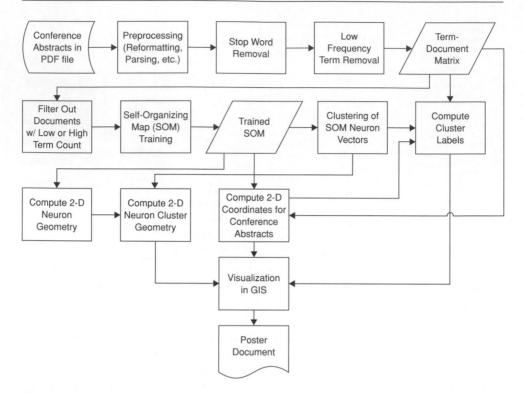

Fig. 4.6 Procedure for deriving a spatialization from AAG conference abstracts
From Skupin 2004

have to be devoted to reorganizing source data into a more suitable form. This can already be surprisingly difficult when dealing with multi-temporal, georeferenced data. Both geographic features and their attributes may be subject to change. For example, census block boundaries may be redrawn, ethnic categories redefined, and so forth. However, the resulting difficulties pale in comparison to source data in which there are no set definitions of what constitutes a feature, how features are separated from each other, or what the attributes should be that become associated with a feature.

What one is faced with here is a distinction between structured and unstructured data. The former is what one almost always encounters in GIS. Unstructured data present wholly different challenges. Consider the case of thousands of conference papers that one might have available in text form in a single file (Figure 4.7). There is no unequivocal separation between different documents nor clear distinction between content-bearing elements (title, abstract, keywords) and context elements (authors, affiliations, email addresses). One could look for certain elements (like end-of-line characters) useful for parsing, but such a procedure will be uniquely tailored to this particular data set, may suffer from inconsistencies in the data, and will require extensive modification to be used for differently organized data.

Semi-structured data are an attempt to address many of these problems by organizing data in accordance with a predefined schema. The extensible markup language

David Aagesen, Department of Geography, State University of New York, Geneseo, NY 14454. E-MAIL: aagesen@geneseo.edu. Still Dividing, Still Conquering: Conflict Over the Ralco Dam in Southern Chile. The Bio-Bío River was the longest free-flowing river in southern Chile until the Pangue Dam was completed in 1997. Construction of a second dam, the Ralco Dam, is currently underway some ten kilometers upstream from the Pangue Dam. The reservoir to be created by the Ralco Dam, which proponents claim is necessary to ensure the efficiency and longevity of the Pangue Dam, will inundate nearly 3400 hectares and require the relocation of 85 indigenous Pehuenche families. The Ralco project has polarized the Pehuenche community in the upper Bío-Bío watershed. On the one hand, many Pehuenche consider Ralco a symbol of progress and embrace the project for its short-term employment opportunities. On the other hand, many Pehuenche view forced resettlement as a gross violation of their constitutional and human rights and have stated publicly that they will die fighting for their land if necessary. This paper outlines the evolution of territorial conflict in the upper Bio-Bío watershed in general, and conflict over the Ralco Dam in particular. It and weaken Pehuenche resistance. Material presented in this paper is based on fieldwork conducted in 1993 and a follow-up visit in July 2000. The paper includes a brief discussion of alternatives to the Ralco Dam that could satisfy energy demand in southern Chile without violating indigenous rights to land and resources.

Keyword: Chile, dams, indigenous geography.

Fig. 4.7 Conference abstract as unstructured text

(XML) is the most prominent solution to this. Figure 4.8 shows an example, in which a schema specifically designed for conference abstracts is applied to previously unstructured data. Such data offer many advantages. This XML file is suitable for human reading and computer parsing alike. From a software engineering point of view, this type of hierarchical, unequivocal structure is also very supportive of object-oriented programming and databases.

Spatialization depends on having data in a form that supports computation of item-to-item relationships in n-dimensional space. For structure-based methods, such as those based on citation links (see Table 4.1) or hypertext links, relationships are already explicitly contained and only have to be extracted to construct network graphs. For content-based analysis, the initial segmentation – for example the segmentation

```
<?xml version="1.0" encoding="iso-8859-1"?>
<CONFERENCE>
 <INDIVCONF>
  <CONFNAME>AAG 2001</CONFNAME>
  <YEAR>2001</YEAR>
  <CONFID>0012001</CONFID>
  <PLACE>New York</PLACE>
  <ABSTRACT>
   <ID>001200100001</ID>
   <TITLE>Still Dividing, Still Conquering: Conflict Over the Ralco Dam in Southern Chile</TITLE>
   <AUTHORINFO>
    <AUTHOR>
     <AUTHORID>00001</AUTHORID>
     <NAME>David Aagesen</NAME>
     <ADDRESS>Department of Geography, State University of New York, Geneseo, NY 14454</ADDRESS>
     <EMAIL>aagesen@geneseo.edu</EMAIL>
    </AUTHOR>
   </AUTHORINFO>
   <KEYWORDS>
    <KEYWORD>chile</KEYWORD>
    <KEYWORD>dams</KEYWORD>
    <KEYWORD>indigenous geography</KEYWORD>
   </KEYWORDS>
   <ABSTEXT> The Bío-Bío River was the longest free-flowing river in southern Chile until the Pangue Dam was completed in
     Construction of a second dam, the Ralco Dam, is currently underway some ten kilometers upstream from the Pangue
     Material presented in this paper is based on fieldwork conducted in 1993 and a follow-up visit in July 2000. The paper
   </ABSTEXT>
  </ABSTRACT>
```

Fig. 4.8 Conference abstract in semi-structured form as part of an XML file

of a photograph or the identification of individual words within a text document –
is followed by significant transformations (see top row in Figure 4.6). For example,
text data may undergo stop word removal and stemming (Porter 1980, Salton 1989),
as illustrated here:

INPUT: The paper includes a brief discussion of alternatives to the Ralco Dam that could
satisfy energy demand in southern Chile without violating indigenous rights to land and
resources
ONPUT: paper includ brief discuss altern ralco dam satisfi energi demand southern chile
violat indigen right land resourc

From this, a high-dimensional vector can then be created for each document, with
dimensions corresponding to specific word stems and values expressing the weight
of a term within a document (Skupin and Buttenfield 1996, Salton 1989, Skupin
2002a).

Dimensionality reduction and spatial layout

The core of any spatialization methodology is the transformation of input data into a
low-dimensional, representational space. In the case of data given as distinct features
with a certain number of attributes one can rightfully refer to the corresponding
techniques as *dimensionality reduction*. *Spatial layout* techniques are typically used
when dealing with explicitly linked features, as in the case of citation networks.

Two popular dimensionality reduction techniques are *multidimensional scaling*
(MDS) and the *self-organizing map* (SOM) method. MDS first requires the computation of a dissimilarity matrix from input features, based on a carefully chosen
dissimilarity measure. Then, the method attempts to preserve high-dimensional dis-
similarities as distances in a low-dimensional geometric configuration of features
(Kruskal and Wish 1978). The popular Themescapes application (Wise, Thomas,
Pennock, et al. 1995) is based on a variant of MDS (Wise 1999). Within GI Science,
spatialization efforts have utilized MDS to create 2D point geometries for sub-
disciplines of geography (Goodchild and Janelle 1988), newspaper articles (Skupin
and Buttenfield 1996, 1997), and online catalog entries (Fabrikant and Buttenfield
2001).

The SOM method is an artificial neural network technique (Kohonen 1995).
It starts out with a low-dimensional (typically 2D) grid of n-dimensional neuron
vectors. N-dimensional input data are repeatedly presented to these neurons. The
best matching neuron to each observation is found and small adjustments are made
to the vector of that neuron as well as to the vectors of neighboring neurons. Over
time, this leads to a compressed/expanded representation in response to a sparse/
dense distribution of input features. Consequently, major topological relationships
in n-dimensional feature space become preserved in the two-dimensional neuron
grid. One can then map n-dimensional observations onto it (left half of Figure 4.4),
visualize individual neuron vector components (right half of Figure 4.4), or compute
neuron clusters (Figure 4.5). SOMs have, for example, been used to spatialize Usenet
discussion groups, Web pages (Chen, Schuffels, and Orwig 1996), image content
(Zhu, Ramsey, and Chen 2000), conference abstracts (Skupin 2002a, 2004), and even

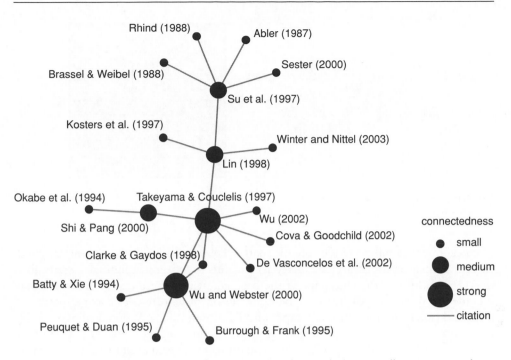

Fig. 4.9 Spring model layout and pathfinder network scaling applied to a small citation network formed by papers in the *International Journal of Geographical Information Science*

a collection of several million patent abstracts (Kohonen, Kaski, Lagus, et al. 1999). *Spring models* are another popular category of dimensionality reduction techniques (Kamada and Kawai 1989, Skupin and Fabrikant 2003).

Pathfinder network scaling (PFN) is a technique used for network visualization, with a preservation of the most salient links between input features. It is frequently applied to citation networks (Chen and Paul 2001). To illustrate this, we computed a PFN solution from the *IJGIS* citation data shown earlier. The result is a network structure consisting of links and nodes. When combined with a geometric layout of nodes derived from a spring model, the citation network can be visualized in GIS (Figure 4.9). Circle sizes represent the degree of centrality a paper has in this network, a measure commonly used in social network analysis (Wasserman and Faust 1999). Note how the centrality of the Takeyama/Couclelis paper derives from it being frequently cited (see Table 4.1), while the Wu/Webster paper establishes a central role because it cites a large number of *IJGIS* papers.

Among spatial layout techniques, the *treemap* method has become especially popular in recent years. It takes a hierarchical tree structure as input and lays portions of it out in a given two-dimensional display space (Johnson and Shneiderman 1991). In the process, node attributes can also be visually encoded (Figure 4.10). For example, when visualizing the directory structure of a hard drive, file size could be encoded as the area size of rectangles. Another important category are *graph layout* algorithms, which attempt to untangle networks of nodes and links in such a manner that crossing lines are avoided as much as possible and network topology is preserved.

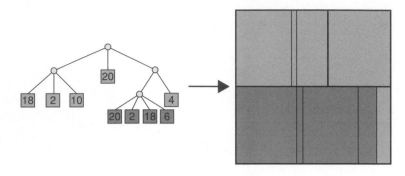

Fig. 4.10 The tree map method
From Skupin and Fabrikant 2003

Once dimensionality reduction or spatial layout methods have been applied, further transformations are necessary to execute the visual design of a spatialization. Depending on the character of the base geometry, these transformations may include the derivation of feature labels, clustering of features, landscape interpolation, and others (Skupin 2002b, Skupin and Fabrikant 2003). When dealing with 2D geometry, much of this can be accomplished in commercial off-the-shelf (COTS) GIS. Many aspects of these transformations remain to be investigated in future research, for instance how scale changes can be implemented as semantic zoom operations (Figure 4.11).

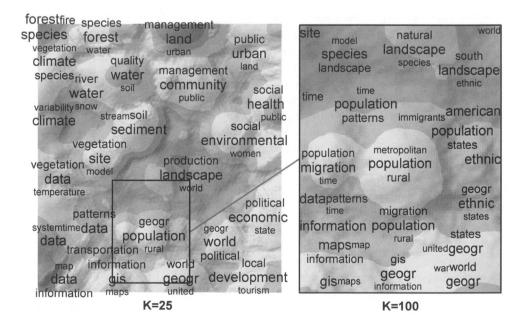

Fig. 4.11 Use of GIS in implementing scale-dependent spatialization of several thousand AAG conference abstracts. Labeling is based on two different k-means cluster solutions
From Skupin 2004

Spatialization geometry can also be linked to attributes that were not part of the input data set. For example, demographic change trajectories (Figures 4.1 and 4.2) could be linked – via symbolization or selection – to voting behavior or public policy decisions (Skupin and Hagelman 2005).

Usability and Cognitive Perspectives

An extensive set of display techniques has been developed for spatialization, and the impressive array of visual forms documents the productivity of this young academic field (Chen 1999). However, few researchers have succeeded in providing empirical evidence to support claims that interactive visual representation tools indeed amplify people's cognition (Ware 2000). Generally, non-expert viewers do not know how spatializations are created and are not told, through legends or traditional map marginalia, how to interpret such aspects of spatialized displays as distance, regionalization, and scale. Of the few existing experimental evaluations in information visualization, most evaluate specific depiction methods or types of software (Chen and Czerwinski 2000, Chen, Czerwinski, and Macredie 2000). While usability engineering approaches are good at testing users' successes in extracting information from a particular visualization, they do not directly assess the underlying theoretic assumptions encoded in the displays, the users' understanding of the semantic mapping between data and metaphor, and between metaphor and graphic variables, or the interaction of graphic variables with perceptual cues.

A fundamental principle in spatialization is the assumption that more similar entities represented in a display should be placed closer together because users will interpret closer entities as being more similar (Wise, Thomas, Pennock, et al. 1995, Card, Mackinlay, and Shneiderman 1999). Montello, Fabrikant, Ruocco, and Middleton (2003) have coined this principle the distance-similarity metaphor. For example, according to the distance-similarity metaphor, US states depicted in Figures 4.3 and 4.4 or conference abstracts shown in Figure 4.5 that are more similar to each other in content are placed closer to one another in the display, while spatialized items that are less similar in content are placed farther apart. In essence, this distance-similarity metaphor is the inverse of Tobler's (1970, p. 236) first law of geography, because similarity typically determines distance in spatializations. Thus we have referred to the "first law of cognitive geography" (Montello, Fabrikant, Ruocco, and Middleton 2003) – people believe that closer features are more similar than distant features. To the extent that this principle is true, it provides theoretical justification for the distance-similarity metaphor as a principle of spatialization design.

In a series of studies relating to point (Fabrikant 2001, Montello, Fabrikant, Ruocco, and Middleton 2003), network (Fabrikant, Montello, Ruocco, and Middleton 2004), region (Fabrikant, Montello, and Mark 2006), and surface display spatializations (Fabrikant 2003) Fabrikant and colleagues have investigated whether the fundamental assumption that spatialization can be intuitively understood as if they represent real-world spaces (Wise, Thomas, Pennock, et al. 1995, Card, Mackinlay, and Shneiderman 1999) is generally true. These studies provide the first empirical evidence of the cognitive adequacy of the distance-similarity metaphor in spatialization.

In these studies, participants have rated the similarity between documents depicted as points in spatialized displays. Four types of spatialization displays have been examined: (1) point displays (e.g., Figures 4.3 and 4.4), (2) network displays linking the points (e.g. Figures 4.1, 4.2 and 4.9), (3) black-and-white regions containing the points (e.g. Figure 4.5), and (4) colored regions containing the points (Figure 4.10). In the point displays, participants based judgments of the relative similarity of two pairs of document points primarily on direct (straight-line or "as the crow flies") metric distances between points, but concentrations of points in the display led to the emergence of visual features in the display, such as lines or clusters, that considerably moderated the operation of the first law of cognitive geography. In the network displays, participants based similarity judgments on metric distances along network links, even though they also had available direct distances across network links and topological separations (numbers of nodes or links connecting points). In the region displays, participants based similarity judgments primarily on region membership so that comparison documents within a region were judged as more similar than documents in different regions, even if the latter were closer in direct distance. Coloring the regions produced thematically-based judgments of similarity that could strengthen or weaken regional membership effects, depending on whether region hues matched or not. In addition, Fabrikant and Montello (2004) also gained explicit information on how similarity judgments directly compare to default distance and direct distance judgments. There are no differences between people's estimates of distance under default (nonspecified) and direct (straight-line) distance instructions for point, network, and region spatializations. Default distance instructions are interpreted as requests for estimates of direct distance in spatializations. They have also found that well-known optical effects such as the vertical (Gregory 1987) and space-filling interval illusion (Thorndyke 1981) affect distance judgments in spatializations and therefore may affect the operation of the first law of cognitive geography.

Without empirical evidence from fundamental cognitive evaluations the identification and establishment of solid theoretical foundations in spatialization will remain one of the major research challenges (Catarci 2000). A solid theoretical scaffold is not only necessary for grounding the information visualization field on sound science, but is also fundamental to deriving valid formalisms for cognitively adequate visualization designs, effective graphical user interface implementations, and their appropriate usability evaluation (Fabrikant and Skupin 2005).

Where Is Spatialization Going?

Spatialization addresses a need to make sense of the information contained in ever-growing digital data collections. There is considerable societal demand for the types of methods discussed in this paper. This includes such obvious applications as counter-terrorism work or the development of improved Web search engine interfaces. Telecommunications companies attempt to find patterns in millions of phone calls through spatialization. Private industry also hopes to use spatialization to detect emerging technological trends from research literature in order to gain a competitive advantage. Funding agencies would like to determine which research grant applications show the most promise. In recent years there have been a growing number of events dedicated to the type of research within which spatialization is prominently featured, organized by the National Academy of Sciences (Shiffrin and Börner 2004), the National Institutes of Health, the National Security Agency, and other public and private entities.

This chapter demonstrates that spatialization may be applicable to both geo-referenced and non-georeferenced phenomena, whenever n-dimensional data need to be investigated in a holistic, visually engaging form. The involvement of GI scientists in spatialization activities does not have to be a one-way street in terms of using spatialization within particular applications. GI Science is also beginning to help answer fundamental questions with regards to how spatializations are con-structed and used (Skupin, Fabrikant, and Couclelis 2002). Our understanding of cognitive underpinnings, usability, and usefulness is still quite incomplete. The com-putational techniques used for spatialization also need further investigation, especially when it comes to developing methods for integrated treatment of the tri-space formed by geographic, temporal, and attribute space. In summary, spatialization is an excit-ing area in which GI Science is challenged to address important issues of theory and practice for many different data and applications.

REFERENCES

Börner, K., Chen, C., and Boyack, K. W. 2002. Visualizing knowledge domains. In B. Cronin (ed.) *Annual Review of Information Science and Technology*. Medford, NJ: Information Today: 179–255.

Card, S. K., Mackinlay, J. D., and Shneiderman, B. 1999. *Readings in Information Visual-ization: Using Vision to Think*. San Francisco, CA: Morgan Kaufmann.

Catarci, T. 2000. What's new in visual query systems? In *Proceedings of the First Inter-national Conference on Geographic Information Science*. Savannah, GA, USA. Santa Barbara, CA: National Center for Geographic Information and Analysis.

Chen, C. 1999. *Information Visualisation and Virtual Environments*. London: Springer.

Chen, C. 2003. *Mapping Scientific Frontiers: The Quest for Knowledge Visualization*. London: Springer.

Chen, C. and Czerwinski, M. 2000. Empirical evaluation of information visualizations: An introduction. *International Journal of Human-Computer Studies* 53: 631–5.

Chen, C., Czerwinski, M., and Macredie, R. D. 2000. Individual differences in virtual envir-onments: Introduction and overview. *Journal of the American Society of Information Science* 51: 499–507.

Chen, C. and Paul, R. J. 2001. Visualizing a knowledge domain's intellectual structure. *IEEE Computer* 34(3): 65–71.

Chen, H., Schuffels, C., and Orwig, R. 1996. Internet categorization and search: A self-organizing approach. *Journal of Visual Communication and Image Representation* 7: 88–102.

Couclelis, H. 1998. Worlds of information: The geographic metaphor in the visualiza-tion of complex information. *Cartography and Geographic Information Systems* 25: 209–20.

Fabrikant, S. I. 2001. Evaluating the usability of the scale metaphor for querying semantic spaces. In R. D. Montello (ed.) *Spatial Information Theory: Foundations of Geographic Information Science*. Berlin: Springer Lecture Notes in Computer Science No. 2205: 156–72.

Fabrikant, S. I. 2003. Distanz als Raummetapher für die Informationsvisualierung (Distance as a spatial metaphor for the visualization of information). *Kartographische Nachrichten* 52: 276–82.

Fabrikant, S. I. and Buttenfield, B. P. 2001. Formalizing semantic spaces for information access. *Annals of the Association of American Geographers* 91: 263–80.

Fabrikant, S. I. and Montello, D. R. 2004. Similarity and distance in information spatializations. In *Proceedings of GIScience 2004*, Adelphi, MD, USA. Santa Barbara, CA National Center for Geographic Information and Analysis, pp, 279–81.

Fabrikant, S. I., Montello, D. R., and Mark, D. M. 2006. The distance-similarity metaphor in region-display spatializations. *IEEE Computer Graphics & Applications* 26: in press.

Fabrikant, S. I., Montello, D. R., Ruocco, M., and Middleton, R. S. 2004. The distance-similarity metaphor in network-display spatializations. *Cartography and Geographic Information Science* 31: 237–52.

Fabrikant, S. I. and Skupin, A. 2005. Cognitively plausible information visualization. In Dykes, J., MacEachren, A. M., and Kraak, M.-J. (eds) *Exploring Geovisualization*. Amsterdam: Elsevier: 667–90.

Goodchild, M. F. and Janelle, D. G. 1988. Specialization in the structure and organization of geography. *Annals of the Association of American Geographers* 78: 1–28.

Gregory, R. L. (ed.). 1987. *The Oxford Companion to the Mind*. Oxford: Oxford University Press.

Johnson, B. and Shneiderman, B. 1991. Treemaps: A space-filling approach to the visualization of hierarchical information structures. In *Proceedings of IEEE Visualization '91*, San Diego, CA, USA. Los Alamitos, CA: Institute of Electrical and Electronics Engineers, pp. 275–82.

Kamada, T. and Kawai, S. 1989. An algorithm for drawing general undirected graphs. *Information Processing Letters* 31: 7–15.

Kohonen, T. 1995. *Self-Organizing Maps*. Berlin: Springer-Verlag.

Kohonen, T., Kaski, S., Lagus, K., Salojärvi, J., Honkela, J., Paatero, V., and Saarela, A. 1999. Self organization of a massive text document collection. In E. Oja and S. Kaski (eds) *Kohonen Maps*. Amsterdam: Elsevier, pp. 171–82.

Kruskal, J. B. and Wish, M. 1978. *Multidimensional Scaling*. London: Sage University Paper Series on Qualitative Applications in the Social Sciences No. 07-011.

Kuhn, W. and Blumenthal, B. 1996. *Spatialization: Spatial Metaphors for User Interfaces*. Vienna, Austria: Technical University of Vienna.

Lakoff, G. and Johnson, M. 1980. *Metaphors We Live By*. Chicago, I: University of Chicago Press.

MacEachren, A. M. 1995. *How Maps Work*. New York: Guilford Press.

Mark, D. M., Skupin, A., and Smith, B. 2001. Features, objects, and other things: Ontological distinctions in the geographic domain. In D. R. Montello (ed.). *Spatial Information Theory: Foundations of Geographic Information Science*. Berlin: Springer-Verlag Lecture Notes in Computer Science No. 2205: 488–502.

McCormick, B. H., Defanti, T. A., and Brown, M. D. 1987. Visualization in scientific computing. *IEEE Computer Graphics and Applications* 7(10): 69–79.

Montello, D. R., Fabrikant, S. I., Ruocco, M., and Middleton, R. 2003. Spatialization: Testing the first law of cognitive geography on point-spatialization displays. In W. Kuhn, M. F. Worboys, and S. Timpf (eds) *Spatial Information Theory: Foundations of Geographic Information Science*. Berlin: Springer-Verlag Lecture Notes in Computer Science No. 2825: 335–51.

Porter, M. F. 1980. An algorithm for suffix stripping. *Program-Automated Library and Information Systems* 14: 130–7.

Salton, G. 1989. *Automated Text Processing: The Transformation, Analysis, and Retrieval of Information by Computer*. Reading, MA: Addison-Wesley.

Shiffrin, R. M. and Börner, K. 2004. Mapping knowledge domains. *Proceedings of the National Academy of Sciences* 101: 5183–5.

Shneiderman, B., Feldman, D., Rose, A., and Grau, X. F. 2000. Visualizing digital library search results with categorical and hierarchical axes. In *Proceedings of the ACM Conference on Digital Libraries (DL2000),* San Antonio, TX, USA. New York: Association of Computing Machinery, pp. 57–65.

Skupin, A. 2000. From metaphor to method: Cartographic perspectives on information visualization. In *Proceedings of the IEEE Symposium on Information Visualization (InfoVis '00),* Salt Lake City, UT, USA. Los Alamitos, CA: Institute of Electrical and Electronics Engineers, pp. 91–7.

Skupin, A. 2002a. A cartographic approach to visualizing conference abstracts. *IEEE Computer Graphics and Applications,* 22: 50–8.

Skupin, A. 2002b. On geometry and transformation in map-like information visualization. In K. Börner and C. Chen (eds) *Visual Interfaces to Digital Libraries.* Berlin: Springer-Verlag Lecture Notes in Computer Science No. 2539: 161–70.

Skupin, A. 2004. The world of geography: Visualizing a knowledge domain with cartographic means. *Proceedings of the National Academy of Sciences* 101: 5274–8.

Skupin, A. and Buttenfield, B. P. 1996. Spatial metaphors for visualizing very large data archives. In *Proceedings of GIS/LIS '96,* Denver, CO, USA. Bethesda, MD: American Society for Photogrammetry and Remote Sensing, pp. 607–17.

Skupin, A. and Buttenfield, B. P. 1997. Spatial metaphors for visualizing information spaces. In *Proceedings of Auto Carto 13.* Bethesda, MD: American Congress of Surveying and Mapping/American Society for Photogrammetry and Remote Sensing, pp. 116–25.

Skupin, A. and Fabrikant, S. I. 2003. Spatialization methods: A cartographic research agenda for non-geographic information visualization. *Cartography and Geographic Information Science* 30: 99–119.

Skupin, A., Fabrikant, S. I., and Couclelis, H. 2002. Spatialization: Spatial metaphors and methods for handling non-spatial data. WWW document, http://www.geog.ucsb.edu/~sara/html/research/ucgis/spatialization_ucsb.pdf.

Skupin, A. and Hagelman, R. 2005. Visualizing demographic trajectories with self-organizing maps. *GeoInformatica* 9: 159–79.

Thorndyke, P. W. 1981. Distance estimation from cognitive maps. *Cognitive Psychology* 13: 526–50.

Tobler, W. 1970. A computer model simulating urban growth in the Detroit region. *Economic Geography* 46: 234–40.

Ware, C. 2000. *Information Visualization: Perception for Design.* San Francisco, CA: Morgan Kaufman.

Wasserman, S. and Faust, K. 1999. *Social Network Analysis.* Cambridge: Cambridge University Press.

Widdows, D. 2004. *Geometry and Meaning.* Stanford: CSLI Publications.

Wise, J. A. 1999. The ecological approach to text visualization. *Journal of the American Society for Information Science* 50: 1224–33.

Wise, J. A., Thomas, A. J., Pennock, K., Lantrip, D., Pottier, M., Schur, A., and Crowet, V. 1995. Visualizing the non-visual: Spatial analysis and interaction with information from text documents. In *Proceedings of the IEEE Symposium on Information Visualization (InfoVis '95),* Atlanta, GA, USA, pp. 51–8.

Zhu, B., Ramsey, M., and Chen, H. 2000. Creating a large-scale content-based airphoto image digital library. *IEEE Transactions on Image Processing* 9: 163–7.

Chapter 5

Uncertainty in Spatial Databases

Ashley Morris

Uncertainty permeates the fabric of spatial data at every level: in the assimilation and storage of geospatial features (which may have uncertain or indeterminate boundaries), in the operations on these features, and in the representation of the results of the operations.

As a spatial database represents features modeled in the real, infinitely complex world, it is not able to be completely faithful to the real features being represented. Uncertainty is often used to describe the differences between what the database captures of the real world, and that which actually exists (Goodchild 1998).

Terms

There is a special vocabulary attached to spatial databases. Some definitions are provided below to ensure that the ambiguity in the verbiage is minimized.

An *error* is simply something that is incorrect. We must know a value to be incorrect for there to be error in our model. Occasionally, we may note that there are missing values for attributes of our objects. We can treat these values as either being in error, or being uncertain. An observed error relates to a single value, whereas accuracy relates to a set of values.

Accuracy may be broken down into *bias* and *precision*. Bias is dependent upon the underlying data model, and is most often predicted by the mean error between a set of known (or actual) values and predicted (or stored) values. Precision is also model-based, but it is often based upon the standard deviation of the error of a set of values. Accuracy can thus be defined as the sum of the precision and the unbias (Foody and Atkinson 2002). Positional inaccuracy is typically caused by limited ability to measure locations on the surface of the earth (Goodchild 1998).

Uncertainty simply means that of which we are not certain; that which is not known. Uncertainty can be divided into ambiguity and vagueness (Klir and Folger 1988). *Ambiguity* is the type of uncertainty most often modeled internally by probability. In probability, we are predicting the chances of an object being a member of a boolean set. So if we were representing the country of Switzerland as a single object in our spatial database, and we had to choose among German, French, Italian, or Romansh as the value for the attribute *language*, we would most likely choose German, as about 64 percent of the population speak German. However, if we went to a higher level of detail and considered Switzerland as being made up of 23 individually modeled cantons, we could represent the predominant language of each canton, and thus end up with less ambiguity (http://www.switzerlandtourism.ch/). Most often, probability is used to express ambiguity when dealing with uncertain data. *Vagueness* refers to data that does not strictly belong to a crisp set.

Unfortunately, we also have ambiguity in the terms being used to denote something of interest in a spatial database. The term *feature* is often used to describe a single entity, although it may refer to a collection of entities. *Entity*, on the other hand, is often ambiguously used, because of the confusion with the term entity in the entity/relationship method of conceptual database modeling. The same problem follows the term *object*, as it is often confused with an object in an object-oriented spatial database. One particularly confusing point is that an object, in object-oriented database terminology, may actually be a *collection* of several objects that is treated as a single object.

We will use the term *feature* to describe the representation of a single physical *thing* within the spatial database, and *entity* to denote the physical thing existing in the real physical world.

Crisp sets, also known as boolean sets or classical sets, consist of unordered collections of unique objects. All members of the set have full (1.0) membership in the set. Objects that are not members of the set have no (0.0) membership in the set.

In classical set theory, there are two fundamental laws: the law of the excluded middle and the law of contradiction. The law of the excluded middle states that every proposition is either true or false, and the law of contradiction states that an element x must either be a member of a set or not be a member of a set (Robinson 2003).

Fuzzy sets introduce the concept of partial membership. In a fuzzy set, objects may have *partial* membership. For example, a certain soil sample may have 0.49 membership in the set of *Loamy Soil*, it may have 0.33 membership in *Sandy Soil*, and it may have 0.18 membership in *Rocky Soil*. Note that this violates both the law of the excluded middle, and the law of contradiction. This will have consequences on the operations that can be performed upon these sets, but it will allow operations to be performed on elements that normally would not be considered as set members.

Within a fuzzy set, we may have objects comprising the *core* (full membership of 1.0 in the set in question) and we may have a *boundary* (the area beyond which they have no or negligible membership in the set). A classic example of the core and boundary problem is determining where a forest begins. Is it determined based on a hard threshold of trees per hectare? This may be the boundary set by management policy but it is likely not to be the natural definition. There are several ways to manage these uncertain boundaries (Fisher 1996, Cheng, Molenaar, and Lin 2001). If our spatial database can represent the outlying trees as being partial members of the forest, then the decision-maker will see these features as being partial members on the display.

In general, the idea of implementing fuzzy set theory as a way to model uncertainty in spatial databases has a long history. In the 1970s, fuzzy set theory was proposed as a technique for geographic analysis (Gale 1972, Leung 1979). However, it is only recently that this idea has taken hold in the modern Geographic Information

System (GIS) (see Chapter 14 by Robinson in this volume for a more detailed discussion of this approach).

What is Stored in Spatial Databases

First, we need to understand what is being stored in a spatial database. It does not store maps; rather it stores *features*, objects with an associated geometry. While a spatial database may store any object with a geometry, it typically stores geo-referenced or geospatial objects. These objects can have attributes answering the questions: who, what, where, when, why, and how.

The *who* refers to the generator of the source material. It may have come from remotely sensed data, bathymetric sidescan sonar, field GPS surveying, or a downloaded shapefile. The *who* often is important when measuring the quality of the data. If a source is not reliable, then we may want to consider the data has an additional degree of vagueness or imprecision.

The *what* is answered by the attributes of the object, the thematic data. Features in a spatial database may have many attributes for thematic data, they may have multiple values for a single attribute of thematic data (like calling the largest mountain in Alaska both *McKinley* and *Denali*), and they might possibly have no thematic attributes at all, other than their system generated primary key or object identifier. Obviously, we may have ambiguity associated with the feature as a result of uncertainty in the thematic attributes.

The *where* is what differentiates spatial data from other kinds of data that may be stored in a database. The spatial attributes not only tell the size and dimensions of an object, but also where that object is positioned in space. This is the area where we will spend most of our effort: looking at the uncertainty that may occur when trying to determine, store, and represent *where* an object may be. The spatial information stored in a spatial database describes the location and shape of the geographic features in terms of points, lines, and areas.

The *when* comes from attempting to manage temporal data. Most databases do not, by default, attempt to manage temporal data. The relational database model is designed to only store a snapshot of the world being modeled at a single point in time. It can be extended to represent the evolution of the modeled world over time better, as can the object-oriented database model. It is slightly easier to model temporality in an object-oriented model, as we will describe in more detail later.

Modeling *why* is beyond the scope of most spatial databases, but may come into play if we are dealing with an advanced GIS that supports modeling and pattern type queries.

How, as it relates to features, describes the scale of the data. When performing queries using features of different scale, we may return imprecise results.

Modeling and Storing Features

GIS users are demanding the ability to represent geographic objects with uncertain boundaries (e.g., Ehlschlaeger and Goodchild 1994, Campari 1996, Goodchild 1998,

Hunter 1998), thus, in modern GIS, there is a need to more precisely model and represent the underlying uncertain spatial data (Zhang and Goodchild 2002). Models have been proposed since 1990 that allow for enriching database models to manage uncertain spatial data (e.g. George, Buckles, and Petry 1992). A major motivation for this is that there exist geographic objects with uncertain boundaries, and fuzzy sets are a natural way to represent this uncertainty (Goodchild and Gopal 1990, Burrough 1996, Burrough and Couclelis 1996, Couclelis 1996, Petry, Cobb and Morris 1999). There are two primary ways to model this data: as individual objects, or as continuous fields. Ideally, we should aim towards a framework that can support both of these.

Feature-based (atomic) models propose that the world can be modeled by representing the smallest single atomic feature, and then layers can be composed of collections of these features. On the other hand, field-based (plenum) models do not have an intrinsic notion of atom, or boundaries. The crisp concept of feature is not applicable; rather each pixel may have a value for one or more attributes. Historically, raster-based GIS have been modeled on the field concept, and vector-based GIS have used feature-based models.

Rather than modeling geospatial features as continuous fields, with perhaps partial membership in every class being modeled, or modeling geospatial features as simple concrete objects, and avoiding the uncertainty altogether, the ideal spatial database should be able to manage both. Nowhere does the problem of real-world complexity cause greater problems than when trying to model the many possible observable abstractions of real-world data.

Relational and object-oriented database management systems (DBMSs)

A problem comes when we have to apply this real-world data model for storage in a DBMS. As we want to store features with uncertain boundaries it follows that a DBMS that supports fuzziness would be a more ideal solution. Buckles and Petry (1982), for example, show that we can easily extend the relational model to support fuzziness. So the problem lies not when we wish to store fuzzy items in a relational DBMS but when we wish to store spatial features in a relational DBMS.

Historically, most spatial databases have stored features in a relational format. In the relational database model, all data must be abstracted into two-dimensional tables. These tables consist of columns of attributes and rows (or tuples) of instances.

There are several problems with the relational database model for storage of spatial entities. One chief problem is that these tables are defined to be "sets of tuples" (Codd 1970). A mathematical *set* is defined to be an unordered collection of objects. This is not a problem when dealing with zero-dimensional points in space. When working with two-dimensional polygons or hulls, and even with one-dimensional lines, this does pose difficulties, as the ordering of the points comprising the boundary of the object makes a definite difference in how the object may be represented (cf. Figures 5.1 and 5.2). So, to store spatial features in a relational DBMS we are violating the basic definition of the relational model by insisting that the spatial attributes be stored in a particular order.

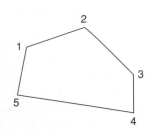

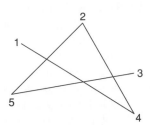

Fig. 5.1 Case in which the points are stored in the database and connected in order from the last point back to the first point

Fig. 5.2 Case in which the segments between the points are connected in no particular order

There are ways to do this, as most GIS have relational databases as their storage mechanism. However, this is one of the reasons why several authors have suggested that the object-oriented data model might be the best available technique to model spatial data (Goodchild and Gopal 1990, Mackay, Band and Robinson 1991, George, Buckles, Petry, et al. 1992, Cross and Firat 2000).

Fuzzy spatial OODBMS

Morris (2003) describes a framework that combines a geographic data model with an object-oriented data model that supports imprecision and uncertainty. In this framework, we are able to incorporate all of the benefits of the object-oriented paradigm into our geographic data framework. This framework is more appropriate for spatial data (Robinson and Sani 1993) as it incorporates fuzziness into both the storage and representation of the spatial features themselves (Burrough 1996, Couclelis 1996, Usery 1996, Morris and Petry 1998).

One of the key elements of this framework is that coverages and/or layers can be stored and represented as a *set* of spatial objects, or features (Morris, Foster, and Petry 1999). This set may be either crisp or fuzzy. By being able to represent a coverage or geographic layer in the object-oriented model as a set we gain the use of all of the normal set operations. Thus, we can use the notion of a set when performing spatial queries or operations. Spatial queries on spatial coverages, as performed by a GIS, are typically akin to set operations on those coverages, and behave as queries on layers or collections (from an OO perspective). Fuzzy queries can also be supported with this framework.

Another advantage of the object-oriented framework is that, internally to the database, the points that comprise a feature may be stored in order. As mentioned before, the relational model does not allow for ordering, as points would be stored as a set, which by definition is unordered. In the OO world, we can store these points using any of the collection types supported by the OODBMS (Cattell and Barry 2000). Since we have these collection types, it is trivial to store the points in a polygon in a particular order, and it is trivial for the rendering engine to connect these points. So there is much less overhead (and conflict with the model) in the OODB model for spatial objects than in the relational database model.

The Open GIS Consortium (OGC; http://www.opengis.org), while not explicitly specifying an OODB format for spatial data, provides a connectivity layer within an object-oriented framework. This will allow users to share existing data sets, regardless of format. The framework of the OGC model is very abstract in its modeling. That is, it makes many demands upon the implementer to manage things such as multiple and partial inheritance.

Multiple inheritance may also introduce uncertainty into object behavior. There may be a conflict in methods when an object inherits from two or more classes. In the fuzzy world, there are a few different ways to manage this. Since an object has a degree of membership in a class, it is possible to simply let the class with the higher membership "win," and the inheritance will be dependent upon the membership value of the direct parent class. In practice, GIS modelers usually take a very hands on approach to the implementation of a particular application, and they will intervene and decide which parent class would be more dominant (Morris 2003).

The OGC framework allows for multiple representations of the same spatial object. First, a feature may be stored in the database as a single object, with several representations of its spatial characteristics. This would allow, for example, the query mechanism to pick the single most appropriate spatial representation of the ones stored for an object. For example, we may have three representations of a feature: one a raster representation at 1:10,000, one a vector representation at 1:24,000, and the third a raster representation at 1:5,000. If we were performing an overlay query with a layer that consisted of vector objects at 1:15,000, our query mechanism would probably choose the 1:24,000 vector representation to use in the query. The GIS modeler could, of course, a priori determine which representation would be the preferred one. The desired representation could also be derived based upon previous user selection (Robinson 2000).

When performing queries on an object-oriented spatial database, we can generalize all collections to the set (Morris, Foster, and Petry 1999). This gives us many advantages. First, since a spatial coverage is by definition a (fuzzy) set (Morris and Petry 1998), when a spatial operation is performed on a collection the order of processing of the elements in the collection is irrelevant, as spatial queries and spatial operations are associative. While the order of processing individual elements of a set may have a drastic effect on performance, it will not affect the final result. Another advantage is that this generalization means we can use any set operation on any spatial data collection, including coverages and layers.

Classes, layers, and membership

The framework described in Morris and Jankowski (2000) and Morris (2003) examines how object collections can be combined in a spatial database so that typical GIS notions and queries are supported and maintained. In practice, the layers used by the GIS often, but not always, correspond to the classes in the underlying object database. This is because the layers are usually organized to represent a theme. Although layers often roughly correspond to classes in an OO model, they are by no means constrained to do so.

To formulate the membership of an object o_j in a class, we must consider the relevance and ranges of the attribute values of the object. With this in mind, the membership of object o_j in class C with attributes *Attr (C)* is defined as:

$$\mu C(o_j) = \mathbf{g}[\mathbf{f}(RLV(a_i,C), INC(rng_c(a_i)/o_j(a_i)))]$$

where $RLV(a_i,C)$ indicates the *relevance* of the attribute a_i to the concept C, INC $(rng_c(a_i)/o_j(a_i))$ denotes the degree of *inclusion* of the attribute values of o_j in the formal range of a_i in class C, **f** represents the function of aggregation over the *n* attributes in the class, and **g** reflects the nature of the semantic link existing between an object and a class (or between a subclass and its superclass). The value of $(RLV(a_i,C)$ may be supplied by the user, or may be calculated by the system, for example, using the weighting method described in (Robinson 2000).

Degree of membership

Using our mountain example, here are some class definitions from our schema:

```
CLASS: Mountain
PROPERTIES: elevation
            avg_snowfall
            avg_alpine_temp
            name
            location (feature_attribute)
            scale (feature_attribute)
END;
CLASS: Skiable_mountain
INHERIT: Mountain
PROPERTIES: trails -> (count(Trails)
                        where mtn = mountain.name)
            lifts -> (count(Lifts)
                        where mtn = mountain.name
                        and operational_flag=TRUE)
            snowmaking_capacity
            longest_run -> (max(Trails.length)
                            where mtn = mountain.name)
            vertical
END;
CLASS: Trail
PROPERTIES: name
            location (feature_attribute)
            difficulty
            length
END;
```

For the *Skiable_mountain* class (abbreviated SM), let us assume the following *typical* attribute ranges:

```
Skiable_mountain attributes:
rng_sm(elevation) = {medium_short, average, medium_tall}
rng_sm(avg_snowfall)= {>4 meters}
rng_sm(avg_alpine_temp)= {slightly_above_freezing,
                    freezing, slightly_below_freezing,
                    below_freezing}
rng_sm(name)          = {X}
rng_sm(location)      = {Y}
rng_sm(trails)        = {a_few, many, a_lot}
rng_sm(lifts)         = {a_few, many, a_lot}
rng_sm(snowmaking_capacity)= {none, limited, moderate, extensive}
rng_sm(longest_run) = {fairly_short, average, fairly_long,
                    long, very_long}
rng_sm(vertical) = {>500m}
```

Note that name, location, and scale are not used in determining class membership.

Calculating membership

Now we can compute the membership of an object in the *Skiable_mountain* class. First, assume the following relevance rules for membership:

```
RLV(elevation, SM)              = 0.25
RLV(avg_snowfall, SM)           = 0.75
RLV(avg_alpine_temp, SM)        = 0.75
RLV(lifts, SM)                  = 0.9
RLV(trails, SM)                 = 0.85
RLV(snowmaking_capacity, SM)    = 0.95
RLV(longest_run, SM)            = 0.8
RLV(vertical, SM)               = 0.5
```

Note that the relevance of lifts and snowmaking capacity are extremely high. If a mountain has snow and has at least one ski lift, the chances are it meets the criteria for a skiable mountain.

Here is an example with Blackcomb mountain in Whistler, Canada:

```
o(elevation)           = {2284m (average)}
o(avg_snowfall)        = {9.14m}
o(name)                = {Blackcomb}
o(avg_alpine_temp)     = {-10C (below_freezing)
o(location)            = {L}
o(trails)              = {>100 (a_lot)}
o(lifts)               = {17 (a_lot)}
o(snowmaking_capacity) = {215 hectares (extensive)}
o(longest_run)         = {11km (very_long)}
o(vertical)            = {1609m}
```

Given these values, and using the *max* function for **g**, we compute the following membership value for *Blackcomb* in the class *Skiable_mountain*:

$\mu C(o_j) = \mathbf{g}[\mathbf{f}(RLV(a_i,C),\ INC(rng_c(a_i)/o_j(a_i)))] =$
$\mu_{SM}\ (Blackcomb) = \mathbf{g}[\mathbf{f}(RLV(a_i,\ Skiable\_mountains),$
$\qquad\qquad\qquad INC\ (rng_c(a_i)/Blackcomb(a_i)))] =$
$\mu_{SM}\ (Blackcomb) = max(0.25\ *\ elevation,\ 0.75\ *\ avg\_snowfall,\ 0.75\ *$
$\qquad\qquad\qquad avg\_alpine\_temp,\ 0.9\ *\ lifts,\ 0.85\ *\ trails,\ 0.95\ *$
$\qquad\qquad\qquad snowmaking\_capacity,\ 0.8\ *\ longest\_run,\ 0.5\ *\ vertical) =$
$\mu_{SM}\ (Blackcomb) = max(0.25*1.0,\ 0.75*1.0,\ 0.75*1.0,\ 0.9*1.0,\ 0.85*1.0,\ 0.95*1.0,$
$\qquad\qquad\qquad 0.8*1.0,\ 0.5*1.0)$
$= \mu_{SM}\ (Blackcomb) = max(0.25,\ 0.75,\ 0.75,\ 0.9,\ 0.85,\ 0.95,\ 0.8,\ 0.5) = 0.95$

Thus one can see that Blackcomb has an extremely high degree of membership (0.95) in the class of skiable mountains, which one would expect.

Modeling within Couclelis's Taxonomy of Object Boundary Uncertainty

As a basic approach for managing uncertainty in spatial databases, we will look at Helen Couclelis's taxonomy of features with ill-defined boundaries (Couclelis 1996), and for each dimensional attribute, we will provide ways for which it may be modeled and stored in a spatial database.

Dimension 1: the empirical nature of the entity

This attempts to distinguish what the nature of the entity is, and whether it lends itself to being well or poorly bounded.

Atomic or plenum   Here, we are simply determining if we will store spatial entities as distinct atomic features, or as a plenum field. If the objects are atomic, their spatial content can be stored as a spatial attribute of the object. If we are managing fields, then we have to parse the region into some form of grid, and each pixel must have a membership value [0.0, 1.0] in all types of features (classes in the OO sense) being stored.

Homogeneous or inhomogeneous   This is managed by providing fuzzy membership values for the class. If we are dealing solely with homogeneous features, then we simply have membership values of 1.0 or 0.0; inhomogeneous features have a membership value somewhere between these points.

Discontinuous or continuous   The management here is somewhat dependent upon whether the features are atomic or plenum. If plenum, then membership values are assigned at the pixel level, and the resulting presentation to the user should be evidence of continuity. If atomic, by providing for multiple alpha cuts in the spatial representation of an object, continuity can be stored and represented in a intuitive way.

Connected or distributed   This can be managed via the object membership value in the *connected* class or collection.

Solid or fluid   Modeling fluid objects is extremely difficult. The best approaches to date are ones that define *core* and *boundary* for the limits of the fluid geometry, or an approach that defines multiple boundaries through temporal modeling.

Two or three-dimensional   Three-dimensional objects may be modeled as easily as two-dimensional objects using the fuzzy spatial object framework. When managing difficult cases (the canopy levels of trees), fuzzy boundaries may be used.

Actual or non-actual   Most object-oriented spatial databases support the concept of *versioning*, where an object may have different numbers of attributes, relationships, and methods, as well as different values for the attributes, at different points in time. Versioning is a nice way to model non-actual objects.

Permanent or variable; fixed or moving   All of these object types may be modeled using versioning.

Conventional or self-defining   The notion of membership can be used to model this as well.

Dimension 2: The mode of observation

Objects may appear to be well or ill bounded, despite their actual nature. Typically, our mode of observation is more important in determining whether the objects should be modeled as well or poorly bounded.

Scale, Resolution, and Perspective   Ideally, we want to store features at the finest level of granularity possible. It will then be the responsibility of the representation tools to either display the finest detail, or the smallest features, so that collections appear to be atomic. In addition, it will be up to the representation tools to manage the viewshed and perspective.

Time   If modeled at all, this may be done by versioning, or another accepted way to model temporality.

Error   As mentioned previously, this must be generalized to uncertainty and modeled via fuzziness.

Theory   Although fuzzy spatial object databases may model atomic or plenum objects, they must still have some concept of boundary, even though the objects may have no core, holes, or an endless boundary. Storing objects using a geostatistical approach will require the representation tools to display the concept of implied or unimplied boundary.

Dimension 3: The user purpose

Often, even though we may know whether or not a feature has a crisp or uncertain boundary, and even though our mode of observation may lend itself to well-defined or ill-defined boundaries, it is immaterial to how we use the system.

It is our opinion that the user purpose should be more clearly defined by the GIS modeler, and thus it is up to the representation software to display features as well, bounded or not.

Queries on Spatial Databases

Regardless of the mechanism used to store features in a spatial database, the queries posed against that data may involve uncertainty to varying degrees (see Chapter 6 of this volume by Brown and Heuvelink for a more detailed treatment of this topic).

Let us consider the query "Display all skiable mountains within 10 km of an airport." There are three terms or phrases in this query that may lead to uncertainty or imprecision: *skiable mountains, airport,* and *within 10 kilometers.* First, the thematic layer, which contains skiable mountains, may consist solely of crisp data. However, if one is an Olympic athlete, then the definition of skiable will differ from the norm. Second, the concept of "airport" may be uncertain as well. One may be looking for a dirt strip, where a tiny two-passenger plane could land, as opposed to a multi-runway tarmac designed for commercial jetliners. A third way uncertainty may exist within this query would be the fuzziness in the semantics. In a crisply modeled world, if one is 10.001 kilometers from an airport, then *that* airport would not satisfy the query. Even though one may ask for skiable mountains within 10 km of an airport, one may wish to know all skiable mountains within walking distance, driving distance, or some other distance depending upon the circumstances. It would be simple for the GIS to display circles around the airports with 10 km radii, but the person posing the query may want to know the mountains within 10 km by road, which is a very different query indeed. What people say is not necessarily what they mean. Ideally, a query system should be robust enough to give the decision-makers more information to help them in their process.

A more classic example of a *fuzzy query* would be to actually include one or more fuzzy terms. An example of this would be: "Display all skiable mountains *near* an airport." This query contains the term "*near,*" which could return a solution set with a degree of membership of 1 for every mountain less than nine kilometers from an airport, and a degree of membership of 0 for every mountain more than 20 km from an airport. Every mountain between nine and 20 km from an airport would have a variable degree of membership.

There are GIS products, namely IDRISI (Jiang and Eastman 2000) that support fuzzy operations on data. Unfortunately, IDRISI does not consider how to store objects with ill-defined boundaries. If we simply allow for fuzziness in the algorithms but not the underlying spatial database (Morris, Petry, and Cobb 1998, Morris, Foster, and Petry 1999), the representation may provide for alterations in visualization quanta, but it will still be represented as discrete data sets.

Representing Query Results

Now that we have stored the uncertain features in our database, how can we represent this uncertainty to the user? Many ways have been proposed. Ehlschlaeger and Goodchild (1997) propose that we can represent positional uncertainty by either blurring features, or making them shake in an animation. Morris (2003) proposed progressive shading, either using a continuous gradient or by using crisp boundaries of fuzzy alpha-cuts. There have been experiments with active interfaces (Morris and Jankowski 2000) and even with sound (Goodchild 1998).

The problem with any method of representing uncertainty is that it is typically counter-intuitive to the user. The user expects the actual physical road to follow the map he or she is holding of the road. The user may even expect there to be crisp red lines denoting state boundaries and the desert to be beige, because that is the color the map shows it to be.

Whenever we want to introduce this uncertainty to the user, to let that user know that this boundary is not necessarily completely precise, then we need to do this in such a way that the user *understands* that this is not necessarily a crisp measurement. The trials of Morris and Jankowski (2000) allowed the user to choose, or even toggle modes from crisp, to uncertain with crisp alpha-cuts, to uncertain with continuous alpha-cuts. These results were much better received than the static presentation, although this did require some education of the users.

Obviously, while this will give added flexibility to the user and modeler, it will also allow the manipulation and distortion of data. For a few details on this, refer to Mark Monmonier's (1996) book entitled *How to Lie with Maps*.

CONCLUSIONS

It is evident to the reader that we feel that the only proper way that uncertainty can be modeled in spatial databases is through the use of fuzzy object-oriented databases. They can provide membership functions to aid in the storage and representation of objects with uncertain boundaries, and they inherit all of the advantages of the object-oriented paradigm. By abstracting to the feature, we are able to store and represent both vector- and raster-based objects. By using multiple alpha-cuts, we are providing an extensible system that may support objects with either crisp or ill-defined boundaries at any level of desired detail.

As users are becoming more sophisticated, the storage, manipulation, and representation of objects with uncertain boundaries is going to become increasingly important, and the fuzzy spatial object framework is available to support it.

REFERENCES

Buckles, B. and Petry, F. 1982. A fuzzy representation of data for relational databases. *International Journal of Fuzzy Sets and Systems* 7: 213–26.

Burrough, P. A. 1996. Natural objects with indeterminate boundaries. In P. A. Burrough and A. Frank (eds) *Geographic Objects with Indeterminate Boundaries*. London: Taylor and Francis: 3–28.

Burrough, P. A. and Couclelis, H. 1996. Practical consequences of distinguishing crisp geographic objects. In P. A. Burrough and A. Frank (eds) *Geographic Objects with Indeterminate Boundaries*. London: Taylor and Francis, pp. 335–7.

Campari, I. 1996. Uncertain boundaries in urban space. In P. Burrough, and A. Frank, (eds) *Geographic Objects with Indeterminate Boundaries*. London: Taylor and Francis, pp. 57–70.

Cattell, R. G. G. and Barry, D. K. (eds). 2000. *The Object Data Standard: ODMG 3.0*. San Francisco, CA: Morgan Kaufman.

Cheng, T., Molenaar, M., and Lin, H. 2001. Formalization and application of fuzzy objects. *International Journal of Geographical Information Science* 15: 27–42.

Codd, E. F. 1970. A relational model of data for large shared data banks. *Communications of the ACM* 13: 377–87.

Couclelis, H. 1996. Towards an operational typology of geographic entities with ill-defined boundaries. In P. A. Burrough and A. Frank (eds). *Geographic Objects with Indeterminate Boundaries*. London: Taylor and Francis: 45–56.

Cross, V. and Firat, A. 2000. Fuzzy objects for geographical information systems. *Fuzzy Sets and Systems* 113: 19–36.

Ehlschlaeger, C. and Goodchild, M. 1994. Uncertainty in spatial data: Defining, visualizing, and managing data errors. In *Proceedings of GIS/LIS'94*. Bethesda, MD: American Congress of Surveying and Mapping, pp. 246–53.

Fisher, P. 1996. Boolean and fuzzy regions. In P. A. Burrough and A. Frank (eds) *Geographic Objects with Indeterminate Boundaries*. London: Taylor and Francis, pp. 87–94.

Foody, G. M. and Atkinson, P. M. 2002. Uncertainty in remote sensing and GIS: Fundamentals. In G. M. Foody and P. M. Atkinson (eds) *Uncertainty in Remote Sensing and GIS*. New York: John Wiley and Sons, pp. 1–18.

Gale, S. 1972. Inexactness, fuzzy sets, and the foundations of behavioral geography. *Geographical Analysis* 4: 337–49.

George, R., Buckles, B., Petry, F., and Yazici, A. 1992. Uncertainty modeling in object-oriented geographical information systems. In Min Tjoa, A. and Ramos, I. (eds) *Database and Expert Systems Applications: Proceedings of the Thirteenth International Conference in Valencia, Spain, 1992*. Vienna: Springer-Verlag, pp. 294–299.

Goodchild, M. F. 1998. Uncertainty: The achilles heel of GIS? *Geo Info Systems* 8(11): 50–2.

Goodchild, M. F. and Gopal, S. (eds). 1990. *The Accuracy of Spatial Databases*. Basingstoke: Taylor and Francis.

Hunter, G. J. 1998. Managing uncertainty in GIS. WWW document, http://www.ncgia.ucsb.edu/giscc/units/u187/u187_f.html.

Jiang, H. and Eastman, J. R. 2000. Application of fuzzy measures in multi-criteria evaluation in GIS. *International Journal of Geographical Information Science* 14: 173–84.

Klir, G. J. and Folger, T. A. 1988 *Fuzzy Sets, Uncertainty, and Information*. Englewood Cliffs, NJ: Prentice-Hall: 246–54.

Leung, J. Y. 1979. Locational choice: A fuzzy set approach. *Geography Bulletin* 15: 28–34.

Mackay, D. S., Band, L. E., and Robinson, V. B. 1991. An object-oriented system for the organization and representation of terrain knowledge for forested ecosystem. In *Proceedings of GIS/LIS'91*. Bethesda, MD: American Congress of Surveying and Mapping, pp. 617–26.

Monmonier, M. 1996. *How to Lie with Maps* (2nd edn). Chicago, University of Chicago Press.

Morris, A., 2003. A framework for modeling uncertainty in spatial databases. *Transactions in GIS* 7: 83–101.

Morris, A. and Jankowski, P. 2000. Combining fuzzy sets and databases in multiple criteria spatial decision making. In H. Larsen, J. Kacprzyk, S. Zadrozny, T. Andreasen, H. Christiansen (eds) *Flexible Query Answering Systems: Recent Advances*. Dordrecht: Kluwer, pp. 103–16.

Morris, A. and Petry, F. E. 1998. Design of fuzzy querying in object-oriented spatial data and geographic information systems. In *Proceedings of Seventeenth International Conference of the North American Fuzzy Information Processing Society*. Pistacaway, NJ: Institute of Electrical and Electronic Engineers, pp. 165–9.

Morris, A., Foster, J., and Petry, F. E. 1999. Providing support for multiple collection types in a fuzzy object-oriented spatial data model. In *Proceedings of the Eighteenth International Conference of the North American Fuzzy Information Processing Society*. Pistacaway, NJ: Institute of Electrical and Electronic Engineers, pp. 824–8.

Morris, A., Petry, F. E., and Cobb, M. 1998. Incorporating spatial data into the fuzzy object-oriented data model. In *Proceedings of Seventh International Conference on Information*

Processing and Management of Uncertainty in Knowledge Based Systems. Paris: Editions EDK: 604–11.

Petry, F., Cobb, M., and Morris, A. 1999. Fuzzy set approaches to model uncertainty in spatial data and geographic information systems. In, L. Zadeh and J. Kacprzyk (eds) *Computing with Words in Information/Intelligent Systems* 2. Heidelberg: Physica-Verlag, pp. 345–67.

Robinson, V. B. 2000. Individual and multi-personal fuzzy spatial relations acquired using human–machine interaction. *International Journal of Fuzzy Sets and Systems* 113: 133–45.

Robinson, V. B. 2003. A perspective on the fundamentals of fuzzy sets and their use in geographic information systems. *Transactions in GIS* 7: 3–30.

Robinson, V. B. and Sani, A. P. 1993. Modeling geographic information resources for airport technical data management using the Information Resources Dictionary System (IRDS) standard. *Computers, Environment, and Urban Systems* 17: 111–27.

Usery, E. L. 1996. A conceptual framework and fuzzy set implementation for geographic features. In P. A. Burrough and A. Frank (eds) *Geographic Objects with Indeterminate Boundaries*. London: Taylor and Francis, pp. 71–86.

Zhang, J. and Goodchild, M. F. 2002. *Uncertainty in Geographical Information*. London: Taylor and Francis.

Chapter 6

On the Identification of Uncertainties in Spatial Data and Their Quantification with Probability Distribution Functions

James D. Brown and Gerald B. M. Heuvelink

Central to understanding and managing social and environmental systems is the need to consider spatial patterns and processes. Spatial data are routinely used to describe, predict, and explain a diverse range of geographical features, including soils, population density, water availability, the spread of disease, and ecological diversity (Fotheringham, Charlton, and Brusden 2000). They are also used routinely in Geographical Information Systems (GIS) to manage social and environmental features (Burrough and McDonnell 1998).

Spatial data are rarely certain or "error free." Rather, in abstracting and simplifying "real" patterns and processes, spatial data contain inherent errors that are insignificant at some spatial scales and for some applications but significant for others (see Chapter 5 by Morris in this volume). Often, decisions are based upon multiple types and sources of data where approximation errors will combine and propagate through spatial models, such as GIS operations (Heuvelink 1998). This may lead to poor decisions about the exploitation and management of social and environmental systems (Harremoës, Gee, MacGarvin, et al. 2002). Understanding the limits and limitations of spatial data is, therefore, essential both for managing social and environmental systems effectively and for encouraging the responsible use of spatial data where knowledge is limited and priorities are varied (Hunter and Lowell 2002, Mowrer and Congalton 2002, Couclelis 2003, Foody and Atkinson 2003).

While it is generally accepted that spatial data are rarely (if ever) "error free," these errors may be difficult to quantify in practice. Indeed, the quantification of error (defined here as a "departure from reality") implies that the "true" nature of the environment or society is known. Yet absolute accuracy is neither achievable nor desirable in scientific research because resources are always limited and must be used efficiently. Rather, in the absence of such confidence, we are uncertain about

the "true" character of the environment and about the errors in our representations of it. However, it may be possible to constrain these errors with a stochastic
model of reality (Ripley 1981), and to explore the impacts of uncertainty on decision-
making through an uncertainty propagation analysis (Heuvelink 1998). The former
corresponds to an assessment of data quality where uncertainties are described (for
example, numerically), while the latter corresponds to an analysis of "fitness for use"
(Veregin 1999, De Bruin and Bregt 2001) where the impacts of uncertainty are
explored in the context of a specific application.

   This chapter focuses on the identification of uncertainties in spatial data (that
is, the assessment of spatial data quality), including the uncertainties associated
with attribute and positional information. Particular emphasis is placed upon the
quantification of uncertainties in spatial attributes with probability distribution
functions (*pdfs*). It does not describe the techniques for propagating uncertainties
through GIS operations, for which reviews can be found elsewhere (for instance,
Heuvelink 1999). The discussion is separated into four sections, namely: (1) an overview of data quality and measures of data quality; (2) the sources of uncertainty in
spatial data; (3) the development of uncertainty models for attribute and positional
information; and (4) a discussion of some major research challenges for improving
the reliability and accessibility of data quality models. In terms of the latter, it is
argued that forward planning of data quality targets, including uncertainty analyses
as well as conventional quality control (error reduction) procedures, will become
increasingly important in the future.

Developing Measures of Spatial Data Quality

Uncertainties about spatial data may refer to a lack of confidence about the quality
of a data set, or about its utility for a particular application (its "fitness for use").
Uncertainty is an expression of confidence about what we know, both as individuals
and communities of scientists, and is, therefore, subjective (Brown 2004). Probability
models are a common approach to describing uncertainties in spatial data. Different
people can reach different conclusions about how probable something is based on
their own personal experiences and world-view, as well as the amount and quality
of information available to them (Cooke 1991, Heuvelink and Bierkens 1992,
Fisher, Comber, and Wadsworth 2002). Indeed, it is widely accepted that stable or
"objective" probabilities cannot be achieved when studying complex, variable, and
poorly sampled environmental systems, as evidenced by the widespread application
of Bayes' theorem to problems of scientific uncertainty in recent years (Beven and
Freer 2001, Greenland 2001).

   While data quality is subjective, the standards used to assess and report uncertainty may be precisely defined (USBB 1947, FGDC 1998, Burkholder 2002). These
standards may refer to the statistical accuracy and precision of data, their numerical
precision, lineage, logical consistency, currency, and "completeness," among others
(Chrisman 1991, Guptill and Morrison 1995). In contrast, the utility or fitness for
use of data cannot be reported with fixed standards, because fitness for use is case-
dependent. For example, an average error of ±1 m in a Digital Elevation Model
(DEM) will be more important for predicting coastal flooding in flat terrain than

for planning an optimal route through mountainous terrain. In principle, therefore, the fitness for use of a data set is conceptually different from its empirical quality. In practice, however, the distinction between data quality and utility may be blurred, both conceptually and operationally. First, if estimates of uncertainty are sensitive to the experiences and judgments of people, the empirical quality of a data set will be partly embedded in the specific applications that led to those experiences. Second, estimates of fitness for use may be relevant for particular classes of application, such as "coastal flood modeling," if not individual applications, such as flooding on the east coast of England. Consequently, they might be stored in spatial databases alongside estimates of empirical quality. In this context, databases might be adapted to allow sampling of larger data sets in order to provide some insight into the fitness for use of data before they are applied in detail (De Bruin and Bregt 2001).

If our definitions of geographic entities, such as "buildings," "electoral wards," and "rivers" are widely accepted (that is, "objective"), uncertainties in data may be reduced to expressions of empirical quality or some proxy for empirical quality, such as trust in the source of information or the methods used to obtain it. In contrast, when our definitions of geographic entities are unclear, or our understandings of reality are varied, assessments on the value of spatial data cannot refer to measures of empirical quality alone. Rather, the failure to distinguish between empirical quality and quality of concepts (for example, clarity of entities) will lead to conflicting observations about what is "real," where geographic variability is wrongly interpreted as observational error (Richards, Brooks, Clifford, Harris, and Lane 1997). These conceptual uncertainties cannot be integrated fully with measures of empirical quality because our constructs of reality may be valued by their adequacy for a particular application rather than any inherent "truth" value. For example, political boundaries may be derived from concepts of "nation states" that only apply during peacetime, and the distinction between a "mountain" and its surroundings may be useful in some situations but not in others.

Finally, states of information on data quality should be reflected in the modes of analyzing and communicating uncertainty, as well as the magnitude of uncertainty itself. For example, *pdfs* imply that all outcomes of an uncertain event are known and that each of their associated probabilities is quantifiable. If probabilities cannot be defined numerically, an arbitrary *pdf* will imply a spurious notion of precision about data quality. When the parameters of a *pdf* cannot be estimated reliably, a spectrum of less precise measures is available for describing uncertainty (Ayyub 2001). These include bounds (for example, binary classifications as certain/unknown) through rough sets or ternary classification (possible/doubtful/unknown), multiple outcomes ranked in order of likelihood, continuous classifications (that is, proportional membership of a "knowledge vector"), "histograms" where outcomes are coarsely graded in numerical frequency, or detailed qualitative information (for example, the pedigree matrices of Funtowicz and Ravetz 1990). Alternatively, scenarios may be preferred if some or all possible outcomes of an uncertain event are known but their individual probabilities cannot be estimated reliably, either in numbers or in narrative (von Reibnitz 1988). The remainder of this chapter focuses on quantitative estimates of probability, as *pdfs* are the most common measure of uncertainty in spatial data.

What are the Key Sources of Uncertainty in Spatial Data?

Sources of uncertainty in spatial data

Social and environmental data are collected within discrete space-time boundaries and are sampled at discrete intervals of space and time within these boundaries. The objects of interest might include people, rivers, houses, mountains, lakes or forests, and their attributes might include "economic wealth," "terrain," "pH," "voting intentions," "ecological diversity," or any other geographic entity derived from a classification of reality. The boundaries may be defined by transitions in physical properties, such as mass, energy, and momentum, or in metaphysical properties (so-called fiat objects), such as political borders, or by a combination of the two. Within these boundaries, individual samples record the aggregate properties of one or many social or physical variables. The samples are collected with instruments that display finite sensitivities to the properties of interest and may not sample these properties directly, but rather infer them from known relationships with other, more easily measurable, properties. Following measurement, the patterns and processes inferred from a sample are implicitly historical. These inferences will often involve predictions about values at unmeasured points (for example, that a measurement is valid beyond the instant it is captured), for which uncertainties are no longer bounded by an observation of the system but rely upon interpolation or extrapolation in space or time. Finally, multiple sources of error and uncertainty are introduced during the registration and transformation of spatial data into digital products that can be used in GIS, including feature extraction and digitization, filtering, vector-to-raster conversion and line simplification.

In practice, these sources of uncertainty are too numerous, and their interactions too complex, to consider in a formal uncertainty analysis (a source of ignorance), and many do not contribute to the quantifiable probability that a value is "correct." Thus, expert judgments are always required to make an assessment of data quality in general, as well as their suitability for making specific decisions (fitness for use). Notwithstanding conceptual uncertainties, a discussion of the major sources of uncertainty in spatial data can usefully be separated into: (1) measurement, sampling, and interpolation; (2) classification (aggregation or dissaggregation of attribute values); and (3) scale and changes between scale (aggregation or dissaggregation in space or time).

Measurement, sampling, and interpolation

Spatial data are often derived from measurements in the field, which introduces measurement uncertainty. Measurement uncertainty originates from a lack of confidence about a *local realization* of the measured variable. It may originate from imperfect knowledge about the accuracy of an instrument, its ability to reconstruct the variable of interest (for example, river discharge from stage measurements), or the control volume for which it is representative (for instance, quantization of light by a remote sensing instrument). While social and environmental parameters vary more or less continuously through space and time, measurements almost always occupy a limited number of space-time points. When exhaustive inputs are required but only partial observations are available they must be interpolated, which leads to interpolation uncertainty (Goovaerts 1997). Uncertainties in sampling and interpolation originate from a lack of confidence about a *distributed realization* of the measured variable. Interpolation uncertainty depends on the sample density, the magnitude and nature of spatial (and temporal) variation in the sampled attribute, and the interpolation algorithm employed. Measurement uncertainties can be estimated through comparisons with more accurate data, laboratory testing of measurement instruments, or repeat measurement with the same instrument. While the former provides an indication of accuracy or "bias," the latter two approaches only indicate precision.

In order to interpolate spatial data and to assess the uncertainties associated with spatial interpolation some assumptions must be made about the behavior of the measured variables at unmeasured locations. A common approach to sampling spatial "fields," such as soil type or land use, involves separating the field site into homogeneous units, sampling these units and calculating a within-unit sample mean and variance (uncertainty). In practice, however, it may not be possible to separate a spatial domain into homogeneous units, but to assume instead that environmental conditions vary continuously in space and time. Geostatistics can be used to interpolate continuous data from partial measurements and to estimate the uncertainties associated with spatial interpolation (Goovaerts 1997).

Classification

Classification leads to uncertainty when our observations of reality cannot be assigned to discrete classes, either because the classes are poorly defined, there are too few classes to capture all of the information available, or there is insufficient information to classify some values (Foody 2002, Stehman and Czaplewski 2003). If the uncertainties associated with classifying entities are to be explored effectively, the entities revealed through classification must be clearly defined and hence "real" rather than "effective" quantities (see below). However, they do not need to be distinct in character, or precisely defined, because many aspects of reality are indistinct, depending upon the space-time scales at which they are observed (Foody 1999). If reality is arbitrarily diffuse, precise definitions may be unhelpful, but definitions of geographic entities can also be insufficiently precise. Hence, the quality of a classification is not implicit in its precision but in its *clarity of meaning*, as well as its *information content* (class "everything" has perfect accuracy) and *empirical accuracy* (all values assigned to the correct classes). In this context, too normative an emphasis on the statistical agreement between remote sensing data and field observations, or on the variance resulting from a geostatistical interpolation, may direct attention away from the epistemic value of the classes themselves. Regardless of the precision with which entities are defined, the scope for confusion about the meaning of geographic entities should be made clear otherwise estimates of uncertainty will be misleading.

Scale and changes between scale

An important consequence of the need for "closure" in geographic research (Lane 2001) is that data and models provide inherently discrete representations of continuous patterns and processes. In particular, they rely upon the specification of dominant patterns and process controls and their discretization over limited areas and with finite control volumes. The need to represent social or environmental systems at one or a combination of scales may lead to uncertainty, because the dominant patterns and processes are not known at all scales or cannot be incorporated practically in models, and the control volumes used to represent them may be fixed at other scales (Hennings 2002).

When spatial data are defined with a different "support" from that required for a given application, these data must be *aggregated* or *disaggregated* to an appropriate support (Heuvelink and Pebesma 1999, Bierkens, et al. 2000). Aggregation or disaggregation of data is commonly referred to as the "change of support problem" (Journel and Huijbregts 1978), the "modifiable areal unit problem" (Cressie 1993) or the "scale problem" (Burt and Barber 1996). In aggregating spatial data it is assumed that the geometry of the control volume can be rescaled without deforming the original data (that is, perfect tessellation) otherwise a "positional" or geometric uncertainty is introduced (Veregin 2000). If the input data are uncertain, these uncertainties will propagate through the rescaling model to the model output, that is, the rescaled quantity.

In practice, space-time aggregation should lead to a reduction in uncertainty and to an increase in spatial autocorrelation, because much of the variability at finer scales is lost and, thus, disappears as a source of uncertainty and spatial divergence (Cressie 1993). Disaggregation of spatial data necessarily leads to uncertainty about the precise value of a given variable at a specific location, because its value is only partially constrained by available information. Specifically, it is constrained by the attribute value at an aggregated level (for example, the block support). Assessments of uncertainty in disaggregated data may be highly subjective, because the spatial variation of the attribute is rarely known at the disaggregated level. In contrast to aggregation, space-time dissaggregation will lead to an increase in uncertainty and a reduction in spatial autocorrelation because the attribute variability is increased at finer scales.

Quantifying Attribute and Positional Uncertainties

Uncertainties about spatial data quality may refer to the space-time domain of a geographic entity, including its absolute and relative position and geometry (*positional uncertainty*), or to an attribute of that entity (*attribute uncertainty*). While assessments of uncertainty in spatial data have traditionally focused on attribute uncertainties (Journel and Huijbregts 1978, Goovaerts 1997; see the following section "Attributes and uncertainties"), studies of positional uncertainty have become increasingly common in recent years (and, following it, the section entitled "Positional uncertainties of geographic objects"). These include studies of positional uncertainties in vector data (Stanislawski, Dewitt, and Shrestha 1996, Kiiveri 1997, Leung and Yan 1998, Shi 1998) and combinations of positional and attribute uncertainties in raster data (Arbia, Griffith, and Haining 1998). For geographic "objects," such as trees, uncertainties in attribute information (for example, tree species) may be separated from uncertainties in positional information (for instance, tree location). In contrast, for geographic fields, such as terrain elevation or land cover, attribute uncertainties interact with positional uncertainties where positional uncertainties may increase attribute uncertainties (Gabrosek and Cressie 2002).

Attribute uncertainties

Uncertainties about the precise value of an attribute at a particular point in space and time may be quantified with a marginal *pdf* (*mpdf*) for that attribute. For a continuous numerical attribute Z, such as "tree height," "rainfall," or "annual income," the *mpdf* is described with the continuous function:

$$F(z) = \text{Prob}(Z \leq z) \qquad -\infty < z < \infty \tag{6.1}$$

For a discrete numerical attribute, such as population size, or a categorical attribute, such as land cover, the *mpdf* is described with a discrete function for that attribute (C):

$$P(c) = \text{Prob}(C = c) \qquad c \in S \tag{6.2}$$

where S represents the set of all possible outcomes for C (e.g. S={urban, arable, forest, water, . . . } for land cover). For pragmatic or theoretical reasons, F and P might be described with an idealized shape function and a parameter set whose values modify that shape for a particular data set. For continuous functions, the *mpdf* may follow a "Gaussian," "uniform," "exponential," "lognormal," or "gamma" distribution, and for discrete functions it may follow a "uniform," "binomial," or "Poisson" distribution (among others). The shape function and parameter set may be derived from expert judgment, or fitted to a sample of differences between more and less accurate data, or modeled from sample data (for example, Kriging). In other cases, Z or C cannot be modeled with an idealized shape function (for example, if the fitting criteria are not met). Here the *mpdf* may be described with an arbitrary "non-parametric" shape providing it satisfies the basic axioms of probability theory.

Uncertainties about the precise value of two or more attributes occupying the same space-time point may be quantified with a joint *pdf* (*jpdf*). For example, the number of road traffic accidents at a motorway junction may depend on the volume of traffic passing through that junction. In this case, they must be described with a *jpdf* because one variable is "statistically dependent" on the other variable. While numerous shape functions are available for the mpdfs in Equations (6.1) and (6.2), few simple models are available for the statistically-dependent *jpdf*. For continuous numerical variables, a common assumption is that the variables follow a joint Gaussian distribution, which may be justified by the Central Limit Theorem. Furthermore, the multi-Gaussian distribution is mathematically simple and requires only a vector of means and a covariance matrix for complete specification.

Statistical dependence may also occur in space and time, both within and between variables, for which a *jpdf* must also be defined. Spatial dependence is common in "field" attributes, such as elevation, rainfall, and land cover, but may also occur in the attributes of spatial "objects," such as tree heights. In modeling the spatial dependence of geographic fields, a common assumption is that the attribute values follow a joint Gaussian distribution and that the correlations or covariances between pairs of locations depend only on their separation distance (Journel and Huijbregts 1978, Goovaerts 1997). Given these assumptions, Webster and Oliver (1992) found that the *jpdf* for geographic fields can be estimated with approximately 80–200 observations. These assumptions may be relaxed if additional observations or auxiliary data are available to improve the model.

For a discrete numerical or categorical variable it is less straightforward to model the statistical dependencies between attribute values at different locations. For example, m^n probabilities must be specified for a categorical variable with m possible outcomes at n locations (for m = 8 and n = 6 this yields more than 250,000 probabilities). In practice, there are few idealized shape functions to model these probabilities and few generally applicable methods for reducing the complexity of the discrete *jpdf*.

In assessing the accuracy of a discrete numerical or categorical variable, an error or "confusion" matrix may be constructed from a sample of more accurate data and assumed valid for a wider population. This approach originates from the classification of land cover with remote sensing imagery (Stehman and Czaplewski 2003, Steele, Patterson, and Redmond 2003). The confusion matrix stores the errors of omission and commission in a contingency table format (Story and Congalton 1986), which allows specific accuracy statistics to be derived for a particular application. Since the confusion matrix is ideally derived from a probabilistic survey design, where each space point has an equal chance of being sampled, it can be used to determine the classification uncertainty of individual points. However, spatial dependence between uncertainties is not included in the confusion matrix, yet spatial dependence may profoundly affect the propagation of uncertainties through GIS operations (Heuvelink 1998).

One approach to modeling statistical dependence in discrete numerical and categorical variables is indicator geostatistics (Goovaerts 1997, Finke, Wladis, Kros, Pebesma, and Reinds 1999, Kyriakidis and Dungan 2001). For example, Finke, Wladis, Kros, Pebesma, and Reinds. (1999) used indicator variograms and cross-variograms to quantify uncertainty in categorical soil maps and land cover maps, and used indicator simulation to generate spatially correlated realizations of these maps for use in an uncertainty propagation analysis. However, indicator geostatistics is inexact (Cressie 1993), and requires a large number of indicator (cross-)variograms to be sampled and modeled, which may be impossible in practice. Other techniques for estimating the *jpdf* of geographic fields, such as land cover or bird counts, include conditional probability networks (Kiiveri and Cacetta 1998), Bayesian Maximum Entropy (Christakos 2000) and Markov Random Fields (Norberg, Rosén, Baran, and Baran 2002). The attributes of spatial objects, such as tree heights, may be treated as "marked point processes" (Diggle 2003, Diggle, Ribeiro, and Christensen 2003, Schlather, Ribeiro, and Diggle 2004) where the spatial positions (points) as well as the attribute values (marks) are modeled as random processes.

Positional uncertainties of geographic objects

Attempts to define positional uncertainties in geographic objects have included partial and full applications of probability theory to vector data where lines or polygons are distorted with an autocorrelated "shock" (Kiiveri 1997), or positional uncertainties are expressed in a confidence region for the end nodes of a line segment (Dutton 1992, Shi 1998). The latter includes an "epsilon (ε) band" approach (Perkal 1966) where the marginal *pdfs* for each node are connected *ex post*, which results in a fixed buffer of radius ε around each line segment, and a non-uniform "error band" model where the end nodes are connected probabilistically through a *jpdf* (Dutton 1992, Shi 1998, Shi and Lui 2000). Models for positional uncertainty also include non-probabilistic approaches where rectangular buffers are used to create a "confidence space" for lines and polygons (Goodchild and Hunter 1997).

Uncertainties about the absolute position or geometry of spatial vectors, such as points and polygons, may lead to uncertainties about the topological relationships between vectors, such as the position of points within polygons (Winter 2000, Shortridge and Goodchild 2002). These uncertainties may be described analytically (Shortridge and Goodchild 2002), or quantified through an uncertainty propagation analysis.

The joint Gaussian distribution is typically assumed in probability models of positional uncertainty (Kiiveri 1997, Leung and Yan 1998, Shi 1994, 1998; Shi and Lui 2000). Here, the uncertainties between locations may be statistically dependent in space for which Shi (1998) and Shi and Lui (2000) introduce the "error-band" and "G-band" models respectively. These models were derived for straight-line segments, but may be extended to other vector shapes and to curved lines (Tong, Shi, and Lui 2003). However, a more fundamental problem arises in the estimation of probability models for curved lines where clear points for computing differences between more and less accurate data do not exist (van Niel and McVicar 2002).

More generally, quantitative probability models (Shi and Lui 2000) may be difficult to estimate if information on positional uncertainty is limited to simple measures of accuracy for whole line segments, rather than two or more points. Furthermore, the joint Gaussian assumption cannot easily be extended to complex features where pre-processing and scale transformations affect some line segments more than others (Goodchild and Hunter 1997). In these cases, simple measures of positional uncertainty, such as non-probabilistic buffers, might be preferred. These measures may be derived from comparisons between more and less accurate data (Goodchild and Hunter 1997, Tveite and Langaas 1999) or internal data geometry (Veregin 2000). They may be extended to probabilistic measures once sufficient data are available and the joint Gaussian assumption can be justified. However, they will ultimately differ from point-based descriptions of positional uncertainty, such as the "G-band" model of Shi and Lui (2000), because points and buffers are conceptually different (van Niel and McVicar 2002).

Some Challenges for Estimating Spatial Data Quality

Understanding the limitations of spatial data is essential both for managing social and environmental systems effectively and for encouraging the responsible use of scientific research where knowledge is limited and priorities are varied. In this context, critical self-reflection about the uncertainties in spatial data must be encouraged alongside attempts to improve data by reducing error. In practice, however, there are many ongoing challenges for the successful application of uncertainty methodologies to spatial data (Brown 2004). This is evidenced by the rapid rate of progress in storing and retrieving "deterministic data" in GIS versus storing and retrieving information about data quality, and the uncertainties associated with spatial data.

While not entirely, or even primarily, a technical challenge, some important technical challenges remain in assessing spatial data quality (Heuvelink 2002). For example, the identification of statistical dependence between uncertain attributes, both in space and between multiple data sets, remains an important technical challenge for applying probability models to spatial data. First, the propagation of uncertainties through GIS operations, such as "spatial overlay," "buffering" and "map algebra," may be highly sensitive to spatial dependencies in the input data (Arbia, Griffith, and Haining 1998, Heuvelink 1998). In addition, environmental models typically rely upon multiple spatial inputs, including both "primary" data (for example, terrain elevation) and data derived from other inputs to the same model (for example, terrain slope), for which spatial dependencies become important. Second, statistical dependencies are more difficult to estimate than an average magnitude of uncertainty (such as a Root Mean Squared Error), and sensitivity testing with a range of possible dependence structures might be preferred in some cases. Third, measurement uncertainties may contain their own space-time dependencies, which are separate from, but complicated by, the real patterns of variation in the measured variable. For example, errors in flow measurements may increase with stream discharge, while current meters may be consistently misused or misinterpreted. However, measurement uncertainties can be obtained from some instruments, such as GPS receivers (temporal correlation of the GPS precision), and remote sensing satellites (Arbia, Griffith, and Haining 1998) and can sometimes be estimated from the data themselves (for instance, through signal processing).

In developing new techniques for assessing uncertainties in spatial data, future research might focus on the identification of quality metrics for groups of objects rather than individual objects, and the interactions between uncertainties in multiple objects and their attributes. It might also focus on the joint modeling of attribute and positional uncertainties, and on the provision of quality metrics that allow fitness for use to be established without an uncertainty propagation analysis (Mowrer and Congalton 2000, Hunter and Lowell 2002). However, the development of new techniques for assessing and representing uncertainties in spatial data must coincide with the development of new concepts for applying groups of techniques to specific problems where multiple data sets, degrees, and sources of uncertainty converge. Here, there is a need for "stochastic information systems" that allow different metrics of uncertainty to be stored in spatial databases and propagated through GIS while educating users about the nature and impacts of uncertainty in spatial data (Brown and Heuvelink 2006). Thus, standardization and automation should be carefully managed in developing uncertainty models. Indeed, in estimating *pdfs*, the balance between model complexity, identifiability, and reliability should be a guided decision by those responsible for assessing uncertainty, and not a predetermined "error button" devoid of specificity or educational value.

Thus, there is a need to provide a flexible, rather than a uniformly simple, infrastructure for retrieving information about spatial data quality from GIS users, and for organizing this information within a database design. In this context, the complexity of an uncertainty model could be linked to the risks and resources associated with decision-making, as well as the state of information on uncertainty. Alongside a flexible framework for assessing and representing uncertainties about spatial data, there is a need for guidelines (and examples) to manage this flexibility in specific cases so that users with limited statistical expertise can develop an appropriate uncertainty model for many applications (obviously, there are limits here). For quantitative descriptions of uncertainty, this might be achieved by grouping and demonstrating the impacts of specific statistical decisions to users.

There are, however, some more fundamental challenges for the development and successful application of uncertainty models to spatial data. These include the social desire for simplicity in applying GIS (from which the "error button" philosophy originates), the time and money required to implement uncertainty methodologies, and the problems of ignorance and indeterminacy in geographic research where, respectively, we do not and cannot know about some aspects of spatial variability (Handmer, Norton, and Dovers 2001, Couclelis 2003). Indeed, too normative an emphasis on quantifying and reducing uncertainty neglects the wider conceptual problems of "representing reality"with spatial data (for example, reality is space *and* time varying), and discourages contingency planning and "openness" more generally. Thus, uncertainty analyses should assist in targeting the causes of uncertainty in GIS, but their success will ultimately rest on the desire for openness and communication about potential errors in spatial data for which some social, institutional, and legal changes, as well as clear evidence on the utility of uncertainty analyses, should further support the widespread application of uncertainty tools in GIS.

REFERENCES

Arbia, G., Griffith, D., and Haining, R. 1998. Error propagation modelling in raster GIS: Overlay operations. *International Journal of Geographical Information Science* 12: 145–67.

Ayyub, B. M. 2001. *Elicitation of Expert Opinions for Uncertainty and Risks*. Boca Raton, FL: CRC Press.

Beven, K. J. and Freer, J. 2001. Equifinality, data assimilation, and uncertainty estimation in mechanistic modelling of complex environmental systems. *Journal of Hydrology* 249: 11–29.

Bierkens, M. F. P., Finke, P. A., and De Willigen, P. 2000. *Upscaling and Downscaling Methods for Environmental Research*. Dordrecht: Kluwer.

Brown, J. D. 2004. Knowledge, uncertainty and physical geography: Towards the development of methodologies for questioning belief. *Transactions of the Institute of British Geographers* 29: 367–81.

Brown, J. D. and Heuvelink, G. B. M. 2006. The Data Uncertainty Engine (DUE): A software tool for assessing and simulating uncertain environmental variables. *Computers and Geosciences* 32: in press.

Burkholder, E. F. 2002. The global spatial data model. In M. F. Goodchild, and A. J. Kimerling (eds) *Discrete Global Grids*. Santa Barbara, CA: National Center for Geographic Information and Analysis (available at http://www.ncgia.ucsb.edu/globalgrids-book/spatialdata/).

Burrough, P. A. and McDonnell, R. 1998. *Principles of Geographical Information Systems.* Oxford: Oxford University Press.

Burt, J. E. and Barber, G. M. 1996. *Elementary Statistics for Geographers.* New York: Guilford Press.

Chrisman, N. R. 1991. The error component of spatial data. In D. J. Maguire, M. F. Goodchild, and D. W. Rhind (eds) *Geographical Information Systems: Principles and Applications.* London: Longman, pp. 164–74.

Christakos, G. 2000. *Modern Spatiotemporal Geostatistics.* Oxford: Oxford University Press.

Cooke, R. M. 1991. *Experts in Uncertainty: Opinion and Subjective Probability in Science.* Oxford: Oxford University Press.

Couclelis, H. 2003. The certainty of uncertainty: GIS and the limits of geographic knowledge. *Transactions in GIS* 7: 165–75.

Cressie, N. A. C. 1993. *Statistics for Spatial Data.* New York: JohnWiley and Sons.

De Bruin, S. and Bregt, A. 2001. Assessing fitness for use: The expected value of spatial data sets. *International Journal of Geographical Information Science* 15: 457–71.

Diggle, P. J. 2003. *Statistical Analysis of Spatial Point Patterns.* Oxford: Oxford University Press.

Diggle, P. J., Ribeiro Jr, P. J., and Christensen, O. F. 2003. An introduction to model based geostatistics. In J. Möller (ed.) *Spatial Statistics and Computational Methods.* Berlin: Springer Lecture Notes in Statistics No. 173: 43–86.

Dutton, G. 1992. Handling positional uncertainty in spatial databases. In *Proceedings of the Fifth International Symposium on Spatial Data Handling*, Charleston, SC, USA. International Geographical Union Commission on GIS, pp. 460–9.

FGDC. 1998. *Geospatial Positioning Accuracy Standards: Part 3, National Standard for Spatial Data Accuracy.* Washington, DC: Federal Geographic Data Committee.

Finke, P. A., Wladis, D., Kros, J., Pebesma, E. J., and Reinds, G. J. 1999. Quantification and simulation of errors in categorical data for uncertainty analysis of soil acidification modelling. *Geoderma* 93: 177–94.

Fisher, P. F., Comber, A. J., and Wadsworth, R. A. 2002. The production of uncertainty in spatial information: The case of land cover mapping. In G. Hunter and K. Lowell (eds) *Accuracy 2002: Proceedings of the Fifth International Symposium on Spatial Accuracy Assessment in Natural Resources and Environmental Sciences.* July 10–12, Melbourne, Australia. Melbourne, Australia: University of Australia, pp. 60–73.

Foody, G. M. 1999. The continuum of classification fuzziness in thematic mapping. *Photogrammetric Engineering and Remote Sensing* 65: 443–51.

Foody. G. M. 2002. Status of land-cover classification accuracy assessment. *Remote Sensing of Environment* 80: 185–201.

Foody, G. M. and Atkinson, P. M. (eds). 2003. *Uncertainty in Remote Sensing and GIS.* London: John Wiley and Sons.

Fotheringham, A. S., Charlton, M. E., and Brunsdon, C. 2000. *Quantitative Geography: Perspectives on Spatial Data Analysis.* London: Sage Publications.

Funtowicz, S. O. and Ravetz, J. R. 1990. *Uncertainty and Quality in Science for Policy.* Dordrecht: Kluwer.

Gabrosek, J. and Cressie, N. 2002. The effects on attribute prediction of location uncertainty in spatial data. *Geographical Analysis* 34: 262–85.

Goodchild, M. F. and Hunter, G. J. 1997. A simple positional accuracy measure for linear features. *International Journal of Geographical Information Science* 11: 299–306.

Goovaerts, P. 1997. *Geostatistics for Natural Resources Evaluation.* New York: Oxford University Press.

Goovaerts, P. 2001. Geostatistical modeling of uncertainty in soil science. *Geoderma* 103: 3–26.

Greenland, S. 2001. Sensitivity analysis, Monte Carlo risk analysis, and Bayesian uncertainty assessment. *Risk Analysis* 21: 579–83.

Guptill, S. C. and Morrison, J. L. 1995. *Elements of Spatial Data Quality*. Oxford: Elsevier.

Handmer, J. W., Norton, T. W., and Dovers, S. R. (eds). 2001. *Ecology, Uncertainty and Policy*. Harlow: Prentice Hall.

Harremoës, P., Gee, D., MacGarvin, M., Stirling, A., Keys, J., Wynne, B., and Guedes, V. S. (eds). 2002. *The Precautionary Principle in the 20th Century: Late Lessons from Early Warnings*. London: Earthscan.

Hennings, V. 2002. Accuracy of coarse-scale land quality maps as a function of the upscaling procedure used for soil data. *Geoderma* 107: 177–96.

Heuvelink, G. B. M. 1998. *Error Propagation in Environmental Modelling with GIS*. London: Taylor and Francis.

Heuvelink, G. B. M. 1999. Propagation of error in spatial modelling with GIS. In P. A. Longley, M. F. Goodchild, D. J. Maguire, and D. W. Rhind (eds) *Geographical Information Systems: Principles, Techniques, Applications, and Management*. New York: John Wiley and Sons: 207–17.

Heuvelink, G. B. M. 2002. Analysing uncertainty propagation in GIS: Why is it not that simple? In G. M. Foody and P. M. Atkinson (eds) *Uncertainty in Remote Sensing and GIS*. Chichester: John Wiley and Sons, pp. 155–65.

Heuvelink, G. B. M. and Bierkens, M. F. P. 1992. Combining soil maps with interpolations from point observations to predict quantitative soil properties. *Geoderma* 55: 1–15.

Heuvelink, G. B. M. and Pebesma, E. J. 1999. Spatial aggregation and soil process modelling. *Geoderma* 89: 47–65.

Hunter, G. and Lowell, K. (eds). 2002. *Proceedings of the Fifth International Symposium on Spatial Accuracy Assessment in Natural Resources and Environmental Sciences (Accuracy 2002)*. Melbourne, Australia: Department of Geomatics, University of Melbourne.

Journel, A. G. and Huijbregts, C. J. 1978. *Mining Geostatistics*. London: Academic Press.

Kiiveri, H. T. 1997. Assessing, representing and transmitting positional uncertainty in maps. *International Journal of Geographical Information Science* 11: 33–52.

Kiiveri, H. T. and Cacetta, P. 1998. Image fusion with conditional probability networks for monitoring the salinization of farmland. *Digital Signal Processing* 8: 225–30.

Kyriakidis, P. C. and Dungan, J. L. 2001. A geostatistical approach for mapping thematic classification accuracy and evaluating the impact of inaccurate spatial data on ecological model predictions. *Ecological and Environmental Statistics* 8: 311–30.

Lane, S. N. 2001. Constructive comments on D Massey "Space-time, 'science' and the relationship between physical geography and human geography." *Transactions of the Institute of British Geographers* 26: 243–56.

Leung, Y. and Yan, J. 1998. A locational error model for spatial features. *International Journal of Geographical Information Science* 12: 607–20.

Mowrer, H. T. and Congalton, R. G. (eds). 2000. *Quantifying Spatial Uncertainty in Natural Resources: Theory and Applications for GIS and Remote Sensing*. Chelsea, MI: Ann Arbor Press.

Norberg, T., Rosén, L., Baran, Á., and Baran, S. 2002. On modeling discrete geological structures as Markov random fields. *Mathematical Geology* 34: 63–77.

Perkal, J. 1966. *On the Length of Empirical Curves*. Ann Arbor, MI: Michigan Inter-University Community of Mathematical Geographers Discussion Paper No. 10.

Richards, K. S., Brooks, S. M., Clifford, N. J., Harris, T. R. M., and Lane, S. N. 1997. Real geomorphology in physical geography: Theory, observation and testing. In D. R. Stoddard (ed.) *Process and Form in Geomorphology*. London: Routledge, pp. 269–92.

Ripley, B. D. 1981. *Spatial Statistics*. New York: John Wiley and Sons.

Schlather, M., Ribeiro, P., and Diggle, P. 2004. Detecting dependence between marks and locations of marked point processes. *Journal of the Royal Statistical Society, Series B* 66: 79–93.

Shi, W. Z. 1994. *Modelling Positional and Thematic Uncertainty in Integration of GIS and Remote Sensing*. Enschede, Netherlands: ITC Publication No. 22.

Shi, W. Z. 1998. A generic statistical approach for modelling error of geometric features in GIS. *International Journal of Geographical Information Science* 12: 131–43.

Shi, W. Z. and Liu, W. 2000. A stochastic process-based model for the positional error of line segments in GIS. *International Journal of Geographical Information Science* 14: 51–66.

Shi, W. Z., Fisher, P. F., and Goodchild, M. F. 2002. Epilogue: A prospective on spatial data quality. In W. Shi, P. F. Fisher, and M. F. Goodchild (eds) *Spatial Data Quality*. London: Taylor and Francis, pp. 304–9.

Shortridge, A. M. and Goodchild, M. F. 2002. Geometric probability and GIS: Some applications for the statistics of intersections. *International Journal of Geographical Information Science* 16: 227–43.

Stanislawski, L. V., Dewitt, B. A., and Shrestha, R. L. 1996. Estimating positional accuracy of data layers within a GIS through error propagation. *Photogrammetric Engineering and Remote Sensing* 62: 429–33.

Steele, B. M., Patterson, D. A., and Redmond, R. A. 2003. Toward estimation of map accuracy without a probability test sample. *Environmental and Ecological Statistics* 10: 333–56.

Stehman, S. V. and Czaplewski, R. L. 2003. Introduction to special issue on map accuracy *Environmental and Ecological Statistics* 10: 301–8.

Story, M. and Congalton, R. G. 1986. Accuracy assessment: a user's perspective. *Photogrammetric Engineering and Remote Sensing* 52: 397–9.

Tong, X., Shi, W., and Liu, D. 2003. An error model of circular curve features in GIS. In *Proceedings of the Eleventh ACM International Symposium on Advances in Geographic Information Systems*, New Orleans, LA, USA. New York, NY: Association of Computing Machinery, pp. 141–6.

Tveite, H. and Langaas, S. 1999. An accuracy assessment method for geographical line data sets based on buffering. *International Journal of Geographical Information Science* 13: 27–47.

USBB. 1947. *United States National Map Accuracy Standards*. Washington, DC: US Bureau of the Budget.

van Niel, T. G. and McVicar, T. R. 2002. Experimental evaluation of positional accuracy estimates from a linear network using point- and line-based testing methods. *International Journal of Geographical Information Science* 16: 455–73.

Veregin, H. 1996. Error propagation through the buffer operation for probability surfaces. *Photogrammetric Engineering and Remote Sensing* 62: 419–28.

Veregin, H. 1999. Data quality parameters. In P. A. Longley, M. F. Goodchild, D. J. Maguire, and D. W. Rhind (eds) *Geographical Information Systems: Volume 1, Principles and Technical Issues*. New York: John Wiley and Sons, pp. 177–89.

Veregin, H. 2000. Quantifying positional error induced by line simplification. *International Journal of Geographical Information Science* 14: 113 30.

von Reibnitz, U. 1988. *Scenario Techniques*. Hamburg: McGraw-Hill.

Webster, R. and Oliver, M. A. 1992. Sampling adequately to estimate variograms of soil properties. *Journal of Soil Science* 43: 177–92.

Winter, S. 2000. Uncertain topological relations between imprecise regions. *International Journal of Geographical Information Science* 14: 411–30.

Part II   Database Trends
and Challenges

The second section of the book explores some of the important and enduring database issues and trends. In Chapter 7, The first chapter in this set, Shashi Shekhar and Ranga Raju Vatsavai describe the relational, object-oriented, and object-relational database management systems (DBMSs) from both the Geographic Information Systems (GIS) and spatial data management systems points of view. The principal characteristics of GIS and spatial database management systems are briefly described and the linkages between these systems and the aforementioned DBMSs are explored. The key concepts of each of these DBMSs are illustrated using simple examples drawn from the standard World database.

The second chapter in this group, Chapter 8 by Michael F. Hutchinson, examines the generation of regular grid digital elevation models from a variety of data sources to support the elevation and landscape shape requirements of environmental modeling over a range of spatial scales. Such models have played an integral role in GI Science since its inception and have directly stimulated new methods for obtaining digital environmental data, new spatial interpolation methods, and new methods for analyzing landscape dependent hydrological and ecological processes. The latter, which are usually performed by various forms of thin plate smoothing splines and geostatistics, demonstrate some of the subtleties and growing importance of multivariate statistical analysis in GI Science.

In Chapter 9, the third chapter in this bundle on database trends and challenges, May Yuan describes the importance of time in geographic inquiry and understanding and some of the conceptual advances that are needed to add time to GIS databases. Separate sections focus on key developments in spatio-temporal ontologies, representation, and data modeling, and support for spatio-temporal queries. The chapter concludes by summarizing the current state of temporal GIS and the prospects for developing temporal GIS that can support spatio-temporal information management, query, analysis, and modeling in the immediate future.

The final chapter of Part II, Chapter 10 by Craig A. Knoblock and Cyrus Shahabi, describes some of their recent work on the extraction and integration of geospatial and related data that go beyond conversion between different products and standard formats for the interoperability of these products. New techniques are introduced to turn online web sources into semi-structured data that can, among other things, be integrated with other geospatial data; automatically and accurately integrate vector data with high-resolution color imagery; integrate online property tax sources and conflated road vector data to identify and annotate buildings on imagery; automatically integrate maps with unknown coordinates and satellite imagery; and efficiently combine online schedules and vector data to predict the locations of moving objects. These examples, while not exhaustive, are used by the authors to illustrate new opportunities for integrating various geospatial and online data sources.

Chapter 7

Object-Oriented Database Management Systems

Shashi Shekhar and Ranga Raju Vatsavai

We are in the midst of an information revolution. The raw material (data) powering this controlled upheaval is not found below the Earth's surface where it has taken million of years to form but is being gathered constantly via sensors and other data-gathering devices. For example, NASA's Earth Observing System (EOS) generates one terabyte of data every day.

Satellite images are one prominent example of spatial data. Extracting information from a satellite image requires that the data be processed with respect to a spatial frame of reference, possibly the Earth's surface. But satellites are not the only source of spatial data, and the Earth's surface is not the only frame of reference. A silicon chip can be, and often is, a frame of reference. In medical imaging the human body acts as a spatial frame of reference. In fact even a supermarket transaction is an example of spatial data if, for example, a zip code is included. Queries, or commands, posed on spatial data are called spatial queries. So, while the query "What are the names of all bookstores with more than ten thousand titles?" is an example of a non-spatial query, "What are the names of all bookstores within ten miles of the Minneapolis downtown?" is an example of a spatial query.

A database is a permanent repository of data which is stored in one or more files and managed by a database management system (DBMS). Databases and the software which manages them are the silent success story of the information age. They have slowly permeated all aspects of daily living, and modern society would come to a halt without them. A DBMS can be characterized by its underlying data model and the query language used for describing and accessing the data. From a database point of view, a data model is a collection of mathematically well-defined concepts for data abstraction. It provides tools for high-level description of database schemas at the conceptual level and hides low-level details of how the data is stored. Traditional DBMSs are generally based on one of the classical data models – hierarchical, network, or relational. However, the focus of attention since the late 1980s has been on relational database management systems (RDBMS), which have gained widespread popularity. More recent data models include object-oriented and object-relational models. The historical evolution of database technology is shown in Figure 7.1.

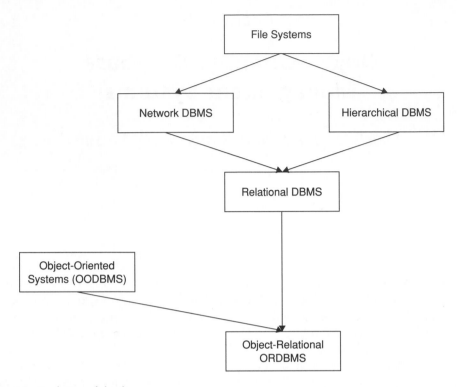

Fig. 7.1   Evolution of databases
After Khoshafian and Baker 1998

Associated with each data model is a database language, which provides two types constructs: one for defining database schemas, and the second for querying and modifying the data in the database. The Structured Query Language (SQL) is the *lingua franca* of the database world, and it is tightly coupled with the relational database model. SQL is a descriptive query language, that is, a user describes what data he or she wants from the database without specifying how to retrieve it. On the other hand, the Object Query Language (OQL) is tightly coupled with the object-oriented model.

Despite their spectacular success, the prevalent view is that a majority of the RDBMSs in existence today are either incapable of managing spatial data or are not user-friendly when doing so. Now, why is that? The traditional role of a RDBMS has been that of a simple but effective warehouse of business and accounting data. Information about employees, suppliers, customers, and products can be safely stored and efficiently retrieved through a RDBMS. The set of likely queries is limited, and the database is organized to answer these queries efficiently. From the business world, the RDBMS made a painless migration into government agencies and academic administrations.

Data residing in these mammoth databases is simple, consisting of numbers, names, addresses, product descriptions, etc. These DBMSs are very efficient for the tasks for which they were designed. For example, a query like "List the top ten customers, in terms of sales, in the year 1998" will be very efficiently answered by a DBMS even if the database has to scan through a very large customer database. Such commands are conventionally called "queries" although they are not questions. The DBMS will not scan through all the customer records; it will use an index, to narrow down the search. By contrast, a relatively simple query such as "List all the customers who reside within 50 miles of the company headquarters" will confound the database. To process this query, the DBMS will have to transform the company headquarters and customer addresses into a suitable reference system, possibly latitude and longitude, in which distances can be computed and compared. Then the DBMS will have to scan through the entire customer list, compute the distance between the company and the customer, and, if this distance is less than 50 miles, save the customer's name. It will not be able to use an index to narrow down the search, because traditional indices are incapable of ordering multi-dimensional coordinate data. A simple and legitimate business query can thus send a DBMS into a hopeless tailspin. The RDBMS, which are designed for business data processing, are capable of managing only simple data types such as numeric, character, and date. They are not suitable to manage complex data types that arise in various emerging application domains such as Geographic Information Systems (GIS), Multimedia, Computer Aided Design (CAD) and Computer Aided Manufacturing (CAM). Therefore, there is an immediate need for databases tailored to handle spatial data and spatial queries, and other complex applications.

The object-oriented paradigm appears to be a natural choice for highly complex application domains such as spatial databases, because it provides a direct correspondence between real-world entities and programming (system) objects. The object-oriented database management systems (OODBMS), which combine object-oriented programming concepts and database management principles, have matured considerably since their appearance in the early 1980s. However, the adoptability of OODBMS has been limited to certain niche markets like e-commerce, engineering, medicine and some complex applications. Recently, hybrid systems, known as object-relational database management systems (ORDBMS), have become popular. The ORDBMS combines the good features of RDBMS (simple data types and SQL) with the good features of OODBMS (complex data types and methods).

This chapter provides an overview of current OODBMS and ORDBMS technologies from a spatial database management point of view. The next section describes the relationship between GIS and Spatial Database Management Systems, and shows the limitations of RDBMS in handling spatial data.

GIS and Spatial Data Management Systems (SDBMS)

GIS are the principal technology motivating interest in SDBMS; they provide a convenient mechanism for the analysis and visualization of geographic data. Geographic data are spatial data whose underlying frame of reference is the Earth's surface. The GIS provide a rich set of analysis functions which allows a user to affect powerful transformations on geographic data. The rich array of techniques which geographers have added to GIS is the reason behind their phenomenal growth and multidisciplinary applications. Table 7.1 lists a small sample of common GIS operations.

Table 7.1   List of common GIS analysis operations (after Albrecht 1998)

Search	Thematic search, search by region, (re-)classification
Location analysis	Buffer, corridor, drainage network
Terrain analysis	Slope/aspect, catchments, drainage network
Flow analysis	Connectivity, shortest path
Distribution	Change detection, proximity, nearest neighbor
Spatial analysis/Statistics	Pattern, centrality, autocorrelation, indices of similarity, topology: hole description
Measurements	Distance, perimeter, shape, adjacency, direction

A GIS provides a rich set of operations over a few objects and layers, whereas an SDBMS provides simpler operations on sets of objects and sets of layers. For example, a GIS can list neighboring countries of a given country (for example, France) given the political boundaries of all countries. However it will be fairly tedious to answer set queries like *list the countries with the highest number of neighboring countries* or *list countries which are completely surrounded by another country.* Set-based queries can be answered efficiently in an SDBMS.

SDBMSs are also designed to handle very large amounts of spatial data stored on secondary devices (for example, magnetic disks, CD-ROMs, jukeboxes, etc.) using specialized indices and query-processing techniques. Finally, SDBMSs inherit the traditional DBMS functionality of providing a concurrency-control mechanism to allow multiple users to simultaneously access shared spatial data, while preserving the consistency of that data. A GIS can be built as the front-end of an SDBMS. Before a GIS can carry out any analysis of spatial data, it accesses that data from a SDBMS. Thus an efficient SDBMS can greatly increase the efficiency and productivity of a GIS.

A typical spatial database consists of several images and vector layers like land parcels, transportation, ecological regions, soils, etc. Let us now consider how census block data can be stored in a DBMS. One *natural* way of storing information about the census blocks (for example, their name, geographic area, population and boundaries) is to create the following table in the database:

```
create table census
blocks (          name         string,
                  area         float,
                  population   number,
                  boundary     polyline);
```

In a (relational) database, all objects, entities, and concepts which have a distinct identity are represented as relations or tables. A relation is defined by a name and a list of distinguishing attributes which characterize the relation. All instances of the relation are stored as tuples in the table. In the preceding code fragment we have created a table(relation) named *census_block*, which has four attributes: name, area, population, and boundary. At table creation time, the types of attributes have to

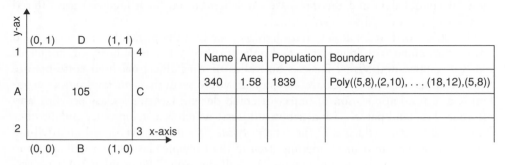

Fig. 7.2   Census blocks with boundary ID:1050

be specified, and here they are: string, float, number, and polyline. A polyline is a data type to represent a sequence of straight lines.

Figure 7.2 shows a hypothetical census block and how information about it can be stored in a table. Unfortunately, such a table is not natural for a traditional relational database because polyline is not a built-in data type. One way to circumvent this problem is to create a collection of tables with overlapping attributes, as shown in Figure 7.3. Another way is to use a stored procedure. For a novice user these implementations are quite complex. The key point is that the census block data cannot be naturally mapped onto a relational database. We need more constructs to handle spatial information in order to reduce the semantic gap between the user's

Census_blocks

Name	Area	Population	Boundary-ID
340	1.58	1839	1050

Polygon

Boundary-ID	Edge-name
1050	A
1050	B
1050	C
1050	D

Edge

edge-name	Endpoint
A	1
A	2
B	2
B	3
C	3
C	3
D	4
D	1

Point

endpoint	x-coor	y-coor
1	0	1
2	0	0
3	1	0
4	1	1

Fig. 7.3   Four tables required in a relational database with overlapping attributes to accommodate the polyline data type

view of spatial data and the database implementation. Such facilities are offered by the object-oriented software paradigm.

The object-oriented software paradigm is based on the principles of user-defined data types, along with inheritance and polymorphism. The popularity of languages like C++, Java, and Visual Basic is an indicator that object-oriented concepts are firmly established in the software industry. It would seem that our land parcel problem is a natural application of object-oriented design: Declare a class polyline and another class land parcel with attribute address, which is a string type, and another attribute *boundary* which is of the type *polyline*. We do not even need an attribute *area* because we can define a method *area* in the polyline class which will compute the area of any land parcel on demand. So will that solve the problem? Are object-oriented databases (OODBMS) the answer? Well, not quite.

The debate between relational versus object-oriented within the database community parallels the debate between vector versus raster in GIS. The introduction of abstract data types (ADTs) clearly adds flexibility to a DBMS, but there are two constraints peculiar to databases that need to be resolved before ADTs can be fully integrated into DBMSs.

- Market adoption of OODBMS products has been limited, despite the availability of such products for several years. This reduces the financial resources and engineering efforts to performance-tune OODBMS products. As a result, many GIS users will use systems other than OODBMS to manage their spatial data in the near future.
- SQL is the *lingua franca* of the database world, and it is tightly coupled with the relational database model. SQL is a declarative language, that is, the user only specifies the desired result rather than the means of production. For example, in SQL the query "*Find all land parcels adjacent to MY_HOUSE.*" should be able to be specified as follows:

```
SELECT    M.address
FROM      land parcel L, M
WHERE     Adjacent(L,M) AND L.address = 'MYHOUSE'
```

It is the responsibility of the DBMS to implement the operations specified in the query. In particular, the function *Adjacent(L,M)* should be callable from within SQL. The current standard, SQL-92, supports user-defined functions, and SQL-3, the next revision, will support ADTs and a host of data structures such as lists, sets, arrays, and bags. Relational databases which incorporate ADTs and other principles of object-oriented design are called object-relational database management systems (ORDBMS).

The current generation of ORDBMSs offers a modular approach to ADTs. An ADT can be built into or deleted from the system without affecting the remainder of the system. While this "plug-in" approach opens up the DBMS for enhanced functionality, there is very little built-in support for the optimization of operations. Our focus will be to specialize an ORDBMS to meet the requirements of spatial data. By doing so, we can extrapolate spatial domain knowledge to improve the overall efficiency of the system. We are now ready to give a definition of SDBMS for setting the scope of this chapter:

1   A spatial database management system is a software module that can work with an underlying database management system, for example, ORDBMS, OODBMS;
2   SDBMSs support multiple spatial data models, commensurate spatial abstract data types (ADTs), and a query language from which these ADTs are callable;
3   SDBMSs support spatial indexing, efficient algorithms for spatial operations, and domain-specific rules for query optimization.

Figure 7.4 shows a representation of an architecture to build an SDBMS on top of an ORDBMS. This is a three-layer architecture. The top layer (from left to right) is the spatial application, such as GIS, MMIS (multimedia information system), or CAD (computer-aided design). This application layer does not interact directly with the ORDBMS but goes through a middle layer which we have labeled "spatial database." The middle layer is where most of the available spatial domain knowledge is encapsulated, and this layer is "plugged" into the ORDBMS. This layered approach explains why commercial ORDBMS products have names like Spatial Data Blade (Illustra), Spatial Data Cartridge (Oracle), and Spatial Data Engine (ESRI).

Let us now summarize the core features that are essential for any DBMS:

1   **Persistence:** The ability to handle both transient and persistent data. While transient data is lost after a program terminates, persistent data not only transcends program invocations but also survives system and media crashes. Further, the DBMS ensures that a smooth recovery takes place after a crash. In database management systems, the state of the persistent object undergoes frequent changes, and it is sometimes desirable to have access to the previous data states.
2   **Transactions:** Transactions map a database from one consistent state to another. This mapping is atomic (that is, it is executed completely or aborted). Typically, many transactions are executed concurrently, and the DBMS imposes an order of execution that can be performed as a series. Consistency in a database is accomplished through the use of integrity constraints. All database states must satisfy these constraints to be deemed consistent. Furthermore, to maintain the security of the database, the scope of the transactions is dependent on the user's access privileges.

Spatial databases can be characterized by a set of spatial data types and the operations permitted on those data types.

Spatial data types

A key issue in the encoding of spatial information is the choice of a basic set of spatial data types required to model common shapes on maps. Many proposals have been made over the years. A consensus is slowly emerging in terms of the OGIS standard (OGIS 1999). Figure 7.5 shows the fundamental building blocks of two-dimensional spatial geometry and their interrelationships in Unified Modeling Language (UML) notation. We will give a brief description of the UML notation in later sections. Let us now look more closely at these building blocks of spatial geometry.

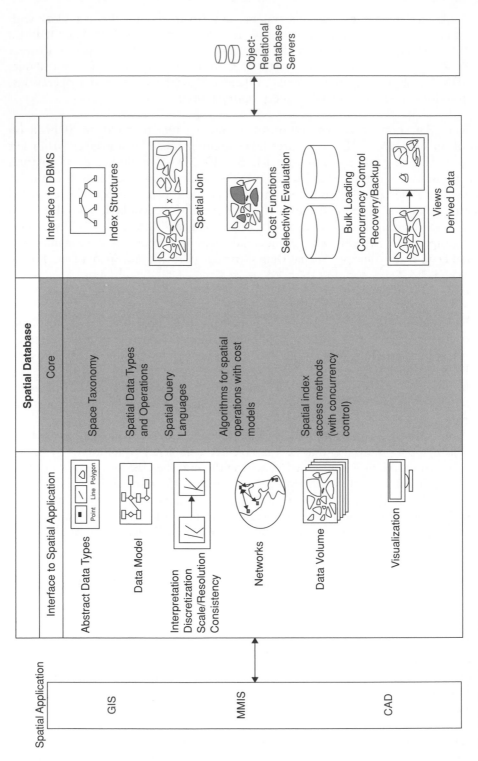

Fig. 7.4 Three-layer architecture

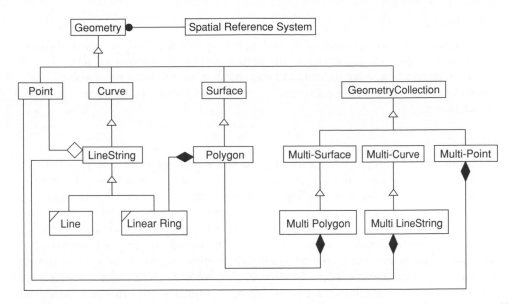

Fig. 7.5 An OGIS proposal for building blocks of spatial geometry in UML notation
After OGIS 1999

The most general shape is represented by "Geometry" described via a "Spatial Representation System" which is a coordinate system like latitude/longitude or some other consensus framework. The "Geometry" is subdivided into four categories, namely *Point, Curve, Surface,* and *GeometryCollection. Point* describes zero-dimensional objects, for example, the city centers in a map of the world. *Curve* describes the shapes of one-dimensional objects, for example, rivers in the map of a world. The *Curve* objects are often approximated by a *LineString,* which is represented by two or more points. The simplest *LineString* is a straight line joining two or more *Points.* The category *Surface* describes the shape of two-dimensional objects, for example, countries on a map of the world. A *Surface* is often modeled by a *Polygon. GeometryCollection* represents complex shapes such as a collection of oil wells, a group of islands, etc. *GeometryCollection* in turn is of three types, namely *Multi-Point, Multi-Curve,* and *Multi-Surface.* The *GeometryCollection* spatial data types provide a "closure" property to OGIS spatial data types under geometric operations such as "geometric-union," "geometric-difference," or "geometric-intersection." For example, if one takes a geometric-difference of the boundaries of Canada and Quebec, the result is a *Multi-Surface* even if Canada and Quebec were of *Surface* spatial data type. This property is useful to support multi-step querying and data processing.

Operations on spatial objects

Let us now briefly look at the typology of embedding space and associated relationships, and some of the common operations defined on spatial objects.

Set-oriented

The simplest and most general of all embedding spaces is called set-oriented space. Common relationships allowable in this setting are the usual set-based relationships of union, intersection, containment, and membership. Hierarchical relationships like a county contained in a state, or a state contained in a country are adequately modeled by set theory.

Topological

For an intuitive feeling of what a topological space is, imagine two polygons which touch (meet) each other and are drawn on a rubber sheet. Now if we deform the rubber sheet by stretching or bending it, but not cutting or folding it, the adjacency of the polygons remains intact. *Meet* is an example of a topological property and the study of transformations (deformations) which preserve topological properties is called topology. Consider a political map of the world that shows boundaries of countries. The neighboring countries meet each other, whether the map is drawn on a sphere or on a flat space. The area of a polygon is clearly not a topological property. In fact, the relative areas of different countries are often not preserved in many maps. Areas of countries near the equator are reduced relative to the areas of countries near the poles in many planar maps. From a spatial/geographic database point of view, topological relationships like *meet*, *within*, and *overlap* are most likely to be queried by a user of a spatial database management system. Is a given land parcel adjacent to a hazardous waste site? Does the river floodplain overlap a proposed highway network? All of these are examples of topological relationships. More detailed information on topological spaces can be found in Egenhofer (1991).

Directional

Directional relationships can be of three types – namely, absolute, object-relative, or viewer based. Absolute directional relationships are defined in the context of a global reference system, for example, North, South, East, West, North-East, etc. Object-relative directions are defined using the orientation of a given object. Example relationships include left, right, front, behind, above, below, etc. Viewer relative directions are defined with respect to a specially designated reference object, called the viewer.

Metric space

Mathematically speaking, a set X is called a metric space if for any pair of points x and y of X, there is an associated real number $d(x,y)$, called the distance (also called a metric) from x to y, with the following properties:

1 $d(x,y) \geq 0$ and $d(x,x) = 0$
2 $d(x,y) = d(y,x)$
3 $d(x,y) \leq d(x,z) + d(z,y)$

for all x,y,z in X. Any function that satisfies these properties is called a metric on X.

In metric spaces the notion of distance is well defined. A distance function can be used to induce a topology on the space, and therefore every metric space is also a topological space. Metric spaces play a crucial role in network or graph settings. Optimal-distance and shortest travel-time queries are ideally handled in a metric space setting.

Euclidean

Let R be the field of real numbers. A *vector space* V over R is a nonempty set V of objects v called *vectors*, together with two operations:

1   Addition: $u + v \in V$ for all $uv \in V$
2   Product: $\alpha u \in V$ for all $\alpha \in R$, $v \in V$.

In addition to the existence of a special vector 0, there are other axioms that the two operations addition and product need to satisfy. For a complete discussion of vector space, see Blythe and Robertson (1999).

If there exists a (minimal) finite set of vectors $\{e_1, e_2, \ldots, e_n\}$ such that any $v \in V$ can be expressed as a linear combination of the e_is, that is, there exists $\alpha_1, \ldots, \alpha_n \in R$ such that:

$$v = \alpha_1 e_1 + \ldots + \alpha_n e_n \tag{7.1}$$

then the vector space is finite-dimensional. In a three-dimensional space the e_is correspond to the familiar x, y, z coordinate axis. If we add the notion of inner-product (angle) to vector space, we get a Euclidean space. In a Euclidean space setting, all spatial relationships including set, topological, metric, and directional (north/south) can be defined.

Dynamic spatial operations

Most of the operations that we have discussed so far have been static, in the sense that the operands are not affected by the application of the operation. For example, calculating the length of a curve has no effect on the curve itself. *Dynamic* operations alter the objects upon which the operations act. The three fundamental dynamic operations are create, destroy, and update. All dynamic operations are variations upon one of these themes (Worboys 1995). The *merge* function, commonly known as "map-reclassification" in many GIS, is an example of the create operation. Several examples of other dynamic spatial operations can be found in cartographic projections and map editing features in a GIS.

RDBMS and SQL

The relational model to represent data, introduced by Codd in 1970, has become one of the most popular logical data models. The power of this model is a consequence of the simplicity of its structure. We explain the terminology of the relational model in the context of a *World database* example.

World database

Suppose we wanted to organize the data of each country in the world. Then we could organize the information about the countries in the form of a table, which we label *Country*, and list the array of available information in columns. For the *Country* table, the associated data consists of six things: the name of the country, the continent to which it belongs, its population and gross domestic product, its life expectancy, and the spatial geometry that holds each country's international boundaries.

The table is called a *relation*, and the columns, *attributes*. Each different instance of the *Country* will be identified with a row in the table. A row is called a *tuple*, and the order in which the rows and columns appear in the table is unimportant. Thus a relation is an unordered collection of tuples. Together the table and column names constitute the *relation schema*, and a collection of rows or tuples is called a *relational instance*. The number of columns is called the *degree of the relation*. The *Country* is a relation of degree six. Similarly, data about, for example, the different cities and major rivers flowing through each country, can be organized as separate tables. Thus, the *World* database consists of three relations or entities: *Country*, *City*, and *River*. The example tables are shown in Table 7.2 and the schema of the database is shown below. Note that an underlined attribute is a primary key. For example, Name is the primary key in Country table.

```
Country (Name: varchar(35), Cont: varchar(35), Pop: integer,
    GDP: integer, Life-Exp: integer, Shape: char(13))
City (Name: varchar(35), Country: varchar(35), Pop: integer,
    Captial: char(1), Shape: char(9))
River (Name: varchar(35), Origin: varchar(35), Length: integer,
    Shape: char(13))
```

The Country entity has six attributes. The *Name* of the country and the continent (*Cont*) it belongs to are character strings of maximum length 35. The population (*Pop*) and gross domestic product (*GDP*) are integer types. The GDP is the total value of goods and services produced in a country in one fiscal year. The *Life-Exp* attribute represents the life expectancy in years (rounded to the nearest integer) for residents of a country. The *Shape* attribute needs some explanation. The geometry of a country is represented in the *Shape* column of Table 7.2. In relational databases, where the data types are limited, the *Shape* attribute is a foreign key to a shape table. In an object-relational or object-oriented database, the *Shape* attribute will be a polygon ADT. Since, for the moment, our aim is to introduce the basics of the SQL, we will not query the *Shape* attribute until the section below "Extending SQL for Spatial Data."

The City relation has five attributes: *Name*, *Country*, *Pop*, *Capital*, and *Shape*. The *Country* attribute is a foreign key into the Country table. *Capital* is a fixed character type of length one; a city is a capital of a country or it is not. The *Shape* attribute is a foreign key into a point shape table. As for the Country relation, we will not query the *Shape* column.

The four attributes of the River relation are *Name*, *Origin*, *Length*, and *Shape*. The *Origin* attribute is a foreign key into the Country relation and specifies the country where the river originates. The *Shape* attribute is a foreign key into a line

Table 7.2 The tables of the World database

(a) Country

COUNTRY	Name	Cont	Pop (millions)	GDP (billions)	Life-Exp	Shape
	Canada	NAM	30.1	658.0	77.08	Polygonid-1
	Mexico	NAM	107.5	694.3	69.36	Polygonid-2
	Brazil	SAM	183.3	1004.0	65.60	Polygonid-3
	Cuba	NAM	11.7	16.9	75.95	Polygonid-4
	USA	NAM	270.0	8003.0	75.75	Polygonid-5
	Argentina	SAM	36.3	348.2	70.75	Polygonid-6

(b) City

CITY	Name	Country	Pop (millions)	Capital	Shape
	Havana	Cuba	2.1	Y	Pointid-1
	Washington, DC	USA	3.2	Y	Pointid-2
	Monterrey	Mexico	2.0	N	Pointid-3
	Toronto	Canada	3.4	N	Pointid-4
	Brasilia	Brazil	1.5	Y	Pointid-5
	Rosario	Argentina	1.1	N	Pointid-6
	Ottawa	Canada	0.8	Y	Pointid-7
	Mexico City	Mexico	14.1	Y	Pointid-8
	Buenos Aires	Argentina	10.75	Y	Pointid-9

(c) River

RIVER	Name	Origin	Length (kilometers)	Shape
	Rio Parana	Brazil	2600	LineStringid-1
	St Lawrence	USA	1200	LineStringid-2
	Rio Grande	USA	3000	LineStringid-3
	Mississippi	USA	6000	LineStringid-4

string shape table. The geometric information specified in the *Shape* attribute is not sufficient, however, to determine the country of origin of a river. The overloading of Name across tables can be resolved by qualifying the attribute with tables using a dot notation table.attribute: Country.Name, City.Name, and River.Name uniquely identifies the Name attribute inside different tables. We also need information about the direction of the river flow.

Basic SQL primer

SQL is a commercial query language first developed at IBM. Since then, it has become the standard query language for RDBMS. SQL is a declarative language; that is, the user of the language has to specify the answer desired, but not the procedure to retrieve the answer.

The SQL language has two separate components: the data definition language (DDL) and the data modification language (DML). The DDL is used to create, delete, and modify the definition of the tables in the database. In the DML, queries are posed and rows inserted and deleted from tables specified in the DDL. We now provide a brief introduction to SQL. Our aim is to provide enough understanding of the language so that readers can appreciate the spatial extensions that we will discuss in under the heading "Extending SQL for Spatial Data." A more detailed and complete exposition of SQL can be found in any standard text on databases (Ullman and Widom 1999, Elmasri and Navathe 2000).

Data definition language

Creation of the relational schema and addition and deletion of tables are specified in the data definition language (DDL) component of SQL. For example, the City schema introduced earlier is defined below in SQL. The Country and River tables are defined in Table 7.3.

```
CREATE TABLE CITY {
        Name VARCHAR(35),
        Country VARCHAR(35),
        Pop INT,
        Capital CHAR(1)
        Shape CHAR(13)
        PRIMARY KEY Name }
```

In the above example, the CREATE TABLE clause is used to define the relational schema. The name of the table is CITY. The table has four columns, and the name of each column and its corresponding data type must be specified. The *Name* and *Country* attributes must be ASCII character strings of less than 35 characters. *Population* is of the type integer and *Capital* is an attribute which is a single character *Y* or *N*. In SQL-92 the possible data types are fixed and cannot be user-defined. We do not give the complete set of data types, which can be found in any text on standard databases. Finally, the *Name* attribute is the primary key of the relation. Thus each row in the table must have a unique value for the *Name* attribute.

Table 7.3 The Country and River schema in SQL

(a) Country schema	(b) River schema
``` CREATE TABLE Country { Name VARCHAR(35), Cont VARCHAR(35), Pop INT, GDP INT Shape CHAR(15) PRIMARY KEY Name } ```	``` CREATE TABLE River { Name VARCHAR(35), Origin VARCHAR(35), Length INT, Shape CHAR(15) PRIMARY KEY Name } ```

Tables no longer in use can be removed from the database using the DROP TABLE command. Another important command in DDL is ALTER TABLE for modifying the schema of the relation.

Data manipulation language

After the table has been created as specified in DDL, it is ready to accept data. This task, which is often called "populating the table," is done in the DML component of SQL. For example, the following statement adds one row to the table River:

```
INSERT   INTO     (Name, Origin, Length)
         River
         VALUES   ('Mississippi', 'USA', 6000)
```

If all the attributes of the relation are not specified, then default values are automatically substituted. The most often used default value is NULL. An attempt to add another row in the River table with Name = 'Mississippi' will be rejected by the DBMS because of the primary key constraint specified in the DDL.

The basic form to remove rows from the table is as follows:

```
DELETE FROM TABLE WHERE < CONDITIONS >
```

For example, the following statement removes the row from the table River that we inserted above:

```
DELETE   FROM     River
         WHERE    Name = 'Mississippi'
```

Basic form of an SQL query

Once the database schema has been defined in the DDL component and the tables populated, queries can be expressed in SQL to extract relevant data from the database. The basic syntax of an SQL query is extremely simple:

```
SELECT   Tuples
FROM     Relations
WHERE    Tuple-constraint
```

SQL has more clauses related to aggregation (for example, GROUP BY, HAVING), ordering results (for example, ORDER BY), etc. In addition, SQL allows the formulation of nested queries. We will illustrate these with a set of examples.

Example queries in SQL

We now give examples of how to pose different types of queries in SQL. Our purpose is to give a flavor of the versatility and power of SQL. All the tables queried are from the WORLD example introduced earlier under the heading "World database." The results of these different queries can be found in Table 7.4.

Table 7.4   Results of example queries

(a) Query 1

Name	Country	Pop (millions)	Capital	Shape
Havana	Cuba	2.1	Y	Point
Washington DC	USA	3.2	Y	Point
Brasilia	Brazil	1.5	Y	Point
Ottawa	Canada	0.8	Y	Point
Mexico City	Mexico	14.1	Y	Point
Buenos Aires	Argentina	10.75	Y	Point

(b) Query 2: Project

Name	Country
Havana	Cuba
Washington DC	USA
Monterrey	Mexico
Toronto	Canada
Brasilia	Brazil
Rosario	Argentina
Ottawa	Canada
Mexico City	Mexico
Buenos Aires	Argentina

(c) Query 3: Life-exp

Name	Life-exp
Mexico	69.36
Brazil	65.60

(d) Query 4

Ci.Name	Co.Pop
Brassilia	183.3
Washington, DC	270.0

(e) Query 5

Ci.Name	Ci.Pop
Washington, DC	3.2

(f) Query 6

Average-Pop
2.2

(g) Query 7

Cont	Continent-Pop
NAM	2343.5
SAM	676.1

(h) Query 8

Origin	Min-length
USA	1200

(i) Query 9

Co.Name
Mexico
Brazil
USA

1   **Query:** List all the cities and the country they belong to in the CITY table.

```
SELECT    Ci.Name,
          Ci.Country
FROM      City Ci
```

2   **Query:** List the names of the capital cities in the CITY table.

```
SELECT    *
FROM      City
WHERE     CAPITAL='Y'
```

3   **Query:** List the names of countries in the Country relation where the life-expectancy is less than 70 years.

```
SELECT    Co.Name,
          Co.Life-Exp
FROM      Country Co
WHERE     Co.Life-Exp < 70
```

4   **Query:** List the capital cities and populations of countries whose GDP exceeds one trillion dollars.

```
SELECT    Ci.Name, Co.Pop
FROM      City Ci, Country Co
WHERE     Ci.Country = Co.Name AND
          Co.GDP > 1000.0 AND
          Ci.Capital= 'Y'
```

Comments: This is the standard way of expressing the join operation. In this case the two tables City and Country are matched on their common attributes Ci.Country and Co.Name. Furthermore, two selection conditions are specified separately in the City and Country table. Notice how the cascading dot notation alleviated the potential confusion that might have arisen as a result of the attribute names in the two relations.

5   **Query:** What is the name and population of the capital city in the country where the St Lawrence River originates?

```
SELECT    Ci.Name, Ci.Pop
FROM      City Ci, Country Co, River R
WHERE     R.Origin = Co.Name AND
          R.Name = 'St. Lawrence' AND
          Ci.Capital= 'Y'
```

Comments: This query involves a join between three tables. The River and Country tables are joined on the attributes Origin and Name. The Country and City tables are joined on the attributes Name and Country. There are two selection conditions on the River and City tables respectively.

6   **Query:** What is the average population of the non-capital cities listed in the `City` table?

```
SELECT    AVG(Ci.Pop)
FROM      City Ci
WHERE     Ci.Capital= 'N'
```

Comments: The `AVG` (Average) is an example of an aggregate operation. Other aggregate operations are `COUNT`, `MAX`, `MIN`, and `SUM`. The aggregate operations expand the functionality of SQL because they allow computations to be performed on the retrieved data.

7   **Query:** For each continent, find the average GDP.

```
SELECT      Co.Cont AVG(Co.GDP) AS Continent-GDP
FROM        Country Co
GROUP BY    Co.Cont
```

Comments: This query expression represents a major departure from the basic SQL query format. The reason is the presence of the `GROUP BY` clause. The `GROUP BY` clause partitions the table on the basis of the attribute listed in the clause. In this example there are two possible values of `Co.cont`: `NAM` and `SAM`. Therefore, the `Country` table is partitioned into two groups. For each group, the average *GDP* is calculated. The average value is then stored under the attribute `Continent-GDP` as specified in the `SELECT` clause.

8   **Query:** For each country in which at least two rivers originate, find the length of the smallest river.

```
SELECT      R.Origin, MIN(R.length) AS Min-length
FROM        River R
GROUP BY    R.Origin
HAVING      COUNT(*) > 1
```

Comments: This is similar to the previous query. The difference is that the `HAVING` clause allows selection conditions to be enforced on the different groups formed in the `GROUP BY` clause. Thus only groups with more than one member are considered.

9   **Query:** List the countries whose GDP is greater than Canada's.

```
SELECT    Co.Name
FROM      Country Co
WHERE     Co.GDP        > ANY ( SELECT  Co1.GDP
                                FROM    Country Co1
                                WHERE   Co1.Name = 'Canada')
```

Comments: This is an example of a nested query. These are queries which have other queries embedded in them. A nested query becomes mandatory when an intermediate table, which does not exist, is required before a query can be evaluated. The embedded query typically appears in the `WHERE` clause, though it can appear, albeit rarely, in the FROM and the SELECT clauses. The ANY is a set comparison operator. Consult a standard database text for a complete overview of nested queries.

SQL summary

Structured query language (SQL) is the most widely implemented database language. SQL has two components: the data definition language (DDL) and data manipulation language (DML). The schema of the database tables are specified and populated in the DDL. The actual queries are posed in DML. We have given a brief overview of SQL. More information can be found in any standard text on databases.

Object-Oriented Database Management Systems

Object-oriented database technology results from the combination of object-oriented programming concepts with database management principles for supporting various non-traditional applications involving complex data types and operations. Let us now briefly review key features offered by most of the OODBMSs available on the market today. These include types, classes, objects, methods, object identity, abstract data types, complex objects, class hierarchies, overloading, and late binding.

1 **Type System:** A rich type system provides mechanisms to construct new types using the base types such as Booleans, integers, reals, and character strings. Many object-oriented programming languages allow constructing record structures and collection types using a feature know as type constructor. A record structure is made of n components, where each component is a duplet consisting of a base type and a field name. A record structure is exactly the same as the "struct" type in the C and C++ programming languages. A collection type is constructed using collection operators, such as arrays, lists, and sets. Thus although a collection type consists of more than one element, all elements are of the same type. Again, these two new types can be applied in the same fashion to construct even more complex types.

2 **Classes:** A class is a description of an object or group of objects with similar properties and behavior. A class is made-up of a type, and one or more functions or procedures known as *methods*.

3 **Objects:** An object is an instance of a class. The simplest objects are just base types, such as Booleans, integers, floats and strings, etc. An object can simply be the value of that class or it can be a variable that holds values of that class, known as immutable and mutable objects respectively. Formally, an object can be thought of as a triple (I,C,V), where I is the object identity (OID), C is a type constructor, and V is the object state (that is, current value). An example is: $o_1 = (i_1, atom, "U.S.A.")$.

4 **Methods:** A method implements a certain behavior or operation for a class. A method takes at least one argument that is an object of that class. Associated with each class is a special method, known as a constructor. Complex objects are built from the simpler objects by applying constructors to them.

5   **Object Identity:** In the object-oriented (OO) paradigm, it is assumed that each object has a unique identity (OID), that is, the identity is independent of the state of the object. It is also assumed that no two objects can have the same OID, and that no object has two different OIDs. As a consequence, OODBMS support two important notions of object equivalence, known as object sharing and object updates.

6   **Abstract Data Types:** ADTs are user defined arbitrary data types that are combinations of atomic data types and the associated methods. ADTs provide data *encapsulation*, that is, they restrict access to objects of a class through a well defined set of functions (methods).

7   **Class Hierarchies:** Generalization and specialization are two important object-oriented concepts that allow sharing or reuse of software. Through generalization we can organize the classes by their similarities and differences. Generalization defines the relationship between a superclass and its subclasses. A subclass inherits all the properties (attributes) and the behavior (methods) of its superclass. In addition, the subclass may have its own properties (attributes and methods). While generalization takes a bottom-up approach, specialization takes a top-down approach – that is, starting with the superclass and then splitting (specializing) into subclasses.

Object-oriented query languages

In the previous section, we have seen that SQL is the standard query language in the RDBMS world. In this section, we present OQL, the object query language that combines the declarative programming feature of SQL with the object-oriented programming. Unlike SQL, OQL is intended to be used as an extension with some object-oriented host language, such as C++ or Java. Though it looks analogous to the way SQL is embedded into a host language, the OQL and object-oriented host language combination provides seamless integration. Objects can be modified by the host language and the OQL without explicitly transferring the values between the two languages, thereby providing an advantage over the embedded SQL. Now let us look at some of the query capabilities of OQL.

OQL supports the basic SELECT-FROM-WHERE expressions of SQL. In addition, OQL supports complex type constructors, such as Set(...), Bag(...), List(...), Array(...), and Struct(...), and path expressions. For simple queries, there would not be any change between SQL and OQL query statements; however, the difference between them becomes obvious if we consider both class hierarchies and methods. In the next section we introduce some examples written in SQL-3 that incorporate some object-oriented features into SQL.

Object Relational Database Management Systems

Despite the advantages offered by object-oriented database management systems, the classical relational model (with various extensions) still dominates the market today. Nevertheless, the motivation that led to the development of OODBMS has also influenced the relational database community. Several database vendors, such as Oracle and IBM (DB2), now offer hybrid systems known as object-relational database management systems (ORDBMS). Though relation is still a core concept in ORDBMS, these newer systems now allow users to store collections of complex objects that encapsulate both data and the behavior. The same is true in the SQL domain, as seen in the recently standardized SQL-3, which brings object-oriented features into the relational world. In this section we briefly review the ORDBMS technology through the OGC and SQL-3 extensions.

Extending SQL for spatial data

Although it is a powerful query-processing language, SQL has its shortcomings. The main one is that this language can handle only simple data types such as integers, dates, and strings. Spatial database applications must handle complex data types like points, lines, and polygons. Database vendors have responded in two ways: They have either used *blobs* to store spatial information, or they have created hybrid systems in which spatial attributes are stored in operating-system files via a GIS. SQL cannot process data stored as blobs, and it is the responsibility of the application techniques to handle data in blob form (Stonebraker and Moore 1997). This solution is neither efficient nor aesthetic because the data depends upon the host-language application code. In a hybrid system, spatial attributes are stored in a separate operating-system file and thus are unable to take advantage of traditional database services like query language, concurrency control, and indexing support.

Object-oriented systems have had a major influence on expanding the capabilities of DBMS to support spatial (complex) objects. The program to extend a relational database with object-oriented features falls under the general framework of object-relational database management systems (ORDBMS). The key feature of ORDBMS is that they support a version of SQL, SQL-3/SQL-99, which supports the notion of user-defined types (as in Java or C++). Our goal is to study SQL-3/SQL-99 enough so that we can use it as a tool to manipulate and retrieve spatial data.

The principal demand of spatial SQL is to provide a higher abstraction of spatial data by incorporating concepts closer to our perception of space (Egenhofer 1994). This is accomplished by incorporating the object-oriented concept of user-defined abstract data types (ADT). An ADT is a user-defined type and its associated functions. For example, if we have land parcels stored as polygons in a database, then a useful ADT may be a combination of the type *polygon* and some associated function (method), say, `adjacent`. The `adjacent` function may be applied to `land parcels` to determine if they share a common boundary. The term *abstract* is used because the end-user need not know the implementation details of the associated functions. All end-users need to know is the interface, that is, the available functions and the data types for the input parameters and output results.

The OGIS standard for extending SQL

The Open GIS Consortium (OGIS) was formed by major software vendors to formulate an industry-wide standard related to GIS interoperability. The OGIS spatial data model can be embedded in a variety of programming languages, for example, C, Java, SQL, etc. We will focus on SQL embedding in this section.

The OGIS is based on a geometry data model shown in Figure 7.5. Recall that the data model consists of a base-class, GEOMETRY, which is non-instantiable (that is, objects cannot be defined as instances of GEOMETRY), but specifies a spatial reference system applicable to all its subclasses. The four major subclasses derived from the GEOMETRY superclass are Point, Curve, Surface, and GeometryCollection. Associated with each class is a set of operations which acts on instances of the classes. A subset of important operations and their definitions are listed in Table 7.5.

Table 7.5   A sample of operations listed in the OGIS standard for SQL (after OGIS 1999)

Basic Functions	SpatialReference()	Returns the underlying coordinate system of the geometry
	Envelope()	Returns the minimum orthogonal bounding rectangle of the geometry
	Export()	Returns the geometry in a different representation
	IsEmpty()	Returns true if the geometry is a null set
	IsSimple()	Returns true if the geometry is simple (no self-intersection)
	Boundary()	Returns the boundary of the geometry
Topological/Set Operators	Equal	Returns true if the interior and boundary of the two geometries are spatially equal
	Disjoint	Returns true if the boundaries and interior do not intersect
	Intersect	Returns true if the geometries are not disjoint
	Touch	Returns true if the boundaries of two surfaces intersect but the interiors do not
	Cross	Returns true if the interior of the surface intersects with a curve
	Within	Returns true if the interior of the given geometry does not intersect with the exterior of another geometry
	Contains	Tests if the given geometry contains another given geometry
	Overlap	Returns true if the interiors of two geometries have non-empty intersection
Spatial Analysis	Distance	Returns the shortest distance between two geometries
	Buffer	Returns a geometry that consists of all points whose distance from the given geometry is less than or equal to the specified distance
	ConvexHull	Returns the smallest convex geometric set enclosing the geometry
	Intersection	Returns the geometric intersection of two geometries
	Union	Returns the geometric union of two geometries
	Difference	Returns the portion of a geometry which does not intersect with another given geometry
	SymmDiff	Returns the portions of two geometries which do not intersect with each other

The operations specified in the OGIS standard fall into three categories:

1. Basic operations applicable to all geometry data types. For example, `SpatialReference` returns the underlying coordinate system where the geometry of the object was defined. Examples of common reference systems include the well-known latitude and longitude system and the often-used Universal Traversal Mercator (UTM) system.

2. Operations which test for topological relationships between spatial objects. For example, `intersect` tests whether the interior of two objects has a non-empty set intersection.

3. General operations for spatial analysis. For example, `distance` returns the shortest distance between two spatial objects.

Limitations of the standard

The OGIS specification is limited to the *object* model of space. However, spatial information is sometimes most naturally mapped onto a field-based model and the OGIS is currently developing consensus models for field data types and operations that will probably be incorporated into a future OGIS standard.

Even within the *object* model, the OGIS operations are limited for simple `SELECT-PROJECT-JOIN` queries. Support for spatial aggregate queries with the `GROUP BY` and `HAVING` clauses does pose problems. Finally, the focus in the OGIS standard is exclusively on basic topological and metric spatial relationships. Support for a whole class of metric operations, namely, those based on the *direction* predicate (for example, North, South, left, front) is missing.

Example queries which emphasize spatial aspects

Using the OGIS data types and operations, we formulate SQL queries in the `World` database which highlight the spatial relationships between the three entities: `Country`, `City`, and `River`. We first redefine the relational schema, assuming that the OGIS data types and operations are available in SQL.

1. **Query:** Find the names of all countries which are neighbors of *USA* in the Country table.

```
SELECT    C1.Name AS "Neighbors of USA"
FROM      Country C1, Country C2
WHERE     Touch (C1.Shape, C2.Shape) = 1 AND
          C2.Name = 'USA'
```

Comments: The `Touch` predicate checks if any two geometric objects are adjacent to each other without overlapping. It is a useful operation to determine neighboring geometric objects. The `Touch` operation is one of the eight topological and set predicates specified in the OGIS Standard. One of the nice properties of topological operations is that they are invariant under many geometric transformations. In particular, the choice of the coordinate system for the `World` database will not affect the results of topological operations.

Table 7.6   Basic data types

	(a)	
CREATE	TABLE	Country(
	Name	varchar(30),
	Cont	varchar(30),
	Pop	Integer,
	GDP	Number,
	Shape	Polygon);

	(b)	
CREATE	TABLE	River(
	Name	varchar(30),
	Origin	varchar(30),
	Length	Number,
	Shape	LineString);

	(c)	
CREATE	TABLE	City (
	Name	varchar(30),
	Country	varchar(30),
	Pop	integer,
	Shape	Point);

Topological operations apply to many different combinations of geometric types. Therefore, in an ideal situation these operations should be defined in an "overloaded" fashion. Unfortunately, many object-relational DBMS do not support object-oriented notions of class inheritance and operation overloading. Thus, for all practical purposes, these operations must be defined individually for each combination of applicable geometric types (Table 7.6).

1   **Query:** For all the rivers listed in the River table, find the countries through which they pass.

```
SELECT    R.Name C.Name
FROM      River R, Country C
WHERE     Cross(R.Shape, C.Shape) = 1
```

Comments: The Cross is also a topological predicate. It is most often used to check for the intersection between LineString and Polygon objects, as in this example, or a pair of LineString objects.

2   **Query:** Which city listed in the City table is closest to each river listed in the River table?

```
SELECT    C1.Name, R1.Name
FROM      City C1, River R1
WHERE     Distance (C1.Shape, R1.Shape) <
              ( SELECT Distance(C2.Shape, R2.Shape)
            FROM City C2, River R2
          WHERE C1.Name <> C2.Name
                AND R1.Name <> R2.Name)
```

Comments: The `Distance` is a real-valued binary operation. It is being used once in the `WHERE` clause and again in the `SELECT` clause of the subquery. The `Distance` function is defined for any combination of geometric objects.

3 **Query:** The St Lawrence River can supply water to cities which are within 300 km. List the cities which can use water from the St Lawrence.

```
SELECT   Ci.Name
FROM     City Ci, River R
WHERE    Overlap (Ci.Shape, Buffer(R.Shape,300)) =1 AND
         R.Name = 'St. Lawrence'
```

Comments: The `Buffer` of a geometric object is a geometric region centered at the object whose size is determined by a parameter in the `Buffer` operation. In the example the query dictates the size of the buffer region. The buffer operation is used in many GIS applications including floodplain management and urban and rural zoning laws. A graphical depiction of the buffer operation is shown in Figure 7.6. In the figure, Cities A and B are likely to be affected if there is a flood on the river, while City C will remain unaffected.

4 **Query:** List the name, population, and area of each country listed in the `Country` table.

```
SELECT   C.Name, C.Pop, Area(C.Shape) AS "Area"
FROM     Country C
```

Comments: This query illustrates the use of the `Area` function. This function is only applicable for `Polygon` and `MultiPolygon` geometry types. Calculating the `Area` clearly depends upon the underlying coordinate system of the `World` database. For example, if the shape of the `Country` tuples is given in terms of latitude and longitude, then an intermediate coordinate transformation must be performed before the `Area` can be calculated. The same care must be taken for `Distance` and the `Length` function.

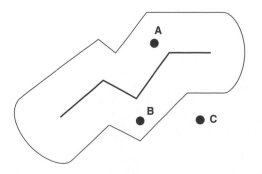

Fig. 7.6 The buffer of a river and points within and outside the resultant buffer

5   **Query:** List the length of the rivers in each of the countries they pass through.

```
SELECT    R.Name, C.Name, Length(Intersection(R.Shape,
          C.Shape)).AS "Length"
FROM      River R, Country C
WHERE     Cross(R.Shape, C.Shape) =1
```

Comments: The return value of the `Intersection` binary operation is a geometry type. The `Intersection` operation is different from the `Intersects` function, which is a topological predicate to determine if two geometries intersect. The `Intersection` of a `LineString` and `Polygon` can either be a `Point` or `LineString` type. If a river does pass through a country, then the result will be a `LineString`. In that case, the `Length` function will return the length of the river in each country it passes through.

6   **Query:** List the GDP and the distance of a country's capital city to the equator for all countries.

```
SELECT    Co.GDP, Distance(Point(0,Ci.y),Ci.Shape) AS
          "Distance"
FROM      Country Co, City Ci
WHERE     Co.Name = Ci.Country AND Ci.Capital = 'Y'
```

Comments: Searching for implicit relationships between data sets stored in a database is outside the scope of standard database functionality. Current DBMS are geared toward online transaction processing (OLTP), while this query, as posed, is in the realm of online analytical processing (OLAP). At the moment the best we can do is list each capital and its distance to the equator (Table 7.7). `Point(0,Ci.y)` is a point on the equator which has the same longitude as that of the current capital instantiated in `Ci.Name`.

7   **Query:** List all countries, ordered by number of neighboring countries.

```
SELECT    Co.Name, Count(Co1.Name)
FROM      Country Co, Country Co1
WHERE     Touch(Co.Shape, Co1.Shape)
GROUP BY  Co.Name
ORDER BY  Count(Co1.Name)
```

Table 7.7   Results of query 7

Co.Name	Co.GDP	Dist-toEq (km)
Havana	16.9	2562
Washington DC	8003	4324
Brazilia	1004	1756
Ottawa	658	5005
Mexico City	694.3	2161
Buenos Aires	348.2	3854

Comment: In this query all the countries with at least one neighbor are sorted on the basis of number of neighbors.

8 **Query:** List the countries with only one neighboring country. A country is a neighbor of another country if their land masses share a boundary.

```
SELECT      Co.Name
FROM        Country Co, Country Co1
WHERE       Touch(Co.Shape, Co1.Shape)
GROUP BY    Co.Name
HAVING      Count(Co1.Name) = 1

SELECT      Co.Name
FROM        Country Co
WHERE       Co.Name IN
            (SELECT Co.Name
            FROM Country Co
            WHERE Touch(Co.Shape, Co1.Shape))
GROUP BY    Co.Name
HAVING      Count(*) = 1
```

Comments: Here we have a nested query in the FROM clause. The result of the query within the FROM clause is a table consisting of pairs of countries which are neighbors. The GROUP BY clause partitions the new table on the basis of the names of the countries. Finally, the HAVING clause forces the selection to be paired to those countries which have only one neighbor. The HAVING clause plays a role similar to the WHERE clause with the exception that it must include aggregate functions like count, sum, max, and min.

9 **Query:** Which country has the maximum number of neighbors?

```
CREATE VIEW   Neighbor AS
SELECT        Co.Name, Count(Co1.Name) AS num neighbors
FROM          Country Co, Country Co1
WHERE         Touch(Co.Shape, Co1.Shape)
GROUP BY      Co.Name

SELECT        Co.Name, num neighbors
FROM          Neighbor
WHERE         num neighbor = (SELECT Max(num neighbors)
FROM          Neighbor
```

Object-relational SQL

The OGIS standard specifies the data types and their associated operations which are considered essential for spatial applications like GIS. For example, for the Point data type an important operation is Distance, which computes the distance between two points. The length operation is not a semantically correct operation on a Point data type. This is similar to the argument that the concatenation operation makes more sense for Character data type than for say, the Integer type.

In relational databases, the set of data types is fixed. In object-relational and object-oriented databases this limitation has been relaxed and there is a built in support for user-defined data types. While this feature is clearly an advantage, especially when dealing with non-traditional database applications like GIS, the burden of constructing syntactically and semantically correct data types is now on the database application developer. To share some of the burden, commercial database vendors have introduced application-specific "packages" which provide a seamless interface to the database user. For example, Oracle markets a GIS specific package called the Spatial Data Cartridge.

The recently standardized SQL-3 allows user-defined data types within the overall framework of a relational database. Two features of the SQL-3 standard which may be beneficial for defining user-defined spatial data types are described below.

A glance at SQL-3

The SQL-3/SQL-99 proposes two major extensions to SQL-2/SQL-92, the current accepted SQL draft.

1   **Abstract Data Type:** An ADT can be defined using a CREATE TYPE statement. Like classes in object-oriented technology, an ADT consists of attributes and member functions to access the values of the attributes. Member functions can potentially modify the value of the attributes in the data type and thus can also change the database state. An ADT can appear as a column type in a relational schema. To access the value that the ADT encapsulates, a member function specified in the CREATE TYPE must be used. For example, the following script creates a type Point with the definition of one member function Distance:

```
CREATE TYPE    Point (
x              NUMBER,
y              NUMBER,
FUNCTION       Distance(:u Point,:v Point)
               RETURNS NUMBER );
```

The colons before u and v signify that these are local variables.

2   **Row Type:** A row type is a type for a relation. A row type specifies the schema of a relation. For example, the following statement creates a row type Point.

```
CREATE ROW TYPE    Point (
x                  NUMBER,
y                  NUMBER );
```

We can now create a table which instantiates the row type. For example:

```
CREATE TABLE Pointtable of TYPE Point;
```

In our own work we emphasize the use of ADT instead of row type. This is because the ADT as a column type naturally harmonizes with the definition of an ORDBMS as an extended relational database.

Object-relational schema

Oracle 8 is an object-relational DBMS introduced by Oracle Corporation. Similar products are available from other database companies such as IBM. Oracle 8 implements a part of the SQL-3 Standard. In this system, the ADT is called the "object type."

Below we describe how the three basic spatial data types – Point, LineString, and Polygon – are constructed in Oracle 8.

```
CREATE     TYPE Point AS OBJECT (
           x NUMBER,
           y NUMBER,
MEMBER     FUNCTION Distance(:u Point,:v Point) RETURN NUMBER,
PRAGMA     RESTRICT_REFERENCES (Distance, WNDS);
```

The Point type has two attributes, x and y, and one member function, Distance. PRAGMA alludes to the fact that the Distance function will not modify the state of the database: WNDS (Write No Database State). Of course in the OGIS standard many other operations related to the Point type are specified, but for simplicity we have shown only one. After its creation the Point type can be used in a relation as an attribute type. For example, the schema of the relation City can be defined as follows:

```
CREATE TABLE    City (
Name            varchar(30),
Pop             int,
Capital         char(1),
Shape           Point);
```

Once the relation schema has been defined, the table can be populated in the usual way. For example, the following statement adds information related to Brasilia, the capital of Brazil, into the database:

```
INSERT INTO CITY('Brasilia', 'Brazil', 1.5, 'Y',
            Point(
            -55.4,-23.2));
```

The construction of the LineString data type is slightly more involved than that of the Point type. We begin by creating an intermediate type, LineType:

```
CREATE TYPE LineType AS VARRAY(500) OF Point;
```

Thus `LineType` is a variable array of `Point` data type with a maximum length of 500. Type-specific member functions cannot be defined if the type is defined as a `Varray`. Therefore, we create another type `LineString`:

```
CREATE TYPE    LineString AS OBJECT (
               Num_of_Points INT,
               Geometry LineType,
               MEMBER FUNCTION Length(SELF IN) RETURN NUMBER,
               PRAGMA RESTRICT REFERENCES(Length, WNDS);
```

The attribute `Num_of_Points` stores the size (in terms of points) of each instance of the `LineString` type. We are now ready to define the schema of the `River` table:

```
CREATE    TABLE    River(
          Name     varchar(30),
          Origin   varchar(30),
          Length   number,
          Shape    LineString );
```

While inserting data into the `River` table, we have to keep track of the different data types involved.

```
INSERT INTO RIVER ('Mississippi', 'USA', 6000, LineString(3,
        LineType(Point(1,1),Point(1,2),Point(2,3)))
```

The `Polygon` type is similar to `LineString`. The sequence of type and table creation and data insertion is given in Table 7.8.

Example queries

1 **Query:** List all the pairs of cities in the `City` table and the distances between them.

```
SELECT    C1.Name, C1.Distance(C2.Shape) AS "Distance"
FROM      City C1, City C2
WHERE     C1.Name <> C2.Name
```

Comments: Notice the object-oriented notation for the `Distance` function in the `SELECT` clause. Contrast it with the predicate notation used in section above, "Example queries which emphasize spatial aspects": `Distance(C1.Shape, C2.Shape)`. The predicate in the `WHERE` clause ensures that the `Distance` function is not applied between two copies of the same city.

2 **Query:** Validate the length of the rivers given in the `River` table, using the geometric information encoded in the `Shape` attribute.

```
SELECT    R.Name, R.Length, R.Length() AS "Derived Length"
FROM      River R
```

Comments: This query is used for data validation. The length of the rivers is already available in the `Length` attribute of the `River` table. Using the `Length()` function we can check the integrity of the data in the table.

Table 7.8 The sequence of creation of the Country table

(a)

```
CREATE  TYPE  PolyType  AS  VARRAY  (500)  OF  Point
```

(b)

```
CREATE    TYPE  Polygon  AS  OBJECT  (
          Num_of_Points  INT,
          Geometry  PolyType,
          MEMBER  FUNCTION  Area(SELF  IN)  RETURN  NUMBER,
          PRAGMA  RESTRICT_REFERENCES  (Length,  WNDS);
```

(c)

```
CREATE    TABLE      Country(
          Name       varchar(30),
          Cont       varchar(30),
          Pop        int,
          GDP        number,
          Life-Exp   number,
          Shape      Polygon       );
```

(d)

```
INSERT  INTO    Country('Mexico',  'NAM',  107.5,  694.3,  1004.0,
                Polygon(23,  Polytype(Point(1,1),  ...,Point(1,1)))
```

3   **Query:** List the names, populations, and areas of all countries adjacent to the USA.

```
SELECT    C2.Name,  C2.Pop,  C2.Area()  AS  "Area"
FROM      Country  C1,  Country  C2
WHERE     *  C1.Name  =  'USA'  AND
          C1.Touch(C2.Shape)  =  1
```

Comments: The Area() function is a natural function for the Polygon ADT to support. Along with Area(), the query also invokes the Touch topological predicate.

CONCLUSIONS

In this chapter we have presented an overview of relational, object-oriented, and object-relational database management systems from a GIS and spatial data management point of view. The relational data model, which has proven to be very successful at solving most business data processing problems, has serious drawbacks when it comes to handling complex information systems. On the other hand, object-oriented database management systems, which are the result of combining good features of object-oriented programming concepts and database management principles, have been limited to certain niche markets. Recent object extensions of RDBMS have resulted in a more flexible hybrid DBMS, known as ORDBMS, and seem to offer the best of both worlds. We have provided key concepts of each of these DBMS through simple examples drawn from the standard World database.

The treatment of this chapter was limited to the basic principles of the concepts needed to understand these DBMS. For additional information the interested reader is directed to the following references. Standard textbooks on DBMS (Ullman and Widom 1999, Elmasri and Navathe 2000) cover a broad spectrum of topics in this field. Readers interested in spatial databases may refer to Rigaux, Scholl, and Voisard (2002) and Shekhar and Chawla (2003), which provide a detailed treatment of DBMS from a spatial data management point of view. Readers wanting additional insights on the intersection of computer science and GIS should consult Worboys (1995); those interested in conceptual data models and their extensions with respect to spatial database applications should consult Tryfona and Hadzilacos (1995), Hadzilacos and Tryfona (1997), Shekhar, Vatsavai, Chawla, and Burk (1999), Brodeur, Bédard, and Proulx (2000), and Bédard, Larrivée, Proulx, and Nadeau (2004). For more on spatial query languages and spatial query processing, there is Orenstein (1986, 1990) and Egenhofer (1994); readers interested in multi-dimensional indexing should refer to Gaede and Günther (1998). Finally, those interested in learning more about standards should consult OGIS (1999).

REFERENCES

Albrecht, J. 1998. Universal analytical GIS operations: A task-oriented systematization of data-structure-independent GIS functionality. In M. Craglia and H. Onsrud (eds) *Geographic Information Research: Transatlantic Perspectives*. London: Taylor and Francis: 557–91.

Bédard, Y., Larrivée, S., Proulx, M.-J., and Nadeau, M. 2004. Modeling geospatial databases with plug-ins for visual languages: A pragmatic approach and the impacts of 16 years of research and experimentations on perceptory. In S. Wang, D. Yang, K. Tanaka, F. Grandi, S. Zhou, E. E. Mangina, T. W. Ling, I.-Y. Song, J. Guan, and H. C. Mayr (eds) *Conceptual Modeling for Advanced Application Domains: Proceedings of the International Conference on Conceptual Modeling (ER 2004)*, Shanghai, China. Berlin: Springer Lecture Notes in Computer Science No. 3289: 17–30.

Blythe, T. and Robertson, E. 1999. *Basic Linear Algebra*. Berlin: Springer-Verlag.

Brodeur, J., Bédard, Y., and Proulx, M.-J. 2000. Modeling geospatial application databases using UML-based repositories aligned with international standards in geomatics. In *Proceedings of the Eighth ACM Symposium on Advances in Geographic Information Systems*, Washington, DC, USA. New York, NY: Association of Computing Machinery, pp. 39–46.

Egenhofer, M. J. 1991. Reasoning about binary topological relations. In O. Günther and H.-J. Schek (eds) *Advances in Spatial Databases: Proceedings of the Second International Symposium, SSD'91*, Zürich, Switzerland. Berlin: Springer Lecture Notes in Computer Science No. 525: 143–60.

Elmasri, R. and Navathe, S. 2000. *Fundamentals of Database Systems*. Boston, MA: Addison-Wesley.

Gaede, V. and Günther, O. 1998. Multidimensional access methods. *ACM Computing Surveys* 30: 170–231.

Hadzilacos, T. and Tryfona, N. 1997. An extended entity-relationship model for geographic applications. *SIGMOD Record* 26: 24–9.

Khoshafian, S. and Baker, A. 1998 *Multimedia and Imaging Databases*. San Francisco, CA: Morgan Kaufmann.

OGIS. 1999. *Open GIS Consortium: Open GIS Simple Features Specification for SQL (Revision 1.1)*. WWW document, http://www.opengis.org/techno/specs.htm.

Orenstein, J. A. 1986. Spatial query processing in an object-oriented database system. In *Proceedings of the ACM SIGMOD International Conference on Management of Data*, Washington, DC, USA. New York: Association of Computing Machinery, pp. 326–36.

Orenstein, J. A. 1990. Comparison of spatial query processing techniques for native and parameter spaces. In *Proceedings of the 1990 ACM SIGMOD International Conference on Management of Data*: Atlantic City, NJ, USA. New York:Association of Computing Machinery, pp. 343–52.

Rigaux P., Scholl, M., and Voisard, A. 2002. *Spatial Databases*. San Francisco, CA: Morgan Kaufman.

Shekhar, S. and Chawla, S. 2003. *Spatial Databases: A Tour*. Upper Saddle Creek, NJ: Prentice Hall.

Shekhar, S., Vatsavai, R. R., Chawla, S., and Burk, T. E. 1999. Spatial pictogram enhanced conceptual data models and their transition to logical data models. In P. Agouris and A. Stefanidis (eds) *Selected Papers from the International Workshop on Integrated Spatial Databases, Digital Images and GIS*. London: Springer Lecture Notes in Computer Science No. 1737: 77–104.

Stonebraker, M. and Moore, D. A. 1997. *Object Relational DBMSs: The Next Great Wave*. San Francisco, CA: Morgan Kaufmann.

Tryfona, N. and Hadzilacos, T. 1995. Geographic applications development: Models and tools for the conceptual level. In *Proceedings of the Third ACM Symposium on Advances in Geographic Information Systems*, Baltimore, MD, USA. New York: Association of Computing Machinery, pp. 19–28.

Ullman, J. and Widom, J. 1999. *A First Course In Database Systems*. Upper Saddle Creek, NJ: Prentice Hall.

Worboys, M. F. 1995. *GIS: A Computing Perspective*. London: Taylor and Francis.

Chapter 8

Adding the Z Dimension

Michael F. Hutchinson

The land surface is the natural context for life on Earth. Above sea-level the land surface plays a fundamental role in modulating land surface and atmospheric processes. Below sea-level it forms an important role in modulating ocean tides and currents. Life depends on these Earth surface processes over a wide range of space and time scales. Conversely, the land surface itself is molded by these processes, also over a wide range of time and space scales. It is thus not surprising that models of the land surface of the Earth have played an integral role in GI Science since its inception. Analyses and representations of the land surface have directly stimulated new methods for obtaining digital environmental data, new spatial interpolation methods and new methods for analysing landscape dependent hydrological and ecological processes (Hutchinson and Gallant 1999). The Z dimension, the elevation above or below sea-level of the land surface, is the primary descriptor of this fundamental layer.

Digital elevation models (DEMs) are used to represent the land surface in different ways, depending on the nature of the application. Visualization of landscapes, and of spatially distributed quantities and entities within landscapes, plays an important role in conceptualization and in providing subjective understanding of surface processes. It can also play an important role in assessing data quality. These applications are discussed extensively in "Optimization of DEM Resolution" below. Of central importance for environmental modeling is the accuracy and spatial coverage that can be achieved by incorporating appropriate dependencies on the land surface. Mesoscale representations of surface climate, particularly temperature and precipitation, have a strong direct dependence on elevation, the Z dimension itself, making such representations truly three-dimensional. Modeling applications at finer scales often depend on representations of the shape of the land surface rather than elevation *per se*. These shape-based applications are largely the subject of terrain analysis as discussed by Wilson and Gallant (2000; see also Chapter 23 by Deng, Wilson, and Gallant in this volume).

This chapter is primarily concerned with the generation of regular grid digital elevation models, from a variety of data sources, to support the elevation and landscape shape requirements of environmental modeling over a range of spatial scales. Issues of data quality and spatial scale naturally arise in this context. An essential shape-based attribute of DEMs for hydrological applications is drainage connectivity. Coupling the process of automatic drainage enforcement to the interpolation and filtering of DEMs, as introduced by Hutchinson (1989), has improved elevation accuracy and directly facilitated hydrological applications. DEM accuracy can also be improved by applying statistical filtering methods that accurately reflect the errors in elevation source data. The issues of drainage connectivity and elevation accuracy are particularly relevant with the advent of fine scale digital elevation data sources.

The chapter also discusses the incorporation of dependencies of environmental variables on elevation and landscape shape. This is largely the domain of multivariate statistical analysis, usually performed by various forms of thin plate smoothing splines and geostatistics. These models normally apply data smoothing, in the same way as can be applied to elevation data, to allow for fine scale variability in the data and produce spatial models with minimal error. The incorporation of dependencies on elevation and landscape shape has played a major role in developing accurate spatial representations of environmental variables such as surface climate and in assessing spatially detailed impacts of projected climate change (Houser, Hutchinson, Viterbo, et al. 2004).

Regular Grid Digital Elevation Models and Spatial Scale

Digital elevation models are commonly based on one of three data structures: triangulated irregular networks, contours, and regular grids. Triangulated irregular networks (TINs) have found most application in visualization where economy of representation is important. This can be achieved if the triangulations are based on well-chosen surface specific points including peaks and points on ridges, streamlines and breaks in slope (Weibel and Heller 1991, Lee 1991). Contour methods tend to be computationally intensive and difficult to apply to larger areas. They are less often used directly as elevation models, but have been used in hydrological applications (Moore, O'Loughlin, and Burch 1988) and in recent derivations of slope (Mizukoshi and Aniya 2002).

On the other hand, regular grid DEMs offer simplicity of representation and of topological relations between points, at the expense of somewhat larger storage requirements than TINs. Regular grid DEMs are readily integrated with remotely sensed environmental data that are normally obtained in regular grid form. The grid spacing can also provide a useful index of scale. If stored with sufficient vertical precision, regular grid DEMs can represent terrain shape in areas of both high and low relief. Thus regular grid DEMs have become the dominant vehicle for environmental applications that depend on elevation and shape of the land surface.

The grid spacing, or spatial resolution, of a regular grid DEM, as well as providing a practical index of scale also provides a measure of information content (Hutchinson 1996). The issue of spatial scale arises at various points in elevation-based environmental analysis and modeling. The scale of source topographic data should guide the choice of grid spacing when generating DEMs from such data. The

Table 8.1 Spatial scales of applications of digital elevation models (DEMs) and common sources of topographic data for generation of DEMs

Scale	DEM Resolution	Common Topographic Data Sources	Hydrological and Ecological Applications
Fine Toposcale	1–50 m	Contour and streamline data from aerial photography and existing topographic maps at scales from 1:5,000 to 1:50,000. Surface specific point and streamline data obtained by ground survey using GPS. Remotely sensed elevation data using airborne and spaceborne radar and laser.	Spatially distributed hydrological modeling. Spatial analysis of soil properties. Topographic aspect corrections to remotely sensed data. Topographic aspect effects on solar radiation, evaporation, and vegetation patterns.
Coarse Toposcale	50–200 m	Contour and streamline data from aerial photography and existing topographic maps at scales from 1:50,000 to 1:200,000. Surface specific point and streamline data digitized from existing topographic maps at 1:100,000 scale.	Broader scale distributed parameter hydrological modeling. Sub-catchment analysis for lumped parameter hydrological modeling and assessment of biodiversity.
Mesoscale	200 m–5 km	Surface specific point and streamline data digitized from existing topographic maps at scales from 1:100,000 to 1:250,000.	Elevation dependent representations of surface temperature and precipitation. Topographic aspect effects on precipitation. Surface roughness effects on wind. Determination of continental drainage divisions.
Macroscale	5–500 km	Surface specific point data digitized from existing topographic maps at scales from 1:250,000 to 1:1,000,000. National archives of ground surveyed topographic data including trigonometric points and bench marks.	Major orographic barriers for general circulation models.

scales of DEMs should also match the natural scales of terrain dependent applications. The determination of appropriate DEM scales for hydrological modeling is an active research issue (for example, Zhang and Montgomery 1994, Blöschl and Sivaplan 1995, Feddes 1995, Sivaplan, Grayson, and Woods 2004). This is particularly relevant in recent years with the advent of high resolution data sources. Incorporation of terrain structure into considerations of spatial scale is also an emerging issue in terrain analysis (Gallant and Dowling 2003).

The range of spatial scales of hydro-ecological applications of DEMs, and the corresponding common primary topographic data sources are indicated in Table 8.1, as adapted from Hutchinson and Gallant (2000). The general trend since the 1980s has been to move from broader continental and regional scales, closely allied to the representation of major drainage divisions (Jenson 1991, Hutchinson and Dowling 1991), to mesoscale representations of surface climate (Hutchinson 1995b, Running and Thornton 1996, Daly, Neilson, and Phillips 1994) and associated flora and fauna (Nix 1986), to finer toposcales suited to the modeling of surface hydrology, vegetation, and soil properties (Moore, Grayson, and Ladson 1991, Quinn, Beven, Chevallier, and Planchon 1991, Zhang and Montgomery 1994, Gessler, Moore, McKenzie, and Ryan 1995, Mackey 1996). This has been accompanied by improvements in methods for representing the fine scale shape and structure of DEMs, supported by the steady increase in resolution of DEM data sources and the capacity of computing platforms. Fine scale processes are often the focus of hydrological applications. However, it should be noted that coarser scale processes, particularly mesoscale climate, can have a significant impact on the spatial distribution of elevation dependent environmental processes.

Of the applications listed in Table 8.1, only representations of surface temperature and rainfall have a direct dependence on elevation. All others depend on measures of surface shape and roughness, as exemplified by the primary and secondary terrain attributes listed in Moore, Grayson, and Ladson (1991). This underlines the importance of DEMs providing accurate representations of surface shape and drainage structure. This is particularly so in low relief areas where elevations must be recorded with sub-meter precision to accurately reflect small elevation gradients.

Though actual terrain can vary across a wide range of spatial scales, in practice, source topographic data are commonly acquired at a particular scale. This places practical limits on the range of DEM resolutions that can be truly supported by a particular source data set. The following section describes the data sources commonly supporting generation of DEMs at each of the scales listed in Table 8.1.

Sources of Topographic Data

Three main classes of source topographic data may be recognized, for which different DEM generation techniques are applicable, as discussed below.

Surface specific point elevation data

Surface specific point elevations, including high and low points, saddle points, and points on streams and ridges make up the skeleton of terrain (Clarke 1990). They are an ideal data source for most interpolation techniques, including triangulation methods and specially adapted gridding methods. These data may be obtained by ground survey and by manually assisted photogrammetric stereo models (Makarovic 1984). They can also be obtained from grid DEMs to construct TIN models (Heller 1990, Lee 1991). The advent of the global positioning system (GPS) has enhanced the availability of accurate ground surveyed data (Dixon 1991, Lange and Gilbert 1998). Such data are mainly obtained for detailed survey of relatively small experimental catchments. They are less often used for larger areas.

Contour and streamline data

Contour and streamline data are still a common terrain data source for larger areas. They have found common application at the toposcale (see Table 8.1). Many of these data have been digitized from existing topographic maps that, until the advent of spaceborne survey methods, were the only source of elevation data for some parts of the world. The conversion of contour maps to digital form has been a major activity of mapping organizations world wide (Hobbs 1995). Contours can also be generated automatically from photogrammetric stereo models (Lemmens 1988), although these methods are subject to error due to variations in surface cover. A sample contour and streamline data set, together with some additional point data, is shown in Figure 8.1. Contours implicitly encode a number of terrain features, including points on streamlines and ridges. The main disadvantage of contour data is that they significantly under sample the areas between contour lines, especially in areas of low relief, such as the lower right hand portion of Figure 8.1. This has

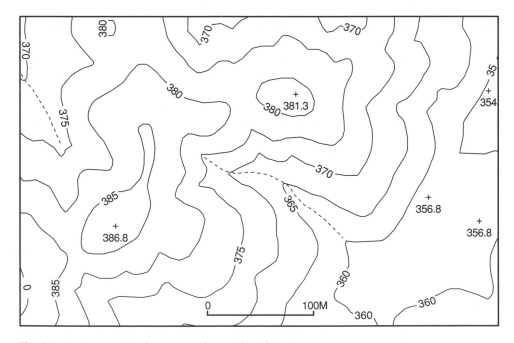

Fig. 8.1   Contour, point elevation, and streamline data

led most investigators to prefer contour specific algorithms over general purpose algorithms when interpolating contour data (Clarke, Grün, and Loon 1982, Mark 1986, Hutchinson 1988, Weibel and Brändl 1995).

Contour data differ from other elevation data sources in that they imply a degree of smoothness of the underlying terrain. When contours are obtained by manually assisted photogrammetric techniques, the operator can remove the effects of obstructions such as vegetation cover and buildings. When coupled with a suitable interpolation technique, contour data can be a superior data source in low relief areas, where moderate elevation errors in remotely sensed data can effectively preclude accurate determination of surface shape and drainage structure (Garbrecht and Starks 1995).

Streamlines are also widely available from topographic maps and provide important structural information about the landscape. However, few interpolation techniques are able to make use of streamline data without associated elevation values. The method developed by Hutchinson (1989) can use such streamline data, provided that the streamlines are digitized in the downslope direction. This imposes a significant editing task, which can be achieved by using a GI System (GIS) with network capabilities.

Remotely sensed elevation data

Gridded DEMs may be calculated directly by stereoscopic interpretation of data collected by airborne and satellite sensors. The traditional source of these data is aerial photography (Kelly, McConnell, and Mildenberger 1977). In the absence of vegetation cover these data can deliver elevations to sub-meter accuracy (Ackermann 1978, Lemmens 1988). Stereoscopic methods have been applied to SPOT imagery (Konecny, Lohmann, Engel, and Kruck 1987, Day and Muller 1988), and more recently to airborne and spaceborne synthetic aperture radar (SAR). Airborne and spaceborne lasers can also provide elevation data in narrow swathes. A major impetus for these developments has been the goal of generating high resolution DEMs with global coverage, recently achieved with the completion of the three-second (90 m) DEM for the globe obtained from the Shuttle Radar Topography Mission (USGS 2005).

Remote sensing methods can provide broad spatial coverage, but have a number of generic limitations. None of the sensors can reliably measure the ground elevations underneath dense vegetation cover. Even in the absence of ground cover, all methods measure elevations with significant random errors that depend on the inherent limitations of the observing instruments, as well as surface slope and roughness (Harding, Bufton, and Frawley 1994, Dixon 1995). The methods also require accurately located ground control points to minimize systematic error. These points are not always easy to locate, especially in remote regions. Best possible standard elevation errors with spaceborne systems currently range between 1 and 10 m, but elevation errors can be much larger, up to 100 m, under unfavorable conditions (Sasowsky, Peterson, and Evans 1992, Harding, Bufton, and Frawley 1994, Zebker, Werner, Rosen, and Hensley 1994, Lanari, Fornaro, Riccio 1997). Averaging of data obtained from multiple passes of the sensor can reduce these errors, but at greater cost.

Airborne laser scanning (ALS) data are an emerging source of fine-scale, irregularly spaced, elevation data with position and elevation errors around 10–20 cm and typical data spacing around 1–2 m (Flood 2001, Maas 2002). Airborne laser can reliably detect the height of the ground surface below significant tree cover. These data offer the potential for a new generation of accurate fine scale DEMs for a wide range of applications (Lane and Chandler 2003). But, as for coarser scale remotely sensed data, careful filtering and interpolation of such data is required to maximize the quality of the resulting DEMs and dependent representations of surface shape and drainage structure.

DEM interpolation methods

Interpolation is required to generate regular grid DEMs from irregularly spaced elevation data. These include surface specific points, contour data and high resolution data obtained by detailed ground survey and by airborne and spaceborne laser. Streamline data can be incorporated into the interpolation process to improve the drainage properties of the interpolated DEM. Since source topographic data sets are usually very large, high quality global interpolation methods, such as thin plate splines, in which every interpolated point depends explicitly on every data point, are computationally impractical. Such methods cannot be easily adapted to the strong anisotropy evidenced by real terrain surfaces. On the other hand, local interpolation methods – such as inverse distance weighting, local kriging, and unconstrained triangulation methods – achieve computational efficiency at the expense of somewhat arbitrary restrictions on the form of the fitted surface. Three classes of interpolation methods are in use. All achieve a degree of *local adaptivity* to anisotropic terrain structure.

Triangulation

Interpolation based on triangulation is achieved by constructing a triangulation of the data points, which form the vertices of the triangles, and then fitting local polynomial functions across each triangle. Linear interpolation is the simplest case, but a variety of higher order polynomial interpolants have been devised to ensure that the interpolated surface has continuous first derivatives (Akima 1978, Sibson 1981, Watson and Philip 1984, Auerbach and Schaeben 1990, Sambridge, Braun, and McQueen 1995). Considerable attention has been directed towards methods for constructing the triangulation. The Delauney triangulation is the most popular method and several efficient algorithms have been devised (for example, Aurenhammer 1991, Tsai 1993).

Triangulation methods have been seen as attractive because they can be adapted to various terrain structures, such as ridge lines and streams, using a minimal number of data points (McCullagh 1988). However, these points are difficult to obtain as primary data. Triangulation methods are sensitive to the positions of the data points and the triangulation needs to be constrained to produce optimal results (Heller 1990, Weibel and Heller 1991). Triangulation methods are known to have difficulties in interpolating contour data. These data tend to generate many flat triangles unless additional structural data points along streams and ridges can be provided (Clarke 1990, Zhu, Eastman, and Toledano 2001).

Local surface patches

Interpolation by local surface patches is achieved by applying a global interpolation method to overlapping regions, usually rectangular in shape, and then smoothly blending the overlapping surfaces. Franke (1982) and Mitasova and Mitas (1993) have used bivariate spline functions in this way. These methods overcome the computational problems posed by large data sets and permit a degree of local anisotropy. They can also perform data smoothing when the data have elevation errors. There are some difficulties in defining patches when data are very irregularly spaced and anisotropy is limited to one direction across each surface patch. Nevertheless, Mitasova and Mitas (1993) have obtained good performance on contour data. An advantage for applications of this method is that topographic parameters such as slope and curvature, as well as flow lines and catchment areas, can be calculated directly from the fitted surface patches, which have continuous first and second derivatives (Mitasova, Hofierka, Zlocha, and Iverson 1996). Local surface patches can also be readily converted into regular grids.

Locally adaptive gridding

Direct gridding methods can provide a computationally efficient means of applying high quality interpolation methods to large elevation data sets. Iterative methods which fit discretized splines in tension, as represented by a finite difference grid, have been described by Hutchinson (1989) and Smith and Wessel (1990). Both methods have their origin in the minimum curvature method developed by Briggs (1974). Computational efficiency can be achieved by using a simple multi-grid strategy which can make computational time optimal, in the sense that it is proportional to the number of interpolated DEM points (Hutchinson 1989). The use of splines in tension is indicated by the statistical nature of actual terrain surfaces (Frederiksen, Jacobi, and Kubik 1985, Goodchild and Mark 1987). It overcomes the tendency of minimum curvature splines to generate spurious surface oscillations in complex areas.

The ANUDEM direct gridding method of Hutchinson (1988, 1989, 2006) is used widely. It has been shown to be superior in terms of elevation accuracy to a variety of local kriging methods (Bishop and McBratney 2002). It has several *locally adaptive* features. It is best described by first defining an appropriate statistical model for the observed elevation data. Each elevation data value z_i at location x_i, y_i is assumed to be given by:

$$z_i = f(x_i, y_i) + \varepsilon_i \qquad (i = 1, \ldots, n) \tag{8.1}$$

where f is an unknown suitably smooth bivariate function of horizontal location represented as a finite difference grid, n is the number of data points and ε_i is a zero mean error term with standard deviation w_i. For accurately surveyed elevation data the standard deviation is dominated by the natural discretization error of the finite difference representation of f. Assuming that each data point is located randomly within its corresponding grid cell, the standard deviation of the discretization error is given by:

$$w_i = hs_i/\sqrt{12} \qquad\qquad (8.2)$$

where h is the grid spacing and s_i is the slope of the grid cell associated with the i th data point (Hutchinson 1996). The function f is then estimated by solving for the regular grid finite difference approximation to the bivariate function f that minimizes:

$$\sum_{i=1,n} [(z_i - f(x_i,y_i))/w_i]^2 + \lambda J(f) \qquad\qquad (8.3)$$

where $J(f)$ is a measure of the roughness of the function f in terms of first and second derivatives (Hutchinson 1989) and λ is a positive number called the smoothing parameter. The smoothing parameter λ is normally chosen so that the weighted residual sum of squares in equation (8.3) is equal to n. This can be achieved with an approximate Newton-Rhapson method coupled with the iterative solution of f (Hutchinson 2000). The spatially varying weights in the residual sum of squares in equation (8.3) is a *locally adaptive* feature that can only be achieved with an iterative interpolation method for which the slopes of the grid cells are available as the iterative solution proceeds.

Former limitations in the ability of general gridding methods to adapt to strong anisotropic structure in actual terrain surfaces, as noted by Ebner, Reinhardt, and Hössler (1988), have been largely overcome by applying a series of *locally adaptive* constraints to the basic gridding procedure. Constraints which have direct relevance for hydrological applications are those imposed by the drainage enforcement algorithm devised by Hutchinson (1989). This algorithm removes spurious depressions in the fitted DEM, in recognition of the fact that sinks are usually quite rare in nature (Band 1986, Goodchild and Mark 1987). This can significantly improve the drainage quality and overall structure of the fitted DEM. It can largely remove the need to modify the interpolated DEM to obtain drainage connectivity. Alternatively, this can be achieved by artificially filling remaining depressions (Jenson and Domingue 1988). A grid carving procedure for removing depressions from DEMs has been recently developed by Soille, Vogt, and Colombo. (2003). Its action is similar to the drainage enforcement algorithm of Hutchinson (1989).

The action of the drainage algorithm can be quite strong in data sparse areas, as illustrated in Figures 8.2 and 8.3 reproduced from Hutchinson (1989). The remaining sinks labeled S1, S2, S3, S4 in Figure 8.2, are removed by systematically identifying the lowest saddle point in the drainage divide surrounding each remaining depression. Thus the point D is the lowest saddle associated with the sink S1. Flow lines on each side of this saddle point are used to enforce approximate linear constraints from S1 down to S2. Other sinks are cleared similarly. Sinks are cleared in order of increasing elevation. This yields a derived drainage network aligned with the actual streamline formation process. Thus, the sink S3 in Figure 8.2 was not initially cleared to the next lowest sink S4, but cleared to join the streamline first inferred between S4 and S1. The drainage enforcement procedure can be made

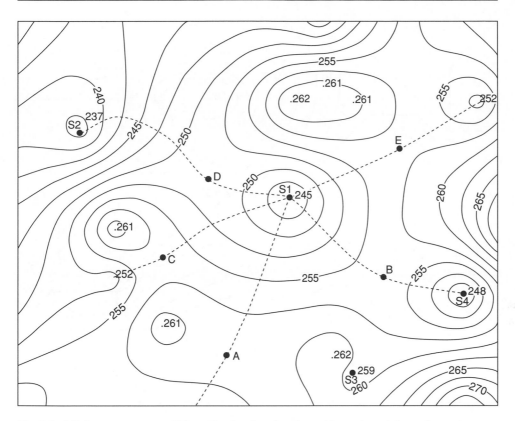

Fig. 8.2 Minimum curvature gridding of point elevation data with spurious sinks or depressions

computationally efficient because both sink points and saddle points can be detected *locally* on the grid DEM as the DEM is being interpolated.

A related *locally adaptive* feature is an algorithm which automatically calculates curvilinear ridge and streamlines from points of locally maximum curvature on contour lines (Hutchinson 1988). This permits interpolation of the fine structure in contours across the area between the contour lines in a more reliable fashion than methods which use linear or cubic interpolation along straight lines in a limited number of directions (Clarke, Grün, and Loon 1982, Oswald and Raetzsch 1984, Legates and Willmott 1986, Cole, MacInnes, and Miller 1990). A partly similar approach, combining triangulation and grid structures, has been described by Aumann, Ebner, and Tang (1992).

The result of applying the ANUDEM program to the contour, streamline and point data in Figure 8.1 is shown Figure 8.4. The inferred stream and ridge lines are particularly curvilinear in the data sparse, low relief portion of the figure, and there are no spurious depressions. The derived contours also closely match the data contours. This locally adaptive method has largely overcome problems formerly encountered by gridding methods in accurately representing drainage structure in low relief areas (Douglas 1986, Carter 1988).

This procedure also yields a generic classification of the landscape into simple, connected, approximately planar, terrain elements, bounded by contour segments and

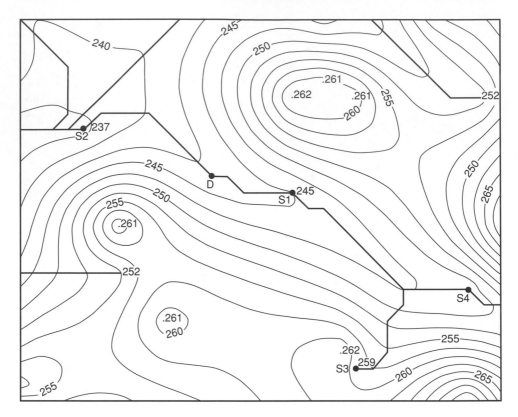

Fig. 8.3 Spurious sinks removed from the surface in Figure 8.2 by the drainage enforcement algorithm

flow line segments. These are similar to the elements calculated by Moore, O'Loughlin, and Burch (1988), but their bounding ridge lines and streamlines are determined in a more stable manner that incorporates uphill searches on ridges and downhill searches in valleys (Hutchinson 1988). A recent development in this elevation gridding method is to include a *locally adaptive* surface roughness penalty defined by profile curvature. This penalty attempts to match fluvial landform processes in a more generic manner and initial results are encouraging (Hutchinson 2000).

Filtering of remotely sensed elevation data

Remotely sensed data can be obtained as regular grid DEMs and as irregularly spaced laser scanned data. Filtering of both forms of data is required to remove errors that can have both random and systematic components. Filtering of DEM data is usually associated with a coarsening of the DEM resolution. Methods include simple nearest neighbor sub-sampling and standard filtering techniques, including median and moving average filtering in the spatial domain, and lowpass filtering in the frequency domain. Several authors have recognized the desirability of filtering remotely sensed DEMs to improve the representation of surface shape.

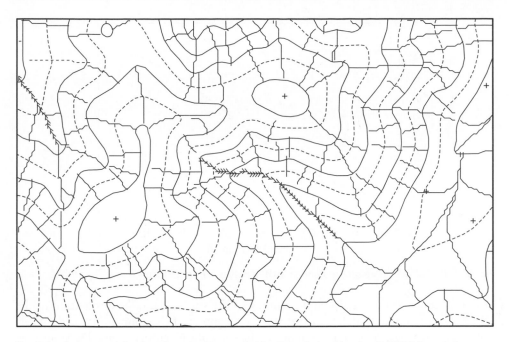

Fig. 8.4 Contours and inferred stream lines and ridge lines derived by the ANUDEM procedure from the topographic data shown in Figure 8.1.

Sasowsky, Peterson, and Evans (1992) and Bolstad and Stowe (1994) used the nearest neighbor method to sub-sample SPOT DEMs, with a spatial resolution of 10 m, to DEMs with spatial resolutions ranging from 20 to 70 m. This generally enhanced the representation of surface shape, although significant errors remained. Giles and Franklin (1996) applied median and moving average filtering methods to a 20 m resolution SPOT DEM. This similarly improved representation of slope and solar incidence angles, although elevation errors were as large as 80 m and no effective representation of profile curvature could be obtained.

Lanari, Fornaro, Riccio (1997) have applied a Kalman filter to spaceborne SAR data obtained on three different wavelengths. Standard elevation errors ranged between about 5 and 80 m, depending on land surface conditions. There is clear potential for the application of smoothing methods that simultaneously maintain sensible morphological constraints, such as connected drainage structure, on the filtered DEM. The locally adaptive finite difference gridding procedure described above is one such method. It can be adapted to DEM data, and to irregularly spaced airborne laser data, by augmenting the error standard deviation described in equation (8.2) to include the standard vertical measurement error in the remotely sensed data.

Optimization of DEM Resolution

Determination of the appropriate resolution of an interpolated or filtered DEM is usually a compromise between achieving fidelity to the true surface and respecting

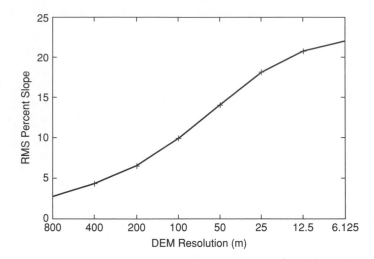

Fig. 8.5 Plot of root mean square slope of a DEM versus DEM resolution

practical limits on the density and accuracy of the source data. Determination of the DEM resolution that matches the information content of the source data is desirable for several reasons. It directly facilitates efficient data inventory, since DEM storage requirements are quite sensitive to resolution. It also permits interpretation of the horizontal resolution of the DEM as an index of information content. This is an important consideration when linking DEMs to other grid data sets and when filtering remotely sensed DEMs. Moreover, it can facilitate assessment of the scale dependence of terrain dependent applications, such as the determination of the spatial distributions of soil properties (Gessler, Moore, McKenzie, and Ryan 1995).

A simple method for matching DEM resolution to source data information content has been developed by Hutchinson (1996). The method is based on the locally adaptive weighting of the residual sum of squares described in equations (8.1) and (8.2) above. The method monitors the root mean square slope of all DEM points associated with elevation data as a function of DEM resolution (Figure 8.5). The optimum resolution is determined by refining the DEM resolution until further refinements produce no significant increase in the root mean square DEM slope. The method is particularly appropriate when source data have been obtained in a spatially uniform manner, such as elevation contours from topographic maps at a fixed scale, or from remotely sensed gridded elevation data.

The root mean square slope criterion appears to be a reliable shape-based way of matching DEM resolution, to within a factor two, to the information content of the source contour and streamline data. This criterion can be refined, especially when source data have positional errors, by examining plots of derived contours and profile curvature, as discussed by Hutchinson and Gallant (2000).

Quality Assessment of DEMs

The quality of a derived DEM can vary greatly depending on the source data and the interpolation technique. The desired quality depends on the application for which the DEM is to be used, but a DEM created for one application is often used for other purposes. Any DEM should therefore be created with care, using the best available data sources and processing techniques. Efficient detection of spurious features in DEMs can lead to improvements in DEM generation techniques as well as detection of errors in source data. Early detection and correction of data errors can avoid expensive reprocessing of DEM dependent applications.

A first measure of DEM accuracy is summary elevation difference, typically the root mean square difference, of reference elevation data from the DEM. This can provide a useful indication of DEM accuracy. However, since many applications of DEMs depend on representations of surface shape and drainage structure, measures of elevation error do not provide a complete assessment of DEM quality (Hutchinson 1989, Wise 2000). A number of graphical techniques for assessing data quality have been developed. These are non-classical measures that offer means of confirmatory data analysis without the use of accurate reference data. Assessments of DEMs in terms of their representation of surface aspect have also been examined by Wise (1998).

Spurious sinks and drainage analysis

Spurious sinks or local depressions in DEMs are frequently encountered and are a significant source of problems in hydrological applications (Mackay and Band 1998). Sinks may be caused by incorrect or insufficient data, or by an interpolation technique that does not enforce surface drainage. They are easily detected by comparing elevations with surrounding neighbors. Hutchinson and Dowling (1991) noted the sensitivity of this method in detecting elevation errors as small as 20 m in source data used to interpolate a continent-wide DEM with a horizontal resolution of 2.5 km. More subtle drainage artifacts in a DEM can be detected by performing a full drainage analysis to derive catchment boundaries and streamline networks, using the technique of Jenson and Domingue (1988).

Views of shaded relief and other terrain attributes

Computing shaded relief allows a rapid visual inspection of the DEM for local anomalies that show up as bright or dark spots. It can indicate both random and systematic errors. It can identify problems with insufficient vertical resolution, since low relief areas will show as highly visible steps between flat areas. It can also detect edge-matching problems (Hunter and Goodchild 1995). Shaded relief is a graphical way of checking the representation of slope and aspect in the DEM. Views of other primary terrain attributes, particularly profile curvature, can provide a sensitive assessment of the accuracy of the DEM in representing terrain shape. Examination of profile curvature can prevent selection of a DEM resolution which is too fine. Overfine resolution can lead to systematic errors in derived primary terrain attributes, as illustrated by Hutchinson and Gallant (2000).

Derived elevation contours

Contours derived from a DEM provide a sensitive check on terrain structure since their position, aspect, and curvature depend directly on the elevation, aspect, and plan curvature respectively of the DEM. Derived contours are a particularly useful diagnostic tool because of their sensitivity to elevation errors in source data. Subtle errors in labeling source data contours digitized from topographic maps are common, particularly for small contour isolations that may have no label on the printed map. Examination of derived contours can prevent selection of a DEM resolution which is too coarse to adequately represent terrain structure, as illustrated by Hutchinson and Gallant (2000).

Frequency histograms of primary terrain attributes

Other deficiencies in the quality of a DEM can be detected by examining frequency histograms of elevation and aspect. DEMs derived from contour data usually show an increased frequency at the data contour elevations in an elevation histogram. The severity of this bias depends on the interpolation algorithm. Its impact is minimal for applications that depend primarily on drainage analyses that are defined primarily by topographic aspect. Frequency histograms of aspect can be biased towards multiples of 45 and 90 degrees by simpler interpolation algorithms that restrict searching to a few specific directions between pairs of data points.

Topographic Dependent Modeling of Environmental Variables

Environmental quantities are naturally distributed over space so accounting for the spatial distribution of environmental variables plays a key role in environmental modeling. Biophysical variables such as climate, soil and terrain are the primary spatially distributed determinants of plant growth. Dependent quantities such as natural vegetation, agricultural productivity, soil erosion and human and animal populations all have spatial dimensions that can be addressed by making explicit links to these contributing spatial biophysical processes. The type of spatial analysis that is discussed here is usually termed geostatistical. It is applied to data that have been sampled at points across the landscape. The broad aim of geostatistical analysis is to identify the nature of the spatial coherence of these data and use it to estimate (or interpolate) the complete spatial distribution from the sampled points.

Geostatistics arose out of efforts by Krige in the 1950s to estimate the spatial distributions of ore bodies. The theory was largely established by Matheron in the 1960s (Chilès and Delfiner 1999). A significant body of related work on multivariate smoothing spline methods also arose in the late 1970s, largely championed by Wahba (1990). The methods are formally equivalent, and have similar accuracy, but tend to differ in practice (Hutchinson and Gessler 1994). Schimek (2000) provides an extensive survey of spline-based approaches to multivariate modeling. A basic two dimensional spatial model underlying both geostatistics and smoothing splines is that there are n measured data values z_i at spatial locations x_i,y_i, given by

$$z_i = f(x_i,y_i) + \varepsilon_i \qquad (i = 1, \ldots, n) \tag{8.4}$$

where f is a function to be estimated from the observations and ε_i is a zero mean error term. In the case of smoothing splines f is assumed to be a smooth unknown function and in the case of geostatistics f is assumed to be a spatially autocorrelated random field.

This spatial model has the same form as equation (8.1) given above for the locally adaptive gridding model for elevation. However, in the case of surface climate data the function can be solved analytically, since climate data sets normally have at most a few thousand points and there is no need to apply locally adaptive constraints to the interpolated function. The degree of data smoothing controls the complexity of the function f and has to be estimated from statistical analysis of the data. This is usually accomplished for smoothing splines by minimizing an estimate of the predictive error of the fitted spline given by the generalized cross validation (GCV) (Wahba 1990). It is often done in geostatistics by variogram analysis (Cressie 1991). Alternatively, maximum likelihood methods can be applied to both splines and geostatistics. Extensions to the basic bivariate model described in equation (8.4) to incorporate dependences on topography are described below in terms of thin plate smoothing splines with their kriging equivalents.

Partial spline model

The simplest way to extend the spline model specified by equation (8.4) to incorporate dependence on topography is to add a parametric linear dependence on elevation. Such a *partial spline* model can be appropriate for representing surface temperature since the environmental lapse rate of temperature is approximately linear. The model can be described as

$$T_i = f(x_i, y_i) + \beta h_i + \varepsilon_i \qquad (i = 1, \ldots, n) \tag{8.5}$$

where T_i is the temperature at location x_i, y_i with elevation h_i above sea-level, n is the number of data points, β is a an unknown but fixed elevation lapse rate and f is an unknown function of horizontal location (Hutchinson 2003). The error term ε_i represents not only measurement error, but also deficiencies in the partial spline model due to fine scale variation in surface temperature below the resolution of the data network. The function f represents sea-level temperature and the scalar β represents the environmental lapse rate, usually around 6.5° C per km. The partial spline model permits simultaneous estimation of f and β with the complexity of the function f determined by minimising the GCV. This corresponds to *detrended kriging* in geostatistics, also called regression kriging. In that case the trend β is usually set by initial linear regression on elevation and the residuals are then spatially interpolated using ordinary kriging.

Jarvis and Stuart (2001a, b) have found partial splines and detrended kriging to perform with similar accuracy in interpolating daily temperature data. A range of topographic predictors have also been used in this manner to predict soil moisture (Famiglietti, Rudnicki, and Rodell 1998, Western, Grayson, Blöschl, Willgoose, and McMahon 1999) and other soil properties (Odeh, McBratney, and Chittleborough 1994, Bourennane, King, Cherry, and Bruand 1996). Additional linear predictors, based on topography and other factors, are easily added to partial spline and detrended kriging models, as illustrated in a spatial analysis of precipitation by Kyriakidis, Kim, and Miller (2001). If there is no significant spatial variation in the dependence on topographic predictors the function f in equation (8.5) may be omitted and the analysis reverts to simple linear regression on the topographic predictors. The spatial variation in the fitted model is then due only to the spatial variation in the topographic predictors.

Trivariate spline model

For variables such as precipitation and soil moisture, with more complex topographic dependences, simple linear regression and partial spline models are not always adequate spatial models. Thus Agnew and Palutik (2000) and Kieffer Weisse and Bois (2001) found elevation detrended kriging analyses of precipitation to perform no better than simple linear regression on elevation. This is because topographic dependencies of precipitation are known to vary over larger areas. Similarly, Qiu, Fu, Wang, and Chen (2001) have noted that relationships between soil moisture and environmental variables can be very variable.

There have been several attempts to introduce *spatially varying* topographic dependences into precipitation analyses. A relatively straightforward way is to divide the region into subregions and perform separate linear regression analyses on elevation for each subregion. Subregions may be defined by partitioning the region into latitude and longitude rectangles (Michaud, Auvine, and Penalba 1995), or more commonly by using a succession of local overlapping neighborhoods, each containing a minimum number of data points (Nalder and Wein 1998). The PRISM method makes a further subdivision based on broad classes of topographic aspect (Daly, Neilson, and Phillips 1994). These methods can be effective. Their main limitations are non-robustness of the local regressions, due to the lack of sufficient numbers of stations at high altitude locations in some subregions, and minimal coherence between analyses for adjoining subregions.

These problems can be addressed by applying a multivariate analysis method that uses all the data to simultaneously incorporate a continuous spatially varying dependence on both horizontal position and elevation. This can be implemented with a trivariate smoothing spline model that can be written

$$R_i = f(x_i, y_i, h_i) + \varepsilon_i \qquad (i = 1, \ldots, n) \tag{8.6}$$

where R_i is the measured precipitation at location x_i, y_i with elevation h_i. A similar trivariate extension can be made to kriging analyses. Trivariate thin plate smoothing spline precipitation analyses have been shown by Hutchinson (1995a) to perform significantly better than both bivariate analyses and elevation detrended bivariate analyses.

The accuracy of these analyses depended critically on exaggerating the scale of elevation in relation to horizontal position by a factor of about 100. If horizontal and elevation coordinates were scaled equally then the trivariate smoothing spline analysis performed no better than a bivariate analysis. This underlines the importance of optimizing the vertical scaling in trivariate spline and kriging analyses. Trivariate spline smoothing has been shown to be superior to a local regression method for the interpolation of both precipitation and temperature data, particularly in data sparse areas (Price, McKenney, Nalder, and Hutchinson 2000). Such analyses can

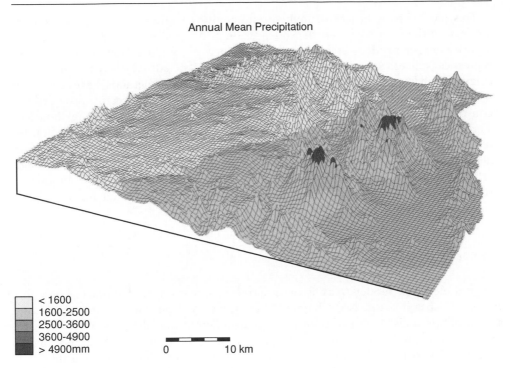

Annual Mean Precipitation

< 1600
1600-2500
2500-3600
3600-4900
> 4900mm

0   10 km

Fig. 8.6   Spline model of annual mean precipitation over topography

also be improved by using elevation data from DEMs at an appropriate horizontal resolution. Several studies have found optimal horizontal resolutions of elevation dependence of precipitation between 4 and 10 km (Schermerhorn 1967, Chuan and Lockwood 1974, Daly, Neilson, and Phillips 1994).

Figure 8.6 shows a plot of annual mean precipitation overlaid on a digital elevation model of northeastern Queensland in Australia. This model has been calculated from a three-dimensional spline model fitted to measured point precipitation data as a function of horizontal location and elevation. The resulting spatial precipitation pattern is quite complex, due in large measure to the complexity of the underlying topography. However, close inspection of the figure also reveals an underlying relatively simple, but spatially varying, dependence on elevation. The relative simplicity of this dependence enables its calibration from data sets of modest size.

Limits to complexity of multivariate spatial models

There is a fundamental limit to the complexity of multivariate spatial models fitted to data. The process indicated above of adding additional predictors, such as elevation and other topographic predictors, to a full multivariate spatial model cannot be continued without limit. Additional predictors rapidly enlarge the space in which the model is fitted. This is called the *curse of dimensionality*. In practice it means that general multivariate functions with no constraints on structure other than surface smoothness cannot be fitted to data sets of typical size if the dimension of the function is more than about four or five.

This has led to a variety of multivariate analysis methods that do impose constraints on the nature of their dependences on additional predictors. These include additive tensor product spline models described by Wahba (1990) and additive regression splines (Sharples and Hutchinson 2004). The latter closely resemble local kriging with external drift. A large variety of cokriging methods have also been developed (Chilès and Delfiner 1999). These methods can be effective provided the constraints on model structure are consistent with the processes being modeled. Thus Phillips, Dolph, and Marks (1992) found that cokriging of precipitation data with elevation peformed no better than ordinary kriging applied to elevation detrended precipitation data. Similarly Odeh, McBratney, and Chittleborough (1994) found cokriging methods to perform less well than regression kriging in predicting soil properties in terms of landform attributes derived from a digital elevation model.

CONCLUSIONS

Digital elevation models play a central role in environmental modeling across a range of spatial scales (see Chapter 23 by Deng, Wilson, Gallant in this volume for additional discussion of numerous environmental modeling applications). The regular grid mode of representation has become the dominant form for digital elevation models used in these applications. This form is directly compatible with remotely sensed geographic data sources and can simplify terrain-based analyses, including assessments of spatial scale. A distinguishing feature for many applications, particularly those that operate at finer scale, is a primary requirement for information about terrain shape and drainage structure, rather than elevation. For this reason, elevation contours and streamlines have remained popular sources of primary topographic data. They can be used to construct fine scale digital elevation models by gridding methods that are locally adaptive to surface shape and drainage structure. Remotely sensed digital topographic data, particularly from airborne sensors, are an emerging source of fine scale digital elevation data. The random errors associated with these data require appropriate filtering, without degrading shape and drainage structure, to maximize the utility of these data in environmental applications, particularly in areas with low relief or with significant surface cover.

Locally adaptive gridding procedures can be used to construct digital elevation models from digital elevation contours, point elevations and streamlines so that the elevation models preserve terrain shape and drainage structure. Grid resolution can be optimized to match the true information content of the source data and to maximize the quality of primary terrain parameters derived from the interpolated DEM. The process of producing a DEM from source data requires careful attention to the accuracy of the source data and the quality of the interpolated DEM. Several shape-based measures of DEM quality, that are readily plotted, can greatly assist in assessing DEM quality and in detecting data errors. These measures do not require the existence of separate reference elevation data. In particular, remaining sinks or depressions in a DEM are an excellent indicator of deficiencies in its representation of terrain shape and drainage structure.

Multivariate smoothing spline and kriging analysis methods of varying complexity have been used to incorporate dependences on topographic predictors to enhance the spatial analysis and mapping of environmental variables. Climate variables dependent on elevation and soil moisture dependent on various shape-based terrain parameters are common applications. The methods include simple regression and partial spline and detrended kriging methods. These often perform as well or better than more complex methods such as various forms of cokriging. Full multivariate spline and kriging models offer the most flexibility and have been successfully applied to the spatial interpolation of precipitation for which there is a spatially varying dependence on elevation. Both the relative vertical scale of elevation and its horizontal resolution need to be chosen carefully in such analyses to produce optimal results. Incorporating additional predictors may require imposition of restrictions on model structure to avoid the instabilities that can arise with higher dimensional analyses. Such methods are still the subject of active investigation. They should be successful if the restrictions on model structure are consistent with the spatial processes being modeled.

REFERENCES

Ackermann, F. 1978. Experimental investigation into the accuracy of contouring from DTM. *Photogrammetric Engineering and Remote Sensing* 44: 1537–48.

Agnew, M. D. and Palutikof, J. P. 2000. GIS-based construction of baseline climatologies for the Mediterranean using terrain variables. *Climate Research* 14: 115–27.

Akima, H. 1978. A method of bivariate interpolation and smooth surface fitting for irregularly distributed data points. *ACM Transactions on Mathematical Software* 4: 148–59.

Auerbach, S. and Schaeben, H. 1990. Surface representation reproducing given digitized contour lines. *Mathematical Geology* 22: 723–42.

Aumann, G., Ebner, H., and Tang, L. 1992. Automatic derivation of skeleton lines from digitized contours. *ISPRS Journal of Photogrammetry and Remote Sensing* 46: 259–68.

Aurenhammer, F. 1991. Voronoi diagrams: A survey of fundamental geometric data structure. *ACM Computing Surveys* 23: 345–405.

Band, L. E. 1986. Topographic partition of watersheds with digital elevation models. *Water Resources Research* 22: 15–24

Bishop, T. F. A. and McBratney, A. B. 2002. Creating field extent digital elevation models for precision agriculture. *Precision Agriculture* 3: 37–46.

Blöschl, G. and Sivaplan, M. 1995. Scale issues in hydrological modelling: A review. *Hydrological Processes* 9: 313–30.

Bolstad, P. V., and Stowe, T. 1994. An evaluation of DEM accuracy: Elevation, slope and aspect. *Photogrammetric Engineering and Remote Sensing* 60: 1327–32.

Bourennane, H., King, D., Cherry, P., and Bruand, A. 1996. Improving the kriging of a soil variable using slope gradient as external drift. *Journal of Soil Science* 47: 473–83.

Briggs, I. C. 1974. Machine contouring using minimum curvature. *Geophysics* 39: 39–48.

Carter, J. R. 1988. Digital representations of topographic surfaces. *Photogrammetric Engineering and Remote Sensing* 54: 1577–80.

Chilès, J. and Delfiner, P. 1999. *Geostatistics: Modelling Spatial Uncertainty*. New York: John Wiley and Sons.

Chuan, G. K. and Lockwood, J. G. 1974. An assessment of topographic controls on the distribution of rainfall in the central Pennines. *Meteorological Magazine* 103: 275–87.

Clarke, A. L., Grün, A., and Loon, J. C. 1982. The application of contour data for generating high fidelity grid digital elevation models. In *Proceedings of Auto Carto 5*, Washington, DC, USA. Bethesda, MD: American Congress on Surveying and Mapping/American Society for Photogrammetry and Remote Sensing, pp. 213–22.

Clarke, K. C. 1990. *Analytical and Computer Cartography*. Englewood Cliffs, NJ: Prentice Hall.

Cole, G., MacInnes, S., and Miller, J. 1990. Conversion of contoured topography to digital terrain data. *Computers and Geosciences* 16: 101–9.

Cressie, N. A. C. 1991. *Statistics for Spatial Data*. New York: John Wiley and Sons.

Daly, C., Neilson, R. P., and Phillips, D. L. 1994. A statistical-topography model for mapping climatological precipitation over mountainous terrain. *Journal of Applied Meteorology* 33: 140–58.

Day, T. and Muller, J.-P. 1988. Quality assessment of digital elevation models produced by automatic stereo matchers from SPOT image pairs. *International Archives of Photogrammetry and Remote Sensing* 27: 148–59.

Dixon, T. H. 1991. An introduction to the global positioning system and some geological applications. *Reviews of Geophysics* 29: 249–76.

Dixon, T. H. 1995. *SAR Interferometry and Surface Change Detection*. Miami, FL, University of Miami RSMAS Technical Report No. 95-003.

Douglas, D. H. 1986. Experiments to locate ridges and channels to create a new type of digital elevation model. *Cartographica* 23: 29–61.

Ebner, H., Reinhardt, W., and Hössler, R. 1988. Generation, management and utilization of high fidelity digital terrain models. *International Archives of Photogrammetry and Remote Sensing* 27: 556–65.

Famiglietti, J. S., Rudnicki, J. W., and Rodell, M. 1998. Variability in surface moisture along a hillslope transect: Rattlesnake Hill, TX. *Journal of Hydrology* 210: 259–81.

Feddes, R. A. 1995. *Space and Time Scale Variability and Interdependencies in Hydrological Processes*. Cambridge, Cambridge University Press.

Flood, M. 2001. Laser altimetry: From science to commercial LiDAR mapping. *Photogrammetric Engineering and Remote Sensing* 67: 1209–18.

Franke, R. 1982. Smooth interpolation of scattered data by local thin plate splines. *Computers and Mathematics with Applications* 8: 273–81.

Frederiksen, P., Jacobi, O., and Kubik, K. 1985. A review of current trends in terrain modelling. *ITC Journal* 1985: 101–6.

Gallant, J. C. and Dowling, T. I. 2003. A multiresolution index of valley bottom flatness for mapping depositional areas. *Water Resources Research* 39: 1347–60.

Garbrecht, J. and Starks, P. 1995. Note on the use of USGS level 1 7.5-minute DEM coverages for landscape drainage analyses. *Photogrammetric Engineering and Remote Sensing* 61: 519–22.

Gessler, P. E., Moore, I. D., McKenzie, N. J., and Ryan, P. J. 1995. Soil-landscape modelling and spatial prediction of soil attributes. *International Journal of Geographic Information Systems* 9: 421–32.

Giles, P. T. and Franklin, S. E. 1996. Comparison of derivative topographic surfaces of a DEM generated from stereographic SPOT images with field measurements. *Photogrammetric Engineering and Remote Sensing* 62: 1165–71.

Goodchild, M. F. and Mark, D. M. 1987. The fractal nature of geographic phenomena. *Annals Association of American Geographers* 77: 265–78.

Harding, D. J., Bufton, J. L., and Frawley, J. 1994. Satellite laser altimetry of terrestrial topography: Vertical accuracy as a function of surface slope, roughness and cloud cover. *IEEE Transactions on Geoscience and Remote Sensing* 32: 329–39.

Heller, M. 1990. Triangulation algorithms for adaptive terrain modelling. In *Proceedings of the Fourth International Symposium on Spatial Data Handling*, Columbus, OH: 163–74.

Hobbs, F. 1995. The rendering of relief images from digital contour data. *Cartographic Journal* 32: 111–6.

Houser, P., Hutchinson, M. F., Viterbo, P., Herve Douville, J., and Running, S. W. 2004. Terrestrial data assimilation. In P. Kabat, M. Claussen, P. A. Dirmeyer, J. H. C. Gash, L. Bravo de Guenni, M. Meybeck, R. S. Pielke, C. J. Vörösmarty, R. W. A. Hutjes, and S. Lütkemeier (eds) *Vegetation, Water, Humans and the Climate: A New Perspective on an Interactive System*. Berlin: Springer-Verlag, pp. 273–87.

Hunter, G. J. and Goodchild, M. F. 1995. Dealing with error in spatial databases: A simple case study. *Photogrammetric Engineering and Remote Sensing* 61: 529–37.

Hutchinson, M. F. 1988. Calculation of hydrologically sound digital elevation models. In *Proceedings of the Third International Symposium on Spatial Data Handling*, Sydney, Australia. Columbus, OH: International Geographical Union, pp. 117–33.

Hutchinson, M. F. 1989. A new procedure for gridding elevation and streamline data with automatic removal of spurious pits. *Journal of Hydrology* 106: 211–32.

Hutchinson, M. F. 1995a. Interpolation of mean rainfall using thin plate smoothing splines. *International Journal of Geographic Information Systems* 9: 385–403.

Hutchinson, M. F. 1995b. Stochastic space–time weather models from ground-based data. *Agricultral and Forest Meteorology* 73: 237–64.

Hutchinson, M. F. 1996. A locally adaptive approach to the interpolation of digital elevation models. In *Proceedings of the Third International Conference/Workshop on Integrating GIS and Environmental Modeling*, Santa Barbara, CA: National Center for Geographic Information and Analysis: CD-ROM.

Hutchinson, M. F. 1998. Interpolation of rainfall using thin plate smoothing splines: II, Analysis of topographic dependence. *Journal of Geographic Information and Decision Making* 2: 168–85.

Hutchinson, M. F. 2000. Optimising the degree of data smoothing for locally adaptive finite element bivariate smoothing splines. *ANZIAM Journal* 42: C774–96.

Hutchinson, M. F. 2003. ANUSPLIN Version 4.3. WWW document, http://cres.anu.edu.au/outputs/anusplin.php.

Hutchinson, M. F. 2006. ANUDEM Version 5.2. WWW document, http://cres.anu.edu.au/outputs/anudem.php.

Hutchinson, M. F. and Dowling, T. I. 1991. A continental hydrological assessment of a new grid-based digital elevation model of Australia. *Hydrological Processes* 5: 45–58.

Hutchinson, M. F. and Gallant, J. C. 1999. Representation of terrain. In P. A. Longley, M. F. Goodchild, D. J. Maguire, and D. W. Rhind (eds) *Geographical Information Systems: Principles, Techniques, Applications, and Management*. New York: John Wiley and Sons, pp. 105–24.

Hutchinson, M. F. and Gallant, J. C. 2000. Digital elevation models and representation of terrain shape. In J. P. Wilson and J. C. Gallant (eds) *Terrain Analysis*. New York: John Wiley and Sons, pp. 29–50.

Hutchinson, M. F. and Gessler, P. T. 1994. Splines: More than just a smooth interpolator. *Geodema* 62: 45–67.

Jarvis, C. H. and Stuart, N. 2001a. A comparison between strategies for interpolating maximum and minimum daily air temperatures: A, The selection of "guiding" topographic and land cover variables. *Journal of Applied Meteorology* 40: 1060–74.

Jarvis, C. H. and Stuart, N. 2001b. A comparison between strategies for interpolating maximum and minimum daily air temperatures: B, The interaction between number of guiding variables and the type of interpolation method. *Journal of Applied Meteorology* 40: 1075–84.

Jenson, S. K. 1991. Applications of hydrologic information automatically extracted from digital elevation models. *Hydrological Processes* 5: 31–44.

Jenson, S. K. and Domingue, J. O. 1988. Extracting topographic structure from digital elevation data for geographic information system analysis. *Photogrammetric Engineering and Remote Sensing* 54: 1593–600.

Kelly, R. E., McConnell, P. R. H., and Mildenberger, S. J. 1977. The Gestalt photomapping system. *Photogrammetric Engineering and Remote Sensing* 43: 1407–17.

Kieffer Weisse, A. and Bois, P. 2001. Topographic effects on statistical characteristics of heavy rainfall and mapping in the French Alps. *Journal of Applied Meteorology* 40: 720–40.

Konecny, G., Lohmann, P., Engel, H., and Kruck, E. 1987. Evaluation of SPOT imagery on analytical instruments. *Photogrammetric Engineering and Remote Sensing* 53: 1223–30.

Kryiakidis, P. C., Kim, J., and Miller, N. L. 2001 Geostatistical mapping of precipitation from rain gauge data using atmospheric and terrain characteristics. *Journal of Applied Meteorology* 40: 1855–77.

Lanari, R., Fornaro, G., Riccio, D., Migliaccio, M., Papathanassiou, K., Moreira, J., Schwäbisch, M., Dutra, L., Puglisi, G., Franceschetti, G., and Coltelli, M. 1997. Generation of digital elevation models by using SIR-C/X-SAR multifrequency two-pass interferometry: The Etna case study. *IEEE Transactions on Geoscience and Remote Sensing* 34: 1097–114.

Lane, S. N. and Chandler, J. H. 2003. The generation of high quality topographic data for hydrology and geomorphology: New data sources, new applications and new problems. *Earth Surface Processes and Landforms* 28: 229–30.

Lange, A. F. and Gilbert, C. 1998. Using GPS for GIS data capture. In P. A. Longley, M. F. Goodchild, D. J. Maguire, and D. W. Rhind (eds) *Geographical Information Systems: Principles, Techniques, Applications and Management*. ChichesterL: John Wiley and Sons, pp. 467–76.

Lee, J. 1991. Comparison of existing methods for building triangular irregular network models of terrain from grid digital elevation models. *International Journal of Geographic Information Systems* 5: 267–85.

Legates, D. R. and Willmott, C. J. 1986. Interpolation of point values from isoline maps. *American Cartographer* 13: 308–23.

Lemmens, M. J. P. M. 1988. A survey on stereo matching techniques. *International Archives of Photogrammetry and Remote Sensing* 27: V11–V23.

Maas, H.-G. 2002. Methods for measuring height and planimetry discrepancies in airborne laserscanner data. *Photogrammetic Engineering and Remote Sensing* 68: 933–40.

Mackay, D. S. and Band, L. E. 1998. Topographic partitioning of watersheds with lakes and other flat areas on digital elevation models. *Water Resources Research* 34: 897–901.

Mackey, B. G. 1996. The role of GIS and environmental modelling in the conservation of biodiversity. In *Proceedings of the Third International Conference/Workshop on Integrating GIS and Environmental Modeling*. Santa Barbara, CA: National Center for Geographic Information and Analysis: CD-ROM.

Mark. D. M. 1986. Knowledge-based approaches for contour-to-grid interpolation on desert pediments and similar surfaces of low relief. In *Second International Symposium on Spatial Data Handling*, Seattle, WA, USA. Columbus, OH: International Geographical Union, pp. 225–34.

Makarovic, B. 1984. Structures for geo-information and their application in selective sampling for digital terrain models. *ITC Journal* 1984: 285–95.

McCullagh, M. J. 1988. Terrain and surface modelling systems: theory and practice. *Photogrammetric Record* 12: 747–79.

Michaud, J. D., Auvine, B. A., and Penalba, O. C. 1995. Spatial and elevational variations of summer rainfall in the southwestern United States. *Journal of Applied Meteorology* 34: 2689–703.

Mitasova, H. and Mitas, L. 1993. Interpolation by regularised spline with tension: I. Theory and implementation. *Mathematical Geology* 25: 641–55.

Mitasova, H., Hofierka, J., Zlocha, M., and Iverson, L. 1996. Modelling topographic potential for erosion and deposition using GIS. *International Journal of Geographical Information Systems* 10: 629–41.

Mizukoshi, H. and Aniya, M. 2002. Use of contour-based DEMs for deriving and mapping topographic attributes. *Photogrammetric Engineering and Remote Sensing* 68: 83–93.

Moore, I. D., O'Loughlin, E. M., and Burch, G. J. 1988. A contour-based topographic model for hydrological and ecological applications. *Earth Surface Processes and Landforms* 13: 305–20.

Moore, I. D., Grayson, R. B., and Ladson, A. R. 1991. Digital terrain modelling: A review of hydrological, geomorphological and biological applications. *Hydrological Processes* 5: 3–30.

Nalder, I. A. and Wein, R. W. 1998. Spatial interpolation of climatic normals: Test of a new method in the Canadian boreal forest. *Agricultural and Forest Meteorology* 92: 211–25.

Nix, H. A. 1986. A biogeographic analysis of Australian elapid snakes. In R. Longmore (ed.) *Atlas of Elapid Snakes of Australia*. Canberra, Australian Flora and Fauna Series No.7: 4–15.

Odeh, I. O. A., McBratney, A. B., and Chittleborough, D. J. 1994. Spatial prediction of soil properties from landform attributes derived from a digital elevation model. *Geoderma* 63: 197–214.

Oswald, H. and Raetzsch, H. 1984. A system for generation and display of digital elevation models. *Geo-Processing* 2: 197–218.

Phillips, D. L., Dolph, J., and Marks, D. 1992. A comparison of geostatistical procedures for spatial analysis of precipitation in mountainous terrain. *Agricultural and Forest Meteorology* 58: 119–41.

Price, D. T., McKenney, D. W., Nalder, I. A., and Hutchinson, M. F. 2000. A comparison of two statistical methods for spatial interpolation of Canadian monthly mean climate data. *Agricultural and Forest Meteorology* 101: 81–94.

Qiu, Y., Fu, B., Wang, J., and Chen, L. 2001. Spatial variability of soil moisture content and its relation to environmental indices in a semi-arid gully catchment of the Loess Plateau, China. *Journal of Arid Environments*. 49: 723–50.

Quinn, P., Beven, K., Chevallier, P., and Planchon, O. 1991. The prediction of hillslope flow paths for distributed hydrological modelling using digital terrain models. *Hydrological Processes* 5: 59–79.

Running, S. W. and Thornton, P. E. 1996. Generating daily surfaces of temperature and precipitation over complex topography. In M. F. Goodchild, L. T. Steyaert, B. O. Parks, C. Johnston, D. Maidment, M. Crane, and S. Glendinning (eds) *GIS and Environmental Modeling: Progress and Research Issues*. Fort Collins, CO: GIS World Books, pp. 93–8.

Sambridge, M., Braun, J., and McQueen, H. 1995. Geophysical parameterization and interpolation of irregular data using natural neighbours. *Geophysical Journal International*: 122: 837–57.

Sasowsky, K. C., Peterson, G. W., and Evans, B. M. 1992. Accuracy of SPOT digital elevation model and derivatives: utility for Alaska's North Slope. *Photogrammetric Engineering and Remote Sensing* 58: 815–24.

Schermerhorn, V. P. 1967. Relations between topography and annual precipitation in western Oregon and Washington. *Water Resources Research* 3: 707–11.

Schimek, M. G. (ed.). 2000. *Smoothing and Regression: Approaches, Computation and Application*. New York: John Wiley and Sons.

Sharples, J. J. and Hutchinson, M. F. 2004. Multivariate spatial smoothing using additive regression splines. *ANZIAM Journal* 45: C676–92.

Sibson, R. 1981. A brief description of natural neighbour interpolation. In V. Barnett (ed.) *Interpreting Multivariate Data*. Chichester: John Wiley and Sons, pp. 21–36.

Sivaplan, M., Grayson, R., and Woods, R. 2004. Scale and scaling in hydrology. *Hydrological Processes* 18: 1369–71.

Smith, W. H. F. and Wessel, P. 1990. Gridding with continuous curvature. *Geophysics* 55: 293–305.

Soille, P., Vogt, J., and Colombo, R. 2003. Carving and adaptive drainage enforcement of grid digital elevation models. *Water Resources Research* 39: 1366–75.

Tsai, V. 1993. Delauney triangulations in TIN creation: An overview and linear-time algorithm. *International Journal of Geographical Information Systems* 7: 501–24.

USGS, 2005. *Shuttle Radar Topography Mission: Mapping the World in Three Dimensions*. WWW document, http://srtm.usgs.gov.

Wahba, G. 1990. *Spline Models for Observational Data*. Philadelphia: SIAM.

Watson, D. F. and Philip, G. M. 1984. Triangle based interpolation. *Mathematical Geology* 16: 779–95

Weibel, R. and Brändl, M. 1995. Adaptive methods for the refinement of digital terrain models for geomorphometric applications. *Zeitschrift für Geomorphologie*, Supplementband 101: 13–30.

Weibel, R. and Heller, M. 1991. Digital terrain modelling. In D. J. Maguire, M. F. Goodchild, and D. W. Rhind (eds) *Geographical Information Systems: Principles and Applications*. Harlow: Longman: 269–97.

Western, A. W., Grayson, R. B., Blöschl, G., Willgoose, G. R., and McMahon, T. A. 1999. Observed spatial organization of soil moisture and its relation to terrain indices. *Water Resources Research* 35: 797–810.

Wilson, J. P. and Gallant, J. C. (eds). 2000. *Terrain Analysis: Principles and Applications*. New York: John Wiley and Sons.

Wise, S. 1998. The effect of GIS interpolation errors on the use of digital elevation models in geomorphology. In S. N. Lowe, K. S. Richards, and J. H. Chandler (eds). *Landform Monitoring, Modelling and Analysis*. Chichester: John Wiley and Sons, pp. 139–64.

Wise. S. 2000. Assessing the quality for hydrological applications of digital elevation models derived from contours. *Hydrological Processes* 14: 1909–29.

Zebker, H. A., Werner, C., Rosen, P. A., and Hensley, S. 1994. Accuracy of topographic maps derived from ERS-1 interferometric radar. *IEEE Transactions on Geoscience and Remote Sensing* 32: 823–36.

Zhang, W. and Montgomery, D. R. 1994. Digital elevation model grid size, landscape representation, and hydrologic simulation. *Water Resources Research* 30: 1019–28.

Zhu, H., Eastman, R., and Toledano, J. 2001. Triangulated irregular network optimization from contour data using bridge and tunnel edge removal. *International Journal of Geographic Information Science* 15: 271–86.

Chapter 9

Adding Time into Geographic Information System Databases

May Yuan

Despite substantial advances, Geographic Information Systems (GIS) technology still lacks the ability to handle geospatial information of all kinds. One of the most significant and long-standing issues is integration of spatial and temporal data and support for spatio-temporal analysis. Nevertheless, time is central to geographic inquiry and understanding, and the significance of adding time to GIS databases cannot be over-stated. John Jakle argues that "it is doubtful . . . that geography can continue its search for spatial understanding by ignoring the integral dictates of time and space as a natural unity; thus have geographers come to focus on the processes of spatial organization through time" (Jakle 1972). While the need to incorporate temporal data is common to many information systems, the challenge is arguably much greater in GIS because space and time in geography are interrelated. Consequently, spatio-temporal information about geography cannot be fully captured by simply adding an attribute field "time" in a GIS database. Much research progress has been made in temporal GIS since the 1980s. However, a temporal GIS that is considered sufficiently robust to support spatio-temporal information management, query, analysis, and modeling has still to be developed.

Challenges to the development of a full-fledged temporal GIS arise deep within the conceptual, computational, and presentational foundations of GIS. Conceptually, we need an ontology to categorize and communicate spatio-temporal concepts; we need GIS representations that can capture and frame these concepts; and, furthermore, we need spatio-temporal data models to organize geographic data so that spatio-temporal concepts can be computed and extracted from temporal GIS databases. Computationally, we need query languages to manipulate spatio-temporal data and retrieve information about space, time, and geographic dynamics; we need spatio-temporal logic to reason about geographic dynamics and their relationships; and we need analytical and modeling frameworks to examine spatio-temporal data and make predictions or retrospections in space and time. At the presentation level, we need means to visually communicate the multi-dimensional and dynamic nature of spatial change through time. The challenges demand a fundamental and comprehensive examination of the underlying design of GIS and innovative ways to integrate spatial and temporal information.

The focus of this chapter is on the conceptual foundations of adding time into GIS databases. A sound conceptual foundation serves as the bedrock for computational and presentational advances because modes of computation and presentation rely heavily upon what data are available in what form and structure. Ontology, representation, and data models constitute the conceptual foundations for any information system. The three conceptual components correspond to three stages of conceptualization, identifying: (1) the elements that need to be considered (ontology); (2) the frameworks in which the identified elements will be best abstracted (representation); and (3) the structures with which the abstracted elements can be best organized (data models). In addition, spatio-temporal query is the means to retrieve data of interest for visualization and analysis. What we can do with an information system depends upon what we can access from the system. Hence, spatio-temporal query is critical to the value of a temporal GIS. The following sections, therefore, focus on key developments in spatio-temporal ontologies, representation, data modeling, and spatio-temporal queries. The concluding section then summarizes the current state of temporal GIS and directions for future research.

Ontologies of Space, Time, and Space-time

Ontology, with its early ties to philosophy and linguistics, is "the metaphysical study of the nature of Being and Existence" (Fellbaum 1998). From the perspective of information science, it is the study of fundamental elements, concrete or abstract, in our world. Since a database is built upon identified views of the world, which can be generic or application-specific, the ontology chosen dictates the kinds of information that will be stored in and made available from the database. A closely related term to ontology is "semantics." Generally speaking, semantic modeling stays close to an identified database application while ontology goes beyond the immediate concern of an application. Ontology describes elements of knowing and the basic modes of description that distinguishes one element from the other (Peuquet 2002). Philosophically, there should be only one ontology since there is only one world. Nevertheless, studies show that ontology is tied to human cognition (Mark, Smith, and Tversky 1999), and people may hold distinct conceptualizations of the world, and formal ontologies that deal with interconnections of things (Smith 1998) play a key role in determining methodological and architectural design of information systems (Guarino 1998).

In GIS, field- and object-based conceptualizations of space assign reality into two divergent sets of concrete and abstract kinds in geography (Couclelis 1992). Ontology of objects appears more intuitive to human cognition (Smith and Mark 1998, Mark, Smith, and Tversky 1999), while ontology of fields seems more compatible to scientific computation and mathematical modeling (Peuquet, Smith, and Brogaard 1998). In addition, testing of human subjects suggests that natural features (for example, *mountain, river, lake, ocean, hill*, etc.) receive higher ontological recognition than artificial features (for instance, *town, city*, etc.); furthermore, adjectives that qualify something as geographic or mappable can influence responses to questions of Being and Existence in geography (Mark, Skupina, and Smith 2001, Smith and Mark 2001). The dichotomy the of field- and object-views of the world is challenged by Yuan, Cova, and Goodchild who argue that many geographic phenomena exhibit both field- and object-like properties (Yuan 2001), and locations in a field can be associated with various objects (Cova and Goodchild 2002). Additional beings: f-objects and o-fields, are thus introduced.

Likewise, elements of Being and Existence in time depend upon how time is conceptualized. Most GIS consider time as instantaneous and discrete beings since GIS data are valid at different instants (Frank 1997): this is the so-called SNAP view of the world (Grenon and Smith 2004). Various time onotologies are possible because time can be bounded or unbounded, absolute or relative, discrete or continuous, of different types (for instance, linear, cyclic), of different dimensions (for example, points, intervals), and of different meanings (such as valid time and transition time) (Worboys 1990, 1994, Frank 1998, Raper 2000).

Moreover, some temporal ontologies have been developed, not based on conceptualization of time *per se*, but based on how time is perceived and manifests itself in reality. Time has long been related to cause and effect (Ullman 1974) and can be considered as the abstraction of all relations of sequence (Feather 1959). Accordingly, Terenziani (1995) develops a causal ontology that accounts for temporal constraints between causes and effects. Another important mode of time is repeatability, which is common to many natural phenomena (Frank 1998). Based on return periods, repeatability can be intermittent or periodic. Terenziani (2002), furthermore, develops another ontology based on first-order logic to deal with user-defined periodicity and temporal constraints about repeated events.

The degree of complexity increases exponentially with the development of an ontology of space and time. Compared to ontologies of non-temporal domains, Frank (2003) argues that an ontology of space and time should be more involved in and has a stronger connection to the intended area of application. There are two alternative ways to handle space and time: SNAP considers successions of instantaneous snapshots of the world and SPAN, which is based on a unified view of the spatio-temporal (Grenon and Smith 2004). SNAP entities are indexed at a certain point in time, which implies that these entities may have existed and will continue to exist for some time. Nevertheless, the SNAP ontology only recognizes existence at a defined instant, and consequently, SNAP entities have no temporal parts. Comparison of two instants results in *change* of entity sets. Currently, a common approach to the treatment of spatio-temporal data in GIS is to follow the SNAP ontology. This is because GIS only recognizes a data object on the data layer on which the object is indexed and considers objects on different data layers as different even if they represent the same geographic feature at different times. A SPAN entity, on the other hand, consists of temporal parts which constitute its history or lifeline. Because the SPAN ontology considers the continuum of an entity over space and time, additional concepts about spatio-temporal dynamics emerge, such as *movement*, *vibration*, *sprawl*, *contraction*, and *deformation*.

While SNAP and SPAN offer a succinct ontological view of spatio-temporal Beings, they do not cope with the ontological complexity that arises from the relations among reality, observation, application (purpose), societal constraint, and cognition. To account for the ontological complexity, Frank (2003) proposes a five-tier ontology to describe spatio-temporal things with uniform spatial properties, such as land parcels. The five tiers correspond to a transition from physical reality that exists externally and objectively towards cognitive agents who extract knowledge from the perceived as listed below (adopted from Table 2.1 in Frank 2003):

1 *Ontological Tier 0: Physical Reality*
   - The existence of a single physical reality
   - Determined properties for every point in time and space
   - Space and time as fundamental dimensions of this reality
2 *Ontological Tier 1: Observable Reality*
   - Properties are observable now at a point in space
   - Real observations are incomplete, imprecise, and approximate
3 *Ontological Tier 2: Object World*
   - Objects are defined by uniform properties for regions in space and time
   - Objects continue in time
4 *Ontological Tier 3: Social Reality*
   - Social processes construct external names
   - Social rules create facts and relationships between them
   - Social facts are valid within the social context only
5 *Ontological Tier 4: Cognitive Agents*
   - Agents use their knowledge to derive other facts and make decisions
   - Knowledge is acquired gradually and lags behind reality
   - Reconstruction of previous states of the knowledgebase is required in legal and administrative processes

The five-tier ontology recognizes the existence of the physical reality independent of observers and human constructs derived from observations, generalization, social manipulation, and cognition. Such recognition allows philosophical integration of positivist views (Tier 0) to post-modern positions (Tier 3), discerning differences and promoting understanding. Within each tier, modes of Being can be defined by considering existence in space and time, measurements of existence, traces of existence, and social settings of existence.

Since each formal ontology of space and time circumscribes the conceptual bound of reality that will be considered in an information system, subscription to an ontology determines what should be represented and coded in a temporal GIS.

Spatio-temporal Representation and Data Modeling

GIS representation has followed the map metaphor that depicts geographic features as a set of static objects residing in a two-dimensional planar space. Hence, all geographic features are represented as static, 2D, and geometrically fixed objects. Comparable to SNAP entities, every object in 2D, static GIS is valid at a certain point in time, and objects indexed at different times are independent from each other even if they represent the same geographic entity in the world. While field-based representation concerns only properties at locations, rather than objects, the fact that properties are valid only at the time of measurement denotes the map-based and SNAP-nature of fields.

Map-based representation also limits the ways that we can analyze data in a GIS. While map-based representation greatly facilitates spatial overlay to reveal spatial relationships among geographic variables, problems occur when we try to perform 3D analysis or dynamic modeling. Three-dimensional visualization techniques cannot fully solve the problem because a true 3D application requires information that can only be derived from analyzing 3D topological relationships beyond simple visualizing 3D volumes of data. For example, a GIS must have capabilities to compute information about adjacency in vertical space to answer a 3D query for areas where sandstone lies on top of shalestone layers. Topological integrity forms the basic operations to manipulate and analyze data in 2D, 3D, or 4D GIS (Hazelton 1998) and cannot be overlooked in spatio-temporal representation. In addition, geographic features must be represented in ways that conform to proper analytic methods (Yuan, Mark, Peuquet, and Egenhofer 2004). For example, routing analysis requires a topologically sound transportation network with nodes representing cities or stops, while distributed modeling requires surface conditions to be represented in a grid to represent properties at regular and pre-defined locations (cells). These two representations have become known as feature and location-based representations, respectively, corresponding to object- and field-based ontologies. In addition, numerous regular and irregular tessellation models are used to represent the geographic space of fields (Frank and Mark 1991). Peuquet (1984) gives a penetrating analysis of conceptual frameworks used in GIS to represent geographic phenomena in a two-dimensional space. Abraham and Roddick (1996) offer a comprehensive review of spatio-temporal databases developed by computer scientists and GI scientists.

However, geographic worlds are neither static nor planar. Incorporation of temporal components into a representation is not a trivial task because space and time have distinctive differences in philosophical, computational, and cognitive concerns, from which grows the complexity of spatio-temporal representation (Peuquet 2002). Since the mid- to late-1980s, researchers in both GIS and database management have been examining ways to incorporate time into information systems (Figures 9.1 and 9.2). In relational databases, the time-stamping method appears to be the most popular treatment of time by attaching time to a table or relation (Gadia and Vaishnav 1985), to a data object or tuple (Snodgrass and Ahn 1986), or to an attribute value or a cell (Gadia and Yeung 1988). In these time-stamp approaches, time is considered an intrinsic part of data and is attached by the information system automatically during data entry (Erwig, Guting, and Schneider 1999). An information system may stamp its data with valid time (historical databases), transaction time (rollback databases), or both (bitemporal databases). Historical databases denote when the data is valid in the real world. On the other hand, transaction databases allow retrospection of when data values were validated in the database so that data can be rolled back for editing or revisions. While the time stamp approach seems to serve non-spatial database systems well, it quickly reaches its limits when situations involve changes in space *and* time. Nevertheless, the time-stamping approach is simple to conceptualize and implement and, therefore, remains popular for adding time to GIS databases.

In GIS, time-stamping techniques have been applied to layers in the snapshot model, to attributes in the space-time composite model (Langran and Chrisman 1988), or to spatial objects in the spatio-temporal objects model (Worboys 1994). Different from the temporal databases discussed earlier, time stamps are given by the user, not the system, to cope with the fact that new data objects may be created from the old

1993

County Population	Avg. Income
1994 | Nixon   17,000   20,000

County Population	Avg. Income
Nixon   20,000   19,800
1995 | Cleveland   35,000   32,000

County Population	Avg. Income
Nixon   20,900   21,000
Cleveland   35,000   32,000
Oklahoma   86,000   28,000

a. Time-stamped tables (Gadia and Vaishnav 1985).

Stock	Price	From	To
IBM	16	10-7-91 10:07am	10-15-91 4:35pm
IBM	19	10-15-91 4:35pm	10-30-91 4:57pm
IBM	16	10-30-91 4:57pm	11-2-91 12:53pm
IBM	25	11-2-91 12:53pm	11-5-91 2:02pm

b. Time-stamped tuples (rows): an ungrouped relation
   (Snodgrass and Ahn 1985).

Name	Salary	Department
[11,60] John	[11, 49] 15K [50, 54] 20K [55, 60] 25K	[11,44] Toys [45, 60] Shoes
[0,20] U [41,51] Tom	[0, 20] 20K [41, 51] 30K	[0, 20] Hardware [41, 51] Clothing
[0,44] U [50, Now] Mary	[0,44] U [50, Now] 25K	[0,44] U [50, Now] Credit

c. Time-stamp values (cells): a group relation
   (Gadia and Yeung 1988). [11, 60] represents
   a period starting at T_{11} and ending at T_{60}.

Fig. 9.1 Examples of representations of temporal information in a relational data model
Adopted from Yuan 1999

objects over time (Erwig, Guting, and Schneider 1999). While the snapshot model presents the simplest way to incorporate time with space, it encounters problems of data redundancy and possible data inconsistency, especially in dealing with large data sets. The space-time composite model can eliminate these problems to a degree, but it has problems keeping spatial object identifiers persistent because updating space-time composites can cause fragmentation of existing spatial objects (Langran and Chrisman 1988). On the other hand, the spatio-temporal object model is able to maintain spatial object identifiers, but it, as in all time-stamping approaches, has difficulty representing dynamic information, such as transition, motion, and processes.

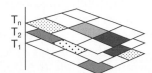

a. Time-stamped layers (Armstrong 1988).

Poly id	T_1	T_2	T_3	T_4
1	Rural	Rural	Rural	Rural
2	Rural	Urban	Urban	Urban
3	Rural	Rural	Urban	Urban
4	Rural	Rural	Urban	Urban
5	Rural	Rural	Rural	Urban

b. Time-stamped attributes (columns): Space-Time Composites
(Langran and Chrisman 1988).

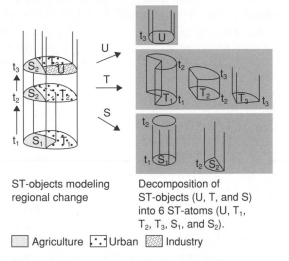

ST-objects modeling regional change

Decomposition of ST-objects (U, T, and S) into 6 ST-atoms (U, T_1, T_2, T_3, S_1, and S_2).

Agriculture ⋮ Urban ▨ Industry

c. Time-stamped space-time objects: the spatiotemporal object model
(Worboys 1994).

Fig. 9.2 Examples of representations of spatio-temporal information in a GIS environment
Adopted from Yuan 1999

Geographic information cannot be extracted from a system where the information cannot be represented. Hence, data models developed using the time-stamping approaches are incapable of supporting spatio-temporal queries about information on the dynamic characteristics of geographic processes, including movement, rate of movement, frequency, and interactions among processes. Geographic representation "*must deal with actual processes, not just the geometry of space-time*" (Chrisman 1998).

The most recent work on GIS representation has emphasized representation of dynamic processes. These models include Peuquet and Duan's (1995) event-based spatio-temporal data model (ESTDM), Raper and Livingstone's (1995) geomorphologic spatial model (OOgeomorph), and Yuan's (1994, 1999) three-domain model. ESTDM is conceptually simple and easily adaptable to other raster-based systems to represent information about locational changes at pre-defined cells along the passage of an event. Central to ESTDM is a chain of vectors that changes at locations (cells) on a raster of a single theme (such as temperature). While the model has shown its efficiency and capability to support spatial and temporal queries in raster systems, it will require a substantial redesign for use with vector-based data layers. On the other hand, OOgeomorph is a vector-based system designed to handle point data of time-stamped locations. The model starts with an object-oriented scheme of geomorphic features of interest. These geomorphic features emerge in the database when an appropriate selection of attributes is made from time-stamped points. For example, data objects of coastlines will be created as all points with an elevation equal to zero are selected. The key idea of OOgeomorph is that instances of spatio-temporal data objects are dynamically derived through selection of attributes, and therefore, object identity results from the interaction of space, time, and attribute. Questions remain as to OOgeomorph's ability to handle spatial objects of higher dimensions and its applicability to systems other than geomorphology.

The three-domain model attempts to capture a broad range of spatio-temporality independent of raster or vector data types and support for spatio-temporal queries (Yuan 1996). While developed as a separate venture, the three-domain model offers a general framework that accounts for histories at locations as in the space-time model, changes at locations along a passage of an event as in the ESTDM, and dynamic creation of object identities through attribute selections as in OOgeomorph. Fundamental to the three-domain model is the idea that geographic semantics (the geographic meaning that a representation attempts to portray), temporal properties, and spatial characteristics are three elements of geographic things to be modeled in a spatio-temporal representation (Figure 9.3). The semantic domain can include specific sets of ontological notations, semantic networks, or conceptual object-oriented models of geographic things, and their temporal and spatial properties are represented in the temporal and spatial domains respectively. Different ways of linking the three elements result in distinct spatio-temporal concepts being represented. For example, linking from spatial objects, through temporal objects, to semantic objects, represents histories at locations; that is how locations change attributes over time, which is the basis of the space-time composite model. When linking from semantic objects through temporal objects to spatial objects, the representation describes how geographic things change locations or geometric properties over time; comparable to what is represented in the ESTDM. As compared to OOgeomorph, the semantic domain corresponds to the object-oriented scheme that defines geographic entities of interest and their relationships. Selection of proper geographic attributes to form instances of geographic entities will be based on the definitions of these geographic entities in the semantic domain. Once attributes are identified, linkages will be made to temporal and spatial data in the other two domains to create spatio-temporal instances.

The three-domain model has shown its abilities to support a wide range of spatio-temporal queries; of particular interest are queries about spatio-temporal behaviors and relationships (Yuan 1999). The model is also implemented in building a database to trace backward information about failure and repairs of electrical transformers

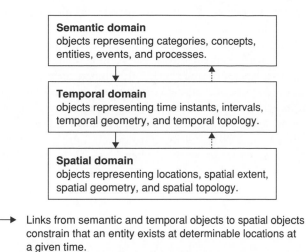

Links from semantic and temporal objects to spatial objects constrain that an entity exists at determinable locations at a given time.

Links from spatial and temporal objects to semantic objects constrain that a location has determinable geographic semantics at a given time.

There are no direct links from semantic to spatial objects because the model assumes that a geographic entity or attribute is contingent on time. A null temporal object can be set to handle data records without temporal measures.

Fig. 9.3 A conceptual framework for a three-domain model
Adopted from Yuan 1999

in a large electric network (Wakim and Chedid 2000) and has shown its ability to model dynamic changes and scenarios. Ideas from the three-domain model are also adopted in the Multidimensional Location Referencing System (MDLRS) data model (Koncz and Adams 2002). By dealing with objects that represent semantics, temporal properties, and spatial properties separately and with links among them to represent dynamic geographic entities in space and time, the MDLRS data model offers the ability and flexibility to integrate data from multiple sources and various transportation needs in multi-dimensional location referencing. Furthermore, the three-domain model is applied to represent dynamics of convective storms based on precipitation data and to develop a prototype system with capabilities to support queries about the development, movement, merger, split, and lifelines of storms in the database (Yuan 2001, Yuan and McIntosh 2003). In the storm application, objects of events, processes, sequences, and states constitute the semantic domain and, as in OOgeomorph, instances of these semantic objects are created by defined threshold values of precipitation.

By using events and processes to integrate space and time, geographic representation can embrace much richer semantics that reflect dynamic characteristics of reality. However, representing events and processes is not a trivial task even at the conceptual level because the interwoven relationships among space, time, and phenomena cannot be fully integrated without a thorough consideration of the fundamentals of geographic systems. Conceptually, it may appear simple to sort out geographic entities and their relationships. Complexity arises, however, from the influences of scale in space and time on entity identification. Consequently, the specification of and sustainability of entity relationships is sometimes challenging to track across spatial and temporal scales. Only since early 2000 has it been recognized that space and time should not always be seen as two orthogonal dimensions. Many researchers advocate using an integrated approach to model geographic reality via events or processes (for example, Peuquet and Duan 1995, Raper and Livingstone 1995, Egenhofer and Golledge 1998, Yuan 2001). Ultimately, events and processes are central to the understanding of geographic worlds. They constitute information of interest to many, and perhaps, the majority of applications and scientific inquiries.

Spatio-temporal Queries

The ability to retrieve data of interest from a massive data set is one of the most fundamental functions in any information system. In fact, it can be argued that query support is the primary driver for developing a database. What data can be retrieved from a database and how efficiently the data can be retrieved depends very much on the chosen ontologies, representation, and data models. For example, if a SNAP ontology is adopted, the GIS will not have the knowledge of lifelines and therefore will not be able to support access to data about lifelines. Likewise, if a space-time composite model is chosen, the GIS will not be able to support information query about movements. At the computational level, evaluation of a query is influenced by its query type and the algorithm designed to process this query type based on storage structure and indexes (Teraoka, Maruyama, Nakamura, and Nishida 1996, Tsotras, Jensen, and Snodgrass 1998).

Accordingly, many spatial or temporal query types identified by computer scientists or geographers are based on indexing or processing needs. In spatial queries, we have query types that center on certain geometry, spatial range, or selection methods. Specifically, spatial queries can be characterized as point query, range query, Boolean query (Knuth 1973, Samet 1989), geometric query (Sourina and Boey 1998), select-by-location, select-by-attribute, spatial joins (Rigaux, Scholl, and Voisard 2002), and query-by-sketch (Egenhofer 1997). Three additional spatial query types are recognized based on GIS computational needs: topological queries, set-theoretical queries, and metric queries (Floriani, Marzano, and Puppo 1993). With historical and rollback databases, temporal query types are categorized as snapshot, timeslice (Verma and Varman 1994), attribute-history, key-history (Verma and Vaishnav 1997), interval intersection (Kanellakis, Ramaswamy, Vengroff, and Vitter 1993), time-range, and bitemporal queries (Kumar, Tsotras, and Faloutsos 1998). As mentioned previously, new ontological beings emerge, such as *movement*, *split*, and *divergence*, when both space and time are under consideration, and the complexity of their organization and relationships grows exponentially (Yuan 2000). Consequently, a systematic way to identify query types for spatio-temporal information is critical to the development of a temporal GIS with adequate support for spatio-temporal queries.

Research on spatio-temporal queries is still in its infancy. An obvious reason is that there is no universal standard spatio-temporal data model, and therefore, full-scale spatio-temporal information systems (that is, temporal GIS) have not yet been developed. Nevertheless, there are several studies that attempt to design search algorithms to support a few specific spatio-temporal query types (Tsotras, Jensen, and Snodgrass 1998). Based on the supply of query predicates, Tsotras, Jensen, and Snodgrass (1998) note that spatio-temporal queries can be categorized as *selection-* based on a single data set or other operations that require more than one data set: *joins*, *unions*, *projections*, *aggregates*, *constraints*, and *differences*.

From around early in 2000, video search and data mining of mobile objects and sensor data have become key drivers in promoting research on spatio-temporal queries. To support searches for objects of interest in video databases, Kuo and Chen (1996) developed a content-based video query language utilizing the spatial and temporal relations of content objects as predicates. Later, algorithms were designed to match trajectory patterns to answer queries about moving objects (*trajectory queries*) in video databases (Li, Ozsu, and Szafron 1997). In addition to the trajectories of moving objects, there is great interest in video queries that return semantic associations among objects under spatial and temporal constraints. A Video Data Base Management System (VSBMS) is developed with a Logical Hypervideo Data Model and query language to support retrieval of video data with specifics to a wide range of spatial and temporal constraints (relations) and semantic descriptions (Jiang and Elmagarmid 1998). Most applications in tracking moving objects consider point-based objects. Querying moving objects that may involve geometric changes over time is very challenging. Attempts have been made to develop extensions to the spatial data model and query language to handle time-dependent geometries (Güting, Erwig, Jensen, et al. 2000).

In addition to trajectories of moving objects, some spatio-temporal queries seek patterns of change in databases, that is to say, spatio-temporal evolution. Djafri, Fernandes, Paton, and Griffiths (2002) referred to these as *evolution queries* because they aim to identify patterns of change throughout histories of the entities. They further classify evolution queries into *sequence queries* for a-spatial data, and *developments queries* for spatial data. With data from the longitudinal study by the UK Office for National Statistics, Djafri, Fernandes, Paton, and Griffiths (2002) develop algorithms to handle evolution queries of individuals over consecutive snapshots. Of significance to the work of Djafri, Fernandes, Paton, and Griffiths (2002) is the consideration of interactions between point-based individuals (persons) and region-based individuals (enumeration units) to seek aligned histories of these objects. Recently, a functional approach is taken to support evolution queries on thematic maps (d'Onofrio and Pourabbas 2003). While these studies collectively address a wide range of spatio-temporal queries, most, if not all, spatio-temporal queries in the literature deal with objects with uniform spatial properties. The assumption can be easily violated in geographic worlds because the properties of some entities may be spatially heterogeneous (Yuan 2001). Examples include storms, heat waves, oil spills, and pollution plumes, to name just a few. These objects have field-like properties (for example, temperature, chemical concentration, etc.), and changes in their field-like properties are central to the evolution of these objects.

With an emphasis on exploring and understanding geographic dynamics as represented by spatio-temporal data, Yuan and McIntosh (2002) have proposed a typology of geographic queries based on the information sought and relevance of spatio-temporal data mining. The typology includes one type of attribute query, three types of spatial query, three types of temporal query, and four types of spatio-temporal query as listed below:

1 *Attribute Queries* seek properties information about specific geographic objects or locations in a field
2 *Spatial Query types* seek geographic objects or locations based on criteria of specific points, ranges, or relations in space
   • Simple spatial queries
   • Spatial range queries
   • Spatial relationship queries
3 *Temporal Query types* seek geographic objects or locations based on criteria of specific points in time, periods, or temporal relations
   • Simple temporal queries
   • Temporal range queries
   • Temporal relationship queries
4 *Spatio-temporal Query types* seek geographic objects or locations based on criteria of specific properties, domains of space and time, spatio-temporal characteristics and spatio-temporal relations
   • Simple spatio-temporal queries
   • Spatio-temporal range queries
   • Spatio-temporal behavior queries
   • Spatio-temporal relationship queries

The most challenging of all queries are those about spatio-temporal behaviors and relationships for spatially heterogeneous (non-uniform) objects as discussed previously. In most cases, geographic objects considered in these queries are complex geographic phenomena, events, and processes (such as wildfires; see Yuan 1997 for further details about the information complexity of wildfire) that exhibit high degrees of geographic dynamics with indeterminate boundaries. Nevertheless, the most challenging query types have the greatest potential to offer new insights into what is embedded in a spatio-temporal database, and furthermore may help us probe for new scientific insight about these geographic worlds. For example, a spatio-temporal query may seek to retrieve events operating at a continental scale with behaviors correlating to droughts in the Midwest of the United States. The results may include some events that are not previously known to be correlates of droughts in the Midwest. Consequently, data mining and knowledge discovery methods can be applied to further validate the correlation in a large database, which may drive new scientific inquiry about the physical causes of the correlation.

CONCLUSIONS

Time is a critical element to geographic study, and GIS technology cannot be fully developed without abilities to handle both spatial and temporal data and to transform these data into spatio-temporal information. While the challenge is grand and multitudinal, a sound conceptual foundation serves as the bedrock upon which a fully fledged temporal GIS can be built.

This chapter has discussed the developments and research needs for ontologies, representation and data modeling, and information queries that constitute the conceptual foundation of a temporal GIS. In summary, research on geographic ontology has resulted in many important advances in our understanding of what constitutes geographies from empirical and cognitive perspectives, and in applications using ontology to improve interoperability of GIS data and feature identification on images. As to representation and data modeling, the general trend directs us to approaches that account for events and processes as integrals of space and time, in contrast to time-stamping approaches. As to spatio-temporal queries, most studies still consider only point-based objects in a 4D space in searching for trajectories of moving objects. Few studies consider two-dimensional spatial objects, like regions, and changes in their geometries over time. New query languages and algorithms have been developed and have shown capabilities to retrieve information about histories of individual spatio-temporal objects and intersections of their histories in space and time.

Common to research on ontology, representation and data modeling, and query is diversity. There are no universally accepted geographic ontologies, representation and data models, or query languages in temporal GIS. The diversity, on the one hand, indicates that this research field is thriving, but on the other hand, we need concerted efforts to ensure sustained progress in temporal GIS. This chapter identified several common threads (i.e. transitions) among the approaches, which may serve as the basis for future integration of the different schools of thoughts. In addition, suggestions have been made for future research in the consideration of complex geographic objects: their ontological implications, representation and data modeling, and information query support.

REFERENCES

Abraham, T. and Roddick, J. F. 1996. *Survey of Spatio-temporal Databases*. Adelaide, University of South Australia, School of Computer and Information Science, Advanced Computing Centre Technical Report No. CIS 96-011.

Chrisman, N. R. 1998. Beyond the snapshot: Changing the approach to change, error, and process. In M. J. Egenhofer and R. G. Golledge (eds) *Spatial and Temporal Reasoning in Geographic Information Systems*. New York: Oxford University Press, pp. 85–93.

Couclelis, H. 1992. People manipulate objects (but cultivate fields): Beyond the raster–vector debate in GIS. In A. U. Frank and U. Formentini (eds) *Theories and Methods of Spatio-temporal Reasoning in Geographic Space*. Berlin: Springer-Verlag Lecture Notes in Computer Science No. 639: 65–77.

Cova, T. J. and Goodchild, M. F. 2002. Extending geographical representation to include fields of spatial objects. *International Journal of Geographical Information Science* 16: 509–32.

Djafri, N., Fernandes, A. A. A., Paton, N. W., and Griffiths, T. 2002. Spatio-temporal evolution: Querying patterns of change in databases. In *Proceedings of the Tenth ACM International Conference on Advances in Geographic Information Systems*, McLean, VA, USA. New York: Association of Computing Machinery, pp. 35–41.

d'Onofrio, A. and Pourabbas, E. 2003. Modeling temporal thematic map contents. *ACM SIGMOD Record* 32(2): 31–41.

Egenhofer, M. J. 1997. Query processing in spatial-query-by-sketch. *Journal of Visual Languages and Computing* 8: 403–24.

Egenhofer, M. J. and Golledge, R. G. 1998. *Spatial and Temporal Reasoning in Geographic Information Systems, edited.* New York: Oxford University Press.

Erwig, M., Guting, R. H., and Schneider, M. 1999. Spatio-temporal data types: An approach to modeling and querying moving objects in databases. *Geoinformatica* 3: 269–96.

Feather, N. 1959. *An Introduction to the Physics of Mass, Length, and Time.* Chicago, IL: Aldine.

Fellbaum, C. (ed.). 1998. *WorldNet: An Electronic Lexical Database for Language, Speech, and Communication.* Cambridge, MA: MIT Press.

Floriani, L. D., Marzano, P., and Puppo, E. 1993. Spatial queries and data models. In A. U. Frank and I. Campari (eds) *Spatial Information Theory: A Theoretical Basis for GIS.* Berlin: Springer-Verlag Lecture Notes in Computer Science No. 716: 113–38.

Frank, A. U. 1997. Spatial ontology: A geographical point of view. In O. Stock (ed.) *Spatial and Temporal Reasoning.* Dordrecht: Kluwer, pp. 135–53.

Frank, A. U. 1998. Different types of "Times" in GIS. In M. J. Egenhofer and R. G. Golledge (eds) *Spatial and Temporal Reasoning in Geographic Information Systems.* New York: Oxford University Press, pp. 40–62.

Frank, A. U. 2003. Ontology for spatio-temporal databases. In M. Koubarakis, T. K. Sellis, A. U. Frank, S. Grumbach, R. H. Güting, C. S. Jensen, N. Lorentzos, Y. Manolopoulos, E. Nardelli, B. Pernici, H.-J. Schek, M. Scholl, B. Theodoulidis, and N. Tryfona (eds) *Spatio-temporal Databases: The Chorochronos Approach.* Berlin: Springer Lecture Notes in Computer Science No. 2520: 9–77.

Frank, A. U. and Mark, D. M. 1991. Language issues for GIS. In D. J. Maguire, M. F. Goodchild, and D. W. Rhind (eds) *Geographical Information Systems: Principles and Applications* Vol. 1. Harlow: Longman, pp. 147–63.

Gadia, S. K. and Vaishnav, J. H. 1985. A query language for a homogeneous temporal database. In *Proceedings of the ACM Symposium on Principles of Database Systems*: New York: Association of Computing Machinery, pp. 51–6.

Gadia, S. K. and Yeung, C. S. 1988. A generalized model for a relational temporal database. In *Proceedings of the ACM SIGMOD International Conference on Management of Data*: New York: Association of Computing Machinery, pp. 251–9.

Grenon, P. and Smith, B. 2004. SNAP and SPAN: Towards dynamic spatial ontology. *Spatial Cognition and Computation* 4: 69–103.

Guarino, N. 1998. Formal ontology and information systems. In N. Guarino (ed.) *Formal Ontology in Information Systems.* Amsterdam: IOS Press: 3–15.

Güting, R. H., Böhlen, M. H., Erwig, M., Jensen, C. S., Lorentzos, N. A., Schneider, M., and Vazirgiannis, M. 2000. A foundation for representing and querying moving objects. *ACM Transactions on Database Systems* 25: 1–42.

Hazelton, N. W. J. 1998. Some operational requirements for a multi-temporal 4D GIS. In M. J. Egenhofer and R. G. Golledge (eds) *Spatial and Temporal Reasoning in Geographic Information Systems.* New York: Oxford University Press, pp. 63–73.

Jakle, J. A. 1972. Time, space and the geographic past: A prospectus. *American History Reviews* 77: 1084–103.

Jiang, H. and Elmagarmid, A. K. 1998. Spatial and temporal content-based access to hyper-video databases. *International Journal on Very Large Data Bases* 7: 226–38.

Kanellakis, P. C., Ramaswamy, S., Vengroff, D. E., and Vitter, J. S. 1993. Indexing for data models with constraints and classes. In *Proceedings of the Twelfth ACM Symposium on Principles of Database Systems*, Washington, DC, USA. New York: Association of Computing Machinery, pp. 233–43.

Knuth, D. E. 1973. *The Art of Computer Programming: Volume1, Fundamental Algorithms.* Reading, MA: Addison-Wesley.

Koncz, N. and Adams, T. 2002. A data model for multi-dimensional transportation applications. *International Journal of Geographical Information Science* 16: 551–69.

Kumar, A., Tsotras, V. J., and Faloutsos, C. 1998. Designing access methods for bitemporal databases. *IEEE Transactions on Knowledge and Data Engineering* 10: 1–20.

Kuo, T. C. T. and Chen, A. L. P. 1996. A content-based query language for video databases. In M. Sakanchi and W. Klas (eds) *Proceedings of the Third IEEE International Conference on Multimedia Computing and Systems*. Los Alamitos, CA, USA. IEEE Computer Society Press: 209–14.

Langran, G. and Chrisman, N. R. 1988. A framework for temporal geographic information. *Cartographica* 25: 1–14.

Li, J. Z., Ozsu, M. T., and Szafron, D. 1997. Modeling of moving objects in a video database. In N. D. Georganas (ed.) *Proceedings of the Fourth IEEE International Conference on Multimedia Computing and Systems*. Los Alamitos, CA, USA. IEEE Computer Society Press: 336–43.

Mark, D. M., Skupina, A., and Smith, B. 2001. Features, objects, and other things: Ontological distinctions in the geographic domain. In D. Montello (ed.) *Spatial Information Theory*. Berlin: Springer Lecture Notes in Computer Science No. 2205: 488–502.

Mark, D. M., Smith, B., and Tversky, B. 1999. Ontology and geographic objects: An empirical study of cognitive categorization. In C. Freksa and D. M. Mark (ed.) *Spatial Information Theory: A Theoretical Basis for GIS*. Berlin: Springer Lecture Notes in Computer Science No. 1661: 283–98.

Peuquet, D., Smith, B., and Brogaard, B. O. 1998. *The Ontology of Fields: Report of the Specialist Meeting held under the auspices of the Varenius Project, Bar Harbor, Maine*. Santa Barbara, CA: National Center for Geographic Information and Analysis.

Peuquet, D. J. 1984. A conceptual framework and comparison of spatial data models. *Cartographica* 21: 66–113.

Peuquet, D. J. 2002. *Representation of Space and Time*. New York: Guilford.

Peuquet, D. J. and Duan, N. 1995. An event-based spatio-temporal model (ESTDM) for temporal analysis of geographical data. *International Journal of Geographical Information Systems* 9: 7–24.

Raper, J. 2000. *Multidimensional Geographic Information Science*. London: Taylor and Francis.

Raper, J. and Livingstone, D. 1995. Development of a geomorphologic spatial model using object-oriented design. *International Journal of Geographical Information Systems* 9: 359–84.

Rigaux, P., Scholl, M., and Voisard, A. 2002. *Spatial Databases with Application to GIS*. San Francisco, CA: Morgan Kaufmann.

Samet, H. 1989. *The Design and Analysis of Spatial Data Structures*. Reading, MA: Addison-Wesley.

Smith, B. 1998. The basic tools of formal ontology. In N. Guarino (ed.) *Formal Ontology in Information Systems*. Amsterdam: IOS Press, pp. 19–28.

Smith, B. and Mark, D. M. 1998. Ontology and geographic kinds. In *Proceedings of the Eighth International Symposium on Spatial Data Handling (SDH'98)*, Vancouver, Canada, pp. 308–20.

Smith, B. and Mark, D. M. 2001. Geographic categories: An ontological investigation. *International Journal of Geographical Information Science* 15: 591–612.

Snodgrass, R. and Ahn, I. 1986. Temporal databases. *IEEE Computer* September: 35–42.

Sourina, L. and Boey, S. H. 1998. Geometric query types for data retrieval in relational databases. *Data and Knowledge Engineering* 27: 207–29.

Teraoka, T., Maruyama, M., Nakamura, Y., and Nishida, S. 1996. The multidimensional persistent tree: A spatio-temporal data management structure suitable for spatial search. *Systems and Computers in Japan* 27: 60–72.

Terenziani, P. 1995. Towards a causal ontology coping with the temporal constraints between causes and effects. *International Journal of Human-Computer Studies* 43: 847–63.

Terenziani, P. 2002. Toward a unifying ontology dealing with both user-defined periodicity and temporal constraints about repeated events. *Computational Intelligence* 18: 336–85.

Tsotras, V., Jensen, C., and Snodgrass, R. 1998. An extensible notation for spatio-temporal index queries. *SIGMOD Record* 27: 47–53.

Ullman, E. L. 1974. Space and/or time: Opportunity for substitution and prediction. *Transactions of the Institute of British Geographers* 62: 125–39.

Verma, R. M. and Vaishnav, J. H. 1997. An efficient multiversion access structure. *IEEE Transactions on Knowledge and Data Engineering* 9: 391–409.

Verma, R. M. and Varman, P. J. 1994. Efficient archivalable time index: a dynamic indexing scheme for temporal data. In *Proceedings of the International Conference on Computer Systems and Education.*

Wakim, T. and Chedid, F. B. 2000. On the reconstruction of old versions from a spatio-temporal database. In *Proceedings of the Twentieth ESRI International User Conference*, San Diego, CA, USA. (Available at gis.esri.com/library/userconf/proc00/professional/papers/PAP458/p458.htm.)

Worboys, M. F. 1990. *Reasoning about GIS Using Temporal and Dynamic Logics.* Santa Barbara, CA: National Center for Geographical Information and Analysis.

Worboys, M. F. 1994. A unified model of spatial and temporal information. *Computer Journal* 37: 26–34.

Yuan, M. 1994. Wildfire conceptual modeling for building GIS space-time models. In *Proceedings of GIS/LIS '94*, Phoenix, AZ, USA, pp. 860–9.

Yuan, M. 1996. Modeling semantical, temporal, and spatial information in geographic information systems. In M. Craglia and H. Couclelis (ed.) *Geographic Information Research: Bridging the Atlantic.* London: Taylor and Francis, pp. 334–7.

Yuan, M. 1997. Knowledge acquisition for building wildfire representation in Geographic Information Systems. *International Journal of Geographic Information Science* 11: 723–45.

Yuan, M. 1999. Use of a three-domain representation to enhance GIS support for complex spatio-temporal queries. *Transactions in GIS* 3: 137–59.

Yuan, M. 2000. Representation of dynamic geographic phenomena based on hierarchical theory. In *Proceedings of the Ninth International Symposium on Spatial Data Handling Spatial Data Handling SDH'00*, Beijing, People's Republic of China.

Yuan, M. 2001. Representing complex geographic phenomena with both object- and field-like properties. *Cartography and Geographic Information Science* 28: 83–96.

Yuan, M., Mark, D., Peuquet, D., and Egenhofer, M. 2004. Extensions to geographic representations. In R. McMaster and L. Usery (eds) *Research Challenges in Geographic Information Science.* Boca Raton, FL: CRC Press, pp. 129–56.

Yuan, M. and McIntosh, J. 2002. A typology of spatio-temporal information queries. In K. Shaw, R. Ladner, and M. Abdelguerfi (eds) *Mining Spatio-temporal Information Systems.* Berlin: Kluwer Academic Publishers, pp. 63–82.

Yuan, M. and McIntosh, J. 2003. Weather intelligence: A GIS approach to enrich weather/climate databases. In *Proceedings of the Eighty-third Annual Meeting of the American Meteorological Society*, Long Beach, CA. New York: Association of Computing Machinery.

Chapter 10

Geospatial Data Integration

Craig A. Knoblock and Cyrus Shahabi

The problem of integrating geospatial data is ubiquitous since there is so much geospatial data available and such a variety of geospatial formats. The commercial world has numerous products that allow one to combine data that is represented in the myriad of geospatial formats and perform the conversion between products in these various formats. In an effort to improve the sharing and interoperability of geospatial information, the Open GIS Consortium (see http://www.opengis.org for additional details) has created the Geographic Markup Language (GML) to support the sharing of geographic information. GML provides an agreed-upon representation for publishing and using the various types of spatial data (for example, maps, vectors, etc.). As the interest in and use of geospatial data continues to grow, these efforts will be critical in exploiting the geospatial data that is available.

However, even in a world where geospatial products can be readily converted between different formats and geospatial information is published using agreed standards for interoperability, the problem is not fully solved. First, there may be sources that are not represented in any geospatial data format. For example, there are libraries of maps on the web that contain maps with only a textual description of the map and no metadata about the location or scale of the map. Second, there are many sources of online information that can be placed in a geospatial context, but the information is only available on a website as HTML pages, not in any standard geospatial format. Finally, even sources that may be available in one of the many standard formats may be difficult to integrate in a meaningful way due to differences in the resolution of the products, differences in the algorithms used to orthorectify the products, or just the lack of metadata on the products.

In this chapter we describe recent work on the extraction and integration of geo-spatial and geospatial-related data that go beyond conversion between different products and standard formats for the interoperability of these products. First, we describe techniques for turning online web sources into more structured sources where the information in these sources can then be integrated with other geo-spatial data. Then we present techniques for accurately and automatically integrating vector data with high-resolution color imagery. Today this type of alignment is performed manually by identifying a set of control point pairs across different products and then using rubber-sheeting techniques (Saalfeld 1993) to align the products, but this approach is very slow and labor intensive. Next, we describe techniques for exploiting online property tax sources and conflated road vector data to identify and annotate the buildings in an image. This goes beyond image processing, which may be able to identify an object as a building, but cannot provide any of the identifying information about the building. Then we present an approach that automatically integrates maps with unknown coordinates with satellite imagery. This makes it possible to exploit the data on a map to help label an image or vice versa. Finally, we present an approach to efficiently combine online schedules with vector data to predict the location of moving objects in the world, such as trains or buses.

Extracting Data from Online Sources

Beyond the traditional types of geospatial data sources, including satellite imagery, maps, vector data, elevation data, and gazetteers, there are many other sources of information available on the web that can be placed in a geospatial context. This includes sources such as property tax sites, telephone books, train, and bus schedules. The amount of such information is large and continues to grow at a rapid rate. The challenge is how to make effective use of all of this information and how to place it in a geospatial context. The first step is to turn the online sources that were intended for browsing by people into sources that can be effectively integrated with the more traditional types of geospatial sources.

To address this challenge, researchers have developed machine learning techniques for rapidly converting online web sources into sources that can be queried as if they were databases (Kushmerick 1997, Hsu and Dung 1998, Knoblock, Lerman, Minton, and Muslea 2003). These techniques greatly simplify the problem of turning web pages into structured data. The user provides examples of the information to be extracted and the system learns a wrapper that can dynamically extract data from an online source or convert an online source into a database. A wrapper is defined by a set of extraction rules that are specific to extracting the data from a particular website. The extraction rules specify how to locate specific types of information from a page and these rules must work over the potentially large number of pages available on a given website. The machine learning techniques developed for this problem are designed to produce highly accurate extraction rules with a minimum number of training examples. The extraction rules for many websites can be learned with just a few examples.

Figure 10.1 shows the property tax site for New York State. This site contains detailed property information such as name, address, lot size, and date of purchase for all of the properties in the state. In order to exploit this information for geospatial data integration we would first need to build a wrapper that provides programmatic access to the data. This is accomplished by providing examples of the data to be extracted from several example pages on this site. The system then learns the set of extraction rules for the site and uses these rules to construct a wrapper tailored to this website. This wrapper can take a request, such as to return the properties of everyone named "Smith" in "Syracuse," and will return the information in a structured format, such as the XML document shown at the bottom of Figure 10.1.

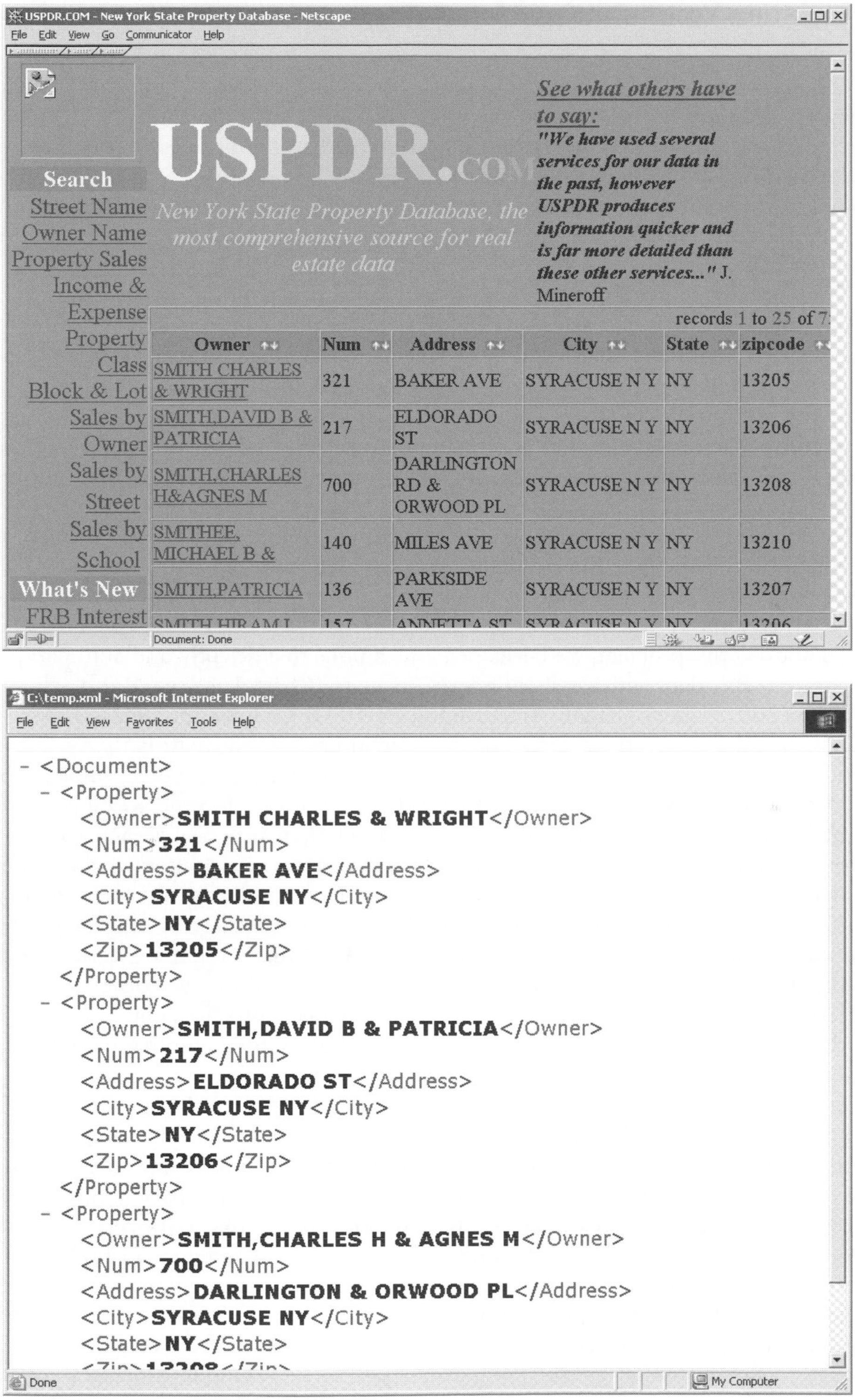

Fig. 10.1 A wrapper converts the New York State Property Database into structured data that can then be integrated with other sources

Integrating Vector Data and Imagery

When combining different geospatial data sources, such as vector data and satellite imagery, a critical problem is that the products do not correctly align. This problem is caused by the fact that geospatial data obtained from various sources may use different projections, may have different accuracy levels, and may have been corrected in different ways. The applications that integrate information from various geospatial data sources must be able to overcome these inconsistencies accurately and for large regions.

Traditionally, this problem has been solved in the domain of image processing and Geographic Information Systems (GIS). The focus of image processing techniques has been on automatic identification of objects in the image (Auclair-Fortier, Ziou, Armenakis, and Wang 2000; see also http://iris.usc.edu/Vision-Notes/bibliography/ contents.html, topic 21, for a comprehensive bibliography of work on automatic extraction of road networks) in order to resolve vector-image inconsistencies. However, these techniques require significant CPU time to process an image in its entirety and still may result in inaccurate results. The primary approach used in most GIS is to require a user to manually identify a set of control point pairs and then to use a technique called conflation (Saalfeld 1993) to align two geospatial data sets. The need to manually identify control point pairs means that this approach does not scale up to large regions.

To address this problem, we developed a technique to efficiently and automatically integrate vector data with satellite or aerial imagery (Chen, Thakkar, Knoblock, and Shahabi 2003, Chen, Shahabi, and Knoblock 2004b). Our approach is based on two important observations. First, there is a great deal of information that is known about a given location beyond the data that is to be integrated. For example, most road-network vector data sets also contain the road direction, road width and even location of the road intersections. Second, rather than processing each source of information in isolation, it is much more effective to apply what is known about these sources to help in the integration of two sources. For example, the information about road directions and road width can be used to help locate the corresponding intersections in the satellite imagery. In the remainder of this section we discuss some of the details of our technique to show how we utilized these two observations to effectively incorporate the image processing techniques into the conflation process so that the resulting approach is both more effective and fully automatic (and hence scalable to large regions).

To explain our approach, we first need to explain the conflation process. The conflation process divides into the following tasks: (1) find a set of conjugate point pairs, termed "control point pairs," in both the vector and image data sets, (2) filter the control point pairs, and (3) utilize algorithms, such as triangulation and rubber-sheeting, to align the remaining points and lines in the two data sets using the control point pairs. Traditionally, human input has been essential to find control point pairs and/or filter control points. Instead, we developed completely automatic techniques to find control point pairs in both data sets and designed novel filtering techniques to remove inaccurate control points. We developed two different techniques to find accurate control point pairs. Our first technique generates control points using localized image processing. The second technique finds control points by querying information from online web sources about known locations in the imagery. Because of lack of space, we only briefly describe the first technique, which relies only on the imagery and vector data for accurate integration.

We first find feature points, such as the road intersection points from the vector data set. For each intersection point, we perform image processing in a small area around the intersection point to find the corresponding point in a satellite image. This is an example of the first observation by exploiting what is known about the location, size, and orientation of the intersections and of the second observation by applying this information during the image processing task to focus the processing on the small area around the location of the intersection on the image (rather than processing the entire image). In addition to the approximate location of intersections, to locate intersections on the images, we also utilize the information inferred from vector data such as road directions, widths and shapes. In particular, we generate a template inferred from all the vector information and then match it against the small area in the image to find the corresponding intersection point on the imagery. This process is then repeated for every candidate intersection. Finally, during an automatic filtering step, we eliminate those intersection point pairs that do not agree with the majority of pairs (that is, outliers). In contrast to techniques for simply extracting road networks from imagery our technique does not need to locate all of the intersection points in order to accurately align the vector data with the imagery. The remaining steps are the same as those of the traditional conflation process.

An example of the results of this technique is shown in Figure 10.2. The running time for this approach is dramatically lower than traditional image processing techniques due to the more focused image processing. Furthermore, the road direction and width information makes detecting edges in the image a much easier problem, thus reducing the running time further.

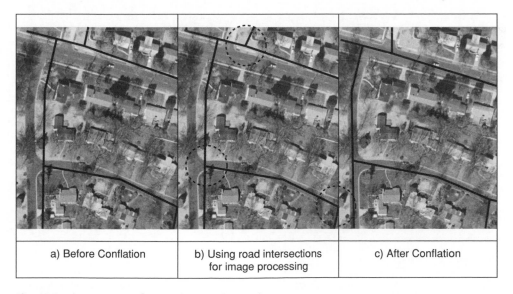

| a) Before Conflation | b) Using road intersections for image processing | c) After Conflation |

Fig. 10.2 Automatic conflation of vector data with imagery

Identifying Structures in Imagery

We can exploit the various online sources of data to identify the structures (that is, buildings) in a satellite image. The online property tax records are an especially rich source of data and, as described above, these can be turned into structured sources that can be integrated with other data sources. In the previous section we described how we can automatically identify the roads in a satellite image. In this section we describe how we can combine this information with the property tax records to identify the buildings in an image.

The traditional approach to locate a house is to use a geocoder, which maps street addresses into latitude and longitude coordinates. Current geocoders perform this mapping by using street vector data that is annotated with the address ranges and then interpolating the address within the range. This approach provides inaccurate results (for example, Ratcliffe 2001) since it assumes that the address ranges are fully populated (that is, there are 50 houses on each side of the street) and that all lots are of equal size, but this is rarely the case. Since the commercial geocoders are not accurate enough to precisely identify the houses in an image, instead we combine the information that is known about the houses on a block to determine the precise identity of each house.

Figure 10.3 illustrates how the various sources of information can be fused to precisely identify the houses in an image (Bakshi, Knoblock, and Thakkar 2004). We can combine the satellite imagery from a source such as terraserver.com, the

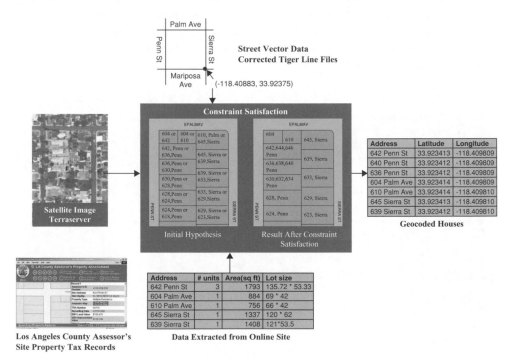

Fig. 10.3 Integrating and reasoning about the property tax data, satellite imagery, and road vector data to identify the structures in an image

street vector data, which has been aligned with the imagery using the techniques described in the previous section, and the property tax records for the given region. The satellite imagery will show where the houses are located on a block. The street vector data will provide the geographic coordinates of the block as well as the overall dimensions of a block. Finally, the property tax site will provide the details on each property including the street address and lot size.

The critical piece of information that is missing from the property tax site is the exact location of the lots. We can tell how many houses are on a street, but the corner lots create a problem since we do not know which street each of the corner houses is listed under. To solve this problem we treat this as a constraint satisfaction problem (Russell and Norvig 1995), which will consider the possible orientations of the corner lots and find the layout that is closest to the actual dimension of the entire block. This is illustrated in the center of Figure 10.3, which shows the initial hypothesis where there is uncertainty about which houses are on the corners and the result after running the constraint satisfaction. Once the exact layout of the block has been determined, the lots can be accurately geocoded and the buildings in the original image can be accurately identified.

Integrating Maps and Imagery

There are a wide variety of maps available from various sources, such as the US Geological Survey, University of Texas Map Library, and various government agencies. These maps include street maps, property survey maps, maps of oil and natural gas fields, and so on. However, for many of these maps, the geographic coordinates and scale of the maps are unknown. Even if this information is known, accurately integrating maps and imagery from different data sources remains a challenging task. This is because spatial data obtained from various data sources may have different projections and different accuracy levels. If the geographic projections of these data sets are known, then they can be converted to the same geographic projections. However, the geographic projection for a wide variety of geospatial data available on the Internet is not known. To address this problem, we built on our previous work on automatic vector to image conflation and developed efficient techniques to the problem of automatically conflating maps with satellite imagery (Chen, Knoblock, Shahabi, Thakkar, and Chiang 2004a).

To tackle this integration task, we continue to rely on the two observations discussed in the previous section: (1) utilizing all of the information known about a given location; and (2) exploiting this information to integrate the products. Using the first observation, we also consider the vector data of the roads in addition to the map and imagery that we want to integrate. With the second observation, we apply the vector data to help locate the intersection points on both the imagery and maps. The common vector data serves as the "glue" to integrate these two sources.

The steps of our approach are illustrated in Figure 10.4. First, we utilize the techniques described above to align the road vector data with the imagery to identify the intersection points on the imagery. Then we apply techniques for identifying the intersections on maps (Sebok, Roemer, and Malindzak 1981, Musavi, Shirvaikar, Ramanathan, and Nekovei 1988), which we have extended to support maps with

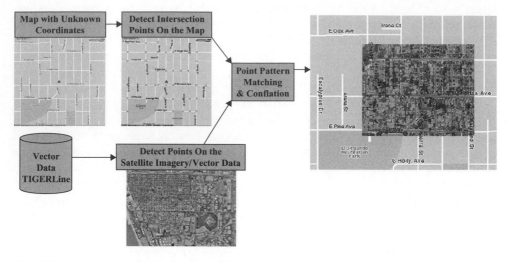

Fig. 10.4 Automatic conflation of maps with imagery

double-lined roads and maps with lots of extraneous data such as on topographic maps. Next, we apply a specialized point matching algorithm (Irani and Raghavan 1999) to compute the alignment between the two sets of intersection points. This matching problem is challenging because of the potential of both missing and extraneous intersection points from the map intersection detection algorithms. Finally, we use the resulting set of control point pairs to automatically conflate the map and image.

Experimental results on the city of El Segundo, California demonstrate that our approach leads to remarkably accurate alignments of maps and satellite imagery. The aligned map and satellite imagery supports inferences that could not have been made from the map or imagery alone. Figure 10.5 shows an example of our results for the city of El Segundo.

Fig. 10.5 Results of conflating MapQuest map with imagery

Integrating Online Schedules (Moving Objects) with Vectors

This final example of geospatial data integration is rather different than the previous cases in that it integrates spatial sources with temporal ones. In particular, we study the integration of vector data sets, for example, train tracks or road networks, with routing schedules such as train or bus schedules. In the database literature, the querying of these specific types of spatio-temporal sources is sometimes referred to as "moving object queries" or "spatio-temporal range queries". In previous work (Shahabi, Kolahdouzan, and Sharifzadeh 2003), we investigated these queries. However, one of our previous studies (Shahabi, Kolahdouzan, Thakka, Ambite, and Knoblock 2001) stands out in that it is the only one we are aware of that focuses on the "integration" challenges in efficient support of queries on spatio-temporal sources.

In Shahabi Kolahdouzan, Thakkar, Ambite, and Knoblock (2001), we assumed a mediator-based web architecture, where some of the information sources contain spatial and temporal data. For example, a temporal source may be a wrapped website providing train schedule information, while a spatial source is a database containing railroad vector data. A spatio-temporal range query would then impose bounds on spatial and temporal attributes and ask for all tuples satisfying the constraints. For example, given a point on the vector data and a time interval, we would like to find all the trains that would pass that point in the given time interval. The user interface of this application is shown in Figure 10.6, where the railroad vector data is drawn on a map and the point on the vector data is shown with an "x." The bottom part of Figure 10.6 lists all of the trains that will pass the point and the estimated time they will reach that point.

There are two main integration challenges with this application. First, as in the previous cases, we need an accurate alignment of railroad tracks with maps and/ or imagery. Here we can utilize the techniques described in previous sections by using the train stations as control point pairs. The other challenge is the efficient integration of spatial and temporal data to answer spatio-temporal queries.

Evaluation of spatio-temporal range queries on distributed sources is time consuming because of the complex computational geometry functions (for example, the shortest path function) that need to be executed on the large volume of vector data as well as the temporal intersections that need to be applied among large sets of time intervals. One solution to reduce the query processing time of spatio-temporal range queries is to pre-compute the required information and materialize it using a moving object data model such as the 3D Trajectory model (Vazirgiannis and Wolfson 2001). This is a feasible approach if we assume that different schedule, railroad, and station information is all local and something over which we have full control. However, with our assumed distributed environment, the sources of information that we would like to access are autonomous and dynamic.

Therefore, we investigated alternative distributed query plans to realize the integration of spatial and temporal information (for example, for the railroad network and train schedules) from distributed, heterogeneous web sources. One approach to this problem is to first look into the spatial source (containing railroads) and filter out only the railroad segments that overlap with the window query (spatial filter).

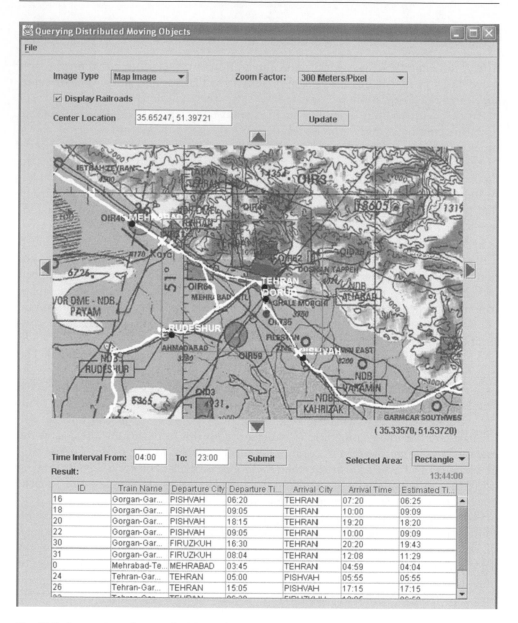

Fig. 10.6 Integration of train schedules with vector data and maps

Next, for those qualifying segments, we check the temporal source (containing train schedules) and find the trains passing through the segments during the query time interval (temporal filter). Another approach is to do the same in the opposite order. We investigated both of these traditional filter+semi-join plans by applying the temporal filter first and then performing the spatial semi-join or vice versa. However, we showed that there are two significant drawbacks with both these plans.

Instead, we introduced a novel spatio-temporal filter (termed deviation filter), which can exploit the spatial and temporal characteristics of the data simultaneously to improve the selectivity.

CONCLUSIONS

In this chapter we described a set of techniques for integrating geospatial data. These techniques are by no means comprehensive, but provide a set of examples of how the wide range of geospatial data can be combined in novel ways. The techniques for extracting data from online sources make available a wide range of data from online sources that can now be integrated with geospatial data sources. The approach to automatically aligning road vector data with imagery makes it possible to accurately identify roads in imagery and shows how we can combine different sources of information (that is, what is known about the vector and what can be extracted from the imagery) to automate difficult tasks. The integration of the constraint satisfaction techniques with the property tax data, imagery, and vector data provides an approach to identify the structures in imagery and shows how the combination of diverse types of data can provide new information. The automatic alignment of street maps with imagery provides an approach to exploiting maps that lack geospatial coordinates and is an example of how seemingly incompatible sources can be fused to provide new insights into the individual data products. Finally, the combination of online schedules with railroad vectors provides an approach to efficiently predicting the location of moving objects and provides another example of inferring new information from diverse sources of geospatial data. In sum, the techniques presented in this chapter are illustrative of the many possible ways of integrating the wide range of geospatial data that are available today.

REFERENCES

Auclair-Fortier, M. F., Ziou, D., Armenakis, C., and Wang, S. 2000. *Survey of Work on Road Extraction in Aerial and Satellite Images*. Sherbrooke, Quebec: Université de Sherbrooke, Département de Mathématiques et d'Informatique Technical Report No. 247.

Bakshi, R., Knoblock, C. A., and Thakkar, S. 2004. Exploiting online sources to accurately geocode addresses. In *Proceedings of the Twelfth ACM International Symposium on Advances in Geographic Information Systems (ACM-GIS '04)*, Washington, DC, USA.

Chen, C. C., Thakkar, S., Knoblock, C. A., and Shahabi, C. 2003. Automatically annotating and interpreting spatial datasets. In T. Hadzilacos, Y. Manolopoulos, J. F. Roddick, and T. Theodoridis (eds) *Advances in Spatial and Temporal Databases: Proceedings of the Eighth International Symposium (SSTD 2003), Santorini Island, Greece*. Berlin: Springer Lecture Notes in Computer Science No. 2750: 469–488.

Chen, C. C., Knoblock, C. A., Shahabi, C., Thakkar, S., and Chiang, Y. Y. 2004a. Automatically and accurately conflating orthoimagery and street maps. In *Proceedings of the Twelfth ACM International Symposium on Advances in Geographic Information Systems (ACM-GIS '04)*, Washington, DC, USA. New York: Association for Computing Machinery.

Chen, C. C., Shahabi, C., and Knoblock, C. A. 2004b. Utilizing road network data for automatic identification of road intersections from high resolution color orthoimagery. In *Proceedings of the Second Workshop on Spatio-Temporal Database Management (STDBM '04)*, Toronto, Canada. New York: Association for Computing Machinery, pp. 17–24.

Hsu, C. N. and Dung, M. T. 1998. Generating finite-state transducers for semi-structured data extraction from the web. *Information Systems* 23: 521–38.

Irani, S. and Raghavan, P. 1999. Combinatorial and experimental results for randomized point matching algorithms. *Computational Geometry* 12: 17–31.

Knoblock, C. A., Lerman, K., Minton, S., and Muslea, I. 2003. Accurately and reliably extracting data from the web: A machine learning approach. In L. A. Zadeh (ed.) *Intelligent Exploration of the Web*. Berkeley, CA: Springer-Verlag: 275–87.

Kushmerick, N. 1997. Wrapper induction for information extraction. Unpublished PhD Dissertation, Department of Computer Science and Engineering, University of Washington.

Musavi, M. T., Shirvaikar, M. V., Ramanathan, E., and Nekovei, A. R. 1988. A vision-based method to automate map processing. *Pattern Recognition* 21: 319–26.

Ratcliffe, J. H. 2001. On the accuracy of TIGER-type geocoded address data in relation to cadastral and census areal units. *International Journal of Geographical Information Science* 15: 473–85.

Russell, S. and Norvig, P. 1995. *Artificial Intelligence: A Modern Approach*. Englewood Cliffs, NJ: Prentice Hall.

Saalfeld, A. 1993. *Conflation: Automated Map Compilation*. College Park, MD: University of Maryland, Center for Automation Research, Computer Vision Laboratory.

Sebok, T. J., Roemer, L. E., and Malindzak, J. 1981. An algorithm for line intersection identification. *Pattern Recognition* 13: 159–66.

Shahabi, C., Kolahdouzan, M. R., Thakkar, S., Ambite, J. L., and Knoblock, C. A. 2001. Efficiently querying moving objects with pre-defined paths in a distributed environment. In *Proceedings of the Ninth ACM International Symposium on Advances in Geographic Information Systems (ACM-GIS)*, Atlanta, GA, USA. New York: Association for Computing Machinery, pp. 34–40.

Shahabi, C., Kolahdouzan, M. R., and Sharifzadeh, M. 2003. A road network embedding technique for k-nearest neighbor search in moving object databases. *Geoinformatica* 7: 255–73.

Vazirgiannis, M. and Wolfson, O. 2001. A spatiotemporal model and language for moving objects on road networks. In *Proceedings of the Seventh International Symposium on Spatial and Temporal Databases (SSTD)*, Redondo Beach, CA, USA. New York: Association for Computing Machinery, pp. 20–35.

Part III   Visualization

This third section of the book consists of seven chapters that explore some of the recent accomplishments and outstanding challenges concerned with the visualization of spatial data. The first chapter, Chapter 11 by William E. Cartwright, starts out by describing the digital pre-web endeavors of designers and cartographers and the various applications of new media (for example, hypermedia, videodisks, CD-ROMs) that led to the establishment of theories and practical methods for interactive multimedia map production. The role of traditional cartographic theory and practice and contributions of computer-assisted cartography and Geographic Information Systems are highlighted along with the challenges that new media pose for those involved in the design and provision of contemporary mapping products. The conclusion describes some of the products available on the Web for both expert and novice map users.

Chapter 12 by William A. Mackaness, the second chapter in this group, examines the role of generalization and scale in this digital age where the database is the knowledge store and the map is the metaphorical window by which geographical information is dynamically explored. The key generalization concepts, methods, and algorithms that have been proposed for creating and evaluating candidate solutions for graphical visualization and multiple representations are described. Mackaness (like Cartwright in the previous chapter) concludes by noting the importance of the art and science of cartography as well as the need for it to keep abreast of the changing environments of map use and analysis and broader developments in visualization methodologies if it is to remain relevant.

In the next chapter, Chapter 13, Nicholas J. Tate, Peter F. Fisher, and David J. Martin looks at some of the opportunities and challenges that are encountered when displaying and analyzing a variety of geographical phenomena as surfaces. The chapter starts with a discussion of the advantages of surface representation and moves quickly to explore the various types of surface model used in GIS given that surface modeling is a necessary precursor to surface analysis and visualization. The authors then move on to examine some of the more popular and powerful visualization tools afforded by surfaces, and they conclude their chapter by summarizing the unique place of surface representation and visualization in GIS.

Vincent B. Robinson describes the basic concepts underlying fuzzy set theory and their relationship to fuzzy classification and mapping in GIS in Chapter 14. The first part of the chapter summarizes the basic approaches to assigning a fuzzy membership value to a location and then visualizing that classification. The two basic approaches – fuzzy classification with a priori knowledge and data-driven fuzzy classification – are described in considerable detail along with visualization strategies and challenges. The latter sections of the chapter discuss the benefits and costs of fuzzy classification-based mapping systems and software availability to support fuzzy classification – topics that provide a nice conduit to the next chapter by A-Xing Zhu on predictive rule-based mapping.

Zhu focuses in Chapter 15 on predictive or rule-based mapping from a knowledge-based perspective that relies on the qualitative knowledge of human experts (that is, performing fuzzy classification with a priori knowledge). The implementation of rules in rule-based mapping under Boolean and fuzzy logic are compared and the chapter concludes by noting some of the enduring challenges and research issues that will need to be solved to propel rule-based mapping to a new era in the years ahead.

In the next chapter in this group (Chapter 16), Mark Gahegan looks at the opportunities for multivariate visualization. The chapter starts with a brief discussion of the need for multivariate analysis and the different approaches that have been utilized to analyze multivariate data. From there, Gahegan notes how the development of multivariate visualization can, and is, approached from a variety of perspectives and he lists a selection of these approaches as a set of motivating questions. The major methods of visualization and a series of visualization examples are then described in considerable detail. The chapter concludes with descriptions of some of the ways in which multivariate visualization might support analysis, the process of multivariate visualization, the technological tools that contribute to one or more geovisualization systems, and the problems that remain to be solved before geovisualization can fulfill its potential.

In Chapter 17, the final chapter in this group, Michael Batty examines the ways in which digital representations of two-dimemsional space can be enriched and augmented through interactivity with users in the third dimension and beyond. This chapter starts out by identifying the salient features of virtual reality (VR) and the ways in which GIS is being extended to embrace the third dimension, the question of time, and the media used to communicate this science to users with very different professional backgrounds and skills. The focus of VR in GI Science is then elaborated by examining different visualizations of 3-dimensional space in terms of geographic surfaces and geometric structures, illustrating the different media in which VR environments are constructed and showing how VR is beginning to form as emergent interface to GIS and potentially to GI Science. Batty concludes by speculating that as the digital solution deepens and matures virtual environments will become recursive, with many different renditions of the same underlying digital representation being used in a variety of ways with the same user interface.

Chapter 11

Mapping in a Digital Age

William E. Cartwright

Cartographers have always striven to make their products more accessible, current, and usable. Access to products was greatly enhanced when printing was embraced as a means for making faithful reproductions of geographic information artifacts. Printing changed how information was "packaged" and delivered, including geographic information. Early digital media trials led to the formulation of theories about how best interactive multimedia mapping products might "work" and how conventional practices might be modified or, in some cases, not used at all. Now the Web has also revolutionized how we address geographic information design and realization and, with users in mind, how these products are procured and utilized. What we do has changed and what we produce has also changed.

When compared to other areas of cartography Web mapping is new, but, relatively quickly, a new genre of mapping product has evolved. This has caused cartographers to assess how they need to design, produce, and deliver these new, innovative products, resulting in a new "mindset" about what they do and how they approach and conduct their activities. A new modus operandi (for mapping) has been established, one that facilitates the provision of new and exciting Web-delivered products.

This chapter addresses mapping in a digital age from the perspective of using New Media for the provision of cartographic artifacts. In doing so it acknowledges the important contributions of computer-assisted cartography and Geographic Information Systems (GIS), and how, in many instances, they have been "harnessed" to interactive multimedia to provide powerful analysis, production, and visualization tools. In order to appreciate the design and delivery task at hand when using the Web, it is appropriate to reflect upon what pre-empted Web mapping and some of the New Media foundations that were laid in the heady experimental days of hypermedia, videodisc, and CD-ROM. Interactive multimedia mapping was new and exciting, and many innovative products were designed and trialed. This chapter begins by describing the digital pre-Web endeavors of designer-cartographers and the various applications of New Media that led to the establishment of theories and practical methods for interactive multimedia map production. It then provides and overview of the range of products available on the Web for both expert and novice map users. Finally, the chapter addresses the challenges that New Media poses those involved in the design and provision of contemporary mapping products.

The Web

With the arrival of the Web, and the use of Berners-Lee's browser-driven information displays, a different, graphical access method to information was made available. The Web is an information discovery system for browsing and searching the Internet's worldwide "web" of digital information. There now exists almost instant access to geographic information, including maps. The extent and the use of the Internet have now matured to such a point that users see it as an everyday commodity or communications device. Education has embraced it for content delivery, face-to-face lecture support, and as a tool for students to keep in touch with academics and peers, as well as to conduct day-to-day administrative and general queries. Industry uses it as a tool for facilitating more effective logistical approaches. Commerce uses it as a means for linking their services, and business views it as a conduit for marketing, selling, and delivering (digital) products.

In *The UCLA Internet Report: Surveying the Digital Future* (Lebo 2003), the extent to which the Internet has been adopted can be seen. This report provides an annual survey of the impact of the Internet on the social, political and economic behavior of users and non-users of the Internet. In general, the report noted that

- Internet access remained generally stable from 2001 to 2002 and online hours increased, as did the use of the Internet at home;
- Use of the Internet spans all age groups;
- The Internet was seen as an important source of information (in 2002 over 60 percent of all users surveyed considered the Internet to be a very important or extremely important source of information);
- The Internet had increased the number of people that respondents communicated with;
- Use of the Internet for making purchases online had declined, but the average number of purchases made this way had increased; and
- There were growing numbers of people using the Internet for business purposes, from e-mails to business-to-business and business-to-customer transactions.

The Web can be "traveled" by following hypertext links from document to document that may reside in any of the many servers in different global locations. In 1993 the first Internet workshop on Hypermedia and Hypertext standards was held. At the end of 1994 there were almost 13 million users of the Internet. By late 1995 this number had risen to 23 million (plus an additional 12 million using electronic mail of various kinds) (Parker 1995). The Web grew one home page every four seconds and doubled every 40 days. It had 40 million plus users worldwide by early 1996. In terms of Web servers it grew from 130 in 1993 to an estimated 660,000 in 1997 (Peterson 1997), by late 1998 servers numbered over 3.5 million and in June 1999 there were just fewer than 6.2 million servers (Netcraft Web Server

Surveys 1998, 1999; see http://news.netcraft.com/ for additional details). In 2001 this grew to over 110 million Internet hosts (see http://navigators.com/statall.gif for additional details). In 2004 there were an estimated 945 million Internet users (Clickz Stats 2004), 1.08 billion in 2005, and with a projected "population" of 1.8 billion in 2010 (Clickz Stats 2006).

Pre-Web

A number of trials for using different media were tried pre-Web. Cartographers strived to "harness" New Media methods that could be employed to facilitate the storage and presentation of geographic information in more effective and efficient ways. This included

- Teletext/Videotext;
- Two-way interactive television;
- Hyperlinked television;
- Hypermedia;
- Videodisc;
- CD-ROM; and
- File transfer using the Internet.

Teletext/videotext

Television provides information via news services, documentaries, reports, live links, and film archives. For the general public it has become one of the most usable information resources that supply on-demand information right in the home. Television was seen as a way in which information could be readily disseminated through the use of teletext in its many forms, like Oracle and CEEFAX using the Prestel (Viewdata) system in the United Kingdom (a one-way system), Antiope in France in the mid-1970s, Telidon in Canada (1975) (a two-way version of the British Prestel system), Captain (1979–81) in Japan and Viatel in Australia. In the USA CBS experimented with a system called Extravision, but the system was never implemented.

In Australia, where distance has always been an important information provision issue, the use of teletext/videotext was trialed in the early days of Viatel as an adjunct to other education resources (Hosie 1985). British television ran very basic maps for the provision of information like weather maps. A typical screen page from CEEFAX is shown in Figure 11.1.

By far the most popular service of this type of information resource was Minitel, which replaced Antiope in France. France Telecom launched the system in February 1984 and it consisted of a low-cost dedicated terminal in the home or office or as an anonymous kiosk charging system. It provided a range of services through the use of coarse text and graphics (compared to today's standards) on color television screens, and among its information pages it included the French telephone directory. Minitel proved to be most popular and it saw a rapid growth in consumer interest. At its peak it had 15 million clients in France. A typical information screen is shown in Figure 11.2. The popularity of the Web supplanted Minitel somewhat, and its traffic

Fig. 11.1   CEEFAX weather map
From Pemberton 2004

Fig. 11.2   Minitel "kiosk"
From http://www.ust.hk/~webiway/content/France/history.html

plateaued around 1994 (see http:/www.ust.hk/~webiway/content/France/history.html for additional details). However, i-Minitel, France Telecom's Web information service provides similar information.

The potential of using videotext to provide geographic information and maps was recognized by Taylor (1984), who saw that the Canadian Telidon system could be a useful conduit of map images. Maps have been provided using Minitel by multimedia provider SGCI Planfax. This company has offered maps via the Minitel system since 1992. Now it provides multimedia maps via the Web, for the French Yellow Pages.

Two-way interactive television

Two-way interaction via television/PC/communications equipment is something that is always being talked about for homes of the future, allowing users/viewers to truly interact with broadcast television packages, including mapping packages. Microsoft developed their Tiger interactive television prototype server software in 1994, and later live television through the Internet using technology like Intel's Intercast, which combined the PC, television, and the Internet, and "streaming" techniques (a method whereby video is compressed as it is sent out over the Internet and then "played" by complementary software on the recipient's computer). Intercast simultaneously delivered text, graphics, video, or data to a PC (equipped with Intercast technology) along with a television signal. Content was created with HyperText Mark-up Language (HTML) and the television signal appeared to the user as a web page. Transmission of computer data would take place during the vertical blanking intervals of the television transmission image. The television needed to have a peripheral input device to receive television signals (cable, broadcast, or satellite), a digitizer to convert analog signals and a telephone-line modem to send data back through the Internet service provider (*Intercast: From Web TV to PC*; Advanced Imaging 1996). Now this is more readily realized with digital television.

Hyperlinked television

Consumers would probably argue that the only geographic information they get from television is weather maps and the occasional simple maps that accompany lifestyle programs and travel documentaries. However, pre-Web dominance of information delivery thinking, the use of broadcast television that provided hyperlinks to other types of information was one possible scenario explored theoretically. Negroponte (1996) considered that one possible future scenario for allowing consumers to interact with information resources was hyperlinked television, whereby "touching" an athlete's image on a television screen would produce relevant statistics, or touching an actor reveal that his tie is on sale this week. This would involve embedding extra information from a central database into broadcast television signals. Television could react according to the information delivery designer's intention when viewed under different circumstances. Negroponte (1996) saw Java contributing to the idea of hyperlinked television.

Hypermedia

Much interest was centered on the production of electronic atlases during the late 1980s and early 1990s, mainly due to the availability of Apple's *HyperCard* software developed for the Macintosh computer and released in 1987 (Raveneau, Miller, Brousseau, and Dufour 1991). Typical of what was developed was Parson's *Covent Garden* area prototype (Parsons 1994a, 1994b, 1995). This particular project presented users with a "through the window" view of the market via a 3D view in perspective. Users could then navigate around the package using conventional cursor controls and mouse clicks on directional arrows indicating movement directions.

Another hypermap product was HYPERSNIGE (Camara and Gomes 1991). This was somewhat different, as it was developed to run on both the Apple *Macintosh* using *Hypercard* and the PC with *Matrix Layout*. HYPERSNIGE was a hypermedia system, which included Portugal's national, regional, and sub-regional maps and information. Nodes in the system maps, text, and spreadsheets and links are used for navigation. Other control structures are numerical, linguistic (logical deductions), and pictorial (overlay). Maps were seen as links to spreadsheets filled with statistical data. They (the maps) could be drawn, imported, and exported. The package was further developed to expand the themes and to incorporate multimedia using Digital Video-Interactive (DV-I).

Videodisc

The first real application was the *Aspen Movie Map Project*, devised and undertaken by the MIT Architecture Machine Group in 1978 (Negroponte 1995). This two-videodisc system was developed to demonstrate the possibilities of providing information using multimedia resources and providing surrogate travel through the city of Aspen, Colorado. This was the first time that the term "multimedia" was used and it is interesting to note that the first multimedia product was in fact a mapping project. A map and screen image from the system is illustrated in Figure 11.3.

A later videodisc product was the *Domesday* project, produced in 1986 by the BBC (British Broadcasting Commission), Acorn Computers, and Philips to commemorate the 900th anniversary of William the Conqueror's tally book (Openshaw and Mounsey 1986, 1987, Openshaw, Wymer, and Charlton 1986). Two videodiscs were produced for the project, one concentrating on national data and the other on community information. The community videodisc included surrogate walks like the Aspen Movie map. A typical screen is shown in Figure 11.4.

(a)

(b)

Fig. 11.3 *Aspen Movie Map* Project
From Allen 2003

Fig. 11.4 Surrogate walk from the Domesday Community disc
From http://www.binarydinosaurs.co.uk/Museum/Acorn/domesday.htm

CD-ROM

The compact disc, more commonly referred to as the CD, was jointly developed by Sony of Japan and Philips of The Netherlands in 1982. Initially, the potential of the large storage capacity of CD-ROMs for the distribution of geographic information fostered interest in publishing digital maps using a new medium (Rystedt 1987, Siekierska and Palko 1986). A wide range of products, initially databases and photographic collections, and later encyclopedia and atlases were made available on this product. As they were made to conform to the ISO 9660 standard they were assured "play" success on all available machines, which ensured that they were widely accepted and used.

An interesting CD-ROM mapping package was The Territorial Evolution of Canada interactive multimedia map-pack (developed from a prototype atlas as part of the National Atlas of Canada from the Geographical Sciences Division, Survey and Mapping Branch, Department of Energy and Resources (Siekierska and Palko 1986, Siekierska and Armenakis 1999). The product provided an innovative overview of Canada and it exploited the use of interactive multimedia in its truest sense. A "screen grab" from the product is shown in Figure 11.5. The use of discrete media like CD-ROM or DVD is still a popular means of distributing cartographic products.

Internet

The Internet was used before the Web to deliver mapping products and data sets. Using the File Transfer Protocol (FTP) files, usually compressed, were distributed in this manner. File transfers were quick but the process was burdened with the overheads of file compression and subsequent decompression and the need to have appropriate display software on the "receiving" computer (Peterson 2001a). Collections of scanned paper maps were constructed and delivered to consumers usually as graphic interchange format (GIF) files. While an efficient means of providing information, almost immediately users still needed to undertake some file manipulations prior to the actual image being displayed. The Web enabled this problem to be eradicated.

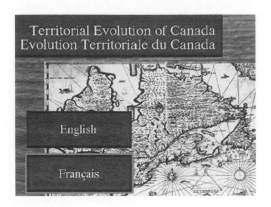

Fig. 11.5   Territorial evolution of Canada
From Siekierska and Armenakis 1999

World Wide Web

The first browser was not all that dissimilar to today's Internet Explorer or Netscape counterpart, and a current-day user of the Web could easily adapt to this original manifestation. Some of the early Web mapping packages used text-heavy interfaces to list the available mapping inventory. The Virtual Atlas, by Ashdowne, Cartwright, and Nevile (1995, 1997) (Figure 11.6) is typical of this genre of geospatial product developed in the "early" days of Web cartography. Once the HTML file was "clicked" the usual means of viewing geographic information was via a collection of scanned maps. The CIA World Fact Book and the PCL (Perry Castaneda Library) Map Collection (University of Texas at Austin) provided excellent collections of scanned maps along with other pertinent information. A most valuable global resource was made available.

Progression of Web Mapping

Along with Web development in the early 1990s map provision via this communications media kept pace. Early implementations were scanned collections of images and maps, and the Corbis collection of images illustrated the wealth of information that could be delivered via the Web. The *CIA World Fact Book*, for example, made available maps of almost any part of the world. And, while the shortcomings of scanned maps must be acknowledged, this site made available, as it still continues to, a plethora of geospatial artifacts and general geographic information. The map downloaded from this site in 1998, Figure 11.7, is typical of the type of early product availability, in this case from the Perry-Castañeda Library. Figure 11.8 shows a similar product from the CIA World Fact Book site. Many simple map access sites were developed and, while powerful media access tools were provided, the reliance of just scanned maps somewhat limited their effectiveness.

Also, quite early, in Web mapping terms, another new genre of "published" map was made available, like products from *MapQuest*. *MapQuest* has probably produced the most impressive product for finding streets and business locations, especially in the

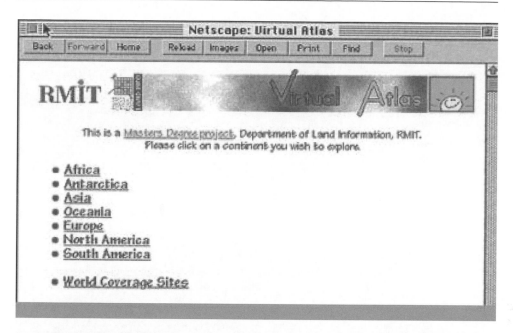

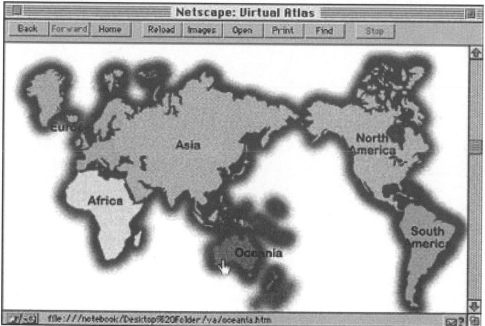

Fig. 11.6 Ashdowne's *Virtual Atlas* – initial text interface
From Ashdowne, Cartwright, and Nevile 1997

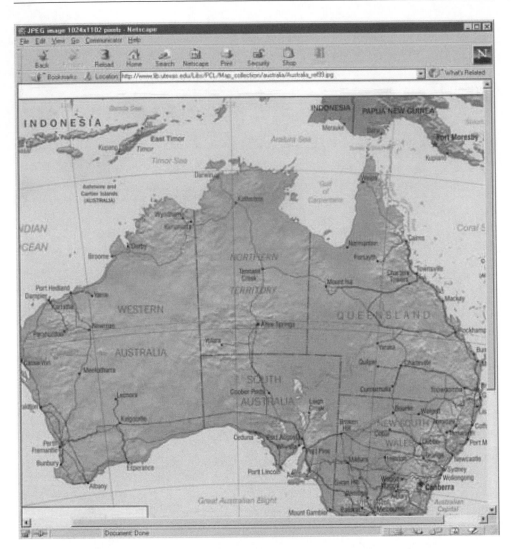

Fig. 11.7   Perry-Castañeda Library, The University of Texas at Austin map of Australia

USA. Using these resource users can pick a country, zoom into part of it, and then down to street level. If two addresses in the USA are known, then both maps and route instructions can be generated and viewed. Whilst a complete coverage of the USA is available, a sparser street database is available for other countries. Nevertheless, the *MapQuest* product is a perfect example of Web-delivered information. Figure 11.9 shows early (1997) *MapQuest* products.

   Problems with scanned maps may include image quality degradation, warping from improper scanning, coarse scanning resolutions, and over-reduction that render many maps unreadable. However, users have accepted these products because of two factors: lower cost (or free) and time (almost immediate delivery of products) (Peterson 2001b).

Fig. 11.8 CIA World Fact Book map of Australia

The Web changed map publishing forever. More maps were made available for free or at modest costs. In addition, collections of valuable maps, once only accessible by a visit to a library or map collection, were now made available to researchers and general map users. Web publishing has become prolific, and Peterson (2001a) noted that by the mid-1990s a single computer operated by Xerox PARC research facility processed over 90,000 Internet requests for maps every day.

Range of Products and Usage

In general terms, mapping services available on the Web include: (1) map and image collections; (2) downloadable data stores; (3) information services with maps; (4) online map-generation services; (5) Web atlases; and (6) hybrid products (Cartwright 2002). These products are briefly described below.

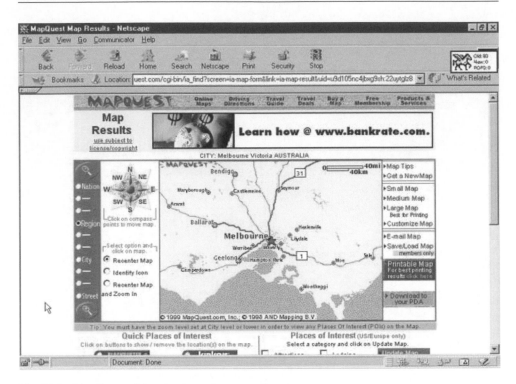

Fig. 11.9 MapQuest – Victoria, Australia
http://www.mapquest.com/cgi-bin/ia_find?screen=ia_find&link=ia_find&uid=a09a4uc0e09edc
accessed July 26, 1997

Map and image collections

The extent of map libraries on the Web can be illustrated by the sheer number listed in the University of Minnesota's Web page (University of Minnesota 2000). It includes details of sites (numerous) in the USA as well as global libraries that provide Web access. A large site to access geospatial information is the Alexandria Digital Library (Andresen, Carver, Dolin, et al. 1997, see http://www.alexandria.ucsb.edu for additional details). It focuses on the provision of spatially indexed information via the World Wide Web (WWW). It contains a collection of geographically referenced materials and services for accessing those collections. The project is being further developed via the Alexandria Digital Earth Prototype (ADEPT), funded for 1999–2004 by the US National Science (Alexandria Digital Library Project 2000).

This type of Web resource is extremely helpful where access to rare or unique maps would otherwise be difficult or impossible. Oxford University's Bodleian Library, a repository of numerous historical artifacts including many related to Oxford and Oxfordshire, makes available via the Web a number of rare map facsimiles. These high-resolution scanned images may be used by scholars in papers without the need to formally request copyright clearance. A typical image is illustrated in Figure 11.10.

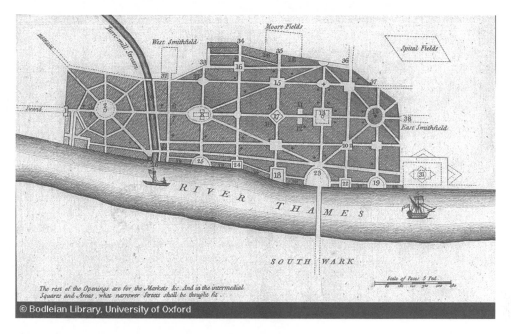

Fig. 11.10　Bodleian Library: Plan for rebuilding . . . London – J Evelyn 1666
From http://www.rsl.ox.ac.uk/nnj/mapcase2.htm

Downloadable data stores

Digital geospatial information files can be accessed and downloaded online. Web repositories have been established by both governmental and private mapping organizations to streamline how these products are marketed, sold and delivered. Typical sites include

- US Geological Survey (http://mapping.usgs.gov/www/products/status.html);
- National Mapping, Australia (formerly AUSLIG) (http://www.auslig.gov.au/index.html);
- Land Victoria – through its LandChannel site; and
- Map Machine (National Geographic Society).

In mid-1998 the Victorian Government, Australia went online with its land-related information and is part of the Electronic Service Delivery program of the State Government of Victoria. Information provided, focusing around the themes of work and home, is related to: (1) buying a property; (2) selling a property; (3) renting and leasing; (4) planning and building; and (5) the land around us. Figure 11.11 illustrates the information provided on this site.

These sites have been developed with the express purpose of making maps more readily available to the general public and professional map users. Most important, they also allow the information to be made available with little cost to the providing organization due to the "hands-free" nature of Web delivery.

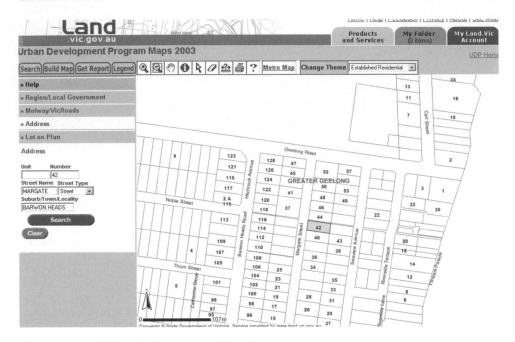

Fig. 11.11   LandChannel, Government of Victoria

Information services with maps

Publishing houses that have traditionally published their information as paper maps and books now use the Web to provide extra information to support their paper publications. The sites are numerous, and they are provided by travel information publishers. In Australia, Melbourne-based Pacific Access Pty Ltd., a Telstra Company, publish their *Whereis* street atlas (http://www.whereis.com.au/). Users gain access to a set of Universal Business Directory's (UBD) scanned street maps by typing-in a street address for Melbourne, Sydney, Brisbane, Gold Coast, Sunshine Coast, Canberra, Perth, and Adelaide. A map is returned along with a UBD map reference. The user is able to navigate to adjoining maps by clicking hot spots on the edges of the displayed maps. Parent company, Pacific Access Pty Ltd of Melbourne also publish paper and Web versions of the Australian White Pages™ directory and Yellow Pages® directory web sites, and the *Whereis* geographical search functionality is built into these sites. Fully interactive *Whereis* street atlas maps can be embedded within Australian corporate web sites. See Figure 11.12 for an illustration of their site.

On-line map generation services

In Australia the first publicly available online environmental information was the Environmental Resources Information Network (ERIN) online service (EOS) (a national facility to provide geographically related environmental information required for planning and decision-making). EOS provides public access to tables, maps, images,

Fig. 11.12 WhereIs map information
From http://www.whereis.com.au

and large databases. Users can undertake their own spatial modeling by passing SQL commands direct to the database. Another product is the Australian Coastal Atlas (ACA) that was initially a component of the Commonwealth Coastal Policy and is now a major component of the Coasts and Clean Seas Initiative. It is an electronic atlas or gateway drawing together the combined data holdings of the Commonwealth in the coastal zone. An Australian Coastal Atlas prototype provides an interactive WWW–GIS interface that uses pre-prepared GIFs of the 250,000-scale Australian National Mapping map sheet with GIS information and allows the user to overlay these GIFs (Blake 1998). Users access AUSLIG maps that form part of the "tiled" mapping resource and also make their own maps (see Figure 11.13).

Web atlases and street directories

Atlas producers, now having to face the realities of the expense of paper publishing and associated distribution costs, increasingly use the Web as a means of providing atlases of countries and regions. Many different configurations have been assembled, from the very simple to the more complex. One of the early products to be placed online was the *Atlas du Quebec et ses Regions*, produced at the University of Quebec at Montreal (see http://www.unites.uqam.ca/atlasquebec/frameSet/fs05.01.html for additional details).

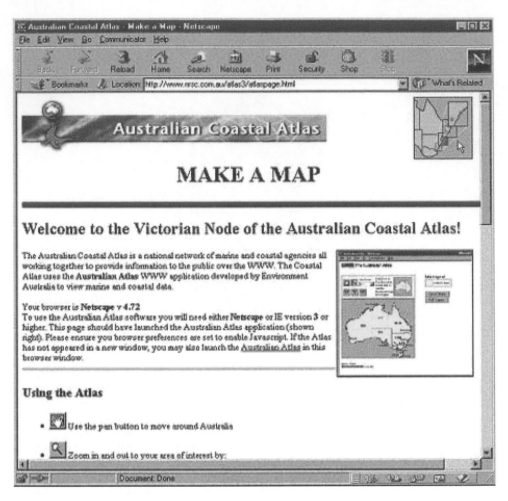

Fig. 11.13 Australian Coastal Atlas
From http://www.environment.gov.au/marine/coastal_atlas/atlaspage.htmle/quickmap.html

Another Canadian product that illustrates the effectiveness of providing atlas products via the Web is the National Atlas of Canada *Quick Maps*. This product provides a number of ready-made maps, as well as the provision for users to "construct" their own maps. The atlas was produced by Natural Resources Canada and it is an excellent example of how atlases can be delivered online. The introductory page is illustrated in Figure 11.14.

Hybrid products

Combined discrete/distributed products that publish on the Web, on CD-ROM and on paper are also being developed. These products include the US Geological Survey's (USGS) national atlas of the United States, published on both CD-ROM and the Web (Guptill 1997). Perhaps one of the most impressive publications of this type

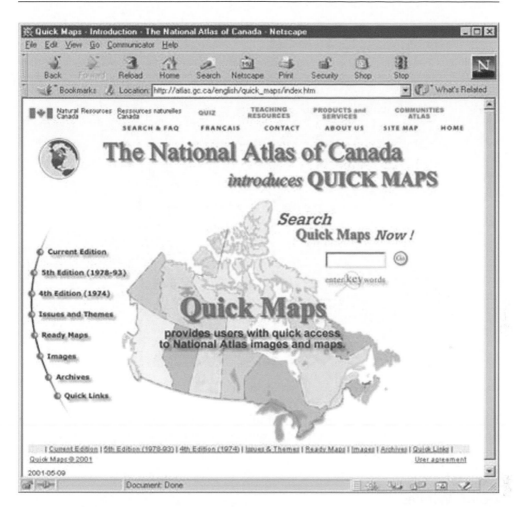

Fig. 11.14 National Atlas of Canada Quick Maps

is the 2001 Atlas of Switzerland (http://www.swisstopo.ch/en/digital/adsi.htm), published on CD-ROM, online via the Web, and as an elegantly bound paper atlas. The atlas has been developed at the Department of Cartography at the ETH in Zurich, which has always developed and produced the atlas on behalf of the Swiss government. Figure 11.15 illustrates a "page" from the atlas.

Travel information

A major Australian-based travel information resource online is Lonely Planet Destinations. Lonely Planet is an Australian company. Currently there are over 200 Lonely Planet titles in print that provide information via walking guides, atlases, phrasebooks, and the "Journeys" series of travel literature. Their Web site (http://www.lonelyplanet.com.au/) provides Destinations – an online guide to travel destinations, Optic Nerve – an online photographic collection, On the Road (extracts

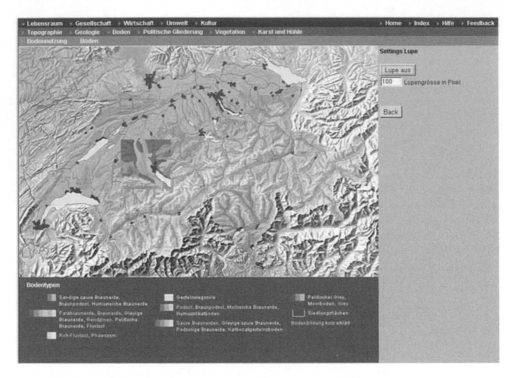

Fig. 11.15   Atlas of Switzerland

from the Journeys book series), The Thorn Tree (an electronic travelers' bulletin board), Postcards (access to individual travelers' thoughts on unique destinations), and Health information.

Going Wireless

The first cellular mobile telephone network was introduced in Japan in 1979. Since then the coverage and the use of mobile devices has grown tremendously. Currently mobile telephones have reached almost saturation point in terms of everyday use. However, using these devices to access geospatial information has not yet been properly exploited. As the information delivery infrastructure is already in place, and because of the fact that the graphics displays on telephone devices are always being improved, they offer the potential for delivering usable geoinformation graphics and support information like sound, or a series of Short Messaging Service (SMS) textual "prompts" to assist navigation. A number of developments that are of interest to geoinformation provision have taken place, including Nokia's Smart Messaging System (SMS; see http://www.forum.nokia.com/ for additional details) and the Wireless Application Protocol (WAP). WAP is simply a protocol – a standardized way that a mobile phone talks to a server installed in the mobile phone network. WAP telephones can deliver images as well as alphanumerics, making it a useful advertising tool (see http://www.mobileaccess.be/pages/products/wap.thm for additional details).

Fig. 11.16 WAP-enabled cellular telephone delivering map information from Webraska

Mapping applications have been developed to deliver mainly traffic information services by the French company, Webraska (see a screen output from an early European application in Figure 11.16).

Wireless is now seen as part of the Internet, and not distinct. It is increasingly being used as a "gateway" to resources provided via the Internet. The provision of geospatial information via these devices has spawned a whole new Location Based Services (LBS) industry that provides information on-site (see Chapter 32 by Brimicombe, this volume, for additional information on this topic). The location services, sometimes referred to as L-commerce, have seen European operators struggling to create LBS revenue models. Short Message Services (SMS) was seen as providing the most dependable revenues from location services over the next several years (Gisler 2001). "Assisted GPS location solutions have been looked-upon favorably by the 'location services' company Snaptrack and telco Sprint, who have conducted a joint case study. The industry sees that the biggest potential money earner is mobile location entertainment, especially amongst teenagers" (Gisler 2001).

A fairly recent example of enterprise using LBS is the Zingo cab service in London (www.zingotaxi.co.uk), which provides a direct connection between available taxis and subscribers. Users call Zingo from a pre-registered mobile telephone, then Zingo's location technology – Cellular and Global Positioning System (GPS) for the user, and GPS for the taxi – links the customer with the nearest available taxi (Zingo 2003). The system pinpoints the potential passenger's location by locking on to the location of the mobile phone. The system operates using the UK's Vodafone, O_2, Orange, Virgin, and Three cellular telephone systems. It works automatically with O_2 and Vodafone. In April 2003 there were 400 cabs using this system, with "several thousand" planned (Rubens 2003).

Obviously the accuracy of the delivery of these services based on location varies between urban areas and rural areas and from country to country, where the saturation of mobile telephone transmission towers might be different. European and Japanese companies were the first to market solutions based on these imprecise location technologies, using the cell ID (see http://www.jlocationservices.com/ for additional details). Accuracy requirements are context and application-specific

– emergency services and navigation requiring the highest degree of accuracy and weather and general information the least (Gisler 2001). The accuracy can be enhanced using GPS, hence the interest in GPS-enabled mobile phones. It is predicted that 95 percent of all handsets sold in the USA by 2005 will be location-enabled.

Ubiquitous computing is also of much interest to computer technologists and also cartographers. Ubiquitous computing has been named as the "third wave" in computing, or "the age of *calm technology*, when technology recedes into the background of our lives" as described by the father of ubiquitous computing, the late Mark Weiser (Weiser 1996). We now see this type of computing in the form of handheld PCs, mobile phones, wireless sensors, radio tags, and Wi-Fi (Baard 2003). Designers of ubiquitous systems envision seeding private and public places with sensors and transmitters that are embedded into objects and hidden from view, providing for the deployment of things like "Audio Tags," which play an infrared sensor-triggered message once a person is within a pre-determined proximity (Wired News 2003). The International Cartographic Association recently formed a Commission on Ubiquitous Cartography to promote the study of this form of information provision.

Advancing Mapping in a Digital Age

It can be argued that maps were in fact the first multimedia products – as they contain text, diagrams, graphics (as ordered symbols), and geographic facts. Paper maps could be considered to be analog Virtual Reality (VR) tools. They have provided the means by which armchair travelers could "go" to places from the comfort of their lounge or study. The rules that govern their design, production, and consumption have evolved over centuries, and the methods of producing maps via the printing press have been established by 500 years of experiment and development. However, multimedia cartography on the Web is relatively new and its use as a geographic visualization tool is still virtually unproven. There is little real information about "best use" with interactive multimedia cartographic tools. Applying old ways to new tools may be the "line of least resistance," but it perhaps does not allow us to properly exploit New Media and new communications systems. In addition, as there are no comprehensive skills packages for teaching how to use interactive multimedia effectively, there exists no "starting point" to decipher if the use of these products is any better than using a conventional paper map.

Cartographers once knew their users: they knew what they wanted and how they intended to use the cartographic artifact produced. It could be argued of conventional paper maps that these products were not considered to be "mainstream" information documents but specialist artifacts to be used by expert users, or users who were willing to "learn the rules" of map use. There now exists a new genre of users. Many may never have used maps before and they consider geographic information in the same way as any other commodity that they can obtain via the Web. Now, with almost instant access to geographic data and graphic products via the Web, it is argued that the general public now considers Web-delivered maps to be just part of what New Media delivers. Geographic information delivered through the use of New Media is seen as part of popular media, rather than scientific documents. As the Web can be considered as an accepted part of popular media, users could consider its employing it in similar ways to those in which they use television, video, movies, books, newspapers, journals, radio, and CD-ROM. The delivery of cartographic artifacts via the Web to naive or inexpert users involves different strategies to traditional map delivery and use. Use strategies need to be developed, tested, and implemented.

CONCLUSIONS

The revolution in information provision prompted by the advent of the Internet, and more particularly the World Wide Web, has changed forever how information products are viewed. They are now wanted, no demanded, almost immediately by users. In newspaper terms this would be described as wanting information "before the ink has dried," but for digital information this is probably best described as wanting information "before the data collection sensor has cooled"! Advances in data collection and telecommunications ensure that collected data is quickly and faithfully transmitted. Processing procedures and equipment, map "construction" and "rendering" software and geographical information delivery systems now provide the ability to deliver on-demand geo-information products in almost real time.

Everything has changed, but the underlying theory and procedural knowledge remain the same. We have powerful tools for the provision of information that has currency, accuracy, and immediacy, but how we apply them depends upon adequate knowledge of cartographic theory and practices. This "new" method of access to and representation of geospatial information is different to formerly used methods and therefore, while New Media applications can be considered to be at a fairly immature stage of development (compared to paper maps), there is a need to identify the positive elements of the media used and to isolate the negative ones, so as to develop strategies to overcome any deficiencies. Web designers and graphic artists can produce elegant information displays for the Web, but the integrity of their map-related products could be questioned. Without adequate grounding in the geospatial sciences, graphical presentations with impact and panache can be produced, but they may be documents of misinformation. We need to ensure that what is delivered via the Web that relates to the provision of geographic information is produced to high standards, of both design and information content, and that users comprehend the underlying structure of the data and the manipulations, or "cartographic gymnastics" that have been employed to ensure that the data is presented in the best possible manner and are thoroughly understood by users.

REFERENCES

Advanced Imaging. 1996. Intercast: From Web TV to PC. April, 52.

Allen, R. 2003. Aspen Movie Map, 1978–80. WWW document, http://rebeccaallen.com/v1/work/work.php?isResearch=1&wNR=18&ord=alph&wLimit=1.

Andresen, D., Carver, L., Dolin, R., Fischer, C., Frew, J., Goodchild, M., Ibarra, O., Kothuri, R., Larsgaard, M., Manjunath, B., Nebert, D., Simpson, J., Smith, T., Yang, T., and Zheng, Q. 1997. The WWW Prototype of the Alexandria Digital Library. WWW document, http://alexandria.sde.ucsb.edu/public-documents/papers/japan-paper.htm.

Ashdowne, S., Cartwright, W. E., and Nevile, L. 1995. Designing a virtual atlas on the World Wide Web. In *Proceedings of AUSWEB '95*, Ballina, Australia (available at http://ausweb.scu.edu.au/aw96/educn/ashdowne/).

Ashdowne, S., Cartwright, W., and Nevile, L. 1997. A virtual atlas on the World Wide Web: Concept, development and implementation. In *Proceedings of the Eighteenth International Cartographic Conference*, Stockholm, Sweden, pp. 663–72.

Baard, M. 2003. A connection in every spot. *Wired News* (October 16; available at http://www.wired.com/news/print/0,1294,60831,00.html).

Blake, S. 1998. Customising maps on the World Wide Web: The Australian Coastal Atlas, an interactive GIS approach. In *Proceedings of the Mapping Sciences '98 Conference*, Fremantle, Australia. Freemantle, Australia: Mapping Sciences Institute, Australia.

Camara, A. and Gomes, A. L. 1991. HYPERSNIGE: A navigation system for geographic information. In *Proceedings of the Second European Conference on Geographical Information Systems (EGIS '91)*, Brussels, Belgium, pp. 175–9.

Cartwright, W. E. 2002. From printing maps to satisfy demand to printing maps on demand. In B. Cope and D. Mason (eds) *Markets for Electronic Book Products*. Melbourne: Common Ground Publishing, pp. 81–96.

Clickz Stats. 2004. Population Explosion! WWW document, http://www.clickz.com/stats/big_picture/geographics/article.php/5911_151151.

Clickz Stats. 2006. Trends & Statistics: The Web's Richest Source. WWW document, http://www.clickz.com/stats/web_worldwide/.

Gisler, M. 2001. Rome mobile location services conference attracts industry leaders. *WLIA Newsletter* (available at www.wliaonline.com/publications/romeconference.html).

Guptill, S. C. 1997. Designing a new national atlas of the United States. In *Proceedings of the Eighteenth International Cartographic Conference*, Stockholm, Sweden. Stockholm: International Cartographic Association, pp. 613–9.

Hosie, P. 1985. Promises, promises: Viatel and education. *Australian Journal of Educational Technology* 1: 39–46.

Lebo, H. 2003. *The UCLA Internet Report: Surveying the Digital Future*. Los Angeles, CA: UCLA Center for Communication Policy (available at http://ccp.ucla.edu/pdf/UCLA-Internet-Report-Year-Three.pdf).

Negroponte, N. 1995. *Being Digital*. Rydalmere: Hodder and Stoughton.

Negroponte, N. 1996. Object-oriented television. *Wired* 3(7): 188.

Netcraft. 1998. Netcraft Web Server Survey. WWW document, http://survey.netcraft.com/Reports/.

Netcraft. 1999. Netcraft Web Server Survey. WWW document, http://survey.netcraft.com/Reports/.

Openshaw, S. and Mounsey, H. 1986. Geographic information systems and the BBC's Domesday interactive videodisk. In *Proceedings of Auto Carto 1*, London, United Kingdom. Bethesda, MD: American Congress of Surveying and Mapping/American Society for Photogrammetry and Remote Sensing, pp. 539–46.

Openshaw, S. and Mounsey, H. 1987. Geographic information systems and the BBC's Domesday interactive videodisk. *International Journal of Geographical Information Systems* 1: 173–9.

Openshaw, S., Wymer, C., and Charlton, M. 1986. A geographical information and mapping system for the BBC Domesday optical discs. *Transactions of the Institute of British Geographers* 11: 296–304.

Parker, H. D. 1995. What does the Internet's future hold? *GIS World* 8(11): 118.

Parsons, E. 1994a. Virtual worlds technology: The ultimate GIS visualization tool. In *Proceedings of AGI '94*, Birmingham, United Kingdom. London: Association for Geographic Information.

Parsons, E. 1994b. Visualisation techniques for qualitative spatial information. In *Proceedings of EGIS '94*, Paris, France. Utrecht: EGIS Foundation, pp. 407–15.

Parsons, E. 1995. GIS visualization tools for qualitative spatial information. In P. Fisher (ed.) *Innovations in GIS 2*. London: Taylor and Francis, pp. 201–10.

Pemberton, A. 2004. Teletext: The Early Years. WWW document, http://www.pembers.freeserve.co.uk/Teletext/Photographs.html.

Peterson, M. P. 1997. Trends in Internet map use. In *Proceedings of Eighteenth International Cartographic Association Conference*, Stockholm, Sweden. Stockholm: International Cartographic Association, pp. 1635–42.

Peterson, M. P. 2001a. Cartography and the Internet: Implications for Modern Cartography. WWW document, http://maps.omaha.edu.NACIS/paper.htm.

Peterson, M. P. 2001b. Cartography and the Internet: Introduction and Research Agenda. WWW document, http://maps.omaha.edu.NACIS/CP26/article1.htm.

Raveneau, J.-L., Miller, M., Brousseau, Y., and Dufour, C. 1991. Micro-Atlases and the diffusion of geographic information: An experiment with Hypercard. In D. R. F. Taylor (ed.) *Geographic Information Systems: The Microcomputer and Modern Cartography*. Oxford: Pergamon, pp. 201–24.

Rubens, P. 2003. How to hail a cab with a mobile phone. *BBC News* (available at http://newsvote.bbc.co.uk/mpapps/pagetools/print/news.bbc.co.uk/1/hi/uk/2946129.stm).

Rystedt, B. 1987. Compact disks for distribution of maps and other geographic information. In *Proceedings of the Thirteenth International Cartographic Association Conference*, Morelia, Mexico. Stockholm: International Cartographic Association, pp. 479–84.

Siekierska, E. M. and Armenakis, C. 1999. Territorial evolution of Canada – An interactive multimedia cartographic presentation. In W. E. Cartwright, M. P. Peterson, and G. Gartner (eds) *Multimedia Cartography*. Heidelberg: Springer-Verlag, pp. 131–9.

Siekierska, E. M. and Palko, S. 1986. Canada's electronic atlas. In *Proceedings of AutoCarto 2*, London. Falls Church, VA: American Congress of Surveying and Mapping-American Society of Photogrammetry and Remote Sensing, pp. 409–17.

Taylor, D. R. F. 1984. The cartographic potential of Telidon. *Cartographica* 19: 3–4, 1830.

University of Minnesota. 2000. Map Libraries on the World Wide Web. WWW document, http://www-map.lib.umn.edu/map_libraries.html.

Weiser, M. 1996. Ubiquitous Computing. WWW document, http://sandbox.xerox.com/hypertext/weiser/UbiHome.html.

Wired News. 2003. Balancing Utility with Privacy. WWW document, http://www.wired.com/news/print/0,1294,60871,00.html.

Zingo. 2003. Get the Upper Hand in the Battle for a Taxi. WWW document, www.zingotaxi.co.uk.

Chapter 12

Generalization of Spatial Databases

William A. Mackaness

"All geographical processes are imbued with scale" (Taylor 2004, p. 214), thus issues of scale are an essential consideration in geographic problem solving. The scale of observation governs what phenomena can be viewed, what patterns are discernible, and what processes can be inferred. We are interested in viewing the precise detail of those phenomena, as well as the broad linkages across regional and global space. Choosing scales of analysis, comparing output at different scales, describing constructions of scale (Leitner 2004) are all common practices in the geosciences. We do this because we wish to know the operational scales of geographic phenomena, how relationships between variables change as the scale of measurement increases or decreases, and we want to know the degree to which information on spatial relationships at one scale can be used to make inferences about relationships at other scales (Sheppard and McMaster 2004; see also Chapter 18 by Brunsdon in this volume). What is always apparent when viewing geographic phenomena is the interdependent nature of geographic processes. Any observation embodies a set of physical and social processes "whose drivers operate at a variety of interlocked and nested geographic scales" (Swyngedouw 2004, p. 129).

Both the scale of observation and of representation reflect a process of abstraction, an instantaneous momentary "slice" through a complex set of spatio-temporal, interdependent processes. Traditionally it has been the cartographer's responsibility to select a scale, to symbolize the phenomena, and to give meaning through the addition of appropriate contextual information. In paper-based mapping, various considerations acted to constrain the choice of solution (the map literacy of the intended audience, map styles, the medium, and choice of cartographic tools). Historically, the paper map reflected the state of geographic knowledge and was the basis of geographic inquiry. Indeed it was argued that if the problem "cannot be studied fundamentally by maps – usually by a comparison of several maps – then it is questionable whether or not it is within the field of geography" (Hartshorne 1939, p. 249). Information technology has not devalued the power of the map, but it has driven a series of paradigm shifts in how we store, represent and interact with geographic information. Early work in automated mapping focused on supporting the activities of the human cartographer who remained central to the map design process. Current research is focused more on ideas of autonomous design – systems capable of selecting optimum solutions among a variety of candidate solutions delivered over the web, in a variety of thematic forms, in anticipation of users who have little or no cartographic skill (see Chapter 11 by Cartwright, this volume). Historically the paper map reflected a state of knowledge. Now it is the database that is the knowledge store, with the map as the metaphorical window by which geographic information is dynamically explored. In these interactive environments, the art and science of cartography (Krygier 1995) must be extended to support the integration of distributed data collected at varying levels of detail, while conforming to issues of data quality and interoperability.

Generalization

At the fine scale, when viewing phenomena at high levels of detail (LoD), we can determine many of the attributes that define individual features (such as their shape, size, and orientation), while at the broad scale, we see a more characteristic view – more particularly the regional context in which these phenomena are situated (for example, their gestaltic and topolgical qualities, and various associations among other phenomena). For example, journey planning requires a broad-scale view in order to gauge timeframes and alternative travel strategies, while a fine-scale, detailed map is required to reach the final point of destination. It is not the case that one map contains less or more information, but that they contain different, albeit inter-related information. Thus maps are required at a range of scales, in a variety of thematic forms, for delivery across a range of media. The term "map generalization" is often used to describe the process by which more general forms of a map can be derived from a detailed form. In the context of twenty-first century technology, there is a vision of a single, detailed database, constantly updated in order to reflect the most current version of a region of the world. For any given National Mapping Agency (such as the Ordnance Survey [OS] of Great Britain or the Institut Géographique National [IGN] of France) that region is defined by its respective national boundaries. In such a context, the process of map generalization entails selecting objects from that detailed database, and representing them in various simplified forms appropriate to the level of detail required, and according to some purpose (or theme). By way of example, Figure 12.1 shows a series of maps at different scales, of Lanvollon in France. The goal remains the creation of automated map generalization techniques that would enable the derivation of such maps from a single, detailed database. This vision is driven by a variety of motivations: data redundancy (maintaining a single, detailed database rather than a set of separate, scale-specific databases; Oosterom 1995); storage efficiency (recording the fine detail of a feature in as few points as possible); exploratory data analysis (MacEachren and Kraak 1997) (being able to dynamically zoom in and explore the data, and to support hypermapping); integration (combining data from disparate databases of varying levels of detail); and paper map production (for traditional series mapping).

Given the strong association of map generalization with traditional cartography it is worth stressing its broader relevance to spatial analysis and ideas inherent in

Fig. 12.1   1:25,000, 1:100,000, and 1:250,000
Copyright of the IGN

visualization methodologies. Though discussion will focus on the cartographic, we are in essence dealing with the generalization of spatial databases (Muller 1991, van Smaalen 2003). In this context we can view the fine-scale, detailed database as the first abstraction of space – often called the Primary Model or Digital Landscape Model (DLM) (Grünreich 1985). As a prerequisite the DLM requires the definition of a schema that will support the explicit storage, analysis, and characterization of all the geographic phenomena we wish to record. A series of secondary models can be derived from this primary model via the process of "model generalization." These abstractions are free from cartographic representational information and could be used to support spatial analysis at various levels of detail. Both primary and secondary models can be used as a basis for creating cartographic products (Digital Cartographic Models) via the process of "cartographic generalization". Figure 12.2 summarizes the relationships between these models and the generalization processes.

Model generalization may involve reduction of data volume, for example via the selection, classification, or grouping of phenomena, or the simplification of phenomena such as network structures. This may be required as a prerequisite to spatial analysis, the integration of different data sets, or for computational efficiency. It is certainly an integral step in the derivation of multi-scaled cartographic products. Though it has important ramifications for cartographic generalization, model generalization does not itself seek to resolve issues of graphic depiction, such as clarity or emphasis in depiction.

Often seen as the complement to model generalization, cartographic generalization describes the process by which phenomena are rendered, dealing with the challenges of creating appropriate symbols and the placement of text within the limited space of the medium (whether on paper or the small screen of a mobile device). The symbology used to represent a geographic feature must be of a size discernible to the naked eye. At reduced scale, less space is available on the map to place the symbols. At coarser scales, the symbols become increasingly larger than the feature they represent. It therefore becomes necessary to omit symbology associated with certain features, to group features, to characterize them in a simpler way, or to choose alternate forms of symbology in response to this competition for space (Mark 1990). Figure 12.3 nicely illustrates this idea, showing The Tower of London and

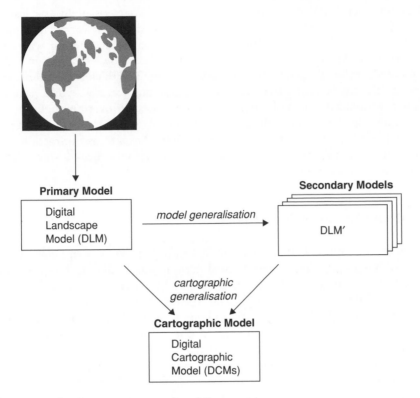

Fig. 12.2 Generalization as a sequence of modeling operations
After Grunreich 1985

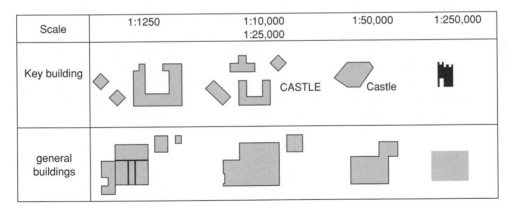

Fig. 12.3 Model and cartographic generalization acting in unison to reveal different qualities about
The Tower of London
Copyright Ordnance Survey

its surroundings at scales of 1:10,000, 1:25,000, and 1:50,000. At the finest level
of detail we can discern individual walls, courtyards, pavements, and trees, and the
buildings are individually named. We can make many inferences drawing on our
understanding and experiences of geographic space, such as the function of buildings

and the components of the various fortifications. At a coarser scale we see less detail, in exchange for more of the context. For example, we discern its strategic importance along the bank of the river Thames, and text is used in a different way to label various features. At the coarse scale of 1:50,000 we see how competition for space has presented further challenges for the cartographer. The thick red symbology used to represent the roads has encroached upon surrounding features, which have had to be slightly "displaced" or made smaller in order to avoid overlapping and causing confusion among the represented features. We can also discern more of a thematic edge to this representation, with the Tower highlighted as a tourist attraction. Overall then, we can discern the processes of model and cartographic generalization at work in the creation of such map designs.

Conceptual Models of Generalization

Initial research in automated cartography began in the 1960s (Coppock and Rhind 1991) and sought to replace the manual scribing tools and techniques used by the human cartographer, with their automated equivalent. Paper-based maps were digitized to create inherently cartographic, vector based databases – in essence the map became a set of points, lines, areas, and text to which feature codes were attached in order to control the symbolization process. But research soon highlighted the limits of this approach, and revealed the art and science of cartographer as a design task involving complex decision making. There was a clear need for conceptual models (such as those presented by Brassel and Weibel 1988 and McMaster and Shea 1992) as a basis for understanding the process of generalization, and developing auto-mated solutions. McMaster and Shea (1992) presented a comprehensive model that decomposed the generalization process into three stages: definition of philosophical objectives (why generalize), cartometric evaluation (when to generalize), and a set of spatial and attribute transformations (how to generalize). A complementary view that reflects the potential of more complete solutions to automated generalization is one in which a variety of candidate solutions are considered (synthesis), based on cartometric and topological analysis (analysis). This is followed by an evaluation phase that selects the most appropriate candidate based on both fine-scale and holistic evaluation techniques (Figure 12.4).

Multi-scale databases

Aligned closely to the topic of map generalization is the idea of "multiple repres-entation," in which various cartographic representations of a single object are stored for viewing or analysis at various levels of abstraction (Goodchild and Yang 1992, Kidner and Jones 1994, Kilpelainen and Sajakoski 1995, Devogele, Trevisan, and Ranal 1996). A specific advantage being that their forms can be pre-cast and imme-diately presented to the user (thus avoiding the time cost associated with creating solutions "on the fly"). Though the DLM (Figure 12.2) remains unchanged, a series of multiple representations can be derived at any time, only needing to be recast when the central database is updated to reflect changes in the real world. There are complicating issues in the management of the database, in particular ensuring

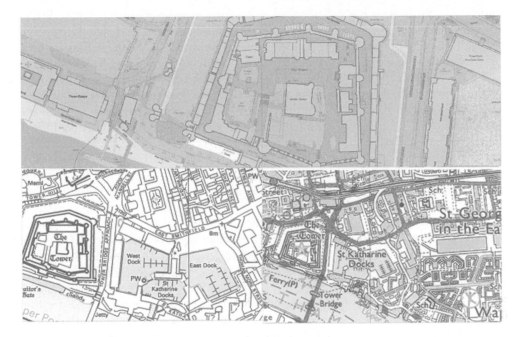

Fig. 12.4 Generalization in the context of automated solutions

the seamless joining together of multiple representation after an update cycle. Ideas of multiple representation mirror the idea of a single, detailed database, from which other databases are derived using map generalization techniques.

Generalization Methods and Algorithms

For any given conceptual framework, it is necessary to precisely define the methods by which we can analyze, synthesize, and evaluate solutions. Early research focused on reverse engineering the design process, observing the human cartographer at work, and, via a process of stepwise refinement, identifying the discrete methods used by the cartographer. In some instances the cartographer would omit selected features, or whole classes of features. Some features were merged and enlarged and (if space allowed, and where symbology overlapped) features were marginally displaced in order to distinguish more easily between features. These and other methods can be divided into two types of transformation: spatial and attribute transformation. The ten spatial transformation methods are: amalgamate, aggregate, collapse, displace, eliminate, enhance, merge, refine, simplify, and smooth. The two attribute transformation methods are classify and symbolize (Weibel and Dutton 1999).

Van Smaalen (2003) argues that in essence map features fall into one of three metaclasses (Molenaar 1998). Classes that contain "network like" objects, such as railways, rivers, and roads; classes of relatively small, often rigid, "island" objects – typically buildings; and a third class of mostly "natural" area objects – often forming exhaustive tessellations of space, for example land parcels, lakes, forested regions, and farms. Each class has different behaviors, and can be characterized in different ways. One can therefore envisage a matrix of these metaclasses against the twelve generalization methods. Each cell in the matrix containing a number of algorithms for modeling transformations of that particular metaclass for varying levels of detail, and for a range of themes. A huge amount of research has been devoted to populating such a matrix – developing methods that can be applied to various classes of objects. By way of illustration, Dutton (1999) and others have worked on methods for generalizing linear features (Buttenfield 1985, Plazanet, Bigolin, and Ruas 1998); finite element analysis and other techniques have been used to model displacement among features (Burghardt and Meier 1997, Hojholt 2000). Considerable effort has been devoted to methods for generalizing buildings (Regnauld 2001, Jiang and Claramunt 2004), while other research has focused on how space-exhaustive tessellations of space can be generalized – as, for example, is found in geological mapping (Bader and Weibel 1997, Downs and Mackaness 2002). Others have researched the problem of attenuating network structures (Richardson and Thomson 1996, Mackaness and MacKechnie 1999) while others have proposed solutions to the problem of text placement (Christensen, Marks, and Shieber 1995).

These methods have been framed in a variety of strategic contexts. For example Molenaar (1998) stratifies these methods under four headings that reflect a need to model both individual and structural characteristics of the map. Importantly he discusses the idea of functional generalization – a generalization technique used to group objects in close proximity, and non-similar objects in order to create meaningful composites (van Smaalen 2003). Figure 12.1 presents a nice example of this whereby the various objects comprising the town of Lanvollon represented at 1:25,000 scale, have been grouped and replaced by a single point symbol at the 1:250,000 scale. Functional generalization is particularly appropriate in the case of significant scale change.

Analysis

A strong recurrent theme in all the research into generalization algorithms has been the need for techniques that make explicit the metric and topological qualities that exist within and between classes of features. Effective characterization of geographic space requires us to make explicit the trends and patterns among and between phenomena, to examine densities and neighborhoods, and to model connectivity and network properties, as well as the tessellation of space. Thus the field draws heavily on spatial analysis techniques such as graph theory (Hartsfield and Ringel 1990), Voronoi techniques (Peng, Sijmons, and Brown 1995, Christophe and Ruas 2002) and skeletonization techniques (Costa 2000). The identification of pattern draws on regression techniques, and automated feature recognition techniques (Priestnal, Hatcher, Morton, Wallace, and Ley 2004). These "supporting" structures (Jones, et al. 1995, Jones and Ware 1998) are used to enrich the database and enable the modeling of topological transitions (Molenaar 1998).

Synthesis and evaluation

Research has also tried to model the process by which a combination of methods is used to synthesize various solutions. For example, a group of islands may be merged,

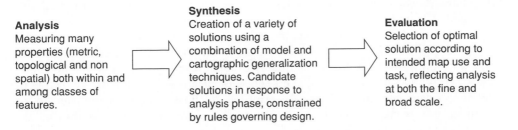

Analysis
Measuring many properties (metric, topological and non spatial) both within and among classes of features.

Synthesis
Creation of a variety of solutions using a combination of model and cartographic generalization techniques. Candidate solutions in response to analysis phase, constrained by rules governing design.

Evaluation
Selection of optimal solution according to intended map use and task, reflecting analysis at both the fine and broad scale.

Fig. 12.5 The choice, sequence, and degree of application of various methods enable synthesis of different solutions, but which one is "correct"?

and enlarged in order to remain visible to the naked eye at smaller scale. The process of enlargement may require marginal displacement to distinguish between the islands. Different results emerge according to the sequence in which the methods are applied, and the degree to which they are applied (Mackaness 1996). The evaluation of candidate solutions must be graded against a set of criteria, themselves defined by the map task. For example, a map intended for tourists may accommodate greater generalization of the characteristic form than a map intended for sea navigation. In Figure 12.6 the two generalized forms (hand drawn) are shown at the same scale as the original (in order to compare), prior to being reduced in size to 30 percent of the original.

Even in the very simple example of Figure 12.6, with a restricted set of considerations, it is easy to imagine a very large set of permutations. But it is possible to define evaluation criteria. For example, shape and area metrics can be used to measure alignments (Christophe and Ruas 2002) or the degree of distortion from the original (Whang and Muller 1998, Cheung and Shi 2004). Topological modeling in surfaces and networks can be used to model neighborhood changes among a group of objects. Density and distribution measures can be used to determine trends in the frequency of occurrence or the degree of isolation of a feature. Distance metrics can be used to assess the perceptibility of an object (is it too small to be represented at the intended scale) and the degree of crowding among objects. Evaluation also includes assessment of non-spatial attributes. For example, is it a rare geological unit relative to the surrounding region (Downs and Mackaness 2002), or a special point of interest in the landscape? Techniques have also been developed to measure the content of a map, and to evaluate levels of content as a function of change in scale (Topfer and Pillewizer 1966, Dutton 1999). Many of the cartometric techniques used to analyze the properties of a map as part of the synthesis of candidate solutions can also be used in this process of evaluation. In effect, each and every one of these techniques makes explicit some property within or between classes of objects.

But a map in its generalized form reflects a compromise among a competing set of characteristics. There is very little in the map that remains invariant over changes in scale. Indeed generalization is all about changing the characteristics of a map in order to reveal different patterns and relationships among the phenomena being mapped. Often the preservation of one characteristic can only be achieved by compromising another. Thus, among a group of buildings do we give emphasis to the "odd one out" because it is significantly larger than the remainder, or preserve the characteristic orientation shared among the group of buildings and the adjoining road? We know that the topology among a set of objects changes if we remove, aggregate, or functionally combine objects. But how do we ensure that the new topology is a "valid" one? And, where we wish to combine data from different sources and scales, how do we validate the quality of any given solution? There is no shortage of techniques for measuring the properties of an object, but the challenge of defining tolerances and collectively prioritizing those characteristics (linked to intended use) remains a significant impediment to development of systems that are more autonomous in their operation.

A Rule-Based Approach

More challenging than the development of generalization methods, has been the formalization of the procedural knowledge required to trigger the use of such methods. At any instant in the design phase, there may exist a range of alternate candidate solutions, whose creation and choice is based on rules of thumb (heuristics), to a goal state that is somewhat hazy and hard to define (Starr and Zeleny 1977). Various attempts have therefore been made to use a rule-based approach to automated map generalization (Richardson and Muller 1991, Heisser, Vickus, and Schoppmeyer 1995, Keller 1995) in which sequences of conditions and actions are matched in order to control the overall process. For example, a small remote building in a rural context has a significance much greater than its counterpart in a cityscape and is therefore treated differently. A solution might be to enlarge the symbology in order that the building remains discernible to the naked eye, according to those conditions:

> IF a building.context = rural AND building.neighborhood = isolated AND building.size = small THEN building.generalization = enlarge.

We can formalise both the <condition> and <action> part of such rules from observation of how features are symbolized on paper maps at various scales. We observe how particular solutions operate over a band of scales (akin to the idea of an "operational scale", Phillips 1997) and that beyond a certain threshold a change in the level of generalization is invoked. Figure 12.7a illustrates the various representational forms of a cathedral and Figure 12.7b shows the scale bands over which those representations might operate. These threshold points are determined by (1) a feature's geometry and size; (2) its non spatial attributes; (3) its distribution and association with other features; (4) its immediate proximity to other features; and (5) the resolution of the device on which the information is being displayed or printed (Glover and Mackaness 1999).

A feature's treatment also depends upon its importance in relation to the intended theme. For example castles and visitor attractions in a tourist map will be given greater emphasis from those buildings deemed more general. Figure 12.7 is based on observations made from paper maps over a range of scales, and shows how key (or special buildings) and general buildings are typically represented.

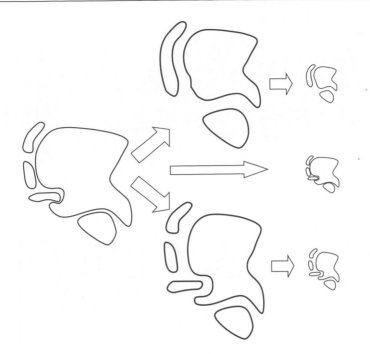

Fig. 12.6 (a) Transformations with decreasing map scale, and (b) corresponding scale bands for a topographic map
Glover and Mackaness 1999

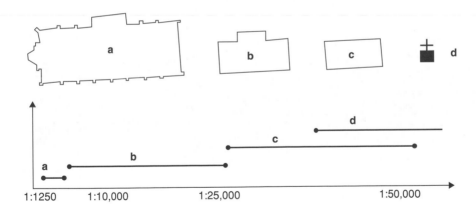

Fig. 12.7 Examples drawn from paper maps of building generalization at various scales

Again from observation we can identify the generalization methods that can be applied at the fine scale to derive these various solutions – that their forms are simplified, or grouped, or collapsed and replaced with an iconic form. For example derivation of the castle representational form at 1:50,000 scale can be formed by placing a minimum bounding rectangle (MBR) around the group of "castle" buildings (so deriving its convex hull), and substituting this form for the group of individual

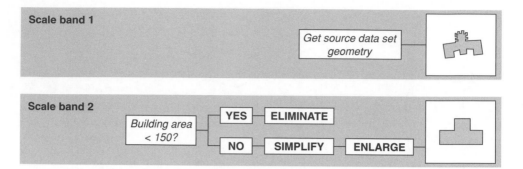
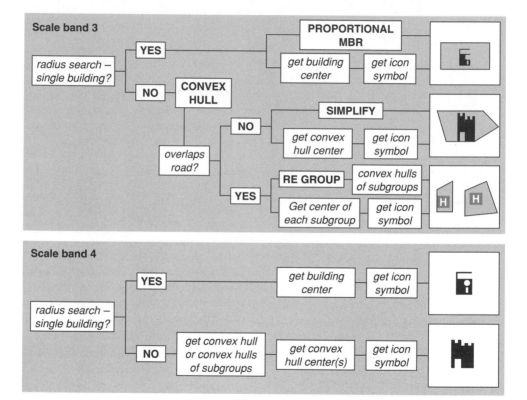

Fig. 12.8 Decision tree for key buildings

buildings. One can envisage a similar process applied to each metaclass, and for each scale band transition point (similar to the one illustrated in Figure 12.6). In this manner we can define a decision tree that incorporates the various generalization methods used, according to: the building type, its association with adjacent features, and the operational scales of the various representational forms. Figure 12.8 is the decision tree for "key" buildings intended for use in urban environments.

These and other decision trees were collectively implemented in a Geographic Information System (GIS) that was able to derive different thematic maps from

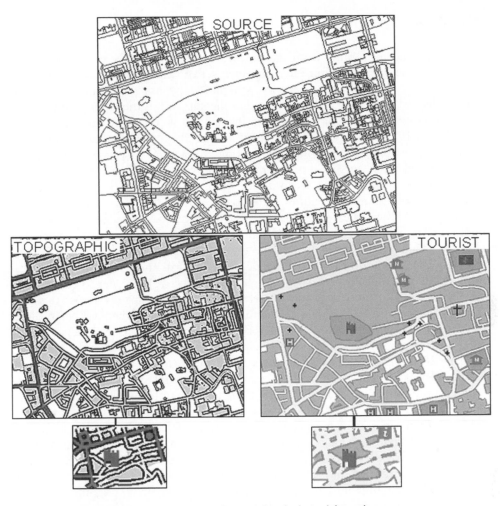

Fig. 12.9 Different products according to theme and scale derived from the same source

a single detailed source (Glover and Mackaness 1999). The results (Figure 12.9) were compared with their manual equivalent, as the basis for identifying future work.

Such a system works quite well for relatively small changes in scale. The system is limited by its inability to generate alternate solutions to a design problem and to automatically evaluate the correctness of the final solution. The work also highlighted the need for cartometric tools capable of analyzing both "local" constraints (imposed by surrounding objects) and "global" constraints (ensuring consistency across the region including preservation of trends). What was required was a system that would enable consideration of alternate designs that took into account a shared view of these and other design constraints. One such approach that has shown great promise in this regard has been in the use of multi-agent systems (MAS).

Multi-Agent Systems

The idea of "agents" came from the observation that complex processes can be modeled as a set of simple but interconnected set of tasks. For example, the complex task of sustaining an ant colony is achieved by assigning ants (agents) to specific, defined tasks that collectively ensure the survival of the colony. In this way quite complex emergent behavior can arise from a set of connected but simple agent tasks (Weiss 1999). Thus one definition of an agent is "a self contained program capable of controlling its own decision making and acting, based on its perception of its environment, in pursuit of one or more objectives" (Luck 1997, p. 309). Where more than one agent exists, we can define what are called multi-agent systems in which several computational entities, called agents, interact with one another (Huhns and Singh 1998). In the context of map generalization, it has been possible to model various characteristics of features and to implement an agent-based approach whereby agents are assigned to manage the generalization process across a geographic region (with a local perspective on the problem), and to communicate with other agents at a more regional scale (a global perspective) in order to ensure consistency in solution, and to ensure preservation of general trends across the map space (Duchêne 2003). This was the methodology utilized in the AGENT project, a European Union funded project, comprised of a consortium of universities and commercial enterprises (Lamy, Ruas, Demazeau, et al. 1999; Barrault, Regnauld, Duchêne 2001). The system was capable of analyzing various properties within and between classes of objects, of synthesizing alternate candidate solutions and evaluating the optimum choice against a set of design constraints. Where a solution was not forthcoming, a more radical or broad-scale solution was proposed and control passed from the local perspective to a more global one. Thus there existed a hierarchical structure of mico, meso and macro agents, which, in effect, modeled both a fine-scale view of design, as well as the more general view of the problem. The project commenced in 1998, and its commercial form is currently manifest in the CLARITY system from Laser Scan (http://www.laser-scan.co.uk), and continues to form the basis of on going research among a consortium of national mapping agencies across Europe under the MAGNET program. Given its adoption by a number of European NMAs it is arguably the best solution to date to the challenges of autonomous map generalization, though a number of challenges remain. The first is in the development of an interface that enables "tuning" of solutions that arise from complex emergent behavior and interactions. The second is in defining the type of information that is passed among the hierarchies of agents, and how this information is utilized in the various stages of decision making.

CONCLUSIONS

Generalization holds an important position in the development of a theoretical framework for handling geographic information "as it deals with the structure and transformation of complex spatial notions at different levels of abstraction" (van Smaalen 2003). As a modeling process, map generalization is about characterizing space in a way that precipitates out the broader contextual relationships that exist among geographic phenomena. It is about making sense of things (Krippendorf 1995) and is intrinsic to geographic ways of knowing.

In essence, a database is a system of relationships – the process of generalization is about abstracting and representing those patterns of relationships inherent among phenomena viewed at different levels of detail (similar to the goals of scientific visualization). The enduring vision is of a single, detailed database from which such multiple views can be automatically derived according to a broad range of tasks.

Over the years a variety of solutions have emerged in response to both a growing understanding of the complexities of automated map design and to the changing context of use arising from developments in information technology. Attempts at automation have highlighted the complexity of this task. It is certainly the case that the design of a map (irrespective of medium) is a hugely challenging task, though the paradigm shift afforded by data modeling techniques has called into question the appropriateness of trying to mimic the human cartographer as a basis to automation.

Developments in the field of generalization continue to advance three key areas: (1) development of algorithms for model generalizations with the focus on spatial data handling and analysis; (2) methods for creating and evaluating candidate solutions for graphical visualization and multiple representation; and (3) development of human computer interaction models that enable integration of these methodologies in both the presentation and exploration of geographic information. Research continues to reveal the subtleties of the art and science of cartography. For it to remain relevant, however, it must keep abreast of the changing environments of map use and analysis (including interoperability requirements), and the broader developments in visualization methodologies.

REFERENCES

Bader, M. and Weibel, R. 1997. Detecting and resolving size and proximity conflicts in the generalization of polygon maps. In *Proceedings of the Eighteenth International Cartographic Association Conference*, Stockholm, Sweden. Sweden: International Cartographic Association, pp. 1525–32.

Barrault, M., Regnauld, N., Duchène C., Haire, K., Baeijs, C., Demazeau, Y., Hardy, P., Mackaness, I., Ruas, A., and Weibel, R. 2001. Integrating multi agent, object oriented and algorithmic techniques for improved autmoated map generalisation. In *Proceedings of the Twentieth International Cartographic Conference*, Beijing, China. Stockholm: International Cartographic Association, pp. 2110–16.

Brassel, K. E. and Weibel, R. 1988. A review and conceptual framework of automated map generalization. *International Journal of Geographical Information Systems* 2: 229–44.

Burghardt, D. and Meier, S. 1997. Cartographic displacement using the snakes concept. In W. Foerstner and L. Pluemer (eds) *Semantic Modelling for the Acquisition of Topographic Information from Images and Maps*. Basel: Birkhaeuser Verlag, pp. 59–71.

Buttenfield, B. 1985. Treatment of the cartographic line. *Cartographica* 22: 1–26.

Cheung, C. K. and Shi, W. 2004. Estimation of the positional uncertainty in line simplification in GIS. *Cartographic Journal* 41: 37–45.

Christensen, J., Marks, J., and Shieber, S. 1995. An empirical study of algorithms for point feature label placement. *ACM Transactions on Graphics* 14: 203–32.

Christophe, S. and Ruas, A. 2002. Detecting building alignments for generalisation purposes. In D. Richardson and P. van Oosterom (eds) *Advances in Spatial Data Handling*. Berlin: Springer-Verlag, pp. 419–32.

Coppock, J. T. and Rhind, D. W. 1991. The history of GIS. In D. J. Maguire, M. F. Goodchild, and D. W. Rhind (eds) *Geographical Information Systems: Principles and Applications*. Harlow: Longman, pp. 21–43.

Costa, L. D. F. 2000. Robust skeletonization through exact Euclidean distance transform and its application to neuromorphometry. *Journal of Real Time Imaging* 6: 415–31.

Devogele, T., Trevisan, J., and Ranal, L. 1996. Building a multi-scale database with scale transition relationships. In *Proceedings of the Seventh International Symposium on Spatial Data Handling*, Delft, Netherlands. Delft: Delft University of Technology, pp. 337–52.

Downs, T. C. and Mackaness, W. A. 2002. Automating the generalisation of geological maps: The need for an integrated approach. *Cartographic Journal* 39: 137–52.

Duchêne, C. 2003. Automated map generalisation using communicating agents. In *Proceedings of the Twenty-first International Cartographic Conference*, Durban, South Africa. Stockholm: International Cartographic Association, pp. 160–9.

Dutton, G. 1999. Scale, sinuosity and point selection in digital line generalisation. *Cartography and Geographic Information Systems* 26: 33–54.

Glover, L. and Mackaness, W. A. 1999. Dynamic generalisation from single detailed database to support web based interaction. In *Proceedings of the Nineteenth International Cartographic Conference*, Ottawa, Canada. Stockholm: International Cartographic Association, pp. 1175–83.

Goodchild, M. F. and Yang, S. 1992. A hierarchical data structure for global geographic information systems. *Computer Vision, Graphics, and Image Processing* 54: 31–44.

Grünreich, D. 1985. Computer-assisted Generalisation. Unpublished Paper, CERCO-Cartography Course, Institut für Angewandte Geodasie, Frankfurt am Main.

Hartsfield, N. and Ringel, G. 1990. *Pearls in Graph Theory: A Comprehensive Introduction*. Boston, MA: Academic Press.

Hartshorne, R. 1939. *The Nature of Geography: A Critical Survey of Current Thought in the Light of the Past*. Washington, DC: Association of American Geographers, p. 249.

Heisser, M., Vickus, G., and Schoppmeyer, J. 1995. Rule-orientated definition of small area selection and combination steps of the generalization procedure. In J.-C. Muller, J. P. Lagrange, and R. Weibel (eds) *GIS and Generalization: Methodology and Practice*. London: Taylor & Francis, pp. 148–60.

Hojholt, P. 2000. Solving space conflicts in map generalisation: Using a finite element method. *Cartography and Geographic Information Science* 27: 65–73.

Huhns, M. N. and Singh, M. P. 1998. *Readings in Agents*. San Francsico, CA: Morgan Kaufmann.

Jiang, B. and Claramunt, C. 2004. A structural approach to the model generalisation of an urban street network. *GeoInformatica* 8: 157–71.

Jones, C. B., Bundy, G. L., and Ware, J. M. 1995. Map generalization with a triangulated data structure. *Cartography and Geographic Information Systems* 22: 317–331.

Jones, C. B. and Ware, J. M. 1998. Proximity relations with triangulated spatial models. *Computer Journal* 41: 71–83.

Keller, S. F. 1995. Potentials and limitations of artificial intelligence techniques applied to generalization. In J. C. Muller, J. P. Lagrange, and R. Weibel (eds) *GIS and Generalization: Methodology and Practice*. London: Taylor and Francis, pp. 135–47.

Kidner, D. B. and Jones, C. B. 1994. A deductive object oriented GIS for handling multiple representation. In *Proceedings of the Sixth International Symposium on Spatial Data Handling*, Edinburgh, Scotland. Delft: Delft University of Technology, pp. 882–900.

Kilpelainen, T. and Sarjakoski, T. 1995. Incremental generalisation for multiple representations of geographical objects. In J. C. Muller, J. P. Lagrange, and R. Weibel (eds) *GIS and Generalisation: Methodology and Practice*. London: Taylor and Francis, pp. 209–18.

Krippendorff, K. 1995. On the essential contexts of artifacts or on the proposition that "design is making sense (of things)." In V. Margolin and R. Buchanan (eds) *The Idea of Design*. Cambridge, MA: MIT Press, pp. 156–84.

Krygier, J. B. 1995. Cartography as an art and a science. *Cartographic Journal* 32: 3–10.

Lamy, S., Ruas, A., Demazeau, Y., Jackson, M., Mackaness, W. A., and Weibel, R. 1999. The application of agents in automated map generalisation. In *Proceedings of the Nineteenth International Cartographic Conference*, Ottawa, Canada. Stockholm: International Cartographic Association, pp. 1225–34.

Leitner, H. 2004. The politics of scale and networks of spatial connectivity: Transnational interurban networks and the rescaling of political governance in Europe. In E. Sheppard and R. B. McMaster (eds) *Scale and Geographic Inquiry: Nature Society and Method*. Malden, MA: Blackwell, pp. 236–55.

Luck, M. 1997. Foundations of multi-agent systems: Issues and directions. *Knowledge Engineering Review* 12: 307–18.

MacEachren, A. M. and Kraak, M. J. 1997. Exploratory cartographic visualization: Advancing the agenda. *Computers and Geosciences* 23: 335–43.

Mackaness, W. A. 1996. Automated cartography and the human paradigm. In C. H. Wood and C. P. Keller (eds) *Cartographic Design: Theoretical and Practical Perspectives*. New York: John Wiley and Sons, pp. 55–66.

Mackaness, W. A. and Mackechnie, G. 1999. Automating the detection and simplification of junctions in road networks. *GeoInformatica* 3: 185–200.

McMaster, R. B. and Shea, K. S. 1992. *Generalization in Digital Cartography: Resource Publication in Geography*. Washington DC: Association of American Geographers.

Mark, D. M. 1990. Competition for map space as a paradigm for automated map design. In *Proceedings of GIS/LIS '90*, Anaheim, CA, USA. Bethesda, MD: American Society of Photogrammetry and Remote Sensing, pp. 97–106.

Molenaar, M. 1998. *An Introduction to the Theory of Spatial Object Modelling for GIS*. London: Taylor and Francis.

Muller, J. C. 1991. Generalisation of spatial databases. In D. J. Maguire, M. F. Goodchild, and D. W. Rhind (eds) *Geographical Information Systems*. London: Longman, pp. 457–75.

Oosterom, P. V. 1995. The GAP-tree: An approach to "on-the-fly" map generalization of an area partitioning. In J. C. Muller, J. P. Lagrange, and R. Weibel (eds) *GIS and Generalization: Methodology and Practice*. London: Taylor and Francis, pp. 120–32.

Peng, W., Sijmons, K., and Brown, A. 1995. Voronoi diagram and Delaunay triangulation supporting automated generalization. In *Proceedings of the Seventeenth International Cartographic Conference*, Barcelona, Spain. Stockholm: International Cartographic Association, pp. 301–10.

Phillips, J. D. 1997. Humans as geological agents and the question of scale. *American Journal of Science* 297: 98–115.

Plazanet, C., Bigolin, N. M., and Ruas, A. 1998. Experiments with learning techniques for spatial model enrichment and line generalization. *GeoInformatica* 2: 315–33.

Priestnall, G., Hatcher, M. J., Morton, R. D., Wallace, S. J., and Ley, R. G. 2004. A framework for automated feature extraction and classification of linear networks. *Photogrammetric Engineering and Remote Sensing* 70: 73–82.

Regnauld, N. 1996. Recognition of building cluster for generalization. In *Proceedings of the Seventh International Symposium on Spatial Data Handling*, Delft, Netherlands. Delft: Delft University of Technology, pp. 185–98.

Regnauld, N. 2001. Contextual building typification in automated map generalisation. *Algorithmica* 30: 312–33.

Richardson, D. E. and Muller, J.-C. 1991. Rule selection for small-scale map generalization. In B. P. Buttenfield and R. B. McMaster (eds) *Map Generalization: Making Rules for Knowledge Representation*. Harlow: Longman, pp. 136–49.

Richardson, D. and Thomson, R. C. 1996. Integrating thematic, geometric and topological information in the generalisation of road networks. *Cartographica* 33: 75–84.

Ruas, A. 1995. Multiple paradigms for automating map generalization: Geometry, topology, hierarchical partitioning and local triangulation. In *Proceedings of Auto Carto 12*, Charlotte, NC, USA. Bethesda, MD: American Congress on Surveying and Mapping, pp. 69–78.

Ruas, A. and Mackaness, W. A. 1997. Strategies for urban map generalization. In *Proceedings of the Eighteenth ICA/ACI International Cartographic Conference*, Stockholm, Sweden. Stockholm: International Cartographic Association, 1387–94.

Sheppard, E. and McMaster, R. B. (eds) 2004. *Scale and Geographic Inquiry: Nature Society and Method*. Oxford: Blackwell.

Starr, M. K. and Zeleny, M. 1977. MCDM: State and future of the arts. In M. K. Starr and M. Zeleny (eds) *Multiple Criteria Decision Making*. New York: North-Holland, pp. 5–29.

Swyngedouw, E. 2004. Scaled geographies: Nature, place, and the politics of scale. In E. Sheppard and R. B. McMaster (eds) *Scale and Geographic Inquiry: Nature Society and Method*. Oxford: Blackwell, pp. 129–53.

Taylor, P. J. 2004. Is there a Europe of cities? World cities and the limitations of Geographical Scale Analyses. In E. Sheppard and R. B. McMaster (eds) *Scale and Geographic Inquiry: Nature, Society, and Method*. Oxford: Blackwell, pp. 213–35.

Topfer, F. and Pillewizer, W. 1966. The principles of selection. *Cartographic Journal* 3: 10–6.

van Smaalen, J. W. N. 2003. *Automated Aggregation of Geographic Objects: A New Approach to the Conceptual Generalisation of Geographic Databases*. Delft: Netherlands Geodetic Commission, p. 1.

Weibel, R. and Dutton, G. 1999. Generalising spatial data and dealing with multiple representations. In P. A. Longley, M. F. Goodchild, D. J. Maguire, and D. W. Rhind (eds) *Geographical Information Systems*. New York: John Wiley and Sons, pp. 125–56.

Weiss, G. (ed.). 1999. *Multiagent Systems: A Modern Approach to Distributed Artificial Intelligence*. Cambridge, MA: MIT Press.

Whang, Z. and Muller, J. C. 1998. Line generalisation based on analysis of shape characteristics. *Cartography and Geographic Information Systems* 25: 3–15.

Chapter 13

Geographic Information Systems and Surfaces

Nicholas J. Tate, Peter F. Fisher, and David J. Martin

Many different geographic phenomena are displayed and analyzed as surfaces. Some are among the most concrete that Geographic Information Systems (GIS) are designed to work with (*bona fide*, in the sense of Smith 2001), while others are among the most abstract. The surfaces concerned are completely different in type, but they all have one thing in common: a surface representation is appropriate under any circumstances where the phenomena being modeled can be thought of as varying *continuously* across space. Indeed, the vector polygon model is a special case of a surface in which the changes in value across space happen abruptly at polygon boundaries. The most tangible surface is the land, the ground under our feet, measured as an elevation above a particular datum, commonly mean sea level, which is conceptualized as a horizontal surface. The more abstract surfaces conceived by geographers include, for example, population density surfaces (the probability that you will meet someone at a particular location), soil pH, and atmospheric air pressure.

Conventional conceptualizations of geographic objects make a distinction between point, line, area, and surface types. The continuous nature of surfaces means that, strictly, they do not embody any topology, although our inability to create truly continuous data structures means that surfaces are represented by various approximations that may include topological information. These include the use of triangulated irregular networks (TINs), digital elevation models (DEMs), and isolines/contours, each of which are considered in more detail below. We conventionally refer to the variable of interest, represented as the height of the surface as the Z variable (and associated Z values), to distinguish it from the familiar X and Y variables of two-dimensional cartographic space.

The purpose of this chapter is to outline the nature and procedures which are available for surface modeling and visualization. In the following section we consider some of the advantages of surface representation. The focus of the second section (Surface Modeling) is on exploration of the various types of surface model used in GIS. Surface concepts provide us with particularly powerful tools for visualization which is the subject of the third section, Surface Visualization.

Surface Representation

Viewing space in terms of continuously varying scalar fields where the Z value is a continuous function of location f(X,Y) is often identified as the *field perspective* (for example, Goodchild 1992, 2003) and is one of the two most frequently used models in GI Science. Fields traditionally comprise regularly/irregularly distributed *points*, tessellations of various regular/irregular *areas*, and *isolines* (Goodchild 2003).

Point phenomena represented as surfaces can be further divided into two categories: *point* and *reference interval* functions (Nordbeck and Rystedt 1970). Point functions are those such as elevation of the land surface above sea level which, given certain assumptions, are measurable at single point locations. However, reference interval functions such as population density cannot be measured in this way and can only be measured in relation to a reference area – for example persons per hectare. In the latter case the Z value at a single point location is not constant but varies according to the size and shape of the reference area. The treatment of their Z values in surface construction varies according to whether they can be directly interpreted as Z values, or quantities from which Z must be derived. The population density at a point in the City of London would be low if measured in relation to its immediate neighborhood which comprises primarily commercial premises, but would rise as surrounding residential areas are included in the reference interval. As we continue to extend the reference interval beyond the build-up area, the density would fall again, and all of these values would be "correct" for that same location. The distinction between point and reference interval functions is more subtle than might first appear, with phenomena such as atmospheric pressure being reference interval functions (expressed as force per unit area), although the reference area is conventionally very small and measured at individual locations.

Cartographic convention has led to some phenomena being more commonly conceptualized as surfaces than others: in general terms, point functions are more readily imagined to be continuously varying surfaces and the convention of representing them by mapping isolines (for example, elevation contours, pressure isobars) that connect locations with the same Z value is long established. The continuously varying distribution of population density by contrast is more often represented in the form of shaded area (choropleth) maps which convey an impression of extensive regions of uniformity separated by sharp and geographically irregular changes at boundaries. This convention owes more to the data collection processes associated with these geographic regions than to any specific advantages that they offer for analysis or visualization. Effectively, a choropleth map of population density is one in which the value at each point is calculated using a uniquely shaped and sized reference interval. This is a manifestation of the modifiable areal unit problem (Openshaw 1984) which affects all area-based data where boundaries are imposed on a continuously varying phenomenon: both the scale of the units and details of boundary placement at a given scale affect the mapped values. Interpretation of choropleth maps of phenomena such as population density is made particularly difficult if the underlying distribution is non-uniform, as the largest population concentrations are afforded only the smallest areas on the map while unpopulated regions dominate the visual image. These types of problem are much diminished if the phenomenon is modeled as a surface, as a continuous model has no spatial units and no boundaries. An elevation matrix approximation provides us with very small, regularly placed units which can overcome many of the problems of choropleth representation. A more extensive comparison between surface and zonal models of population is presented in Martin (1996).

Key considerations in the selection of a surface representation should be its fitness for the analysis and visualization methodologies to be employed (as discussed in the following two sections). Analysis of runoff from a land surface or identification of discrete settlements from a population density surface require representations which capture surface characteristics not directly calculable from isoline or choropleth mapping and are more appropriately addressed through surface modeling. For example, Thurstain-Goodwin (2003) explores the advantages of socio-economic surface representations in a policy making context.

Surface Modeling

Surface modeling is a necessary preliminary to surface analysis or visualization. Mathematically we can conceive of a surface as a bivariate scalar field $Z = f(X,Y)$. If we can sample Z at a sufficient intensity/number of discrete points in space, we can define a discrete point model of a surface also known as a 'height field'. Similarly, if we obtain Z as a set of isolines we define a discrete line model of a surface. In the context of terrain modeling these would be identified as a Digital Evaluation Model (DEM) and a contour model respectively. In both cases the geometric continuity between the points/lines is only *implied* (Schneider 2001a). In the context of terrain modeling Schneider (2001b) and Hugentobler (2004) have argued that such surface models are less useful than models where the continuity between points is *explicitly* modeled and the surface reconstructed as a set of piecewise functions in the form of a polygonal mesh. Using this distinction we will discuss the methods of surface modeling below.

Height Fields, DEMs, and contours

The direct measurement of the Z variable at point locations is the most direct, often the most accurate, and the most costly method of surface construction since it requires the most measured data. Surface modeling in this form requires measured samples of the Z variable at sufficient spatial locations in order to be able to characterize the surface, which in turn necessarily involves choices to be made with respect to spatial sampling, the scale and pattern of variation, and the actual process of physical measurement. There are a variety of spatial sampling frames that might be suitable for the construction of surfaces by direct measurement; however, by far the most common are those that employ a regular square grid form, although irregularly distributed points can also be collected. Regular grids offer a variety of advantages for the digital/computer representation of surfaces; for example, both position and neighbors are explicit (Mark and Smith 2004). Ideally, the choice of frame is determined by the pattern of spatial variation that exists in the Z variable in reality, as well as the form and the resolution of surface model that is required. Real surfaces often manifest complex spatial patterns of scaling (dependent or fractal) and directionality (isotropic or anisotropic), and a suitable sampling pattern will need to honor these properties. Various methods might be employed to vary the density of sampling: one example used in the photogrammetric construction of terrain models is a technique based on the progressive density increases in areas of complex spatial patterns (Petrie 1990a). Once an appropriate sample frame has been established, measurement of the Z variable is then obtained for each sample location – a process that might require fieldwork, or lab-based measurement, and which may prove cost-prohibitive. The grid-based frame for sampling soil fertility characteristics at Broom's Barn Experimental station is shown in Figure 13.1. A more irregular sample frame of Light Detection and Ranging (LiDAR) points used to construct a DEM for a section of Ribble Catchment in NW England is shown in Figure 13.2.

An inherent feature of most surface-type geographic phenomena is that we often cannot comprehensively capture their form by a measured sample alone. Surface modeling is therefore nearly always based on some combination of measurement

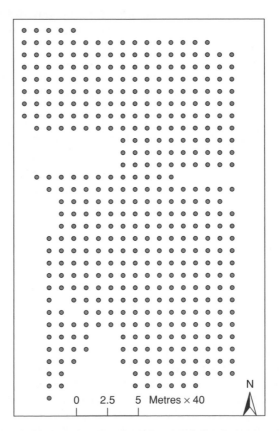

Fig. 13.1 Sample frame for soil fertility characteristics recorded at Broom's Barn Experimental Station, Suffolk
Webster and McBratney 1987, Webster and Oliver 2001

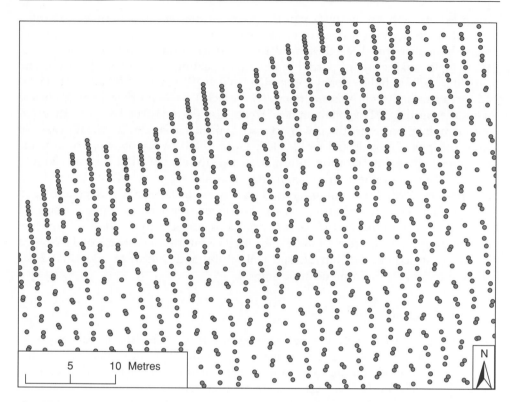

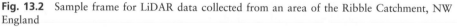

Fig. 13.2 Sample frame for LiDAR data collected from an area of the Ribble Catchment, NW England
This map is derived from data Copyright Environment Agency Science Group – Technology (2004)

and statistical modeling. In the case of point functions the former may be known values of the surface at measured locations, but for reference interval functions these will be estimates based on specific reference areas, which may themselves be irregular in size and shape. With known values at measured locations but at an insufficient spatial density/intensity, the statistical model required is some form of interpolation, that is, the estimation of Z at unmeasured locations. A great variety of generic point-based as well as surface-specific interpolation methods can be employed, and there are numerous surveys and classifications of interpolation methods relevant to surface construction (see Lam 1983, Oliver and Webster 1990; Petrie 1990a, Watson 1992, Hutchinson 1993, Myers 1994, Mardia, Kent, Goodall, and Little 1996, Mitas and Mitasova 1999).

The construction of surfaces – particularly terrain surfaces – from contours/isolines has been a topic of interest for some time in the GIS community, where paper maps were often the only convenient source of surface information (Legates and Willmott 1986, Yoeli 1986, Petrie 1990b; see also Hutchinson, Chapter 8 in this volume for additional discussion of this topic). As noted by Hormann, Spinello, and Schroeder (2003) contours can be treated as sets of points and subjected to standard interpolation methods, although Gold (1994) has observed that artifacts

– such as zero slope of the surface at the position of each sample point – are often introduced into surfaces derived from such methods. Various purpose-designed algorithms have therefore been developed to construct surfaces from contour maps, for example making use of partial differential equations (Hormann, Spinello, and Schroeder 2003), bidirectional Hermitian splines (Legates and Willmott 1986) and "area stealing" from Voronoi tilings (Gold 1994). TINs (further discussed below under the heading "Poygonal models") can also be constructed from digitized contour points; however, these will often contain flat triangles in certain cases of peaks and pits (Thibault and Gold 2000, Gold 2003). Attempts have been made to augment the point elevation information in TIN construction by using graph-based methods developed from computer graphics approaches (Amenta, Bern, and Eppstein 1998). Here, graphs derived both from the Delaunay triangulation and Voronoi/Thiessen polygonization of a set of digitized contours, which in a connected form are termed a *crust* and *skeleton/medial* axis respectively (using the jargon of Amenta, Bern, and Eppstein 1998), can be used to inform the process of TIN-based surface construction (see Thibault and Gold 2000 and Dakowicz and Gold 2002 for examples in a terrain modeling context).

Polygonal models

As noted above, surfaces can be represented in terms of discrete models of points and lines such as a DEM and contour map. These models can be estimated to higher resolution point models using some form of interpolation. However, as noted above, continuity is only *implied* between individual points. An alternative is to reconstruct the surface *explicitly* in continuous form. Surface reconstruction from scattered and often noisy point samples has been of considerable interest in computer graphics (for example, Hoppe, DeRose, Duchamp, McDonald, and Stuetzle 1992, Amenta, Bern, and Eppstein 1998, Xie, Wang, Hua, Qin, and Kaufman 2003) where the aim is to represent and model a variety of solid 3D objects by using only partial information about the original surface (Hoppe, DeRose, Duchamp, et al. 1994). Although much of this work often addresses more mathematically complex 3D *manifold* surfaces than the simpler 2.5D surfaces encountered in GIS, many of the algorithms developed have found application in a GIS context (for example, Thibault and Gold 2000, Gold 2003). Surface reconstruction methods are often classified into those that construct *piecewise* (or patchwise) *polygonal surfaces* between data points and those that employ *approximation techniques* which fit functions to the data (Hoppe, DeRose, Duchamp 1992, Xie, Wang, Hua, Qin, and Kaufman 2003).

In a GIS context polygonal surfaces are less frequently encountered in a regular square grid form but more frequently encountered in a triangular/TIN form (Figure 13.3). De Floriani, Marzano, and Puppo (1996) have defined this polygonal surface representation[1] as a combination of geometric domain partition (for example, a set of triangles or squares) and the function defined on the partition.

Often the functions that are used to model each triangle (or square) are simply *piecewise linear* (or *bilinear*) as in Figure 13.3. However, the use of nonlinear functions such as Coons patches, NURBS, Clough-Tocher and Bezier splines – long popular in Computer Aided Geometric Design (CAGD) for smooth modeling (for example, Foley, Van Dam, Feiner, and Hughes 1990, Barnhill 1993, Hoppe, DeRose, Duchamp 1994, Farin 1997), and long cited in the GIS and terrain modeling

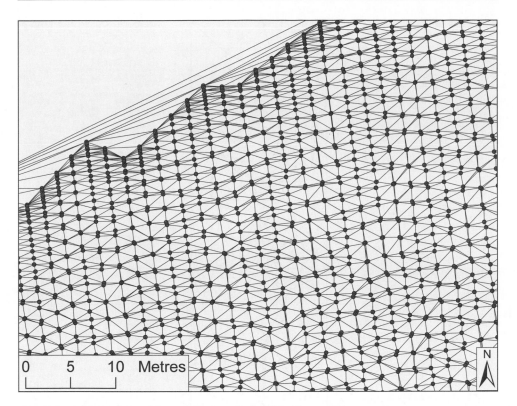

Fig. 13.3 A TIN generated from the points displayed in Figure 13.2
This map is derived from data Copyright Environment Agency Science Group – Technology (2004)

literature (see Gold 1979, McCullagh 1990, Watson 1992, Mark and Smith 2004, and references therein) – are becoming more popular within a GIS context (Schneider 2001b, Hugentobler 2002, Hugentobler 2004). Importantly, such non-linear functions allow the expression of varying degrees of continuity.

Mathematically, continuity is defined as geometric continuity G^n (or parametric continuity C^n), where n indicates the order of the derivative (Foley, Van Dam, Feiner, and Hughes 1990). In this context we can define G^0 where the surface function itself varies continuously, G^1 and G^2 which indicate continuity in first and second derivatives, that is, surface slope and surface curvature respectively (Hugentobler 2004). For example, a simple bilinear function that might be used to model each square of a regular square grid allows the surface function to vary continuously with no gaps – it is therefore G^0 continuous – but the breaks of slope between each cell that will occur will preclude G^1 and G^2 continuity. Different piecewise functions will impart different degrees of continuity to the surface (Hugentobler 2004) and a different visual appearance (McCullagh 1990). Certain applications of surfaces – for example, the use of terrain surfaces for the modeling of flows in geomorphology (Mark and Smith 2004) as well as for modeling erosion and deposition (for example, Mitasova, Hofierka, Zlocha, and Iverson 1996) – require local continuity. However, many natural surfaces are non-differentiable (fractal) at certain scales, contradicting the usual assumption of differentiability made in much modeling work

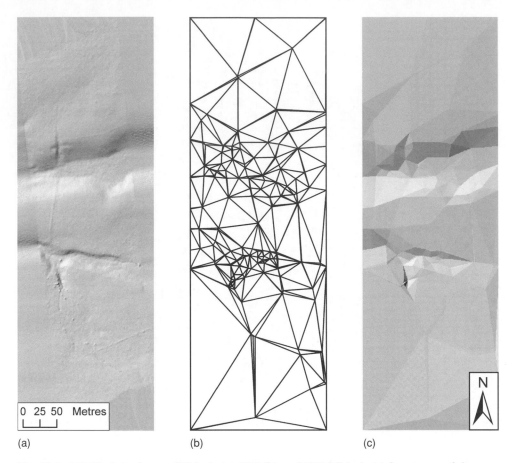

(a)   (b)   (c)

Fig. 13.4 LiDAR derived raster hillshade (a), TIN (b), and TIN hillshade (c) for an area of the Ribble Catchment, NW England
This map is derived from data Copyright Environment Agency Science Group – Technology (2004)

(Mark and Smith 2004). In a GIS context the degree of continuity required, and the scales that continuity applies to determine an accurate representation of a given surface are not always straightforward to determine.

Although a polygonal model in a regular square grid form is a logical progression from the points collected in the form of a height field/DEM, TINs are often preferred primarily because they allow more efficient representations of surfaces and are easily adapted (De Floriani, Marzano, and Puppo 1996) in proportion to the complexity of the surface (Mark and Smith 2004). A TIN (with hillshade) is visible in Figure 13.4 alongside the original hillshaded DEM.

Such efficiencies are also relevant in the visualization of surfaces where a smaller scale or lower level of detail (LOD) is required; for example, in rendering more distant objects in a scene (Luebke 2001). Views and their associated TINs with a differing LOD are displayed in Figure 13.5 where Figure 13.5a indicates the visualization and TIN with a constant LOD and Figure 13.5b indicates the same view/TIN with an adaptive LOD proportional to the viewpoint.

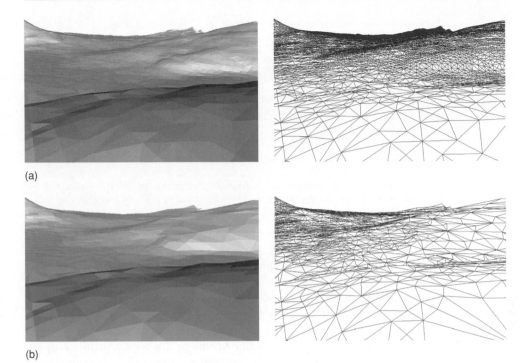

(a)

(b)

Fig. 13.5 Perspective surface views and TINS with (b) and without (a) adaptive LOD proportional to the viewpoint
From Garland 1999, used with permission

There has been considerable effort in computer graphics and GIS/terrain modeling to develop efficient structures for multi-resolution TINs, and algorithms for TIN simplification both direct from the original height field/DEM (for example, Fowler and Little 1979, Lee 1991, Little and Shi 2001) and from an existing TIN (for example, Puppo and Scopigno 1997, Garland 1999, Kidner, Ware, Sparkes, and Jones 2000, Luebke 2001, Danovaro, De Floriani, Magillo, Mesmoudi, and Puppo 2003).

Reference interval functions

A different approach is required when we wish to model surfaces from reference interval functions, for example population counts for areas. Each count relates to an (initially irregular) reference area. In modeling this type of information in surface form we may either produce a density value at a reference point for each area and interpolate between these points, or explicitly redistribute the count associated with each area into a fine grid approximating the required surface. The latter introduces the possibility of having zero-value (unpopulated) regions within the output surface and is suitable with small areas and high resolution grid references. Both approaches require the presence of reference points for each area. For interpolation a simple geometric centroid may be sufficient. The density value for each area is assigned to its centroid and an appropriate interpolation algorithm applied as for point reference functions.

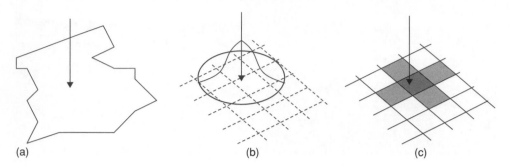

(a)   (b)   (c)

Fig. 13.6 Kernel estimation of a reference-interval function: (a) original data collection area with population count and centroid location; (b) local kernel superimposed at centroid location; and (c) population count redistributed onto grid according to kernel function

For the redistribution approach a weighted centroid will produce better results, and if this is not directly available in the input data it may be possible to create one by overlaying ancillary data onto the original areas in order to estimate such centroids. An example would be the overlay of remotely sensed land use data onto census areas in order to identify centroids of populated areas to which the population counts can be assigned. Redistribution algorithms of this type typically employ a spatial kernel which is focused over each centroid in turn and a distance decay model for the redistribution of the count into the immediate neighborhood applied. The process is illustrated by Figure 13.6.

In Figure 13.6a a population count is associated with the area centroid location. In Figure 13.6b a kernel function is centered on the centroid in order to estimate weights associated with each location on a fine-resolution grid. If the size of this kernel is varied according to, for example, the number of other centroids encountered, this is a form of adaptive kernel estimation (Bailey and Gatrell 1995). In Figure 13.6c the weights associated with each cell are used to guide redistribution of the original population count onto the surface model. The form of the surface is strongly determined by the choice of kernel function: larger kernels will result in smoother surface representations. Some cells will not fall within the kernel of any centroids and thus remain unpopulated, while others will receive counts from several centroids, allowing the surface model to reconstruct aspects of the underlying settlement geography. This is the method employed by Martin (1996) and Martin, Tate, and Langford (2000) in population density surface construction from census area centroid locations.

Surface Visualization

Rendering

Surface representation provokes a number of possible visualization strategies. First among them is the classic contour map (Figure 13.7).

Historically this has been the most important representation for two reasons: first, it is depicted by a set of lines in Cartesian space and as such it is easy to construct

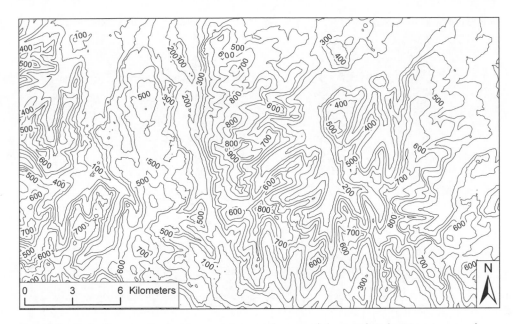

Fig. 13.7 An isoline visualization or contour map of a part of the British Lake District centered on the Helvellyn range
The map is derived from the Ordinance Survey 50 m DEM, Panorama product © Crown copyright/database right 2006. An Ordnance Survey/EDINA supplied service

and relatively inexpensive to reproduce; and second, it leaves much space in any map empty and therefore available for other information, particularly in topographic mapping. If these other information (place names, data values, etc.), or reproduction costs are still a concern then the contour map remains the most effective representation. The contour map, however, leaves much to the skill of the map reader, and the understanding of contours is not necessarily as intuitive as those who can appreciate them may believe. Furthermore, contouring is about the least informative visualization method.

The use of a color fill between contours is the next simplest method (Figure 13.8), and embeds an ordering in the colors used between contours.

This has been less-favored due to the cost of production and reproduction, but, with computer technology and the gridded data in particular, it is actually easier than most other methods for visualization of surface data in most GIS. The choice of colors, however, needs to respect the properties of color space, and blend through a limited part of that space. A number of conventions for coloring surfaces exist, but can be confusing. The classic method is to use a monochrome representation varying the lightness or intensity of the color. For example, when terrain is visualized, the highest ground should normally be shown in the palest color (white in grey scale image; Figure 13.8), but when the subject is some other environmental or social parameter, it is more usual to show the largest values with the darkest color. Terrain can also be illustrated with a multi-hue color scheme as is done in atlases, with yellow or brown through green to blue and white being a typical color scheme for increasing altitude.

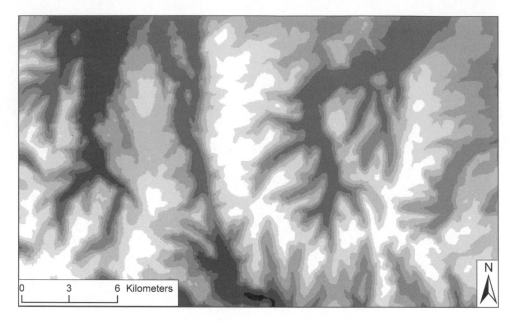

Fig. 13.8   A filled contour visualization of the same area as is shown in Figure 13.7
© Crown copyright/database right 2006. An Ordnance Survey/EDINA supplied service

Also now commonplace are 2.5D or pseudo-3D views (Figure 13.9), so-called to distinguish them from true 3D navigable views discussed in the section on navigable views below. Again, the cost and skill required for constructing these views was a deterrent to their reproduction in the past, but in computer systems it is trivial to generate perspective views of the surface. There are two principal types, *isometric*, and *perspective*, where the former ignores the complexity of perspective. However, there is little difference, when viewed from a distance, typical in cartographic representation. The problem of these views is that hills in the foreground will conceal parts of the terrain rather than revealing it as is usually the intention in mapping.

Derivatives and drapes

It is relatively simple to derive first and second order derivatives of any surface, including the gradient or slope of the surface, the aspect and curvature (plan and profile). These too can be visualized. Curvature and slope are continuous variables and can be illustrated with simple monochrome colorings, such as a grey-scale from white to dark grey (Figure 13.10), but aspect is measured as an angular bearing and so should use a color scheme where 360° is colored the same as 0°.

Color hue can be used to visualize aspect and is sometimes specified in Intensity–Hue–Saturation schemes by 0°–360° (Brown and Feringa 2003). Moellering and Kimerling (1990) and Brewer and Marlow (1993) (see also the cover illustration of MacEachren and Fraser-Taylor [1994]), have suggested combining hue to indicate aspect and lightness (or intensity) to indicate slope. The resulting representations are very striking and informative.

(a)

(b)

Fig. 13.9 A psuedo-3D view of the same general area as is shown in Figure 13.7, showing (a) a filled contour drape and (b) a hillshaded drape
© Crown copyright/database right 2006. An Ordnance Survey/EDINA supplied service

Another striking effect for surface visualization is produced when hillshading is used as a grey-scale background (lightness), and hue and saturation are used to show elevation. This scheme is used in many atlases and topographic mappings, although the hue-saturation scheme for elevation varies.

Contouring and color filled contours are appropriate for viewing the gross properties of a surface, but the detail of that surface will only appear when the derivatives are visualized (Wood and Fisher 1993, Wood 1994). To see detail, the most effective methods for visualization are entirely based on the derivatives. Mapping slope will show areas of steepness occurring in bands in a DEM where there are no bands of steep slope in the field (they are ghosts of the digitized contour lines from which some DEMs are created). Perhaps hillshading is the most effective (Figure 13.11) of all, showing many different artifacts of a DEM where contour and other mapping do not.

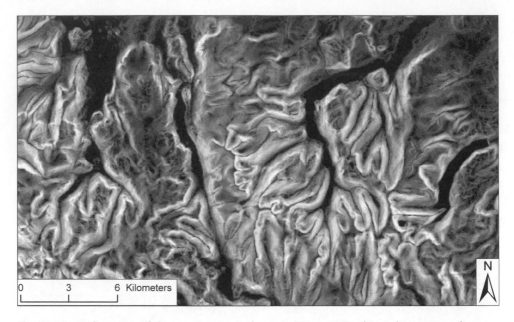

Fig. 13.10   A slope map of the same area as is shown in Figure 13.7; white indicates steep slopes and black flat
© Crown copyright/database right 2006. An Ordnance Survey/EDINA supplied service

Over any surface it is also possible to drape some other theme. Many such drapes are used, common ones being an aerial photograph or satellite image of the area, and surface derivatives. The information has to be georegistered to the same reference framework, but otherwise the process is quite straightforward in most modern GIS.

Navigable views

The ultimate form of visualization of a surface is as a navigable "Virtual Reality" (VR) model. The difference between the 2.5D view and the VR model is simply that the user has the ability to navigate around the VR model, they can immerse themselves in the 3D model, and they can traverse it (Brodlie, Dykes, Gillings, et al. 2002; Figure 13.12). VR models are all grounded in the same basic technology but come in many different versions, from the simple "through-the-window" VR on the standard computer screen, to VR theaters, caves, and headsets which all give increasing amounts of personal immersion in the environment (Brodlie and El-Khalili 2002; see also Chapter 17 by Batty, in this volume for additional discussion of VR and GI Science).

VR models bring with them the desire to create photo-realistic views, which look as the actual landscape might look. If the user intends to hover over the landscape, then it can be sufficient to drape an aerial photograph over the landscape to give a realistic-looking context (Miller, Dunham, and Chen 2002), but if the viewer is to be embedded in the landscape it is not. Ultimately the VR view is based on a

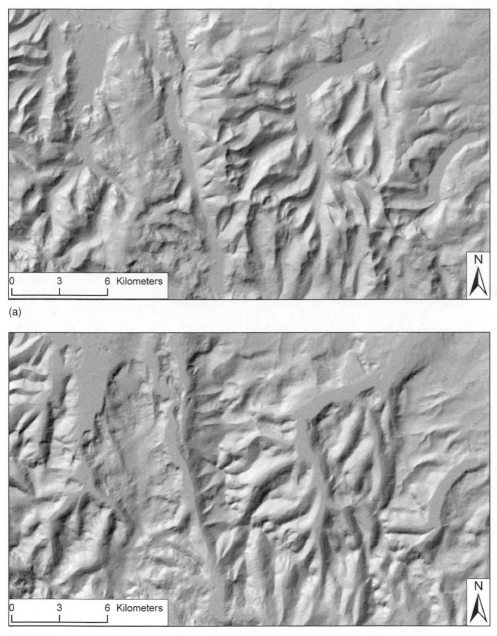

(a)

(b)

Fig. 13.11 Hillshaded maps of the same area as is shown in Figure 13.7; (a) is illuminated from the northwest and (b) from the southeast

© Crown copyright/database right 2006. An Ordnance Survey/EDINA supplied service

Fig. 13.12 A navigable 3D view of the area around the University of Leicester. Buildings are
extruded from the land surface as prisms using only the number of floors for height without accurate
roof lines or photographic drapes for detail or the elevations
Some Ordnance Survey Landline data was used © Crown copyright/database right 2006. An
Ordnance Survey/EDINA supplied service

sampling or abstraction of the landscape (elevations, houses, trees, etc.) and so the
challenge of generating a photo-realistic view is considerable (Ervin and Hasbrouck
2001). The principal problem with realistic rendering of scenes is that so much moves
in the real world, and if it does not move in the VR model the model looks like
what it is, an abstraction (Gillings 2002).

CONCLUSIONS

In this chapter we have reviewed the unique characteristics of surface-type geographic
phenomena. These are characterized by continuous variation in some variable of
interest over geographic space and cover a wide range of application areas. The
challenge of surface representation in GIS concerns selection of an appropriate tradeoff between costly measurement and complex modeling. Nevertheless, a number of
standard approaches are available which allow us to achieve close approximations
to true surface properties and these support a range of enormously powerful visualization and analysis functions.

ACKNOWLEDGEMENTS

The data representations in Figure 13.1 were kindly provided by Richard Webster and Rothamsted Research. The data used in Figures 2, 3, and 4 were provided courtesy of the Environment Agency Science Group. Figure 13.5 is courtesy of Michael Garland. The assistance of Kate Moore in assembling the database for Figure 13.11 is also acknowledged.

ENDNOTE

1 This they term a "Digital Surface Model" which contrasts with the use of the term in GIS/Photogrammetry as an elevation model which includes vegetation/building in addition to the bare earth (Weibel 1997).

REFERENCES

Amenta, N., Bern, M., and Eppstein, D. 1998. The crust and the β-skeleton: Combinatorial curve reconstruction. *Graphical Models and Image Processing* 60: 125–35.

Bailey, T. C. and Gatrell, A. C. 1995. *Interactive Spatial Data Analysis*. Harlow: Longman.

Barnhill, R. E. 1983. A survey of the representation and design of surfaces. *IEEE Computer Graphics and Applications* 3: 9–16.

Brewer, C. A. and Marlow, K. A. 1993. Color representation of aspect and slope simultaneously. In *Proceedings of Auto Carto 11*, Bethesda, MD, USA. Bethesda, MD: American Society for Photogrammetry and Remote Sensing, pp. 328–37.

Brodlie, K., Dykes, J., Gillings, M., Haklay, M. E., Kitchin, R., and Kraak, M. J. 2002. Geography in VR: Context. In P. F. Fisher and D. J. Unwin (eds) *Virtual Reality in Geography*. London: Taylor and Francis, pp. 7–16.

Brodlie, K. and El-Khalili. 2002. Web-based virtual environments. In P. F. Fisher and D. J. Unwin (eds) *Virtual Reality in Geography*. London: Taylor and Francis, pp. 35–46.

Brown, A. and Feringa, W. 2003. *Colour Basics for GIS Users*. Englewood Cliffs, NJ: Prentice-Hall.

Dakowicz, M. and Gold, C. M. 2002. Extracting meaningful slopes from terrain contours. In P. M. A. Sloot, C. J. K. Tan, J. J. Dongarra, and A. G. Hoekstra (eds) *Computational Science*. Berlin, Springer-Verlag: Lecture Notes in Computer Science No 2329: 144–53.

Danovaro, E., De Floriani, L., Magillo, P., Mesmoudi, M. M. and Puppo, E. 2003. Morphology-driven simplification and multi-resolution modeling of terrains. In *Proceedings of the Eleventh International Symposium on Advances in Geographical Information Systems (ACM-GIS 2003)*, New York, USA. New York: Association for Computing Machinery, pp. 63–70.

De, Floriani. L., Marzano, P., and Puppo, E. 1996. Multiresolution models for topographic surface description. *The Visual Computer* 12: 317–45.

Ervin, S. M. and Hasbrouck, H. H. 2001. *Landscape Modeling: Digital Technique for Landscape Visualization*. NewYork: McGraw-Hill.

Farin, G. 1997. *Curves and Surfaces for Computer-Aided Geometric Design: A Practical Guide* (4th edn). San Diego, CA: McGraw-Hill.

Foley, J. D., Van Dam, A., Feiner, S., and Hughes, J. 1990. *Computer Graphics: Principles and Practice* (2nd edn). Massachusetts: Addison-Wesley.

Fowler, R. J. and Little, J. J. 1979. Automatic extraction of irregular network digital terrain models. *Computer Graphics* 13: 199–207.

Garland, M. 1999. Multi-resolution modeling: Survey and future opportunities. In *Proceedings of Eurographics '99*, Milan, Italy. Aire-la-Ville, Switzerland: European Association for Computer Graphics, pp. 111–31.

Gillings, M. 2002. Virtual archaeologies and the hyper-real. In P. F. Fisher and D. J. Unwin (eds) *Virtual Reality in Geography*. London: Taylor and Francis, pp. 17–34.

Gold, C. M. 1979. Triangulation based terrain modeling: Where are we now? In *Proceedings of Auto Carto 4*, Baltimore, MD, USA. Bethesda, MD: American Society for Photogrammetry and Remote Sensing, pp. 104–11.

Gold, C. M. 1994. An object-based method for modeling geological surfaces containing linear data. In *Proceedings of the Annual Meeting of the International Association for Mathematical Geology*, Mont Tremblant, Quebec. Kingston, Ontario: International Association for Mathematical Geology, pp. 141–6.

Gold, C. M. 2003. But is it GIS? *Journal of Geospatial Engineering* 5: 11–26.

Goodchild, M. F. 1992. Geographical data modeling. *Computers and Geosciences* 18: 401–8.

Goodchild, M. F. 2003. The nature and value of geographic information. In M. Duckham, M. F. Goodchild, and M. F. Worboys (eds) *Foundations of Geographic Information Science*. London: Taylor and Francis, pp. 19–32.

Hoppe, H., DeRose, T., Duchamp, T., Halstead, M., Jin, H., McDonald, J., Schweitzer, J., and Stuetzle, W. 1994. Piecewise smooth surface reconstruction. In D. Schweitzer, A. Glassner, and M. Keeler (eds) *Proceedings of the Twenty-first Annual Conference on Computer Graphics and Interactive Techniques*. New York: ACM Press, pp. 295–302.

Hoppe, H., DeRose, T., Duchamp, T., McDonald, J., and Stuetzle, W. 1992. Surface reconstruction from unorganized points. In J. J. Thomas (ed.) *Proceedings of the Nineteenth Annual Conference on Computer Graphics and Interactive Techniques*. New York: ACM Press, pp. 71–8.

Hormann, K., Spinello, S., and Schroeder, P. 2003. C^1 continuous terrain reconstruction from sparse contours. In *Proceedings of the Eighth International Workshop on Vision, Modeling, and Visualization*, Munich, Germany, pp. 289–97.

Hugentobler, M. 2002. Interpolation of continuous surfaces for terrain modeling with coons patches. In *Proceedings of the Tenth GIS Research UK Annual Conference*, Sheffield, United Kingdom. Sheffield, United Kingdom: GISRUK, pp. 13–15. (Available on-line via http://www.shef.ac.uk/gisruk/.)

Hugentobler, M. 2004. Terrain Modeling with Triangle Based Free-Form Surfaces. Unpublished PhD Dissertation, University of Zurich.

Hutchinson, M. F. 1993. On thin plate splines and kriging. In M. E. Tarter and M. D. Lock (eds) *Computing and Science in Statistics 25*. Berkeley, CA: Interface Foundation of North America, pp. 55–62.

Kidner, D. B., Ware, J. M., Sparkes, A. J., and Jones, C. B. 2000. Multiscale terrain and topographic modeling with the implicit TIN. *Transactions in GIS* 4: 379–408.

Lam, N. S. N. 1983. Spatial interpolation methods: A review. *American Cartographer* 10: 129–49.

Lee, J. 1991. A comparison of the existing methods for building TIN models of terrain. *International Journal of Geographical Information Systems* 5: 267–85.

Legates, D. R. and Willmott, C. J. 1986. Interpolation of point values from isoline maps. *American Cartographer* 13: 308–23.

Little, J. J. and Shi, P. 2001. Structural lines, TINs and DEMs. *Algorithmica* 30: 243–63.

Luebke, D. P. 2001. A developer's survey of polygonal simplification algorithms. *IEEE Computer Graphics and Applications* 21: 24–35.

McCullagh, M. J. 1990. Digital terrain modeling and visualization. In G. Pertrie and T. J. M. Kennie (eds) *Terrain Modeling in Surveying and Civil Engineering*. London: Whittles Publishing/Thomas Telford, pp. 128–51.

MacEachren, A. M. and Fraser-Taylor, D. 1994. *Visualization in Modern Cartography*. Oxford: Elsevier.

Mardia, K. V., Kent, J. T., Goodall, C. R., and Little, J. A. 1996. Kriging and splines with derivative information. *Biometrika* 81: 207–21.

Mark, D. M. and Smith, B. 2004. A science of topography: From qualitative ontology to digital representations. In M. P. Bishop and J. Shroder (eds) *Geographic Information Science and Mountain Geomorphology*. Chichester: Springer-Praxis: 75–100.

Martin, D. 1996. An assessment of surface and zonal models of population. *International Journal of Geographical Information Systems* 10: 973–89.

Martin, D., Tate, N. J., and Langford, M. 2000. Refining population surface models: Experiments with Northern Ireland census data. *Transactions in GIS* 4: 343–60.

Miller, D. R., Dunham, R. A., and Chen, W. 2002. The application of VR modeling in assessing potential visual impacts of rural development. In P. F. Fisher and D. J. Unwin (eds) *Virtual Reality in Geography*. London: Taylor and Francis, pp. 131–43.

Mitas, L. and Mitasova, H. 1999. Spatial interpolation. In P. A. Longley, D. J. Maguire, M. F. Goodchild, and D. W. Rhind (eds) *Geographical Information Systems: Principles and Applications*. New York: John Wiley and Sons, pp. 481–92.

Mitasova, H., Hofierka, J., Zlocha, M., and Iverson, J. 1996. Modeling topographic potential for erosion and deposition using a GIS. *International Journal of Geographic Information Systems* 10: 629–41.

Moellering, H. and Kimerling, A. J. 1990. A new digital slope-aspect display process. *Cartography and Geographic Information Systems* 17: 151–9.

Myers, D. E. 1994. Spatial interpolation: An overview. *Geoderma* 62: 17–28.

Nordbeck, S. and Rystedt, B. 1970. Isarithmic maps and the continuity of reference interval functions. *Geografiska Annaler* 52B: 92–123.

Oliver, M. A. and Webster, R. 1990. Kriging: A method of interpolation for geographical information systems. *International Journal of Geographic Information Systems* 4: 313–32.

Oliver, M. A. and Webster, R. 1991. How geostatistics can help you. *Soil Use and Management* 7: 206–17.

Openshaw, S. 1984. *The Modifiable Areal Unit Problem: Concepts and Techniques in Modern Geography*. Norwich: Geo Books.

Petrie, G. 1990a. Modeling, interpolation and contouring procedures. In G. Petrie and T. J. M. Kennie (eds) *Terrain Modeling in Surveying and Civil Engineering*. London: Wittles Publishing/Thomas Telford, pp. 112–27.

Petrie, G. 1990b. Terrain data acquisition and modeling from existing maps. In G. Petrie and T. J. M. Kennie (eds) *Terrain Modeling in Surveying and Civil Engineering*. London: Wittles Publishing/Thomas Telford: 85–111.

Puppo, E. and Scopigno, R. 1997. Simplification, LOD and multi-resolution: Principles and applications. In *Proceedings of the Eighteenth Annual European Conference on Computer Graphics (Eurographics '97)*, Budapest, Hungary. Aire-la-Ville, Switzerland: European Association for Computer Graphics.

Schneider, B. 2001a. On the uncertainty of local shape of lines and surfaces. *Cartography and Geographic Information Science* 28: 237–47.

Schneider, B. 2001b. Phenomenon-based specification of the digital representation of terrain surfaces. *Transactions in GIS* 5: 39–52.

Smith, B. 2001. Fiat objects. *Topoi* 20: 131–48.

Thibault, D. and Gold, C. M. 2000. Terrain reconstruction from contours by skeleton construction. *GeoInformatica* 4: 349–73.

Thurstain-Goodwin, M. 2003. Data surfaces for a new policy geography. In P. A. Longley and M. Batty (eds) *The CASA Book of GIS*. Redlands, CA: ESRI Press, pp. 145–70.

Watson, D. F. 1992. *Contouring: A Guide to the Analysis and Display of Spatial Data*. Oxford: Pergamon.

Webster, R. and McBratney, A. B. 1987. Mapping soil fertility at Broom's Barn by simple Kriging. *Journal of the Science of Food and Agriculture* 36: 97–115.

Webster, R. and Oliver, M. A. 2001. *Geostatistics for Environmental Scientists*. Chichester: John Wiley and Sons.

Weibel, R. 1997. Digital terrain modeling for environmental applications: A review of techniques and trends. In *Progress Seminar on Developments and Applications of Digital Elevation Models in Environmental Modeling*, Vienna, Austria.

Wood, J. 1994. Visualizing contour interpolation accuracy in digital elevation models. In H. M. Hearnshaw and D. J. Unwin (eds) *Visualization in Geographical Information Systems*. Chichester: John Wiley and Sons, pp. 168–80.

Wood, J. D. and Fisher, P. F. 1993. Assessing interpolation accuracy in elevation models. *IEEE Computer Graphics and Applications* 13: 48–56.

Xie, H., Wang, J., Hua, J., Qin, H., and Kaufman, A. 2003. Piecewise C^1 continuous surface reconstruction of noisy point clouds via local implicit quadric regression. In *Proceedings of the IEEE Conference on Visualization*, Seattle, WA: 91–8.

Yoeli, P. 1986. Computer executed production of a regular grid of height points from digital contours. *The American Cartographer* 13: 219–29.

Chapter 14

Fuzzy Classification and Mapping

Vincent B. Robinson

The mapping process is basically concerned with determining "what" is "where" and representing that "what is where" knowledge in a form that can be stored for later retrieval, display, and analysis. With the advent of computerized Geographic Information Systems (GIS) and related technologies, the digital representation of information from the mapping exercise led to the development of spatial databases that permit the representation, manipulation, and display of geographical phenomena with greater rigor, detail, and consistency than ever before. However, owing to the characteristics of the mapping methodology and/or the nature of phenomenon being mapped it is often difficult to be absolutely certain of "what is where." For many phenomena that are distributed over space there are no crisp boundaries that can be identified to differentiate classified geographic zones (Cheng, Molenaar, and Lin 2002). For examples, the boundary between beach and foreshore, between woodland and grassland, and between urban and rural areas may be gradual rather than defined by a sharp boundary. Furthermore, when we use a remotely sensed imagery to extract objects of interest, there are pixels that may contain sub-pixel objects, trans-pixel objects, boundary pixels and/or natural intergrades (Foody 1999). The mixture of spectral information at the sub-pixel scale can lead to uncertain classification and indeterminate boundaries. These problems of uncertainty have led many to use classification techniques based on fuzzy set theory.

As researchers and practitioners wrestled with how to represent, manipulate, and manage geographic data within the developing field of GIS, issues of error and uncertainty began to be recognized (Robinson and Frank 1985), and the potential of fuzzy set theory (Zadeh 1965) in GIS began to be addressed. McBratney and Odeh (1997), in their perspective piece on fuzzy sets in soil science, note that the inadequacy of traditional Boolean logic for the design of spatial databases for GIS has been identified since the 1980s (Robinson and Strahler 1984). In his essay on the principles of logic for GIS, Robinove (1989) discusses the potential value of fuzzy set theory. Thus, the relevance of fuzzy sets to GIS was recognized early on by a small group of researchers.

This chapter first presents some basic concepts of fuzzy set theory and how they are related to fuzzy classification and mapping for GIS. A more in-depth presentation of many of the topics summarized in this chapter can be found in other works (for example, Klir and Yuan 1995, Burrough and Frank 1996, McBratney and Odeh 1997, Burrough and McDonnell 1998, Robinson 2003). The reader is encouraged to consult them as well as the references cited.

Fuzzy Classification: The Basics

In set theory the membership of an element in a particular set is defined by a characteristic function. Nonfuzzy classification uses characteristic functions that result in a location being classified as either a member of a set or not. For example, in GIS applications percent slope is often calculated for each location in a study area. We may use a characteristic function that says that all locations where the percent slope is between 5 and 17 will be classified as *gentle slope*. Now consider the question of at which value of percent slope specifically a location goes from being gentle to not gentle?" Our rule implies that locations with a percent slope of 17.01 are classed as *not gentle* while locations with 16.98 are *gentle*. Perhaps a slope of 17.01 is simply not as gently sloping as one of 16.98, hence locations with a slope of 17.01 might be considered both gentle and not gentle but to differing degrees. This is the fundamental proposition upon which fuzzy set theory is based. In other words, the characteristic function indexes the degree to which a location is a member of a set with larger values denoting higher degrees of set membership. Such a function is referred to as a *membership function*. The set defined by such a membership function is a *fuzzy set*.

Assigning Membership Values

It is common in the GIS literature to describe the approaches to assigning fuzzy membership values as following either the Semantic Import (SI) or Similarity Relation (SR) model (Robinson 1988, Burrough and McDonnell 1998). The SI model is based on using *a priori* knowledge to construct fuzzy membership functions with which individuals can be assigned a membership grade. If the process of assigning membership values depends on similarity relations then it fits the SR model. The essential aspect of the SR model is that membership values of elements are a function of how similar/ dissimilar an element is to some ideal. The essential difference between the SI and SR models is the degree to which an approach is dependent on similarity measures to derive a function that assigns membership values to individual elements. So rather than organize the methods of fuzzy classification along SI/SR lines, the approach in this chapter is to first consider use of models that rely heavily on a priori knowledge to arrive at a classification. Then the important technique of fuzzy clustering will be considered since it focuses on extracting fuzzy classification 'rules' from the data itself.

Fuzzy Classification with a priori Knowledge

One common approach to specifying a membership function is to use a plausible standard function (for example, triangular or sigmoidal) (Robinson 2003). This presumes some degree of a priori knowledge. No matter which standard function is used there remains the necessity of having the function be applicable to the problem at hand. In other words, parameters must be specified so the membership function makes sense in the context of the problem. On the other hand, there have been cases where membership functions have been constructed that are unique to the problem (for example, Brown 1998; Wang and Hall 1996).

Although solitary membership functions have been used to provide linguistic interpretations of quantitative data, a fuzzy model is usually composed of many membership functions in combination. The techniques of combining the membership functions are grouped into aggregation models and rule-based models.

Aggregation models

Aggregation operations on fuzzy sets are operations that combine several fuzzy sets in a desirable manner to produce a single fuzzy set (Klir and Yuan 1995). The several fuzzy sets that are being combined are defined by membership functions applied to a number of different map layers. This process is referred to as fuzzification. For example, application of membership functions may result in map layers defining the fuzzy sets *good_slope*, *close_to_road*, and *far_from_stream*. Aggregation operators would be used to then combine them into a single fuzzy set such as *good_location*. The term Joint Membership Function (JMF) conveys the idea that the fuzzy membership at a map location is a joint function of several fuzzy sets (Burrough and McDonnell 1998) combined using some model of aggregation. There are many aggregation operators from which to choose. The most commonly used operators are the intersection (min) and union (max) operators first proposed by Zadeh (1965).

One problem with the max/min operators is that they confer equal weight to each map. Some studies have incorporated preferences by formulating a weighted aggregation model. Oberthur, Dobermann, and Aylward (2000) used a convex combination in their study to combine sets of soil properties into indices of soil quality. Jiang and Eastman (2000) used a modified ordered weighted averaging operator (OWA) and showed different aggregation approaches can yield strikingly different results.

Rule-based models

Fuzzy rule-based models are another class of a priori knowledge-based models seen in GIS applications and are covered in more detail by Zhu in Chapter 15 of this volume. Fuzzy rules define the connection between input and output fuzzy sets (that is, membership functions). The rules have the general structure of the form: **IF** (antecedent) **THEN** (consequent). The antecedent and consequent conditions are defined by fuzzy sets. If the fuzzy system is used as a classifier the consequent can become a crisp value (or label). The specific form of the antecedent fuzzy sets can be based on standard membership functions, as was done in a fuzzy rule-based approach to map comparison (Power, Simms, and White 2001). To evaluate the rule base and arrive at an answer requires the application of an inference, or implication, method. The use of rule-based fuzzy models in GIS include applications for the conflation of vector maps (Cobb, Chung, Foley, Petry, and Shaw 1998), real estate evaluation (Zeng and Zhou 2001), and land fill location (Charnpratheep, Zhou, and Garner 1997).

Specifying membership functions

Regardless of whether an aggregation or rule-based model is being used, the specification of the membership functions is fundamental to model formulation. It is the specification and tuning of membership functions that has been a source of much criticism leveled at the fuzzy logic approach in general (Yen 1999) and more specifically in GIS (Goodchild 2000). Often the parameters of standard membership functions are described as being chosen by experts (for example, Stefanakis, Vazirgiannis, and Sellis 1999, MacMillan, Pettapiece, Nolan, and Goddard 2000, DeGenst, Canters, and Gulink 2001). Rigorous specification of membership functions from experts can be a difficult task (Zhu 1999). The challenging proposition of acquiring fuzzy memberships from domain experts has been addressed by automating the process to some extent in systems such as the spatial relations acquisition station (SRAS) (Robinson 2000), SOLIM (Zhu 1999; Zhu, Hudson, Burt, Lubich, and Simonson 2001) and others (Foley, Petry, Cobb, and Shaw 1997). Each uses a different, yet formal approach to acquiring fuzzy membership functions from experts. Another emerging class of approaches seeks to supplement, or supplant, the input of human expertise with automated techniques for parameterizing fuzzy models (Huang, Gedeon, and Wong 1998, Mackay, Samanta, Ahl, et al. 2003).

Data-Driven Fuzzy Classification

So what happens when we do not have sufficient a priori knowledge to construct a fuzzy model? One of the most commonly used class of methods is fuzzy clustering. Less often seen is the use of neural networks to extract fuzzy rule bases or as the basis for fuzzy classification directly (Carpenter, Gjaja, Gopan, and Woodstock 1997, Foody and Boyd 1999, Zheng and Kainz 1999).

Neural networks and fuzzy classification

In the mapping of land cover over large regions remote sensing techniques are used and the application of traditional neural network methods are adjusted so that the output is a fuzzy membership of a pixel in a land cover class. Thus, it is used to directly produce a fuzzy classification (Foody and Boyd 1999). Other GIS applications may use neural networks to extract a rule base from the data (Zheng and Kainz 1999). In either case, the general architecture of the artificial neural network is basically the same. These networks are composed of a set of simple processing units, or nodes, that are interconnected by some predefined architecture which can be trained. The processing nodes are arranged in a layered architecture typically composed of three layers. The first layer is the input, or fuzzification, layer where there is one node per input variable. The implication layer is comprised of a number of processing units. These are the processing nodes that do most of the "thinking" of the neural network. The outcomes of processing at this layer flow to the output layer. In general, there is one output node associated with each class. Although these networks are typically used to arrive at a crisp classification, they can be designed to provide a fuzzy classification (Foody and Boyd 1999). Alternatively, the process of generating classifications can be exploited to extract a fuzzy rule base (Zheng and Kainz 1999).

A related technique is the adaptive neuro-fuzzy inference system (ANFIS) (Jang 1993). Using a given input/output data set the objective is to construct a fuzzy inference system whose membership functions best suit the data set. Using either a back-propagation algorithm and/or a least-squares method, the membership parameters are tuned in a training exercise similar to regular neural networks. ANFIS has been used for map revision (Teng and Fairbairn 2002) and land cover classification.

The neural network approach has advantages and disadvantages. The advantages of neural networks include an ability to learn from past experience and they can handle noise and incomplete data. Once trained, a neural network can respond to a new set of data instantly. However, some of the disadvantages remain significant hurdles for application in GIS. They can take a long time to train, especially since training is still largely a trial and error situation further complicated by the fact that incomplete training data can cause the network to give wrong results, as can the incorrect network target outputs in a supervised learning algorithm. Perhaps the most important disadvantage is that it is difficult to explain the reasoning that led to the output product. It is perhaps for this reason that the neuro-fuzzy techniques are not quite as commonly used as another class of data-driven fuzzy classification method, namely fuzzy clustering.

Fuzzy clustering

Originally developed by Dunn (1973) and later generalized by Bezdek (1981) the fuzzy c-means (FCM) algorithm (also known as the fuzzy k-means algorithm) remains one of the most popular techniques for allocating individuals to fuzzy sets (that is, assigning membership values). The basic FCM algorithm seeks to minimize an objective function with respect to the membership functions and centroids of the clusters. As input, the FCM algorithm requires that number of clusters be specified. When the number of classes is not known a priori, the FCM algorithm is run using different numbers of clusters and the number of clusters is chosen based on an index, such as the fuzzy performance index (Roubens 1982), partition coefficient, classification entropy (Bezdek, Ehrlich, and Full 1984), or Xie-Beni validity index (Xie and Beni 1991), that provides the best results.

FCM did not receive much attention for GIS applications until the publication of the algorithm in Bezdek, Ehrlich, and Full (1984). Shortly after Bezdek, Ehrlich, and Full's (1984) publication of FCM, Robinson and Thongs (1986) demonstrated how it could be used to obtain a fuzzy classification of land cover from Landsat data. It was subsequently shown how the results of a fuzzy classification of land cover could be represented in a geographical database using concepts from the fuzzy database field (Robinson 1988). Since that early exploratory work, FCM has been used in a variety of mapping problems. It has been used to develop classifications of land cover in suburban areas (Fisher and Pathirana 1994, Zhang and Foody 1998).

In studies of vegetation FCM has been used to model vegetation using remote sensing data as input (Foody 1996). It has been combined with geostatistical techniques to model soil variability from hyperspectral remote sensing data as a means of improving soils maps (Ahn, Baumgardner, and Biehl 1999). There has also been an exploration of the utility of incorporating terrain and other data with vegetation or remote sensing data to improve on landscape characterizations (Burrough, van Gaans, and MacMillan 2000, 2001).

Although FCM (FKM) figures prominently in their papers on new approaches for physical geographers (Wilson and Burrough 1999, Scull, Franklin, Chadwick, and McArthur 2003), it is but one approach to fuzzy clustering, albeit the one most commonly found in GIS applications. There are now several methods for fuzzy clustering that might prove useful when applied to GIS-related problems. For example, Yao, Dash, Tan, and Liu (2000) have described an entropy-based fuzzy clustering (EFC) algorithm that requires the specification of fewer parameters. One of the limitations of FCM has been its reliance on the use of ratio/interval data. The development of a mixed-variable fuzzy c-means (MVFCM) algorithm seeks to address that shortcoming (Yang, Hwang, and Chen 2004). Like FCM, MVFCM is capable of assigning membership values to each element.

Spatial Interpolation and Fuzzy Classification

When data are not present at every location in an area the technique of spatial interpolation is often used to estimate what the attribute might be at a particular location based on the spatial distribution of attributes surrounding that particular location generally using either a local or global approach. For example, the output from FCM may provide fuzzy memberships for only some, not all, points (or cells) in a region. In this case spatial interpolation techniques have been used to interpolate fuzzy membership functions for the remainder of the region (McBratney and de Gruijter 1992; Brown 1998; Bragato 2004). Dragicevic and Marceau (2000) have extended this idea of interpolation of membership values to the temporal dynamics of spatial land use change.

Walvoort and de Gruijter (2001) consider the basis and additive log-ratio methods in comparison with their compositional kriging approach to spatial interpolation of fuzzy membership values. Another approach to the use of kriging with fuzzy data has been to fuzzify the variogram with either the application of the extension principle or with the use of fuzzy arithmetic (Bardossy, Bogardi, and Kelly 1989). In this fuzzy kriging formulation, the use of fuzzy data and a fuzzy variogram results in fuzzy numbers as estimates and fuzzy numbers as estimation variances for each point.

It has been suggested that a fuzzy membership function represents a form of interpolation. Thus, in geographical applications the fuzzy membership function can be considered a spatial, temporal or spatio-temporal interpolator (Anile, Furno, Gallo, and Massolo 2003). In many of these approaches fuzzy set theory is an intrinsic part of the interpolation technique. For example, Gedeon, Wong, Wong, and Huang (2003) integrate fuzzy reasoning with techniques of local and Euclidean interpolation that makes for a computationally efficient spatial interpolation technique resulting in a crisp number at a location. Thus, the goal is not to interpolate membership values but to use fuzzy set theory to better interpolate attributes themselves as was done using cubic splines in Lodwick and Santos (2003).

Building on their work using B-splines to construct fuzzy terrain models, Anile, Furno, Gallo, and Massolo (2003) report an algorithm for constructing inter-visibility maps from a fuzzy terrain model. They show how a fuzzy terrain model that can be coupled to a GIS is constructed using fuzzy numbers, intervals, and B-splines (Anile, Furno, Gallo, and Massolo 2000).

Visualizing Fuzzy Classifications

Maps depicting membership in a single class are easily visualized by grey, or pseudo-color, scales that are related to strength of membership. However, when there is more than one class to which elements are mapped then each class is associated with a separate map (Figure 14.1). The potentially large number of separate maps does not allow the user to visualize the membership maps as a whole. Visualization of fuzziness and uncertainty is important as it allows users to explore it and investigate

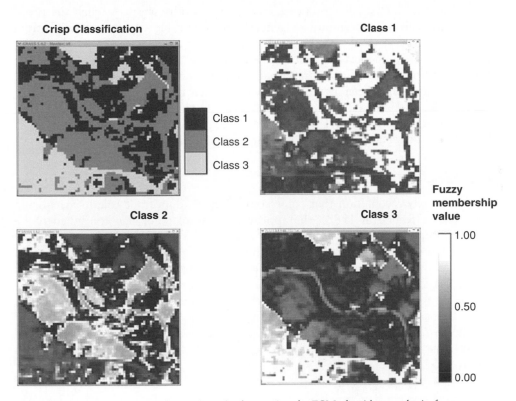

Fig. 14.1 A simple example of typical results from using the FCM algorithm to obtain fuzzy classification of land cover as a function of four bands from Landsat Thematic Mapper data. The crisp classification map is based on allocating a cell to that class which has the highest membership value

the effects of different decisions in the classification (MacEachren and Kraak 1997). Typically, defuzzification of the fuzzy maps allocates a spatial element to a single class thus collapsing the number of maps from many to one (see crisp classification in Figure 14.1) where each spatial element is depicted as belonging to a single, crisp, class. The most common defuzzification method is to produce a color map where each spatial element is assigned to the class of which it has the highest membership value. Clearly, this does not provide a means of visualizing the information contained in the underlying fuzzy classification.

Use of measures such as entropy, ignorance, and ignorance uncertainty (Zhu 1997) illustrate a strategy of visualizing the fuzzy information contained in multiple maps (as in Figure 14.1) by using a measure that collapses multiple maps into a single map. For example, the confusion index (CI), a ratio of the dominant and first sub-dominant membership value for each spatial element, has been mapped to visualize which parts of a landscape are characterized by an abrupt, or gradual, spatial change in classes (Burrough, van Gaans, and Hootman 1997, 2001). The presentation of CI is often done in tandem with other visual information such as a map of the maximum membership values or class map (DeBruin and Stein 1998, Burrough, Wilson, van Gaans, and Hansen 2001, Bragato 2004).

The problem of visualizing fuzzy classifications is generally subsumed within the larger literature of visualizing uncertainty (fuzzy or otherwise). As such the problems of visualizing uncertainty in general are applicable to fuzzy classification. For instance, FCM would produce c maps, one for each cluster. Thus, presenting those maps simultaneously in a 2D space completely fills the 2D space, saturating it and potentially providing more visual stimulus than a viewer can comprehend. Furthermore, it has been noted that to map onto the 2D space supplementary information relating to the uncertainty of that 2D view increases clutter and confuses viewers (MacEachren 1992, van der Wel, Van der Gaag, and Gorte 1998). A number of strategies have been suggested to address these problems.

Visualizing with color

Since the early recognition that color could play an important role in visualizing fuzzy maps (Robinson 1988), the appropriate use of color has continued to be explored (Hengl 2003). A color mixture with fuzzy-metric legend methodology has been proposed to more effectively portray the results of fuzzy classification (Hengl, Walvoort, Brown, and Rossiter 2004). This method is related to earlier work on the *pixel mixture* (PM) technique developed for visualizing membership maps by including all membership values in the representation. PM randomly assigns pixels to a sub-pixel grid with a probability proportional to the (normalized) membership in the class. It is argued that this technique provides a visual impression of both the possible classes and their confusion (de Gruijter, Walvoort, and Van Gaans 1997).

Dynamic visualization

With the development of computer graphics and related technology researchers began to consider ways of taking advantage of the computational/graphical environment in which GIS users now operate. Thus, one approach to exploiting the potential of computer-controlled displays constructs dynamic displays which can toggle between a primary view and a view containing more information about fuzziness, or uncertainty (Evans 1997, van der Wel, Van der Gaag, and Gorte 1998). Although past studies have tended to focus on comparing a classified map with its associated memberships, this visualization technique has also been incorporated into a fuzzy GIS query system so that users could toggle between the continuous map of results and a discrete representation (Morris 2003).

Animation has been used to portray uncertainty in GIS applications (Fisher 1993, Davis and Keller 1997, Bastin, Fisher, and Wood 2002). There are two major categories of animation that have been reported: serial animation (Bastin, Fisher, and Wood 2002) and the moving shutter method (Blenkinsop, Fisher, Bastin, and Wood 2000). The pace at which the animation proceeds is one problem that has been noted as well as choice of hue in color animations (Davis and Keller 1997).

In random animation a population of possible land cover maps is produced based on (normalized) fuzzy memberships derived for every pixel. When the images are shown in succession, color at any point will change so that the mixing of the land covers and uncertainty is identifiable (Bastin, Fisher, and Wood 2002).

An interesting adjunct to visualizing fuzzy classifications was suggested by Fisher (1994). He discussed the use of sound by relating tone and/or rhythm to the level of uncertainty (fuzziness) associated with a particular location, or mapped unit. Although the efficacy of audio remains to be demonstrated, it is an intriguing elaboration of the visualization problem to a multimedia environment.

Typically these techniques do not allow for a user to interact with the classification algorithm itself as part of the visualization process. This could potentially be a valuable tool, especially in the classification of remotely sensed data. One prototypical method uses a feature space plot dynamically linked with an image display and the classification result to allow the user to adapt class clusters in the feature space so as to change the parameters of a fuzzy classification algorithm (Lucieer and Kraak 2004).

Human subjects and visualization strategies

Although anecdotal evidence has been offered for the effectiveness of the toggle strategy (Morris 2003), human subject testing suggests it can be an effective strategy (Evans 1997). In one study that included expert and novice users of land cover classification information, it was found that both were highly successful at determining classification uncertainty among pixels when shown grey-scale images and histograms (Blenkinsop, Fisher, Bastin, and Wood 2000). In addition, results suggest that users can extract uncertainty information from random animations (Evans 1997, Blenkinsop, Fisher, Bastin, and Wood 2000).

Issues of Cost and Fuzzy Mapping

Results of some applications of fuzzy methods in mapping and GIS indicate that use of these techniques can be cost-effective. Using the soil-land inference model

(SoLIM) approach (Zhu 1997, Zhu, Hudson, Burt, Lubich, and Simonson 2001) it has been estimated that the accuracy of a map product is about 80 percent as compared to 60 percent for traditional map products and is fifteen times faster with the cost being about one-third of the conventional approach (Zhu 2004). MacMillan, Martin, Earle, and McNabb (2003) report dramatically lower mapping costs when using a fuzzy-based methodology. Thus, it is clear that incorporating fuzzy classification in mapping methodologies holds promise of increasing accuracy while reducing costs.

Recent developments in mapping systems that integrate Global Positioning Systems (GPS), mobile computing, GIS, and one or more sensors to take physical measurements (Arvanitis, Ramachandran, Brackett, Abd-El Rasol, and Xu 2000) suggest that it is now realistic to think in terms of spatial data collection agents that use fuzzy logic to adapt their spatial sampling strategy in real-time as they move along a transect network. Simulation results of a prototypical system indicate that a fuzzy adaptive sampler may be able to reduce the cost of collecting field data while simultaneously improving overall data quality (Graniero and Robinson 2003).

Even though many studies stress the utility of the additional information afforded by fuzzy classification, it is also noted by many, often the same, researchers that a fuzzy classification can be subject to serious limitations (Oberthur, Dobermann, and Aylward 2000, Jiang and Eastman 2000, Guneralp, Mendoza, Gertner, and Anderson 2003). Most often the problem arises from the large potential parameter space that accompanies any fuzzy classification methodology and the subjectivity that may be inherent in the choice of parameters. Thus, it is advisable to take care in developing the model upon which a fuzzy classification is based.

CONCLUSIONS

This chapter has summarized the basic approaches to assigning a fuzzy membership value to a location and then visualizing that classification. Hopefully the extensive set of references will allow the reader to pursue selected topics in more detail.

There remain many areas for further development and refinement of fuzzy classification in mapping. For example, using fuzzy classification for assessing the accuracy of thematic maps is perhaps the first stage in developing intelligent map update, or change detection, systems. As applications of GIS become more intelligent and/or embedded in other systems where automated decision making is important, fuzzy classification (that is, fuzzy set theory) may provide a cost-effective set of tools to address these increasingly complex problems of mapping. Although much progress has been made in the last few decades, many challenges remain.

ACKNOWLEDGEMENTS

Partial support of a grant from the Natural Sciences and Engineering Research Council (NSERC) of Canada is gratefully acknowledged. The comments of Linda Robinson and Phil Graniero improved the manuscript.

REFERENCES

Ahn, C.-W., Baumgardner, M. F., and Biehl, L. L. 1999. Delineation of soil variability using geostatistics and fuzzy clustering analyses of hyperspectral data. *Soil Science Society America Journal* 63: 142–50.

Anile, A. M., Falcidieno, B., Gallo, G., Spagnuolo, M., and Spinello, S. 2000. Modeling uncertain data with fuzzy B-splines. *Fuzzy Sets and Systems* 113: 397–410.

Anile, M. A., Furno, P., Gallo, G., and Massolo, A. 2003. A fuzzy approach to visibility maps creation over digital terrains. *Fuzzy Sets and Systems* 135: 63–80.

Arvanitis, L. G., Ramachandran, B., Brackett, D. P., Abd-El Rasol, H., and Xu, X. S. 2000. Multi-resource inventories incorporating GIS, GPS and database management systems: A conceptual model. *Computers and Electronics in Agriculture* 28: 89–100.

Bardossy, A., Bogardi, I., and Kelly, W. E. 1989. Geostatistics utilizing imprecise (fuzzy) information. *Fuzzy Sets and Systems* 31: 311–28.

Bastin, L., Fisher, P. F., and Wood, J. 2002. Visualizing uncertainty in multi-spectral remotely sensed imagery. *Computers and Geosciences* 28: 337–50.

Bezdek, J. C. 1981. *Pattern Recognition with Fuzzy Objective Function Algorithms.* New York: Plenum Press.

Bezdek, J. C., Ehrlich, R., and Full, W. 1984. FCM: The fuzzy c-means clustering algorithm. *Computers and Geosciences* 10: 191–203.

Blenkinsop, S., Fisher, P. F., Bastin, L., and Wood, J. 2000. Evaluating the perception of uncertainty in alternative visualisation strategies. *Cartographica* 37: 1–14.

Bragato, G. 2004. Fuzzy continuous classification and spatial interpolation in conventional soil survey for soil mapping of the lower Piave plain. *Geoderma* 118: 1–16.

Brown, D. G. 1998. Classification and boundary vagueness in mapping pre-settlement forest types. *International Journal of Geographical Information Science* 12: 105–29.

Burrough, P. A. and Frank, A. U. 1996. *Geographic Objects with Indeterminate Boundaries.* London: Taylor and Francis.

Burrough, P. A. and McDonnell, R. A. 1998. *Principles of Geographical Information Systems.* New York: Oxford University Press.

Burrough, P. A., van Gaans, P. F. M., and Hootsman, R. J. 1997. Continuous classification in soil survey: Spatial correlation, confusion and boundaries. *Geoderma* 77: 115–35.

Burrough, P. A., van Gaans, P. F. M., and MacMillan, R. A. 2000. High resolution landform classification using fuzzy k-means. *Fuzzy Sets and Systems* 113: 37–52.

Burrough, P. A., Wilson, J. P., van Gaans, P. F. M., and Hansen, A. J. 2001. Fuzzy k-means classification of topo-climatic data as an aid to forest mapping. *Landscape Ecology* 16: 523–46.

Carpenter, G. A., Gjaja, M. N., Gopal, S., and Woodcock, C. E. 1997. ART neural networks for remote sensing: Vegetation classification from Landsat TM and terrain data. *IEEE Transactions on Geoscience and Remote Sensing* 35: 308–25.

Charnpratheep, K., Zhou, Q., and Garner, B. 1997. Preliminary landfill site screening using fuzzy geographical information systems. *Waste Management and Research* 15: 197–215.

Cheng, T., Molenaar, M., and Lin, H. 2002. Formalizing fuzzy objects from uncertain classification results. *International Journal of Geographical Information Science* 15: 27–42.

Cobb, M. A., Chung, M. J., Foley, I. H., Petry, F. E., and Shaw, K. B. 1998. A rule-based approach for the conflation of attributed vector data. *Geoinformatica* 2: 7–35.

Davis, T. J. and Keller, C. P. 1997. Modelling and visualizing multiple spatial uncertainties. *Computers and Geosciences* 23: 397–408.

DeBruin, S. and Stein, A. 1998. Soil-landscape modelling using fuzzy c-means clustering of attribute data derived from a digital elevation model (DEM). *Geoderma* 83: 17–33.

DeGenst, A., Canters, F., and Gulink, H. 2001. Uncertainty modeling in buffer operations applied to connectivity analysis. *Transactions in GIS* 5: 305–26.

de Gruijter, J. J., Walvoort, D. J. J., and Van Gaans, P. F. M. 1997. Continuous soil maps: A fuzzy set approach to bridge the gap between aggregation levels of process and distribution models. *Geoderma* 77: 169–95.

Dragicevic, S. and Marceau, D. J. 2000. An application of fuzzy logic reasoning for GIS termporal modeling of dynamic processes. *Fuzzy Sets and Systems* 113: 69–80.

Dunn, J. C. 1973. A fuzzy relative of the ISODATA process and its use in detecting compact well-separated clusters. *Journal of Cybernetics* 3: 32–57.

Evans, B. J. 1997. Dynamic display of spatial data-reliability: Does it benefit the map user? *Computers and Geosciences* 23: 409–22.

Fisher, P. F. 1993. Visualizing uncertainty in soil maps by animation. *Cartographica* 30: 20–7.

Fisher, P. F. 1994. Animation and sound for the visualization of uncertain spatial information. In H. H. Hearnshaw and D. J. Unwin (eds) *Visualization in Geographical Information Systems*. New York: John Wiley and Sons, pp. 181–5.

Fisher, P. F. and Pathirana, S. 1994. The evaluation of fuzzy membership of land cover classes in the suburban zone. *Remote Sensing of Environment* 34: 121–32.

Foley, H., Petry, F., Cobb, M., and Shaw, K. 1997. Using semantic constraints for improved conflation in spatial databases. In *Proceedings of the Seventh International Fuzzy Systems Association World Congress*, Prague, Czech Republic. International Fuzzy Systems Association, pp. 193–7.

Foody, G. M. 1996. Fuzzy modelling of vegetation from remotely sensed imagery. *Ecological Modelling* 85: 3–12.

Foody, G. M. 1999. The continuum of classification fuzziness in thematic mapping. *Photogrammetric Engineering and Remote Sensing* 65: 443–51.

Foody, G. M. and Boyd, D. S. 1999. Fuzzy mapping of tropical land cover along an environmental gradient from remotely sensed data with an artificial neural network. *Journal of Geographical Systems* 1: 23–35.

Gedeon, T. D., Wong, K. W., Wong, P., and Huang, Y. 2003. Spatial interpolation using fuzzy reasoning. *Transactions in GIS* 7: 55–66.

Goodchild, M. F. 2000. Introduction: Special issue on uncertainty in geographic information systems. *Fuzzy Sets and Systems* 113: 3–5.

Graniero, P. A. and Robinson, V. B. 2003. A real-time adaptive sampling method for field mapping in patchy, heterogeneous environments. *Transactions in GIS* 7: 31–54.

Guneralp, B., Mendoza, G., Gertner, G., and Anderson, A. 2003. Spatial simulation and fuzzy threshold analyses for allocating restoration areas. *Transactions in GIS* 7: 325–43.

Hengl, T. 2003. Visualisation of uncertainty using the HSI colour model: Computations with colours. In *Proceedings of the Seventh International Conference on GeoComputation*, Southampton, United Kingdom. Southampton: School of Geography, University of Southampton, pp. 8–17.

Hengl, T., Walvoort, D. J. J., Brown, A., and Rossiter, D. G. 2004. A double continuous approach to visualisation and analysis of categorical maps. *International Journal of Geographical Information Science* 18: 183–202.

Huang, Y., Gedeon, T., and Wong, P. 1998. Spatial interpolation using fuzzy reasoning and genetic algorithms. *Journal of Geographic Information and Decision Analysis* 2: 223–33.

Jang, J.-S. 1993. ANFIS: Adaptive-network-based fuzzy inference systems. *IEEE Transactions on Systems, Man and Cybernetics* 23: 665–85.

Jiang, H. and Eastman, J. R. 2000. Application of fuzzy measures in multi-criteria evaluation in GIS. *International Journal of Geographical Information Science* 14: 173–84.

Klir, G. J. and Yuan, B. 1995. *Fuzzy Sets and Fuzzy Logic: Theory and Applications*. Upper Saddle River NJ: Prentice-Hall.

Lodwick, W. A. and Santos, J. 2003. Constructing consistent fuzzy surfaces from fuzzy data. *Fuzzy Sets and Systems* 135: 259–77.

Lucieer, A. and Kraak, M. J. 2004. Interactive and visual fuzzy classification of remotely sensed imagery for exploration of uncertainty. *International Journal of Geographical Information Science* 18: 491–512.

McBratney, A. B. and de Gruijter, J. J. 1992. A continuum approach to soil classification by modified fuzzy k-means with extragrades. *Journal of Soil Science* 43: 159–75.

McBratney, A. B. and Odeh, I. O. A. 1997. Application of fuzzy sets in soil science: Fuzzy logic, fuzzy measurements and fuzzy decisions. *Geoderma* 77: 85–113.

MacEachren, A. M. 1992. Visualizing uncertain information. *Cartographic Perspectives* 13: 10–9.

MacEachren, A. M. and Kraak, M. J. 1997. Exploratory cartographic visualization: Advancing the agenda. *Computers and Geosciences* 23: 335–44.

Mackay, D. S., Samanta, S., Ahl, D. E., Ewers, B. E., Gower, S. T., and Burrows, S. N. 2003 Automated parameterization of land surface process models using fuzzy logic. *Transactions in GIS* 7: 139–53.

MacMillan, R. A., Pettapiece, W. W., Nolan, S. C., and Goddard, T. W. 2000. A generic procedure for automatically segmenting landforms into landform elements using DEMs, heuristic rules, and fuzzy logic. *Fuzzy Sets and Systems* 113: 81–109.

MacMillan, R. A., Martin, T. C., Earle, T. J., and McNabb, D. H. 2003. Automated analysis and classification of landforms using high-resolution digital elevation data: Applications and issues. *Canadian Journal of Remote Sensing* 29: 592–606.

Morris, A. 2003. A framework for modeling uncertainty in spatial databases. *Transactions in GIS* 7: 83–103.

Oberthur, T., Dobermann, A., and Aylward, M. 2000. Using auxiliary information to adjust fuzzy membership functions for improved mapping of soil qualities. *International Journal of Geographical Information Science* 14: 431–54.

Power, C., Simms, A., and White, R. 2001. Hierarchical fuzzy pattern matching for the regional comparison of land use maps. *International Journal of Geographical Information Science* 15: 77–100.

Robinove, C. J. 1989. Principles of logic and the use of digital geographic information systems. In W. J. Ripple (ed.) *Fundamentals of GIS: A Compendium*. Washington, DC: American Society for Photogrammetry and Remote Sensing: 61–79.

Robinson, V. B. 1988. Some implications of fuzzy set theory applied to geographic databases. *Computers, Environment, and Urban Systems* 12: 89–97.

Robinson, V. B. 2000. Individual and multipersonal fuzzy spatial relations acquired using human-machine interaction. *Fuzzy Sets and Systems* 113: 133–45.

Robinson, V. B. 2003. A perspective on the fundamentals of fuzzy sets and their use in geographic information systems. *Transactions in GIS* 7: 3–30.

Robinson, V. B. and Frank, A. U. 1985. About different kinds of uncertainty in collections of spatial data. In *Proceedings of the Seventh International Symposium on Automated Cartography (Auto Carto 7)*, Baltimore, MD, USA. Bethesda, MD: American Congress on Surveying and Mapping / American Society for Photogrammetry and Remote Sensing, pp. 440–50.

Robinson, V. B. and Strahler, A. H. 1984. Issues in designing geographic information systems under conditions of inexactness. In *Proceedings of the Tenth International Symposium on Machine Processing of Remotely Sensed Data*, West Lafayette, IN, USA. New York: Institute of Electrical and Electronics Engineers, pp. 179–88.

Robinson, V. B. and Thongs, D. 1986. Fuzzy set theory applied to the mixed pixel problem of multispectral landcover databases. In B. K. Opitz (ed.) *Geographic Information Systems in Government.* Hampton, VA: A. Deepak Publishing, pp. 871–85.

Roubens, M. 1982. Fuzzy clustering algorithms and their cluster validity. *European Journal of Operational Research* 10: 294–301.

Scul, P., Franklin, J., Chadwick, O. A., and McArthur, D. 2003. Predictive soil mapping: A review. *Progress in Physical Geography* 27: 171–97.

Stefanakis, E., Vazirgiannis, M., and Sellis, T. 1999. Incorporating fuzzy set methodologies in a DBMS repository for the application domain of GIS. *International Journal of Geographical Information Science* 13: 657–75.

Teng, C. H. and Fairbairn, D. 2002. Comparing expert systems and neural fuzzy systems for object recognition in map data set revision. *International Journal of Remote Sensing* 23: 555–67.

van der Wel, F., Van der Gaag, L. C., and Gorte, B. G. H. 1998. Visual exploration of uncertainty in remote-sensing classification. *Computers and Geosciences* 24: 335–43.

Walvoort, D. J. J. and de Gruijter, J. J. 2001. Compositional kriging: A spatial interpolation method for compositional data. *Mathematical Geology* 33: 951–66.

Wang, F. and Hall, G. B. 1996. Fuzzy representation of geographical boundaries in GIS. *International Journal of Geographical Information Science* 10: 573–90.

Wilson, J. P. and Burrough, P. A. 1999. Dynamic modeling, geostatistics, and fuzzy classification: New sneakers for a new geography? *Annals of the Association of American Geographers* 89: 736–46.

Xie, X. L. and Beni, G. 1991. A validity measure for fuzzy clustering. *IEEE Transactions on Pattern Analysis and Machine Intelligence* 13: 841–7.

Yang, M.-S., Hwang, P.-Y., and Chen, D.-H. 2004. Fuzzy clustering algorithms for mixed feature variables. *Fuzzy Sets and Systems* 141: 301–17.

Yao, J., Dash, M., Tan, S. T., and Liu, H. 2000. Entropy-based clustering and fuzzy modeling. *Fuzzy Sets and Systems* 113: 381–8.

Yen, J. 1999. Fuzzy logic: A modern perspective. *IEEE Transactions on Knowledge and Data Engineering* 11: 153–65.

Zadeh, L. A. 1965. Fuzzy sets. *Information and Control* 8: 338–53.

Zeng, T. Q. and Zhou, Q. 2001. Optimal spatial decision making using GIS: A prototype of a real estate geographical information system (REGIS). *International Journal of Geographical Information Science* 15: 307–21.

Zhang, J. and Foody, G. M. 1998. A fuzzy classification of sub-urban land cover from remotely sensed imagery. *International Journal of Remote Sensing* 19: 2721–38.

Zheng, D. and Kainz, W. 1999. Fuzzy rule extraction from GIS data with a neural fuzzy system for decision making. In *Proceedings of the Seventh ACM International Symposium on Advances in Geographic Information Systems*, Kansas City, MS: 79–84.

Zhu, A.-X. 1997. Measuring uncertainty in class assignment for natural resource maps under fuzzy logic. *Photogrammetric Engineering and Remote Sensing*: 1195–202.

Zhu, A. X. 1999. A personal construct-based knowledge acquisition process for natural resource mapping. *International Journal of Geographical Information Science* 13: 119–41.

Zhu, A. X., Hudson, B., Burt, J., Lubich, K., and Simonson, D. 2001. Soil mapping using GIS, expert knowledge, and fuzzy logic. *Soil Science Society of America Journal* 65: 1463–72.

Chapter 15

Rule-Based Mapping

A-Xing Zhu

Rule-based mapping is a predictive approach to mapping. By predictive it is meant that the status of the geographic entity to be mapped at a given point is inferred from the conditions of other variables which are related to or influence the existence and status of the entity. Examples of rule-based mapping include the prediction of the spatial distribution of potential customers for a retail store using census data and the identification of habitat areas for a specific wildlife using a set of physical landscape conditions. The set of variables used in the prediction are collectively referred to as predictive variables. Data on the spatial variation of these predictive variables are often easier to obtain than the status or condition of the entity to be mapped. Predictive mapping is based on the concept that there is a relationship between the phenomenon to be mapped and a set of predictive variables as expressed in the following equation:

$$S_{ij} = f_{ij}(E_{ij}) \tag{15.1}$$

where S_{ij} is the status or property value of a geographic entity at location (i,j), E_{ij} is a set of predictive conditions $(E_{ij}^1, E_{ij}^2, \ldots, E_{ij}^v, \ldots, E_{ij}^m)$ at the location, and f_{ij} is a set of relationships $(f_{ij}^1, f_{ij}^2, \ldots, f_{ij}^v, \ldots, f_{ij}^m)$ between the entity and the set of predictive conditions for that location and is often expressed in the form of rules. Often we assume that f_{ij} is stationary over a small area or over areas where geographic processes are similar across locations. Under this assumption, Equation (15.1) can be simplified to

$$S_{ij} = f(E_{ij}) \tag{15.2}$$

Information on the spatial variation of E is often derived using Geographic Information System (GIS)/remote sensing techniques. The development of GI Science and technology has made the acquisition and compilation of predictive conditions (E) over large areas not only possible but also easier. Thus, rule-based mapping is now very easy to implement in a GIS setting. In fact, many of the GIS applications, particularly those related to suitability analysis, are in the form of rule-based mapping (see Scott, Davis, Cusuti, et al. [1993] for rule-based mapping in Gap Analysis Program [GAP] analysis and Zhu, Hudson, Burt, Lubich, and Simonson [2001] in soil survey).

The generation of rules is one of the most important parts of rule-based mapping. In a broad sense rule-based mapping can be classified into three major categories based on the form and level of sophistication of the rules: physical process-based, knowledge-based, and statistical. The physical process-based approaches represent rules as physical and mechanistic processes governed by a set of mechanistic equations and parameters. Process-based modeling (such as climate, ecosystem, and hydrological modeling) belongs to this category. For example, the Penman-Monteith combination equation (Monteith 1965) is often used to model (predict) the transpiration of a forest canopy. This equation expresses the relationships between daily canopy transpiration and a set of variables (parameters) describing the environmental conditions (such as vapor pressure and temperature) and tree physiology (such as stomatal conductance and leaf area index) (Ehleringer and Field 1993, Waring and Running 1998). Examples of urban and planning applications can be found in Simpson (2001). The deterministic nature of these equations requires us to have a thorough understanding of these processes and the role that each parameter plays in the respective processes so that these processes can be reasonably quantified using mathematical equations.

The statistical approaches, to the contrary, assume little prior knowledge of the relationships between the phenomenon to be mapped (dependent variable) and the predictive (independent) variables. Rather, this type of approach often first extracts the relationships (rules) between dependent variable and predictive variables from sample locations and then predicts (maps) the value of the dependent variable at unvisited sites using the extracted relationships (rules). Examples of these approaches include linear regression models (Hastie, Tibshirani, and Friedman 2001) and generalized linear models (Hastie and Pregibon 1992). Statistical clustering can also be considered as rule-based because it first develops class signatures (a set of rules in the attribute domain) and then classifies individual locations based on the established class signatures (Lillesand, Kiefer, and Chipman 2004). Statistical simulation techniques (that is, cellular automata; Couclelis 1997, Batty 1998), agent-based modeling (O'Sullivan and Haklay 2000), and spatial analytical techniques (Goodchild and Haining 2004) can be considered in this group as well.

Knowledge-based approaches to rule-based mapping sit between the above two extremes in terms of the required level of understanding about geographic systems. Often, we understand geographic systems at some level that is not sufficient for prescribing physical processes using a set of deterministic equations but that is more than enough to make the statistical approaches not worthwhile. In these circumstances, knowledge-based approaches to rule-based mapping are very useful in exploiting this type of human expertise and in providing accurate mapping of geographic phenomena.

This chapter focuses on rule-based mapping from a knowledge-based perspective, particularly in terms of making use of the qualitative knowledge of human experts. The other two types of mapping approaches (that is, the physical process-based and statistical approaches) are not discussed further because they are extensively discussed in the existing literature.

Under a GIS framework and with a knowledge-based approach to rule-based mapping, Equation (15.2) can be perceived to be that shown in Figure 15.1. The required environmental conditions can be characterized using GIS/RS techniques (Zhu, Band, Dutton, and Nimlos 1996, Wilson and Gallant 2000). The relationship between the feature to be mapped and the related environmental conditions is approximated with descriptive knowledge of human experts (Zhu 1999). The descriptive knowledge and the characterized environmental conditions are then combined to predict the spatial distribution of the phenomenon through a set of inference techniques such as spatial overlay and raster map algebra.

The unique aspect of this knowledge-based approach is that the relationship is now expressed as descriptive knowledge in the form of rules, rather than mechanistic equations or statistical relations. Figure 15.2 shows, as an example, the descriptive knowledge used to map a soil series, Basco, over Pleasant Valley in Wisconsin, USA (this example will be used throughout this chapter to illustrate the implementation of rule-based mapping under GIS; readers can replace soil series with number of potential customers or habitat of particular wildlife). The generation, encoding of descriptive

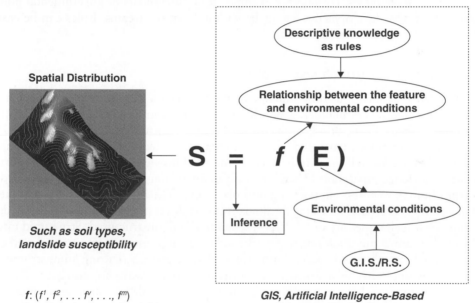

Fig. 15.1   Knowledge-based approach to rule-based mapping under GIS

Fig. 15.2   Example of descriptive knowledge expressed as a rule for rule-based mapping

knowledge, and inference using descriptive knowledge are the key issues in rule-based mapping. The following sections will address each of these issues respectively.

Obtaining Rules

There are three major sources from which rules can be derived: existing documents, human experts, and spatial data. The following subsections describe each of these three sources.

Extracting rules from existing documents

Existing documents often contain information about relationships between phenomena or regulations about certain geographic activities. For examples, an insurance company may have documentation about how social and economic conditions of family are related to the likelihood of property insurance purchase. The ecology of an endangered species can provide us the rules needed to map the spatial distribution of its habitats. For another example, legal documents or governmental policy state that new developments cannot be within 20 m of streams. Rules can be easily defined using this type of information.

Extracting rules from human experts

Knowledge acquisition from human experts has been a subject of study in the artificial intelligence community, a sub-discipline of computer science, for decades (for example, Hart 1986, Boose and Gaines 1988, Fensel and Studer 1999, Liao 2005). There are often two parties involved in human knowledge acquisition: the domain expert whose knowledge is to be elicited and the knowledge engineer who performs the knowledge elicitation. There are two main strategies in acquiring knowledge from human experts: interviewing and observing. The interview techniques take a questioning approach by asking the domain expert direct questions to elicit knowledge. Interview techniques are of two types: structured and unstructured. Structured interviews organize the questions to be asked. These questions are interrelated and are presented in a particular order during the interview process. During an unstructured interview questions being asked are not organized in any way. An unstructured interview is often used for initial information gathering while the structured interview is much more suited for extracting the specific rules.

The observing techniques elicit knowledge by studying the domain experts while they are in the problem solving process. There are two main groups: prototype analysis and ethnographic methods. The former requires experts to think aloud and report what they do when solving a prototype problem but they do not need to justify their actions. Prototype analysis techniques lead to a verbal protocol which the knowledge engineer can use to build a model of problem solving. The latter focuses on the actions of a domain expert by observing the behaviors of the domain expert in the workplace. This requires the knowledge engineer to be at the location where the domain expert solves actual problems. The observing techniques are more suited for knowledge related to operational or analytical procedures.

Knowledge on geographic relationships is often not well formulated. As a result, it is often very challenging to formally extract rules on spatial distribution of geographic phenomenon/features. Many of these knowledge acquisition techniques developed in the artificial intelligence community need to be revised when they are applied in geography. In fact, there exist few knowledge extraction techniques specifically designed to elicit knowledge on relationships between geographic phenomena/features (Zhu 1999, Yamada, Elith, McCarthy, and Zerger 2003).

Extracting rules from existing spatial data

Descriptive knowledge can also be derived from existing data (such as previously obtained field observations and/or existing maps). The explosive growth in geospatial data since the late 1980s, as a result of the advent of remote sensing and other geospatial data collection technologies, has also made it viable to extract descriptive knowledge by linking field observation data or existing maps with these remotely sensed data and other geospatial data derived from GIS technology. This emerging field is referred to as the geospatial data mining and knowledge discovery (Miller and Han 2001, Buttenfield, Gahegan, Miller, and Yuan 2002). Qi and Zhu (2003) provided a detailed example on how to use geospatial data mining techniques to extract knowledge on soil-landscape relationships embedded in existing soil maps. See Miller in Chapter 19 of this volume, for a more detailed treatment of this topic.

Encoding Rules in Rule-Based Mapping

Rules as descriptive knowledge cannot be used directly in GIS-based mapping. Descriptive rules need to be encoded into a numeric form so that they can be applied in a computer environment. In other words, the question of how to use these descriptive rules to approximate f in Figure 15.1 needs to be addressed. A rule typically consists of several components with each component describing the required condition for a given predictive variable. For example, the rule shown in Figure 15.2 consists of four components: geology, slope gradient, profile curvature, and plan curvature. The implementation of a descriptive rule needs to address the following two issues: the encoding of each component into a function form (the f^v in Figure 15.1) and the interaction of these components during inference (the "=" in Figure 15.1). The inference is a process of determining the values from functions (components of a rule) for a given location and then integrating these values to derive the final value on the status of the mapped entity. The implementation of descriptive rules is discussed below under two logic frameworks (Boolean and fuzzy logic).

Rule-based mapping under Boolean logic

Boolean encoding of rule components

Under Boolean logic each component of a descriptive rule is encoded as a step function. The function produces a value of 1 (or true) if the stated condition in the component is met, 0 (false) otherwise. For a component describing a nominal/categorical variable, even for an ordinal variable, the step function is easy to define. For example, the step function for the component on geology in the rule shown in Figure 15.2 can be expressed as

$$^G(G) = 1 \text{ if } G = \text{"Jordan Sandstone"} \tag{15.3a}$$

$$^G(G) = 0 \text{ otherwise} \tag{15.3b}$$

For interval and ratio variables, a numeric range or threshold needs to be established in order to define the step function. With some rules or components of a rule, the range or threshold is embedded in the descriptive term of the knowledge while others may just contain a linguistic term without any indication of what the range or threshold is. An example of the former is the slope gradient component and an example of the latter is the profile curvature component. For the former, the step function can be expressed as:

$$f^{slope}(\text{slope}) = 1 \text{ if slope} < 20\% \tag{15.4a}$$

$$f^{slope}(\text{slope}) = 0 \text{ otherwise} \tag{15.4b}$$

For the components which contain only linguistic values (such as the profile component in Figure 15.2) the numeric meaning (such as range or threshold) of these linguistic values needs to be defined before the step function can be formulated. The numeric meaning of the linguistic values can be defined during the initial knowledge acquisition process or through a secondary knowledge acquisition process specifically designed for this purpose. An example of the former is the preparation interview in the iterative knowledge acquisition process developed by (Zhu 1999). Liu (2004) developed a knowledge acquisition process specific for obtaining the numeric meaning of geographic linguistic values (terms) (such as "convex," "linear," and "concave"). During the process the domain expert was presented with a set of landscapes using a three dimensional visualization tool (3dMapper, see http://TerrainAnalytics.com/ for additional details) and asked to delineate areas where the expert thinks the definition of the linguistic terms (such as "convex surface" or "linear surface" or "concave surface") are well met. The delineated areas are then analyzed together with the data layer of the variable (such as profile curvature) for determining the numeric meaning of the term in the domain of the variable. Once numeric meanings are obtained, the step function for each of these linguistic terms can be defined in the same way as the components with known ranges or thresholds.

Boolean integration of rule components

Under Boolean logic the values from the step functions are binary (0 or 1) and the integration of these values are often illustrated via a truth table (see below). The integration of these values depends on the logic operator joining these components in formulating the rule. There are two basic logic operators that can be used to join the components depending on the nature of a rule: the logic AND and the logic OR. Logic AND is used when all components of a rule need to be satisfied

Table 15.1 The truth table for mapping Soil Series "Basco" by integrating the Geology component with the Slope component using logic AND under Boolean logic

Location	Conditions	Geology Component (= "Jordan")	Slope Component (Slope < 20%)	Final Value
1	Geology = Oneota, Slope = 25%	0	0	0
2	Geology = Jordan, Slope = 25%	1	0	0
3	Geology = Oneota, Slope = 18%	0	1	0
4	Geology = Jordan, Slope = 18%	1	1	1

before the entire rule is met. Integration of components using logic AND is often accomplished using the minimum operator. As its name indicates, the final value of the rule is the minimum of the values from the components. A truth table for integrating the values from the components of the rule in Figure 15.2 is shown in Table 15.1. For simplicity, only the first two components are used in this table.

To produce a map of Basco the inference process will evaluate the conditions at every pixel and determine the final value using the integration process as illustrated here across the landscape. Figure 15.3 shows the predicted distribution of Basco based on the rule shown in Figure 15.2. One unique aspect of the prediction produced under Boolean logic is that it is binary (1 or 0). In this case, it is either Basco or not. There is no between.

Logic OR is used when a rule is met when only some of its components are true. For example, in landslide susceptibility mapping several combinations of conditions can cause landslides to occur but each of these combinations impact landslide occurrences independently (Zhu, Wang, Qiao, et al. 2004). When assessing landslide susceptibility these combinations should be considered separately and landslide susceptibility should be considered high when any of these combinations of conditions exists. Thus, the OR operator should be used to integrate these combinations. Integration under Logic OR is often accomplished using the maximum operator, that is, the final value is the maximum of the values from the components of the rule. Table 15.2 shows the truth table for integrating the components of the following rule: "*Landslides are most likely to occur when the Geology is 'Shale' or when the slope is steep (exceeding 50 percent)*".

Common sense also tells us that Location 4 in Table 15.2 is most susceptible because each condition can cause landslides independently. However, the OR implementation of the rule does not reflect this situation. In addition to this problem, the Boolean encoding of components is often inappropriate or too rigid. For example, the step function used in encoding the impact of slope gradient on landslide susceptibility does not consider the difference in impact between a location with slope of 55 percent and a location with slope of 90 percent. Clearly the latter location is more susceptible to landslide than the former.

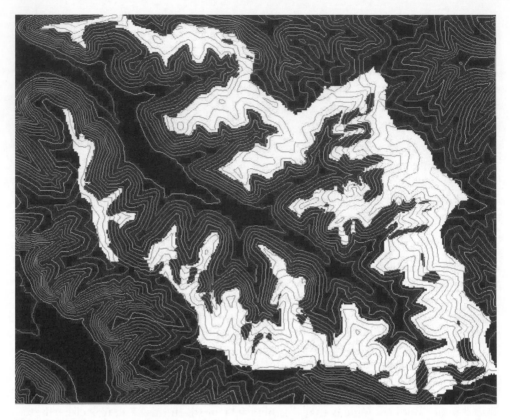

Fig. 15.3 A distribution of Soil Series Basco in Pleasant Valley, Wisconsin, USA based on a Boolean integration of the rule in Figure 15.2. White indicates areas of Basco soil (the thin lines are contours) and black areas are non-Basco soils

Table 15.2 The truth table for mapping landslide susceptibility by integrating the Geology component with the Slope component using logic OR under Boolean logic

Location	Conditions	Geology Component (= "Shale")	Slope Component (Slope > 50%)	Final Value
1	Geology = Oneota, Slope = 30%	0	0	0
2	Geology = Shale, Slope = 30%	1	0	1
3	Geology = Oneota, Slope = 60%	0	1	1
4	Geology = Shale, Slope = 60%	1	1	1

Rule-based mapping under fuzzy logic

Fuzzy encoding of rule components

Fuzzy logic allows the encoding of components to reflect the degree of suitability. In other words, under fuzzy logic the level at which a component is satisfied can be expressed. The level is often referred to as *membership value* (grade) ranging from 0 to 1, with 0 meaning no membership (not satisfying the component of a rule at all) and 1 meaning full membership (fully satisfying the component). The function used to express (encode) the level of satisfaction is referred to as *membership function* (see Zimmermann [1985] for details on fuzzy logic; Zhu (1997a, 2005) for application of fuzzy logic in rule-based soil mapping). For rule-based mapping the membership function for each component is defined using what is referred to as the *Semantic Import Model* (SI) (see Burrough 1996, McBratney and Odeh 1997, Robinson 2003 for details), which means that the definition of a membership function is based on the knowledge or experience of one or more domain specialists. Fuzzy membership functions for categorical/nominal variables are the same as that under Boolean logic. For ordinal variables membership functions can take a function with "multiple steps" with each ordinal value corresponding to one membership grade. The association of a membership grade to a given ordinal value depends on the preference assigned to the specific ordinal value. The more preferred the value the higher the membership grade.

The definition of fuzzy membership functions for interval and ratio variable types typically requires three critical pieces of information: the form of the membership curve, the optimal value, and crossover point(s) (Figure 15.4). There are three basic forms of membership curves which are often used in fuzzy mathematics (Figure 15.5): the bell-, Z-, and S-shaped. The bell-shaped curve describes that there

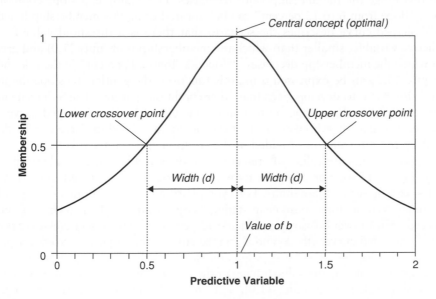

Fig. 15.4 The metrics of a fuzzy membership function

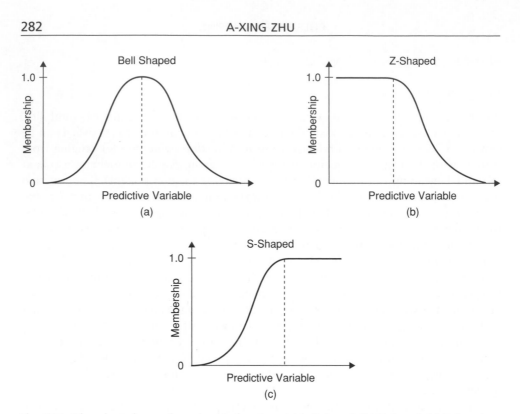

Fig. 15.5   Three basic forms of membership functions: (a) bell-shaped; (b) Z-shaped; (c) S-shaped

is an optimal attribute value or range over which membership is at unity (1.0) and as the attribute of the predictive variable deviates from this value or moves away from this range the membership value decreases. For example, a slope component stating "slope from 6 to 12 percent" can be captured using this membership function. The Z-shaped curve describes the scenario that there is a threshold value for the predictive variable, smaller than which the membership is at unity (1.0) and greater than which the membership decreases. "Slope less than 20 percent" in the rule shown in Figure 15.2 can be expressed using this function (the gentler the slope the more suitable for Basco to develop given that other conditions permit). The S-shaped curves define the relationships opposite to that characterized by the Z-shaped curves. As indicated through this discussion, the form of the membership curve can be determined by the information provided in the component of a given rule.

Crossover points are values of predictive variables at which the membership value is 0.5 (Figure 15.4). For components with numeric limits specified, the crossover point(s) can be set to these limits. For example, the crossover points of a membership function for a slope component stating "slope from 6 to 12 percent" can be set to 6 percent (for lower crossover point) and 12 percent (for the upper crossover point). For the one-sided curves (the S- and Z-shaped curves) the respective crossover points can be set to the "greater than" value and the "less than" value, respectively.

The optimal value is the value (range) of the predictive variable at (over) which the membership value is at unity (1.0). For the bell-shaped membership function, the optimal value can be a single point or can be a range. For one-sided curves the optimal value is a point from which the membership decreases (Z-shaped) from or increases (S-shaped) to unity (1.0) as the attribute value increases. Determining the optimal value (range) is more difficult than determining the crossover points because the descriptive components in a rule may not contain this information. Thus, this information often has to be obtained through knowledge acquisition. For components with numeric limits specified the optimal values (range) can be obtained from domain experts using the approach developed by Zhu (1999).

For components with linguistic terms as values (such as curvature variables with "convex," "linear," "concave" terms) the determination of optimal and crossover points is difficult. Researchers have explored the use of geovisualization techniques and frequency analysis techniques to define these values (see Robinson (2000) and Liu (2004) for additional details on these techniques).

Once the form of the curve and the optimal and crossover points are determined the membership function can be formulated quite easily. There are many mathematical functions that can be used in this respect (see Robinson (2003) for additional details). Burrough (1989) use the following function to depict a symmetrical bell-shaped membership function:

$$\mu_{fuzzy}(x) = \frac{1}{1 + \left(\dfrac{x - b}{d}\right)^2} \qquad (15.5)$$

where x is the value of a predictive variable, b is the attribute value representing the central concept (optimal value), and d is the width of the bell-shaped curve between one of the cross-over points and b (Figure 15.4). The S-shaped curve can be created by setting the membership value to 1 for x greater than b while the Z-shaped curve can be created by setting the membership value to 1 for x less than b. Equation (15.5) can also be modified to model a membership function that has a range over which the membership value is at unity (Shi, Zhu, Burt, and Simonson 2004).

Fuzzy integration of rule components

As we can see from the above-mentioned discussion on fuzzy encoding of rule components, for a given location (i,j) each membership function will produce a fuzzy membership value for its respective component. The integration of these fuzzy membership values produces a fuzzy membership value describing how much the rule is satisfied at this location. The operator used to integrate the component values under fuzzy logic is problem specific. There are a number of operators that can be used to accomplish the integration: the common ones are the logic AND and the logic OR depending on the nature of joining these components in formulating the rule. There are a number of implementations for each of these two operators. The commonly used implementation of logic AND is what is termed the *fuzzy minimum operator*. The theoretical basis of the fuzzy minimum operator is the limiting-factor principle in ecology, which states that the formation or development of an ecological feature, such as vegetation or soil, is controlled or limited by the least favorable factor in its environment. The integration under the fuzzy minimum operator can be expressed as follows:

$$S_{i,j} = \min_{v=1}^{m} [f^v(E_{i,j}^v)] \qquad (15.6)$$

Table 15.3 The membership values for Soil Series "Basco" by integrating the Geology component with the Slope component using the AND operator under fuzzy logic

Location	Conditions	Geology Component (= "Jordan")	Slope Component (Slope < 20%)	Final Value
1	Geology = Oneota, Slope = 25%	0	0.2	0
2	Geology = Jordan, Slope = 25%	1	0.2	0.2
3	Geology = Oneota, Slope = 18%	0	0.74	0
4	Geology = Jordan, Slope = 18%	1	0.74	0.74

where $S_{i,j}$ is the outcome of the rule at location (i,j) and m is the number of components in the rule. Table 15.3 shows the fuzzy equivalent of Table 15.1 given that a Z-shaped curve (Equation 15.7) based on Equation (15.5) is used to encode the slope component (with $b = 15\%$ and crossover point at 20%, $d = 20\% - 15\%$):

$$\mu_{slope}(x) = \frac{1}{1 + \left(\dfrac{x - 15}{5}\right)^2} \qquad \text{for } x > 15\% \tag{15.7}$$

$$\mu_{slope}(x) = 1 \qquad \text{otherwise}$$

The clear cut situation expressed in Table 15.1 no longer exists in Table 15.3. Location 2 now bears some similarity to soil series Basco. It is very possible that the slope condition at Location 2 is not steep enough to preclude the soil from bearing high similarity to Basco. At Location 4 the slope condition is in the transition zone (15~20%) and we expect the soil to bear high similarity to Basco but may not fit the typical (central) concept of Basco exactly. This situation is reflected in its membership value being close to 1 but not 1. If Basco were the soil type which is suitable for a septic tank location, users would be able to gain a better appreciation of how suitable a particular location is for septic tanks with these small but important differentiations provided under fuzzy logic.

The predicted distribution of Basco over the Pleasant Valley of Wisconsin based on the fuzzy implementation of the rule in Figure 15.2 is shown in Figure 15.6. The difference between Figures 15.4 and 15.6 is that in the latter the predicted distribution of Basco is portrayed as memberships ranging from 1 (white area, for fully meeting the condition of the rule) to 0 (the black area, for not meeting the condition of the rule at all). Basco is a soil that occurs on ridgetops and as one moves away from ridges down the slope the typicality of Basco decreases which is captured by this fuzzy soil map.

The commonly used implementation of logic OR is the *fuzzy maximum operator* and can be expressed as:

$$S_{i,j} = \max_{v=1}^{m} [f^v(E_{i,j}^v)] \tag{15.8}$$

Fig. 15.6 The predicted distribution of Soil Series Basco in Pleasant Valley, Wisconsin, USA based on a fuzzy implementation of the rule in Figure 15.2. The lighter the grades the more typical of Basco (thin lines are contours)

Given that Equation (15.9) is used to encode the slope component in the landslide susceptibility example, after applying Equation (15.7) to the membership values from the components of the rule, we have the outcome from the rule as shown in Table 15.4.

$$\mu_{slope}(x) = \frac{1}{1 + \left(\dfrac{x - 70}{20} \right)^2} \qquad \text{for } x < 70\% \qquad (15.9)$$

$$\mu_{slope}(x) = 1 \qquad \text{otherwise}$$

Boolean versus fuzzy

Much of the difference between Boolean and fuzzy logic for rule-based mapping is related to how the rule components are encoded. When the minimum and maximum operators are used for logic AND and logic OR, respectively, the integration of values from components are the same for both Boolean and fuzzy logic. However,

Table 15.4 The membership values for landslide susceptibility by integrating the Geology component with the Slope component using the maximum operator under fuzzy logic

Location	Conditions	Geology Component (= "Shale")	Slope Component (Slope > 50%)	Final Value
1	Geology = Oneota, Slope = 30%	0	0.2	0.2
2	Geology = Shale, Slope = 30%	1	0.2	1
3	Geology = Oneota, Slope = 60%	0	0.8	0.8
4	Geology = Shale, Slope = 60%	1	0.8	1

the difference in the values from rule components under different logic frameworks drives the difference in final results. Under Boolean logic the outcome from a rule is always either 1 (true) or 0 (false). However, the outcome from a rule implemented under fuzzy logic has a value between 0 and 1 (inclusive), although under some circumstances the upper bound can be greater than 1. This flexibility of fuzzy logic in encoding and integrating the rule components greatly enhances our ability to capture and map the spatial gradation a of geographic entity (Zhu 1997a) (cf. Figures 15.3 and 15.6).

The flexibility under fuzzy logic also allows rule-based mapping to be more complete spatially and more informative than under Boolean logic. Figure 15.7 shows the soil map of Pleasant Valley in Wisconsin based on Boolean logic using the rules shown in Figure 15.8. The rigidity of the Boolean implementation means that the conditions at some of the locations in the area do not meet any of the rules specified for the soil series in the area. As a result these areas are not mapped as any of the soil types. Typically these missed locations are in the transitional areas between soil types. This means that under the Boolean implementation the soils in transitional areas are completely missed out. On the contrary, fuzzy implementation retains the level of satisfaction of each rule and this level of satisfaction can be used to map the areas where local conditions do not fully meet the conditions of the rules. Figure 15.9 shows the soil map of the same area using the same set of rules but under fuzzy logic. The soil map in Figure 15.9 is produced by hardening the fuzzy soil membership maps produced for soil types using the respective rules. Hardening is a process of assigning a pixel (location) to a soil type in which the location has the highest membership value (Zhu and Band 1994). The transitional areas are now mapped. More importantly, the hardening process also produces uncertainty information associated with assigning a pixel to a soil class to which the soil at the pixel does not fully belong (Zhu 1997b). Figure 15.10 shows the uncertainty map associated with the soil map in Figure 15.9. The level of whitening in a color for a location indicates the level of uncertainty in assigning the local soil to the given class (represented by the color). It is clearly shown that the locations of whitening in colors are in the transitional areas which are the missed areas in Figure 15.7.

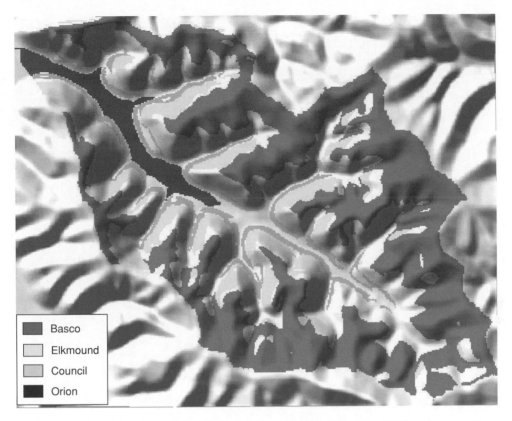

Fig. 15.7 Predicted distribution of soil series in Pleasant Valley, Wisconsin, USA based on a Boolean implementation of the rules in Figure 15.8. Light toned areas in the watershed are areas that do not meet the conditions of any of the rules

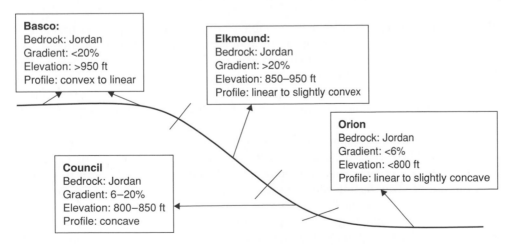

Fig. 15.8 Descriptive rules for soils in Pleasant Valley, Wisconsin, USA

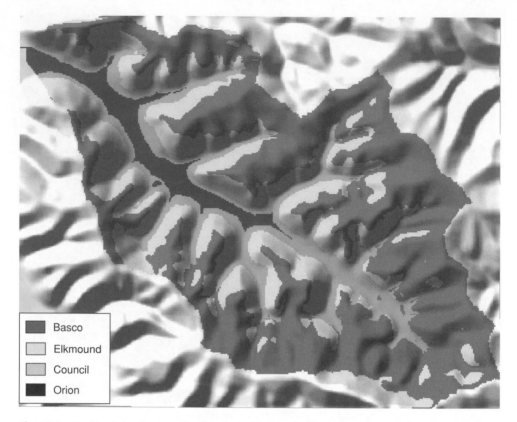

Fig. 15.9 Predicted distribution of soil series in Pleasant Valley, Wisconsin, USA based on a fuzzy implementation of the rules in Figure 15.8. No areas are missed out

Challenges and Research Issues

Rule-based mapping provides a very viable option to physical process-based and statistical approaches for mapping variation of geographic phenomena/features, particularly when we know more than what is needed in statistical approaches but less than what is needed for physical process-based mapping. The development of the fuzzy logic concept, coupled with geographic information processing techniques makes rule-based mapping not only practical but also accurate in portraying spatial gradation of geographic entities.

However, rule-based mapping still faces many challenges and many research issues. Examples of these challenges and research issues are: the extraction of rules, the weighting of rule components in inference, the definition of membership functions, and the use of fuzzy membership values. Much of the knowledge needed for rule-based mapping exists in the form of human expertise. Methods are needed to extract and document this knowledge. One important research issue is how to integrate knowledge from different experts and sources (such as human expertise with knowledge that exists in other forms).

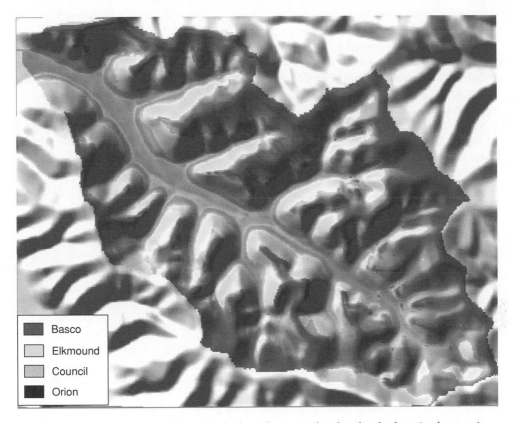

Fig. 15.10 Uncertainty map associated with the soil map produced under the fuzzy implementation of the rules for the Pleasant Valley area. The purer the color the more certain for the area to be labeled as the soil represented by the color

Most of the inference techniques treat rule components (variables in a rule) equally, not for simplicity of inference, but more importantly for the lack of knowledge on how these components (variables) interact. Assigning weights to variables is not difficult in the implementation of an inference engine or overlay process but what weights should be assigned is a challenge. Is there knowledge from domain experts which can be used for determining the weights? Is spatial data mining an approach to this problem?

Definition of fuzzy membership functions has come a long way (from general functions to domain knowledge-based functions). There are still challenges ahead in this arena, particularly with variables which take linguistic terms as values. The following questions need to be answered before the meaningful definition of membership functions for these variables can be found: What does a given linguistic value mean in the numeric domain of data layers? How can we determine the numeric meaning of a given linguistic value? How does this meaning change from one person to another, from one landscape to another and from one application to another?

Fuzzy logic has been well received by researchers in geography, particularly GI scientists (see Robinson, Chapter 14 of this volume, for a more detailed treatment of fuzzy classification and mapping). However, the use of fuzzy membership values in decision making and other management practice has a long way to go. For example, people are used to seeing maps showing the distribution of soil classes and associated properties with these classes. How would a person associate the required soil properties with a fuzzy membership map? What do fuzzy membership maps mean in decision making?

Research in the areas mentioned above will almost certainly propel rule-based mapping to a new era in the years to come. Information produced from rule-based mapping is then more likely to be meaningfully used in decision making and other forms of geographic analysis.

ACKNOWLEDGEMENTS

The author is grateful to the support from the "One-Hundred Talents" program of the Chinese Academy of Sciences. The assistance provided by Mr Rongxun Wang in preparing the maps used in this chapter is greatly appreciated.

REFERENCES

Batty, M. 1998. Urban evolution on the desktop: Simulation with the use of extended cellular automata. *Environment and Planning A* 30: 1943–67.

Boose, J. H. and Gaines, B. R. (eds). 1988. *Knowledge Acquisition Tools for Expert Systems.* London: Academic Press.

Burrough, P. A. 1989. Fuzzy mathematic methods for soil survey and land evaluation. *Journal of Soil Science* 40: 447–92.

Burrough, P. A. 1996. Natural objects with indeterminate boundaries. In P. A. Burrough and A. U. Frank (eds) *Geographic Objects with Indeterminate Boundaries.* London: Taylor and Francis: 3–28.

Buttenfield, B., Gahegan, M., Miller, H. J., and Yuan, M. 2002. Geospatial Data Mining and Knowledge Discovery. WWW document, http://www.ucgis.org/priorities/research/research_white/2000%20Papers/emerging/gkd.pdf.

Couclelis, H. 1997. From cellular automata to urban models: New principles for model development and implementation. *Environment and Planning B* 24: 165–74.

Ehleringer, J. R. and Field, C. B. (eds). 1993. *Scaling Physiological Processes: Leaf to Globe.* San Diego, CA: Academic Press.

Fensel, D. A. and Studer, R. 1999. *Knowledge Acquisition, Modeling and Management: Proceedings of the European Knowledge Acquisition Workshop (EKAW-99).* New York: Springer Lecture Notes in Artificial Intelligence No. 1621.

Goodchild, M. F. and Haining, R. P. 2004. GIS and spatial data analysis: Converging perspectives. *Papers in Regional Sciences* 83: 363–85.

Hart, A. 1986. *Knowledge Acquisition for Expert Systems.* New York: McGraw-Hill.

Hastie, T. J. and Pregibon, D. 1992. Generalized linear models. In J. M. Chambers and T. J. Hastie (eds) *Statistical Models in S.* Pacific Grove, CA: Wadsworth and Brooks, pp. 195–248.

Hastie, T. J., Tibshirani, R. J., and Friedman, J. 2001. *The Elements of Statistical Learning: Data Mining, Inference and Prediction.* New York: Springer.

Liao, S. H. 2005. Expert system methodologies and applications: A decade review from 1995 to 2004. *Expert Systems with Applications* 28: 93–103.

Lillesand, T. M., Kiefer, R. W., and Chipman, J. W. 2004. *Remote Sensing and Image Interpretation* (5th edn). New York: John Wiley and Sons.

Liu, J. 2004. Mapping with words: A new approach to knowledge-based digital soil mapping. Unpublished MSc Thesis, Department of Geography, University of Wisconsin at Madison.

McBratney, A. B. and Odeh, I. O. A. 1997. Application of fuzzy sets in soil science: Fuzzy logic, fuzzy measurements and fuzzy decisions. *Geoderma* 77: 85–113.

Miller, H. J. and Han, J. (eds). 2001. *Geographic Data Mining and Knowledge Discovery*. London: Taylor and Francis.

Monteith, J. L. 1965. Evaporation and environment. In G. E. Fogg (ed.) *The State and Movement of Water in Living Organisms*. San Diego, CA: Academic Press: 205–34.

O'Sullivan, D. and Haklay, M. 2000. Agent-based models and individualism: Is the world agent-based? *Environment and Planning A* 32: 1409–25.

Qi, F. and Zhu, A. X. 2003. Knowledge discovery from soil maps using inductive learning. *International Journal of Geographical Information Science* 17: 771–95.

Robinson, V. B. 2000. Individual and multi-personal fuzzy spatial relations acquired using human–machine interaction. *Fuzzy Sets and Systems* 113: 133–45.

Robinson, V. B. 2003. A perspective on the fundamentals of fuzzy sets and their use in geographic information systems. *Transactions in GIS* 7: 3–30.

Scott, M. J., Davis, F., Cusuti, B., Noss, R., Butterfield, B., Groves, C., Anderson, H., Caicco, S., D'Erchia, F., Edwards, T. C., Ulliman, J., and Wright, R. G. 1993. GAP analysis: A geographic approach to protection of biological diversity. *Wildlife Monographs* 123: 1–41.

Shi, X., Zhu, A. X., Burt, J., Qi, F., and Simonson, D. 2004. A case-based reasoning approach to fuzzy soil mapping. *Soil Science Society of America Journal* 68: 88–94.

Simpson, D. M. 2001. Virtual reality and urban simulation in planning: A literature review and topical bibliography. *Journal of Planning Literature* 15: 359–72.

Waring, R. H. and Running, S. W. 1998. *Forest Ecosystem Analysis at Multiple Scales*. San Diego, CA: Academic Press.

Wilson, J. P. and Gallant, J. C. (eds). 2000. *Terrain Analysis: Principles and Applications*. New York: John Wiley and Sons.

Yamada, K., Elith, J., McCarthy, M., and Zerger, A. 2003. Eliciting and integrating expert knowledge for wildlife habitat modeling. *Ecological Modeling* 165: 251–64.

Zhu, A. X. 1997a. A similarity model for representing soil spatial information. *Geoderma* 77: 217–42.

Zhu, A. X. 1997b. Measuring uncertainty in class assignment for natural resource maps under fuzzy logic. *Photogrammetric Engineering and Remote Sensing* 63: 1195–202.

Zhu, A. X. 1999. A personal construct-based knowledge acquisition process for natural resource mapping. *International Journal of Geographical Information Science* 13: 119–41.

Zhu, A. X. 2005. Fuzzy logic models. In S. Grunwald and M. E. Collins (eds) *Environmental Soil-Landscape Modeling: Geographic Information Technologies and Pedometrics*. Boca Rotan, FL: CRC Press, pp. 215–39.

Zhu, A. X. and Band, L. E. 1994. A knowledge-based approach to data integration for soil mapping. *Canadian Journal of Remote Sensing* 20: 408–18.

Zhu, A. X., Band, L. E., Dutton, B., and Nimlos, T. J. 1996. Automated soil inference under fuzzy logic. *Ecological Modelling* 90: 123–45.

Zhu, A. X., Hudson, B., Burt, J., Lubich, K., and Simonson, D. 2001. Soil mapping using GIS, expert knowledge, and fuzzy logic. *Soil Science Society of America Journal* 65: 1463–72.

Zhu, A. X., Wang, R. X., Qiao, J. P., Chen, Y. B., Cai, Q. G., and Zhou, C. H. 2004. Mapping landslide susceptibility in the Three Gorge Area, China, using GIS, expert systems and fuzzy logic. In Y. Chen, K. Takara, I. D. Cluckie, and F. H. De Smedt (eds) *GIS and Remote Sensing in Hydrology, Water Resources and Environment*. Wallingford, UK, International Association of Hydrological Sciences Publication No. 289: 385–91.

Zimmermann, H. J. 1985. *Fuzzy Set Theory and Its Applications*. Boston, MA: Kluwer-Nijhoff.

Chapter 16

Multivariate Geovisualization

Mark Gahegan

Research into the physiological aspects of human perception points to three primary channels by which visual information passes from the retina to the visual cortex; these essentially parallel channels carry encoded signals describing color, movement, and form (structure) within the scene (see Livingstone and Hubel 1988, for a more detailed account). The signals they carry are integrated during the perception process,[1] and it is their joint interpretation, along with ocular disparity derived from stereo vision and stored experiences and knowledge that enables us to make sense of the complexities of the visual stimulus presented to us with such consummate ease. Using these properties we "see" motion, perspective, speed, distance, pattern, surface properties (shiny, dull), illumination characteristics (position and intensity of the light source), and many other visual qualities. It is therefore a regrettable fact that the following description of the wonders of multivariate visualization is in a book, where color, movement, and stereo vision are not supported and the available bandwidth for visual communication of information is rather drastically restricted to the use of position and shape, with some possible further encoding of information using shades of grey.[2] Thus, many of the important benefits of visualization are lost, and it is a challenge to offer the reader any convincing proof of concept.

I therefore humbly ask, gentle reader, that you try and imagine color, depth, and movement in the rude images included in this chapter. Or if you prefer, visit the GeoVISTA website (http://www.geovista.psu.edu) where these visual properties are available in full measure.

Why Do We Need So Many Variables?

Before discussing the details of multivariate geovisualization – concurrent visual display of many variables or dimensions – it is worth considering why the use of many variables is called for. There seem to be two sets of forces at work here, offering a "push" and a "pull" factor. On the push side, two specific technologies present us with ever more complex data sets. Firstly, measuring instruments from census to satellite are becoming more complex and elaborate, gathering more and more variables; secondly, as many longstanding problems of interoperability and data access are solved it becomes possible to integrate once-separate data sets into highly multivariate collections (Miller and Han 2001).

On the pull side, trying to understand, as geographers are wont to do, the complexities of physical and human systems usually demands that we consider many factors simultaneously; systems tend to be unbounded or "open," and interrelationships are complex, multi-faceted, and often accompanied with significant spatial and temporal variation. For example, understanding cancer incidence and prevalence with respect to healthcare accessibility, demographics, and physical environment might reveal important trends that could form the basis of improved screening programs and public education campaigns.

Multivariate data visualization is one means by which humans can deal with this kind of data complexity, and there are a number of well-established visual methods for displaying several variables concurrently (for example, McLeod and Provost 2001), as we will see later. However, continuing to add in more variables does not necessarily produce a more coherent picture; perceptual and cognitive limitations govern the effectiveness of these multivariate visualizations. What is not always clear is exactly *how* these human limitations govern effectiveness. So, although there seems by now little doubt that multivariate visualizations are a useful means to explore and portray complex data, there are still many unanswered questions regarding perceptual limitations, appropriateness, and effectiveness. Multivariate visualization is therefore not a panacea for large or complex data sets, but just like statistics and data mining, it offers some techniques to address them, while introducing limitations and problems of its own. Good analysis needs to integrate the best that each of these approaches has to offer.

Different Ways to Analyze Multivariate Data

To elaborate, when faced with the task of understanding some complex system – a snapshot of which is captured in a multivariate data set – there are a variety of tools and methodologies that could be used. Multivariate statistics provides a vast array of tried and tested analysis tools (for example, Mardia, Kent, and Bibby 1979); machine learning and data mining supply a newer set of tools that are sometimes more flexible and computationally efficient, but are also less well tested and understood (Gahegan 2003; see also Chapter 19 by Miller in this volume). Table 16.1 contrasts some of the basic tenets of these three different approaches.

Any form of analysis is a collaboration between researchers and their tools and methodologies; but this collaboration is most central to visualization, as it relies on human perceptual and cognitive systems to observe and interact with data, as well as interpret the results. Visualization is also usually interactive, so analysis can be regarded as a form of communication (perhaps even a conversation; MacEachren, Gahegan, and Pike 2004) between geographer and data, and as such offers the potential advantage of bypassing the formal language of mathematics or the theories of machine learning in order to communicate with the researcher. The language is visual as opposed to mathematical, statistical, or logical, and the conversation is often aided

Table 16.1 A brief summary of the differences between statistical, machine learning, and visual processes to analysis, a more detailed version appears in Buttenfield, Flewelling, Gahegan, Miller, and Yuan (2005)

Method	Data analyzed using	Testing, rigor
Statistical analysis	Parametric function fitting, inferential methods	Significance testing, power, goodness of fit
Machine learning	Inductive learning methods, data mining methods	Validation experiments that test generality and repeatability
Visualization	Graphical methods, mapping data to visual variables	User studies, focus groups, protocol analysis

by interactive tools that support a dynamic exploration of the data by allowing the analyst to keep refocusing their attention on different regions of the data set. But because the human is such an integral part of the visualization process, the outcomes are more subjective in nature and hence the evaluation of findings becomes more problematic (Cleveland 1993). We each see and understand differently, so what works well for one person may be unsuitable for another, based on factors as simple as color vision deficiencies and as complex as the difference in prior experiences of analysts (McGuinness 1994, Nyerges, Moore, Montejano, and Compton 1998).

Ways to Approach Multivariate Geovisualization

The development of multivariate visualizations can be, and is, approached from a wide variety of perspectives, a selection of which appear below as a list of motivating questions.

1 **Graphic Design**: How can I make displays that people want to look at – that are intriguing and appealing, that capture the eye? (Arheim 1974, Karssen 1980, Tufte 1990, Landa 2004).
2 **Perception and Cognition**: How can I use the different visual variables at my disposal, such as color, transparency, movement, size, in ways that are known to be effective? (Bertin 1967, 1981, Koch and Ullman 1985, Treisman 1986, Livingstone and Hubel 1988; Grinstein and Levkowitz 1995).
3 **Statistics**: How can I capture the variance of all the salient data attributes so that the underlying trends and patterns can be better understood? (Asimov 1985, Cleveland and McGill 1988, Haslett, Bradley, Craig, Unwin, and Wills 1991, Cook, Majure, Symanzik, and Cressie 1996).
4 **Computer Science**: How can I improve the computational representation and manipulation of graphical data? (Noll 1967, Sammon 1969, Glassner 1989, Spitaleri 1993, Ribarsky, Katz, and Holland 1999).
5 **User-centered design**: Who are the users, and what are they familiar with? (Hix and Hartson 1993, Shneiderman 1998, Vredenburg, Isensee, and Righi 2002, Kosara, Healey, Interrante, Laidlaw, and Ware 2003).

6 **Cartography:** How can I extend the familiar map to include additional useful and relevant information? (Monmonier 1989, 1990 MacDougall 1992, MacEachren 1994; Dykes 1996).

All of these approaches have merits of their own and all of the fields represented above have made significant contributions to the emerging field of information visualization. Historically, perhaps the most commonly used approaches to geovisualization are 6 and 2, with a little of 2, 3, and 5. By contrast, within advertising the emphasis instead is on 1 and 5.

It is not surprising, then, that, as the field of geovisualization develops, approaches and techniques are drawn from an increasingly wide variety of sources. Obviously, cartography remains a major influence, and provides not only a range of symbolization and thematic techniques such as perceptually valid color schemes (Brewer 1997, Green and Horbach 1998) and cartograms (Dorling 1994) but also important early theoretical and conceptual contributions such as those of Taylor (1991), DiBiase (1990) and MacEachren (1995), and a rich history and commitment to testing and evaluation by experiments with users (Wood 1990, McGuinness 1994, Slocum, Blok, Jiang, et al. 2001, Andrienko, Adrienko, Voss, et al. 2002). The work of Bertin (1967, 1981, 1983) in describing low-level graphical primitives (retinal or visual variables) and making recommendations as to how they should be used effectively has become central to geovisualization, as it has to cartography. Efforts to extend our list of useful visual variables as they relate to dynamic, interactive displays are increasingly relevant (for example, Mackinlay 1986, MacEachren 1994).

At the present time, in the early twenty-first century, most experimental systems built for multivariate geovisualization rely on a series of different data displays, including maps and other graphical techniques that are kept in step with each other by a technique called "linking and brushing"; selecting items in one display causes the same items to be selected in another display. This technique seems to have first arisen within the statistical community (Cleveland and McGill 1988) and was later applied to interactive mapping by Monmonier (1989, 1990), where it has become very popular as a means to explore how patterns in geographic space are reflected in attribute space, and vice versa. Linking and brushing, thematic classification, and data exploration form the basis of many environments for visualization (Egbert and Slocum 1992, Dykes 1996, Gahegan 1998, Andrienko and Andrienko 1999, North and Shneiderman 1999, Gahegan, Takatsuka, Wheeler, and Hardisty 2002).

Types of Visualization Methods

"Seeing" into high dimensional spaces typically involves reducing or re-portraying the space into forms that humans can better understand, and that can be physically displayed in a smaller number of dimensions, or in a more compact form, on what is usually a small display device such as a typical computer monitor. Many different approaches have been taken to increase visual effectiveness and to add in further dimensions, producing a wide variety of techniques as a result (Robertson 1997). Of course, it is usually the case for geographic inquiry that the spatial dimensions, and often the temporal dimensions, within the data are of great significance and because of this they are often privileged in that they form the basis (for example, the axes) around which other data dimensions are structured. No surprise, then, that geovisualization is typically biased toward map-based displays.

More recently, information visualization – having become an established field of research in its own right – has supplied a large number of newer techniques: graph-based methods such as the parallel coordinate plot (Inselberg 1985, 1997) and bagplot (Rousseeuw, Ruts, and Tukey 1999), hierarchical methods such as cone-trees (for example, Robertson, Mackinlay, and Card 1991), symbol-based methods such as iconographs (Pickett and Grinstein 1988), techniques for making "information landscapes," for analyzing multi-thematic information sources such as newspapers and scientific literature archives (for example, Miller, Wong, Brewster, and Foote 1999, Fabrikant 2000), and methods for visualizing concept graphs and ontologies (Mutton and Golbeck 2003, Pike and Gahegan 2003). Figure 16.5 shows an example of a parallel coordinate plot, Figure 16.9 a generalization hierarchy, Figure 16.4 an iconic visualization, Figure 16.8 a demographic "information landscape," and Figure 16.10 a concept graph. A useful summary of some of the more established information visualization methods is given by McLeod and Provost (2001).

Various attempts have been made to categorize these different methods according to the kinds of graphics they employ (Hinnenberg, Keim, and Wawryniuk 1999) or to the kinds of inference that they support (Gahegan, Wachowicz, Harrower, and Rhyne 2001). However, any distinctions between different methods are fast becoming moot as they borrow freely from each other, fusing together different graphical and statistical approaches.

Different methods for visualizing multivariate data include

- Multiple linked views: graphs shown in separate (usually linked) displays (multiples or matrices), as in Figures 16.6 and 16.7;
- Symbol construction or graphical mark (glyph) composition: data are mapped to different aspects of a chosen symbol or glyph, such as color, transparency, and shape resulting in a symbol that encodes two or more data variables, as shown in Figures 16.1, 16.2, and 16.4;
- Data projection and/or reduction: data are transformed via some function into a smaller number of variables (for example, projection pursuit, Cook, Buja, Cabrera, and Hurley 1995) sometimes followed by re-projection to a surface (for instance, using a Self Organizing Map as shown in Figure 16.8 and the visual data mining display in Figure 16.11);
- Animation: since the eye is so sensitive to changes in a display, movement or other forms of change can be very effective for getting trends in the data noticed, or, if badly used, can be very distracting (Figures 16.2 and 16.4 animate when viewed on the web);
- Hierarchies and node-edge graphs: these kinds of displays show taxonomies, concept graphs, and ontologies, that encode relationships such as "is a kind of," and "is a part of" between entities or concepts. Figures 16.9 and 16.10 are examples;
- Using additional sensory modes (haptics, sound, etc.).

Choosing a suitable visual representation for data depends heavily on the type of data under consideration, on the number of variables (p) and the number of records or objects (*n*) but also on the nature of the task to be undertaken, such as exploration or communication, as discussed in the next section.

Geovisualization Examples

The following set of geovisualizations exemplify many of the ideas described in this chapter, and are all available as color images or animations on the accompanying website, where the reader is advised to examine them if possible (animated displays require a Flash Plugin available freely from http://www.macromedia.com/software/flashplayer/). Note that as the number of variables increases, the visualizations become more abstract with less of the details in the original data being preserved. For exploring a highly multivariate data set, one might begin with these more abstract techniques, using them to identify lower dimensional sub-spaces that show patterns of interest, then investigating these further using the simpler techniques.

As a starting point, the first image is a simple thematic (choropleth) map, showing just a single attribute value, human population in the year 2000 by county for the contiguous 48 states of the USA (Figure 16.1). As is common in choropleth mapping, the population values are grouped into classes (six in this case) to provide discrete color values since this generally makes the map easier to interpret.

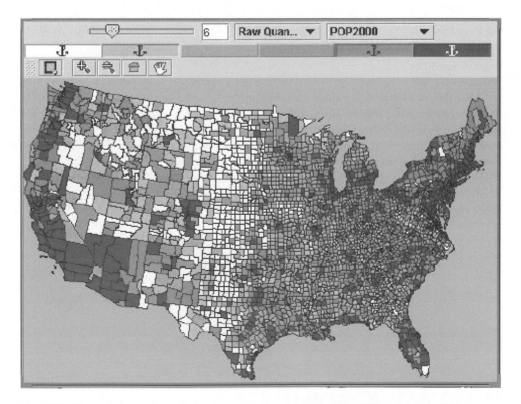

Fig. 16.1 Thematic map of population by county for the conterminous 48 states of the USA

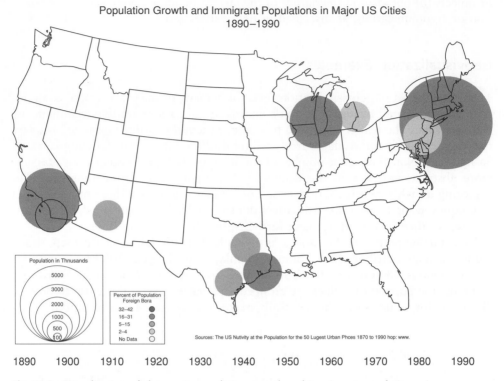

Population Growth and Immigrant Populations in Major US Cities
1890–1990

Fig. 16.2 Visualization of changes in population growth and immigrant population using proportional circles

The map in Figure 16.2 shows two variables concurrently. Proportional circles are centered on the major cities, with the size of each circle encoding population and the shade describing the proportion of foreign-born people who live there. Change over time is also shown in the accompanying applet, as a sequence of maps representing 10-year values from 1890–1990.

Encoding data (such as population) with a visual variable (such as circle size) is never straightforward. A good deal of our visual interpretation is relative, depending in large part on the immediate context within the display. As an example, Figure 16.3

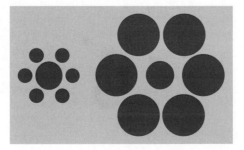

Fig. 16.3 Which inner circle seems bigger? A visual illusion concerning the size of circles (see text for details)

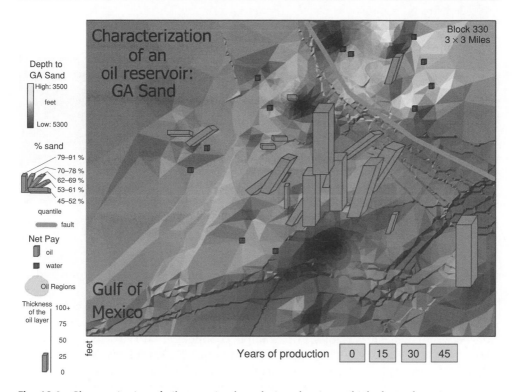

Fig. 16.4   Characterization of oil reservoirs through time showing multiple data values via a combination of symbols or glyphs

below shows a well-known visual illusion in which we see two inner circles of exactly the same size as being of different sizes because of the surrounding context. Many other such examples illustrate similar problems when interpreting orientation, motion, length, color, straightness, in fact just about all possible visual variables. The website: http://www.michaelbach.de/ot/ has several fascinating examples.

The more complicated animated map reproduced in Figure 16.4 shows details of an oil reservoir in terms of each drilled well. Many variables are encoded: the color of the background encodes interpolated depth to reservoir; hypothesized oil regions are shown as a semi-transparent shaded (green) overlay, the bar positions show bore hole location, their tilt angle encodes the percentage sand encountered at that location, the length of the bar shows the thickness of the oil layer; the square dots show the added presence of water. When using the animated version on the website, the bars change as the oil reservoirs become depleted over time.

Figure 16.5 shows a parallel coordinate plot (PCP), a technique proposed by Inselberg (1985) to visualize multiple data dimensions concurrently. The axes are drawn parallel to each other and appear as vertical lines. The data set shown describes various demographic properties for the 48 conterminous US states, hence there are 48 polylines or "strings" shown. If the values for each state are marked off on the axes, then joined with a line, the result is a web of strings that can show trends and clustering in the data. The arrows point to the trace for California (shown light

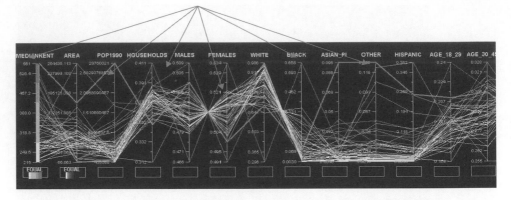

Fig. 16.5 A parallel coordinate plot (PCP) showing demographic and related information for the 48 conterminous US states. The arrows point to the "string" representing California

green on the website) which stands out as being rather anomalous compared to the general trend.

Looking between each pair of axes provides some insight into the correlation structure; where strings do not cross each other much there is a positive correlation between adjacent variables (for example, "population 1990" and number of "households"), where they cross more often than not, the correlation is likely to be negative (for example, % "male" and "female", % "white" and "black"). Color can be added to the display to impose categories onto the strings. Variants of the PCP are often used to examine temporal trends in data, in which case the axes usually represent the same attribute measured at different times and arranged in sequence. Positive and negative trends through time are then shown by the slope of the strings.

The display reproduced in Figure 16.6 provides multiple bivariate views onto a data set describing cancer incidence and prevalence, demographics, and healthcare accessibility in the Appalachian region (West Virginia, Kentucky, most of Pennsylvania). Five variables are explored here, with each variable being assigned to both a row and column, hence five rows and five columns. The cells in the matrix show bivariate views onto the data, that is, the conjunction of one row and one column variable. The upper right cells show a scatterplot for each pair of variables, and the lower left cells show the corresponding map. See MacEachren, Dai, Hardisty, Guo, and Lengerich (2003) for more details. The diagonal cell elements show the data distribution (histogram) for each individual variable – on the website you can also see that each scatterplot and map is colored using a bivariate color scheme to show the relationship between a pair of specially-chosen variables imposed on the graphics in each cell of the matrix.

The visualization in Figure 16.7 shows the results of using a decision tree classifier on a six-dimensional land cover data set using the variables: height, flow accumulation (surface), shape (surface morphometry), slope-degrees, TM-band-2 and TM-band-4 (both from the Landsat sensor). Although the diagram appears complex at first, it conveys a lot of useful information if carefully studied.

The scatterplot matrix display (as in the previous figure) graphs each variable against every other variable, that is to say, the set of all possible bivariate scatterplots for

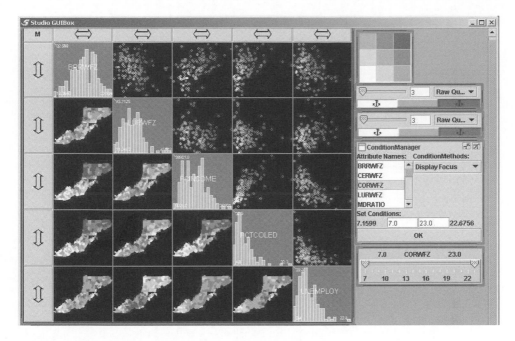

Fig. 16.6 Multiple scatterplots and bivariate maps of demographics and cancer-related data organized in a matrix

the six-dimensional data set. Since it makes no sense to plot a variable against itself, the diagonal cells in the matrix instead show a binned frequency histogram for each individual variable, and so indicate the general shape of the distribution. For all other cells, each cloud of points gives an indication of the degree of association between any pair of variables, and sometimes can also show clusters and outliers, depending on the distribution. If the user wishes, the points can be color coded (according to some other data values such as the land cover class each point indicates, for example). The horizontal and vertical lines represent the output from a decision tree classifier using Quinlan's (1993) C4.5 algorithm. So the horizontal lines across the top row of the matrix show the number and position of decision rules that were constructed using the height attribute to classify the data into different land cover types. In this case three decision rules were used. Vertical lines work the same, so the lines in the rightmost column show decision rules using Landsat TM band 4. The lines can be color coded too, so that we can see which rules help to construct which classes. Many lines in a row or column indicate a variable that is very useful in constructing the classification scheme (formally reduces entropy) or that the variable has a complex relationship to the categories sought, no lines indicate a variable that is redundant and adds no qualifying power to the analysis. Of course, such a display becomes even more powerful when linked to maps, images, and other multivariate displays such as the parallel coordinate plot.

The visualization of the conterminous 48 states of the USA shown in Figure 16.8 was constructed by clustering states based on their similarities or differences computed over 25 census variables describing population profiles in terms of raw count,

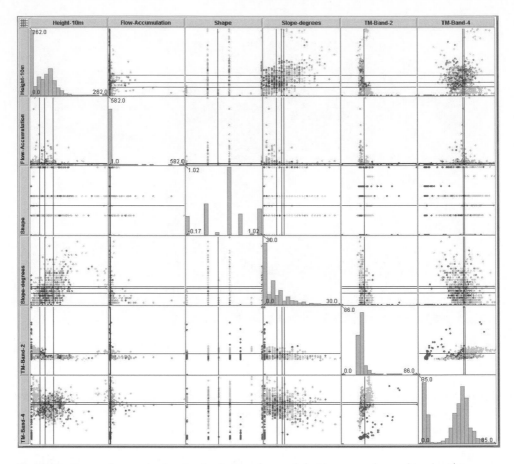

Fig. 16.7 Scatterplot matrix showing land cover data and decision tree rules used in classifying the data

ethnicity, proportions male and female, age, and the cost of renting accommodation. Hence, it encodes a large number of variables so that their dominant trends are preserved although the original values are lost. Clustering was performed using the Kohonen self-organizing map or SOM (Kohonen 1997), and in fact the visualization is a representation of the internal state of the SOM after training; the white dots represent neurons, the labels show the relative location of each state and the background surface is a measure of difference between states that can be interpreted just like a topographical surface; the actual distance between concepts being calculated as the path over the terrain between them, rather than as a straight line. Hence, New Mexico is similar to Nevada, Illinois is similar to Ohio and Pennsylvania, but California (represented by the huge spike bottom left) is very dissimilar to its nearest neighbors, New York, Virginia, and Texas. For contrast, the same data set was shown earlier in the parallel coordinate plot in Figure 16.8.

The example reproduced in Figure 16.9 shows the use of geovisualization to explore historical data. The map above shows the city of London as it was during

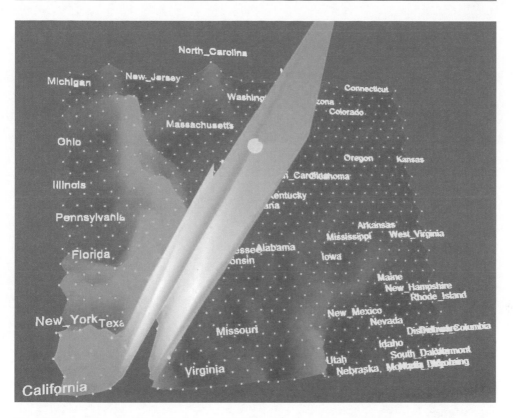

Fig. 16.8 A surface produced by clustering US states based on their demographic properties. California stands out as very anomalous

the seventeenth century, with streets containing significant merchant premises overlaid. Color is used to show the density of merchants. The hierarchical visualization tool (based on the Taxa tool; Graham, Watson, and Kennedy 2002) gives several alternative perspectives onto the same information, organized by market sectors to which merchants can belong. The selected merchant, Wetherell Janaway, is used to highlight relevant branches in the connected hierarchies of: commodities, retail, wholesale, tradesmen, and warehouses; for example *retail-precious metals-goldsmith-Wetherall Janaway* and *production-precious metal-jewelers-Wetherell Janaway*.

The concept map in Figure 16.10 depicts a series of concepts and relationships that together describe 10 significant events that have helped to shape the current physical landscape of central Pennsylvania. The events themselves are shown as pale diamond-shaped nodes and specific consequences and examples are shown as rectangles. Linking all of these together are a series of directed and undirected edges that describe the relationships between these concepts; for example, growth of industry was facilitated by railroad building, forest use and management gave rise to fire suppression and deforestation, and affected forest pathogens. A live version of this application (*ConceptVista*) dynamically reconfigures the graph according to the user's current interest (the concept "in focus"). This concept map also has ancillary materials, such as text and photographs, linked into it for use as an interactive

Density of Merchants

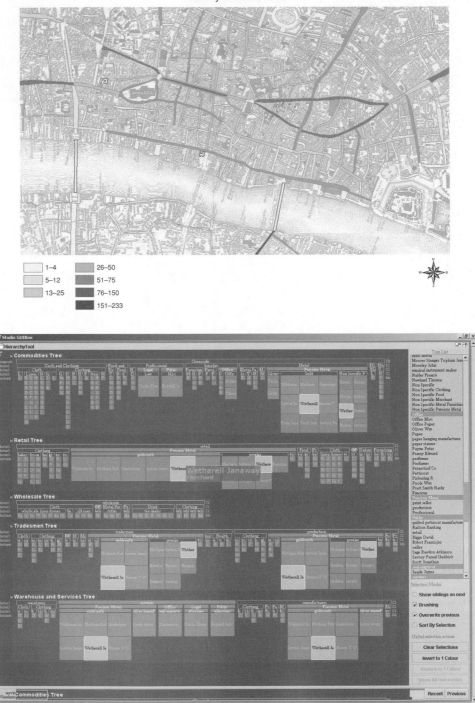

Fig. 16.9 The city of London during the seventeenth century, showing the density of merchants along various street segments (top) and a breakdown of the various hierarchies of goods and services that the data set contains (bottom)

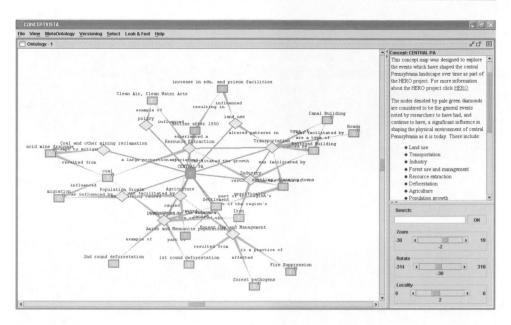

Fig. 16.10   A concept map depicting events that have shaped the landscape of central Pennsylvania

learning application. Tools to visualize and explore concept maps and ontologies are likely to become increasingly important in the future as a means to represent and exchange understanding between collaborating researchers.

The visualization reproduced in Figure 16.11 shows a matrix of around 100 census and disease variables and is used to gain a synoptic overview of possible significant relationships between pairs of variables. Two measures of similarity are used, linear correlation (upper right triangle, the lighter squares show higher correlation) and entropy (lower left triangle, the lighter colors show lower entropy). The diagonal line is kept black since this would represent each variable compared to itself. Lighter areas in the display are candidates for further investigation if the relationships shown are not already known; tools such as scatterplots, parallel coordinate plots, and maps might be used with the identified smaller subsets of data.

The Nature of Analysis: A Variety of Tasks

Multivariate visualization can help support a number of different activities; the various stages in the process of science provide a number of roles that visualization can play, including (but not limited to)

* Exploring data, formulating hypotheses
* Learning patterns, synthesis, making generalizations
* Performing analysis – supporting analytical activities
* Validating results – comparison, verification, error assessment
* Communicating findings

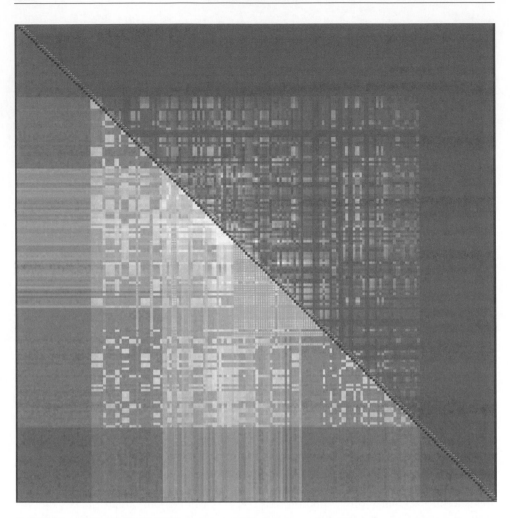

Fig. 16.11 A visualization of 100 census variables showing correlation and entropy

Each of these activities is likely to require a different approach to visualization, and possibly unique tools. For example, formulating a new hypothesis might involve trying to find a hitherto unknown or unexplained pattern in the data, whereas validating results might involve comparing several modeled outcomes to known reference data. For finding a pattern, techniques that help the user explore the data from many different perspectives are likely to be effective, such as scatterplot animation methods or automated clustering, whereas comparison of results against known data needs a method that emphasizes the similarities and differences between two sets of values.

Understanding the nature of the question being asked is vital to the success of visualization. Visualization techniques are often chosen pragmatically because they are easily available or they can support the types and formats of the data under investigation. Three things are needed to improve this situation: (1) wider availability of different techniques; (2) better inter-operation between techniques and other software to be used; and (3) more research on which techniques work best for which tasks (Fabrikant and Skupin 2005, Plaisant 2005, Gahegan 2005).

Creating Multivariate Visualizations

The process of creating a multivariate visualization involves a number of stages via which data are mapped to visual variables, which can then be rendered in a display. The process starts with the data, which will be of a specific type (for example, point, line, region, non-spatial attribute), be discrete or continuous, and measured according to a specific statistical scale. These factors must then be matched first to appropriate visualization geometry or symbols, (the sign vehicle) then to specific visual variables that those vehicles support. After these choices have been made, the range of values in the data must be transformed to a range of values supported by the visual variable; for example, a set of population values for census regions containing between 1,000 and 10,000 people might be matched to the color of a circle, from purple through to green in five discrete steps. The relationship between visual variable and data can be complex; several data variables may be combined to produce one or a few visual variables as in the surface shown in Figure 16.8. Conversely, a single data value might be encoded by more than one visual variable to give it added prominence in the display; a common example is to color a surface according to the "height" value thus using height twice – so called redundant encoding. This specific encoding is also used in Figure 16.8.

The entire process is repeated for each spatial object or attribute to be visualized. This abstraction process is much the same as for traditional cartography (MacEachren 1995, Dent 1998), but with additional data types and visual variables, and sometimes many sophisticated computational methods by which to transform the data into visual variables. Table 16.2 summarizes this process, after Muehrke, Muehrke, and Kimerling (2001) and Fabrikant and Skupin (2005). For simplicity, only static visual variables are considered in the table, though dynamic variables used for animation, and variables for sound and haptics could also be added.

In the table, two examples are given of the arrangements used to support visualization methods: available choices for the production of a scatterplot are shown shaded grey, and those used for a choropleth map have a heavier border; hence, a scatterplot visualizes attribute (non-spatial) data, that can be discrete or continuous, measured on an interval or ratio scale, transformed with a linear function, rendered with a point symbol that supports visual variables for x, y, and z position, color via hue, saturation, and value and size, and is useful for detecting structures and characterizing distributions. By contrast, a choropleth map uses attribute and area data, that is continuous and (usually) differentiable, measured on an ordinal, interval, or ratio scale, transformed using a classification method (equal interval in this case), displayed using polygons that support visual variables for x and y position and color, and used to examine a spatial pattern.

Supporting Multivariate Visualization

Since a number of different disciplines have independently developed the means to visualize multivariate information, it is not surprising that a number of quite different and usually incompatible visualization systems are now available. Choosing the right system depends on how you want to visualize the data and what you hope

Table 16.2 The abstraction process used to construct visualizations from data, showing two specific examples of a scatterplot (grey shading) and a choropleth map (heavy border line)

Data type	Smoothness	Statistical Scale	Transformation function	Sign vehicle	Visual Variable	Activity
Attribute	Discrete	Nominal	linear	Point (circle)	X position	Structure detection
Point	Continuous/differentiable	Ordinal	logarithmic	Line	Y position	Characterization (learn pattern A)
Line	Continuous non-differentiable	Interval	Equal intervals classification	Polygon	Z position	Differentiation (A from B)
Area		Ratio	Self organizing map	Surface	Hue	Examine spatial pattern
Surface			k-means clustering	Volume	Saturation	Examine temporal pattern
Volume				Sphere	Value	Find similar
				Arrow	Transparency	Validate results
					Size	
					Orientation	
					Length	

to achieve by so doing. Is animation important, or visual representation of 3D geometry, or perceptually sound use of color? In practice, the kinds of techniques and functionality available depends very much on the visualization system being used; at the time of writing commercial Geographic Information Systems (GIS) provide sophisticated map rendering, with appropriate color schemes, proportional symbols, and dynamic reclassification, often backed up by a scatterplot with dynamic linking and brushing between the two. For many kinds of multivariate visualization, these capabilities may suffice, but compared to many of the specialized visualization systems available, both commercial and academic, they are somewhat limited. The factors that typically separate GIS visualization capabilities from information visualization systems are:

- Dynamics – temporal animation is either very limited or absent in most current GIS;
- Support for true 3D geometry – most GIS can render one or more surfaces, but not a solid three-dimensional shape;
- Control of light and viewpoint (the user's position) in the 3D space of the scene – though increasingly GIS and remote sensing systems do offer some kind of "sun angle shading" algorithm to mimic the illumination from a single light source (usually the sun) at a given position above the geographical plane;
- Control of textures and properties of surfaces, for instance, surfaces can be made to appear more or less reflective to give objects in the scene a more realistic appearance;
- A rich set of data visualization methods perhaps including some of those described above under the heading "Ways to Approach Multivariate Geovisualization" and as shown in the example figures.
- The ability to "spatialize" data; that is to treat any data attribute as if it defined a geographic coordinate (GIS typically have to be "tricked" into spatializing non-spatial data).

Compared to the comprehensive lists of visual variables proposed by Bertin (1981) and extended by MacEachren (1994) and others, GIS can often support only a limited subset. Furthermore, in GIS there is usually an enforced connection between data type and sign vehicle (refer back to Table 16.2); a point must be rendered as a point, or point symbol, a polygon as a region and so forth. By contrast, most information visualization systems make no assumptions about how different data types will be signified, even if the mappings make no intuitive sense. For example, two data values a and b could be "spatialized" as a point x, y (as in a scatterplot) or a geographic point x, y could be rendered as the size and transparency of a circle.

Returning for a moment to the fundamental visual pathways described in the introduction, (shape, color, movement), GIS provide good facilities for describing color, reasonable control over shape and position – but only for explicitly spatial data, and little or no capability to support movement or dynamics, so there is room for further development, assuming we can determine how to make good use of these extra capabilities. To this end, many cartographers and information visualization specialists have experimented with a plethora of additional variables, from animation to sound to haptics (tactile feedback) and even taste and smell in order to add to the number of concurrent variables that can be studied, or the effectiveness of how they are perceived by the user.

Video graphics production environments such as Flash deserve a special mention as they are playing an increasingly important part in the study and production of dynamic maps, because of the ease by which temporal behavior can be scripted for the individual map components. Figures 16.2 and 16.4 are derived from Flash animated maps. Capacity to handle spatial data remains primitive, but improvements in the programming interface allow the intrepid user to make their own provision, with easy access to the temporal variables being the incentive.

You Usually Do Not Need To See It All

Most of the visualization examples shown above portray a small number of variables that can be visualized quite effectively using existing methods and computing techno-logy. But as the dimensionality grows it becomes difficult to show all variables clearly without exceeding the perceptual and cognitive abilities of the user (Maeder 1997), or the performance of the computer; some mechanism must be used to either drop some variables, reduce the detail shown, provide a summary view or add windowing capabilities (zoom and pan) onto the display. For example, the simplistic approach of assuming that all variables should be shown, and pair-wise related one to another as shown in Figure 16.6 will not scale to the highly multivariate data sets that are increasingly becoming available; to illustrate the point, the AVIRIS 224 band satellite data would require a scatterplot matrix to have 224*223/2 = 24,976 cells!

As argued by Scott (1992), if a highly multivariate space is treated as a hyper-cube, with each data variable having its own dimensions then the vast majority of the contained space is likely to be empty, containing no data values, and much of the remainder may contain no patterns of interest. It therefore makes sense to utilize methods that help users to locate interesting areas of lower dimensionality (sub-spaces), as shown in Figure 16.11, or to project data to a lower dimensional form that is more compact, as in Figure 16.8.

Current Geovisualization Research Needs

Much progress has been made, but many problems remain to be solved before geo-visualization can fulfill all of its potential (Gahegan 1999, Foley 2000). Some specific problems follow:

1 **Perceptual and cognitive limitations** There are certain to be some perceptual and cognitive limits as to how much information a human observer can take in. However, these limits depend very much on the experience of the observer, the types of data, the techniques in use, the tasks to be performed, and even the capabilities of the display environment. A good deal of research has been directed towards studying the effects of the above aspects in isolation, but little is under-stood concerning how these limits apply *together* in determining how much data is too much.

2 **Lack of an overarching theory** It is very difficult to combine the results of independent research to ascertain any global truths about the perceptual and cognitive aspects of visualization. This is not to say that no useful theories or frameworks exist – they do, and several examples are given above. However, as with quantum physics, all of these guidelines do not add up to a unified theory that holds true in all situations. So, rather like great art, there are exceptions that flaunt theory yet appear to work well. Hence there is great need for further research.

3 **Subjectivity of results** When any new understanding is generated as a result of using visualization it is of an informal nature and can be difficult to represent and communicate in an objective way. Contrast this with statistical inference where quantifiable results are produced, often with a known significance and reliability that can be communicated to others in a commonly understood, universal language. However, this may be a strength as well as a weakness because the findings are not restricted to what can be said with statistical methods, but instead by what can be observed and understood by the user.

The International Cartographic Association (ICA) Commission on Geovisualization divides its research initiatives into four thematic areas: representation, knowledge discovery, interfaces and cognition-usability (http://www.ica.org). Details of progress and barriers are provided by MacEachren and Kraak (2001) and Dykes, MacEachren, and Kraak (2005).

The Future

The future will no doubt see many new multivariate visualization techniques proposed, and hopefully accompanied by improvements in tool availability and integration. The following list of developments also seems probable at the time of writing.

1 Increased use of immersive visualization environments such as caves and data walls (Isakovic, Dudziak, and Köchy 2002) that can provide a good deal more display real-estate and hence can visualize greater amounts of data. These environments are usually large enough to accommodate groups of researchers who can then collaborate in the shared visual spaces;

2 Personal augmented reality (glasses, retinal projection), coupled with access to wireless geographic databases will allow visualization to move out of the laboratory and into the real world (Hedley, Billinghurst, Postner, May, and Kato 2004);

3 Further integration of visualization tools within the process of data mining, to improve knowledge discovery (for example, Keim and Kriegel 1996, Ribarsky, Katz, and Holland 1999, Guo, Peuquet, and Gahegan 2003).

4 Bigger screens and better computers may appear to provide access to more data, but the pressing need to consider other limiting factors, such as perception, cognition, and the lack of formal models must be considered concurrently.

5 Better integration with other analysis approaches, such as machine learning, statistics, spatial analysis and GIS. There is already a trend to include visualization tools in GIS, and map rendering capabilities in visualization, statistics and mathematical software environments.

6 Development of standards to encode how geovisualization and map displays appear, so that appearance can be saved and shared more effectively. The Open GIS Consortium has already made significant progress in this regard with their Styled Layer Description (SLD) standard (OpenGIS 2002).

7 Increasing development and use of on-demand geovisualization tools that are highly adaptable to the current needs of users, rather than relying on visualizations that are custom-made, highly specialized and expensive to produce (Kraak 1998).

ACKNOWLEDGEMENTS

The visualization examples shown are all the work of undergraduates, graduates, researchers, and faculty within the GeoVISTA Center at Penn State (http://www.geovista.psu.edu). The research leading to their production was funded in part by grants from the National Science Foundation, National Cancer Institute, and Centers for Disease Control; support from these organizations is gratefully acknowledged.

ENDNOTES

1 Exactly how this happens is known as the "binding problem." For an informal description of binding and a most intriguing optical illusion see the online article Tricks the Brain Plays, (http://www.npr.org/features/feature.php?wfId=1902700), from the radio show "All Things Considered", May 19 2004, National Public Radio (NPR), USA.

2 It is possible to use animation in a book by placing a series of images at the outside corners of pages, one to each page. If images progressively differ from each other in subtle ways, then flipping through the pages can convey the notion of a "movie." See Pilkey (1999) for another form of manual animation using books.

REFERENCES

Andrienko, G. L. and Andrienko, N. V. 1999. Interactive maps for visual data exploration. *International Journal of Geographic Information Science* 13: 355–74.

Andrienko, N., Andrienko, G., Voss, H., Bernardo, F., Hipolito, J., and Kretchmer, U. 2002. Testing the usability of interactive maps in common GIS. *Cartography and Geographic Information Science* 29: 325–42.

Arnheim, R. 1974. *Art and Visual Perception*. Berkeley, CA: University of California Press.

Asimov, D. 1985. The grand tour: A tool for viewing multidimensional data SIAM. *Journal of Science and Statistical Computing* 6: 28–143.

Bertin, J. 1967. *Semiologie Graphique*. Paris: Mouton.

Bertin, J. 1981. *Graphics and Graphic Information Processing*. Berlin: Walter de Gruyter.

Bertin, J. 1983. A new look at cartography. In D. R. F. Taylor (ed.) *Progress in Contemporary Cartography: Graphic Communication and Design in Contemporary Cartography*. New York: John Wiley and Sons, pp. 69–86.

Brewer, C. A. 1997. Evaluation of a model for predicting simultaneous contrast on color maps. *Professional Geographer* 49: 280–94.

Buttenfield, B., Flewelling, D., Gahegan, M., Miller, H., and Yuan, M. 2004. Geospatial data mining and knowledge discovery. In R. B. McMaster and E. L. Usery (eds) *A Research Agenda for Geographic Information Science*. Boca Rotan, FL: CRC Press, pp. 365–88.

Cleveland, W. S. 1993. *Visualizing Data*. Summit, NJ: Hobart Press.

Cleveland, W. S. and McGill, M. E. 1988. *Dynamic Graphics for Statistics*. Belmont, CA: Wadsworth & Brookes/Cole.

Cook, D., Buja, A., Cabrera, J., and Hurley, C. 1995. Grand tour and projection pursuit. *Computational and Graphical Statistics* 4: 155–72.

Cook, D., Majure, J. J., Symanzik, J., and Cressie, N. 1996. Dynamic graphics in a GIS: Exploring and analyzing multivariate spatial data using linked software. *Computational Statistics* 11: 467–80.

Dent, B. D. 1998. *Cartography: Thematic Map Design* (5th edn). New York: McGraw Hill.

DiBiase, D. 1990. Visualization in the earth and mineral sciences. *Bulletin of the College of Earth and Mineral Sciences, Penn State University* 59: 13–8.

Dorling, D. 1994. Cartograms for Visualizing Human Geography. In D. J. Unwin and H. M. Hearnshaw (eds) *Visualization in Geographical Information Systems*. London: John Wiley and Sons: 85–10.

Dykes, J. A. 1996. Dynamic maps for spatial science: A Unified approach to cartographic visualization. In D. Parker (ed.) *Geographical Information Systems 3*. London: Taylor and Francis, pp. 171–81.

Dykes, J. A., MacEachren, A. M., and Kraak, M. J. (eds). 2005. *Exploring Geovisualization*. Amsterdam: Elsevier.

Egbert, S. L. and Slocum, T. A. 1992. ExploreMap: An exploration system for choropleth maps. *Annals of the Association of American Geographers* 82: 275–88.

Fabrikant, S. I. 2000. Spatialization browsing in large data archives. *Transactions in GIS* 4: 65–78.

Fabrikant, S. I. and Skupin, A. 2005. Cognitively plausible information visualization. In J. A. Dykes, A. M. MacEachren, and M. J. Kraak (eds) *Exploring Geovisualization*. Amsterdam: Elsevier, pp. 667–90.

Foley, J. 2000. Getting There: The Ten Top Problems Left. WWW document, http://www.computer.org/cga/articles/topten.htm.

Gahegan, M. N. 1998. Scatterplots and scenes: Visualisation techniques for exploratory spatial analysis. *Computers, Environment and Urban Systems* 21: 43–56.

Gahegan, M. N. 1999. Four barriers to the development of effective exploratory visualization tools for the geosciences. *International Journal of Geographical Information Science* 13: 289–310.

Gahegan, M. N. 2003. Is inductive machine learning just another wild goose (or might it lay the golden egg?). *International Journal of Geographical Information Science* 17: 69–92.

Gahegan, M. N. 2005. Beyond tools: Visual support for the entire process of GIScience. In Dykes, J. A., MacEachren, A. M., and Kraak, M. J. (eds) *Exploring Geovisualization*. Amsterdam: Elsevier, pp. 83–99.

Gahegan, M., Wachowicz, M., Harrower, M., and Rhyne, T. 2001. The integration of geographic visualization with knowledge discovery in databases and geocomputation. *Cartography and Geographic Information Science* 28: 29–44.

Gahegan, M., Takatsuka, M., Wheeler, M., and Hardisty, F. 2002. GeoVISTA Studio: A geocomputational workbench. *Computers, Environment and Urban Systems* 26: 267–92.

Glassner, A. S. 1989. *An Introduction to Ray Tracing*. San Fransisco, CA: Morgan Kaufmann.

Graham, M., Watson, M. F., and Kennedy, J. B. 2002. Novel visualisation techniques for working with multiple, overlapping classification hierarchies *Taxon* 51: 351–8.

Green, D. R. and Horbach, S. 1998. Colour: Difficult to both choose and use in practice. *Cartographic Journal* 35: 169–80.

Grinstein, G. and Levkowitz, H. 1995. *Perceptual Issues in Visualisation*. Berlin: Springer-Verlag.

Guo, D., Peuquet, D., and Gahegan, M. 2003. ICEAGE: Interactive clustering and exploration of large and high-dimensional geodata. *GeoInformatica* 7: 229–53.

Haslett, J., Bradley, R., Craig, P., Unwin, A., and Wills, G. 1991. Dynamic graphics for exploring spatial data with application to locating global and local anomalies. *American Statistician* 45: 234–42.

Hedley, N. R., Billinghurst, M., Postner, L., May, R., and Kato, H. 2004. Explorations in the use of augmented reality for geographic visualization. *Presence: Teleoperators and Virtual Environments* 11: 119–33.

Hinneburg, A., Keim, D., and Wawryniuk, M. 1999. HD-Eye: Visual mining of high dimensional data. *IEEE Computer Graphics and Applications* 19(5): 22–31.

Hix, D. and Hartson, H. R. 1993. *Developing User Interfaces: Ensuring Usability through Product and Process*. New York: John Wiley and Sons.

Inselberg, A. 1985. The plane with parallel coordinates. *The Visual Computer* 1: 69–97.

Inselberg, A. 1997. Multidimensional detective. In *Proceedings of IEEE Conference on Visualization (Visualization '97)*, Los Alamitos, CA, USA. Los Alamitos, CA: Institute of Electrical and Electronics Engineers.

Isakovic, K., Dudziak, T., and Köchy, K. 2002. X-Rooms: A PC-based immersive visualization environment. *In Proceedings of the Web3D Symposium*, Tempe, AZ, USA. New York: Association of Computing Machinery.

Karssen, A. J. 1980. The artistic elements in map design. *Cartographic Journal* 17: 124–7.

Keim, D. and Kriegel, H.-P. 1996. Visualization techniques for mining large databases: A comparison. *IEEE Transactions on Knowledge and Data Engineering* 8: 923–8.

Koch, C. and Ullman, S. 1985. Shifts in selective visual attention: Towards the underlying visual circuitry. *Human Neurobiology* 4: 219–27.

Kohonen, T. 1997. *Self-Organizing Maps*. Berlin: Springer-Verlag.

Kosara, R., Healey, C. G., Interrante, V., Laidlaw, D. H. V., and Ware, C. 2003. User Studies: Why, how, and when? *IEEE Computer Graphics and Applications* 23(4): 20–5.

Kraak, M. J. 1998. The cartographic visualization process: From presentation to exploration. *Cartographic Journal* 35: 11–5.

Landa, R. 2004. *Advertising by Design: Creating Visual Communications with Graphic Impact*. New York: John Wiley and Sons.

Livingstone, M. and Hubel, D. 1988. Segregation of form, colour, movement and depth: Anatomy, physiology and perception. *Science* 240: 740–9.

MacDougall, E. B. 1992. Exploratory analysis, dynamic statistical visualisation and geographic information systems. *Cartography and Geographical Information Systems* 19: 237–46.

MacEachren, A. M. 1994. Time as a cartographic variable. In H. M. Hearnshaw and D. J. Unwin (eds) *Visualisation in Geographical Information Systems*. Chichester: John Wiley and Sons: 115–30.

MacEachren, A. M. 1995. *How Maps Work*. New York: Guilford Press.

MacEachren, A. M. and Kraak, M. J. 2001. Research challenges in geovisualization. *Cartography and Geographic Information Science* 28: 3–12.

MacEachren, A. M., Dai, X., Hardisty, F., Guo, D., and Lengerich, G. 2003. Exploring high-d spaces with multiform matrices and small multiples. In *Proceedings of the International Symposium on Information Visualization*, Seattle, WA, USA. Los Alamitos, CA: Institute of Electrical and Electronics Engineers.

MacEachren, A. M., Gahegan, M., and Pike, W. 2004. Geovisualization for constructing and sharing concepts. *Proceedings of the National Academy of Science* 101, Supplement 1: 5279–86.

Mackinlay, J. D. 1986. Automating the design of graphical presentations of relational information. *ACM Transactions in Graphics* 5: 110–41.

Maeder, A. 1997. Human understanding limits in visualisation. In T. Caelli, P. Lam, and H. Bunke (eds) *Spatial Computing: Issues in Vision, Multimedia and Visualisation Technologies*. Singapore: World Scientific, pp. 229–37.

Mardia, K. V., Kent, T., and Bibby, J. M. 1979. *Multivariate Analysis*. London: Academic Press.

McGuinness, C. 1994. Expert/Novice Use of Visualization Tools. In A. M. MacEachren and D. R. F. Taylor (eds) *Visualization in Modern Cartography*. New York: Elsevier, pp. 185–99.

McLeod, A. I. and Provost, S. B. 2001. Multivariate data visualization. In A. H. El-Shaarawi and W. W. Piegorsch (eds) *Encyclopedia of Environmetrics*. Somerset, NJ: John Wiley and Sons, pp. 1333–44.

Miller, N. E., Wong, P. C., Brewster, M., and Foote, H. 1999. Topic Islands™: A wavelet-based text visualization system. In *Proceedings of the IEEE Symposium on Information Visualization (Infoviz 2000)*, Salt Lake City, UT.

Miller, H. and Han, J. (eds). 2001. *Knowledge Discovery with Geographic Information*. London: Taylor and Francis.

Monmonier, M. 1989. Geographic brushing: Enhancing exploratory analysis of the scatterplot matrix. *Geographical Analysis* 21: 81–4.

Monmonier, M. S. 1990. Strategies for the interactive exploration of geographic correlation. In *Proceedings of the Fourth International Symposium on Spatial Data Handling*, Zurich, Switzerland.

Muehrcke, P., Muehrcke, J., and Kimerling, A. 2001. *Map Use: Reading, Analysis and Interpretation* (4th edn). Madison, WI: JP Publications.

Mutton, P. and Golbeck, J. 2003. Visualization of semantic metadata and ontologies. In *Seventh International Conference on Information Visualization (IV03)*, London, United Kingdom: 300–5.

Noll, A. M. 1967. A computer technique for displaying n-dimensional hyperobjects. *Communications of the Association for Computing Machinery* 10: 469–73.

North, C. and Shneiderman, B. 1999. *Snap-together Visualization: Coordinating Multiple Views to Explore Information*. College Park, MD: University of Maryland, Computer Science Department Technical Report No. CS-TR-4020.

Nyerges, T. L., Moore, T. J., Montejano, R., and Compton, M. 1998. Interaction coding systems for studying the use of groupware. *Journal of Human–Computer Interaction* 13: 127–65.

Open GIS Consortium. 2002. *Styled Layer Descriptor Implementation Specification*. WWW document, http://www.opengis.org/docs/02-070.pdf.

Pickett, R. M. and Grinstein, G. G. 1988. Iconographic displays for visualizing multidimensional data. In *Proceedings of the IEEE Conference on Systems, Man and Cybernetics*, Piscataway, NJ, USA.

Pike, W. and Gahegan, M. 2003. Constructing semantically scalable cognitive spaces. In W. Kuhn, M. Worboys, and S. Timpf (eds) *Spatial Information Theory: Foundations of Geographic Information Science*. Berlin: Springer-Verlag Lecture Notes in Computer Science No. 2825: 332–48.

Pilkey, D. 1999. *Captain Underpants and the Attack of the Talking Toilets*. New York: Scholastic Books.

Plaisant, C. 2005. Information visualization and the challenge of universal access. In J. Dykes, A. M. MacEachren, and M. J. Kraak (eds) *Exploring Geovisualization*. Amsterdam: Elsevier, pp. 53–82.

Quinlan, R. 1993. *C4.5: Programs for Machine Learning*. San Mateo, CA: Morgan Kaufmann.

Ribarsky, W., Katz, J., and Holland, A. 1999. Discovery visualization using fast clustering. *IEEE Computer Graphics and Applications* 19(5): 32–9.

Robertson, G., Mackinlay, J., and Card, S. 1991. Cone trees: Animated 3D visualizations of hierarchical information. *In Proceedings of Human Factors in Computing Systems (CHI'91)*, New York.

Robertson, P. K. 1997. Visualizing spatial data: The problem of paradigms. *International Journal of Pattern recognition and Artificial Intelligence* 11: 263–73.

Rousseeuw, P. J., Ruts, P. J., and Tukey, J. W. 1999. The bagplot: A bivariate boxplot. *American Statistician* 53: 382–7.

Sammon, J. W. 1969. A nonlinear mapping for data structure analysis. *IEEE Transactions on Computers, Series C* 18: 401–9.

Scott, D. W. 1992. *Multivariate Density Estimation: Theory, Practice and Visualization*. New York: John Wiley and Sons.

Shneiderman, B. 1998. *Designing the User Interface: Strategies for Effective Human–Computer Interaction*. Reading: MA, Addison-Wesley.

Slocum, T. A., Blok, C., Jiang, B., Koussoulakou, A., Montello, D., Fuhrmann, S., and Hedley, N. R. 2001. Cognitive and usability issues in geovisualization. *Cartography and Geographic Information Science* 28: 61–75.

Spitaleri, R. 1993. Reference models for computational visual simulations. In P. Palamidese (ed.) *Scientific Visualization: Advanced Software Techniques*. Chichester: Ellis-Horwood, pp. 3–14.

Taylor, D. R. F. 1991. Geographic information systems: The microcomputer and modern cartography. In D. R. F. Taylor (ed.) *Geographic Information Systems: The Microcomputer and Modern Cartography*. Oxford: Pergamon, pp. 1–20.

Treisman, A. 1986. Features and objects in early vision. *Scientific American*: 114B–25.

Tufte, E. R. 1990. *Envisioning Information*. Cheshire, CT: Graphics Press.

Vredenburg, K., Isensee, S., and Righi, C. 2002. *User-Centred Design*. Upper Saddle River, NJ: Prentice-Hall.

Wood, M. 1990. Map perception studies. In C. R. Perkins and R. B. Parry (eds) *Information Sources in Cartography*. London: Bowker-Saur: 441–52.

Chapter 17

Virtual Reality in Geographic Information Systems

Michael Batty

Virtual reality (VR) emerged as part of the extension of computing into graphics which began to accelerate with the advent of the microprocessor and its widespread use in personnel computers, workstations, and supercomputers. Sometime in the mid-1980s, Jaron Lanier coined the term to describe the kinds of virtual environments popping up everywhere in which users had begun to hook themselves into computers in such as way that what they saw, heard, and touched was a product of the machine's simulation (Rheingold 1991). The visual was the most important of these sensory experiences, and, approaching 2010, the typical portrait of VR is still the 3D world in which the user is immersed and able to navigate amongst movable objects. But as computers have fused with networks, the concept of VR has blurred, and no longer does the term imply total immersion in the machine's simulation. A more general definition from the computer and internet encyclopedia *Webopedia* is "an artificial environment created with computer hardware and software and presented to the user in such a way that it appears and feels like a real environment" (see http://www.webopedia.com/TERM/V/virtualreality.html for additional details). Today the term is used to describe many experiences in environments which exist on everything from the desktop to the net.

We need to be a little more specific about how we are beginning to use the term in GI Science for the complete panoply of environments, media, and electronic peripherals used to generate as many sensory experiences as possible (fully-fledged VR) have not yet been widely exploited in our field. In fact GI Science is largely graphical, rooted in the 2D space, the map, extending into 3D, and even into the temporal dimension, but with little use of more esoteric virtual environment hardware and media. The main distinction we must make at the outset is one of abstraction: VR technologies can be and are being exploited in two directions, first with respect to how we interact with analytical conceptions of space, with surfaces and their properties in terms of their statistics, and, second, with respect to much more realistic conceptions of space in terms of landscapes where the quest for simulated realism is to the fore. In a sense, this is the difference between *geographic* and *geometrical analysis*, the difference between navigating in mathematical space and its various transforms in contrast to interacting with the Euclidean space of the three-dimensional world in which we live. Of course, GI Science spans the route between both conceptions of space, but the development of VR along this continuum marks out rather difference styles and methods that we need to consider.

In essence, VR here is limited to that constellation of users and technologies that enable interactions within a visual environment. Indeed, a good working definition of VR is "visualization + interaction + motion" where motion implies that user moves and consciously manipulates the environment in some way. Interaction is more than passive in that users continuously respond to cues within the environment which is configured for some particular task ranging from entertainment to scientific research. In this sense, interactivity is the watchword of VR, and it is not surprising that this domain has been dominated by the development of computer-aided environments oriented to design and problem-solving. In this sense, spatial, particularly fine-scale design, such as computer-aided architecture design, has figured prominently in the development of VR.

The environments we will present here begin with two-dimensional space which can be treated geographically and/or geometrically. Before we begin to define the shape and content of these environments, it is worth charting the range of possibilities and in this, we will define two continua – from real to virtual space and from real to virtual users. Real space is the environment in which we live, while virtual space, is its digital equivalent. In fact this continuum does not specifically pick up the level of abstraction in which real space is transformed to digital and it tends to exclude those fictional virtual realities that have no equivalent in the real world. Nevertheless, we will assume that virtual space covers all types of digital abstractions based on varying degrees of fiction. Real users are ourselves while their virtual equivalents can be both virtual renditions of ourselves – so called avatars of which we show examples below: agents that mirror the requirements of real users, or robots that are under our control. We picture these continua in Table 17.1 where the grey tones indicate that the two kinds of environment involve virtual space which is populated by real users and their virtual equivalents. In GI Science, these virtualities exist at different levels of abstraction from relatively realistic renditions of landscape and urban scenes to more abstract geographic models of how those same spaces function and operate.

Table 17.1 Virtual environments defined in terms space and users

	Real space	Real and virtual space	Virtual space
Real users	Environments in which We Live	....	*Digital Environments in which Real Users Interact*
Real and virtual users	. . .	Augmented Realities	. . .
Virtual users	Populated by Robots, e.g., Hazardous, Remote Environments	....	*Real Users As Avatars, or Real Users Acting Through Agents*

When VR systems were first developed in the 1980s, the virtual worlds were the exclusive domain of single users but two developments have exploded this into multi-user worlds. First, there are specialist hardware/software environments in which more than a single individual can engage immersively such as VR theaters, CAVES,[1] and so on. Second, the convergence of computers with networks has opened a veritable Pandora's Box of possibilities for many users to interact with each other through shared environments in remote locations. We will note these different media below, but this array of possible interactive environments also leads to the idea that multiple virtual worlds can be arranged recursively where real users can take on a variety of virtual personas in building interactions with on another across many levels.

Virtual realities are always constructed with some purpose in mind, and thus the way users interact with the environment, and with each other, is essential to the way such realities are engaged. In GI Science, navigation through the virtual space is crucial – but navigation is simply a means to an end that involves some task which enables the properties of elements and objects that comprise the virtual reality to be extracted. Virtual environments are usually constructed either for design or for understanding in circumstances in which analysis is an important component of their construction. For example, flying though the virtual equivalent of a real landscape might be geared to exploring the properties of landscape not only in immediate visual terms but in terms of statistical relationships and patterns. Such environments may mix many kinds of space – from geographic to geometric but also various transforms of these spaces into non-traditional structures no longer represented as 2D or 3D map-like spaces. Frequently, such worlds are composed of many windows with only a limited number of these representing the 2D and 3D map-like spaces. Frequently such worlds are composed of many windows with only a limited number of these representing the 2D map or 3D extrusion from the map. Graphs, flow charts, and related iconography may be equally important as maps in such worlds where many of the graphic and statistical tools presented elsewhere in this book, can be exploited.

In the remainder of this chapter, we will examine three related aspects of VR for GIS. First, we will look at typical environments that use the third dimensionas their essential organizing concept for visual interaction. Second, we will look at different media – hardware and software for VR, noting the cornucopia of environments form the desktop to the headset and from the theater to the net. Third, we will look at the way VR can be fashioned as an interface to GIS and show how the concept is changing as graphics and other sensory media are fast becoming routine in human–computer interaction. These three facets of VR lead to new possibilities for using digital simulations in diverse ways for design. We will conclude with some suggestions as to how such virtual realities can be grounded in more "concrete" realities, thus giving back to analysis some of the tangibility that the digital domain removes.

Environments for VR: The Third Dimension and Beyond

The 3D geometric environment – rooms, buildings, cities, natural landscapes, and so on – represent the quintessential examples of VR. Many definitions of the term reflect the fact that the third dimension is all important in enabling users to truly immerse themselves in the digital environment. Most of the action to date, however, in urban environments has come from computer-aided architectural design. Computer-aided design (CAD) software has been increasingly used to manufacture such environments although since the mid-1990s, as multimedia techniques have exploded, such environments have been built using much more *ad hoc* techniques which mix many types and flavors of software. In terms of GI Science, GIS have slowly moved to embrace the third dimension with most propriety packages now offering quite elaborate surface generating techniques based on spatial interpolation. Since the late 1990s, 3D extrusion from the map to the cityscape or landscape has become possible, thus providing geometric frameworks in which spatial data can be queried and displayed in 3D, just as such techniques have become the workhorse of routine 2D GIS applications. In fact, there are extremely rapid developments in 3D GIS at present, as we will note in a later section. There is now little doubt that the functionality that such systems offer in building such environments is considerably richer than the rather vacuous CAD methods that emphasize rendering rather than geometric database construction to which spatial attributes can be associated (Batty, Dodge, Doyle, and Smith 1998).

In a sense, the 3D graphic environments that can now be produced easily using GIS do not constitute virtual reality. They may be the basis for constructing virtual realities but such realities in the narrow sense depend upon how the user is immersed in the environment and this depends on the hardware and software used to make such environments work. In a broad sense of course, flying through a 3D model on the desktop might be considered part of VR (and increasingly is so as the definition continues to broaden) but it is worth keeping such graphics separate from the media of communication and interaction. In the past, there have been quite severe limits on what 3D GIS can handle, although these are fast being resolved. In particular, while moving through the model, detailed rendering has either not been possible or has been only primitive whether on-the-fly or using predetermined flight paths. In current generations of software, all these limits are being relaxed while the size of the environment that can be handled is being massively increased through better hardware and software. Embedding other kinds of multimedia into such environments such as panoramas has become possible (Shiffer 2001) and for the first time, it looks at through 3D GIS is becoming the preferred medium for such model construction.

Our own example of this is "Virtual London." This is based on 20 km² of central London where the 45,000 buildings within the Ordnance Survey's *MasterMap* product are extruded to average building heights using LiDAR data from InfoTerra. The model sits on a digital terrain layer and currently data for the buildings reflecting land use, tenure, floorspace, rental levels, and so on are being ported into the model as layers. The block model is built in *ArcScene* but more detailed renditions of certain buildings are produced photogrammetrically and then embedded in the model as we show in Figures 17.1(a) and (b). In fact this is only on part of the virtual environment for much additional multimedia ranging from digital panoramas to fast zoomable and clickable maps are associated with the project. This information is not delivered to the user as a 3D GIS for this would require fast, memory intensive hardware well beyond the resources of the casual user. Instead, various

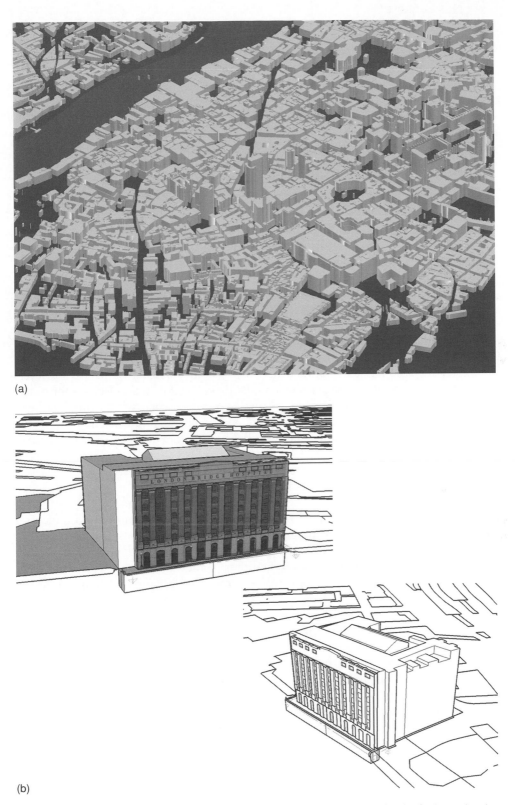

(a)

(b)

Fig. 17.1 (a) The basic geometric model built in *ArcScene* and (b) an example of a finely rendered building imported into the scene

3D products are being manufactured from the base model. Moreover the software used is not restricted to GIS but embraces standard CAD and diverse rendering and photogrammetric techniques. The preferred medium in which to deliver Virtual London to users is over the net, and thus net-based CAD, which is still rooted in Virtual Reality Modeling Language (VRML)-type languages, also plays a part.

We shall return later to ways in which we can make this model "virtual" in the narrow sense but we must now consider more abstract renditions of 2D virtual space which are more strongly linked to GI Science than to GIS. Since the late 1980s, one of the main structures within GIS has been the notion of the layer. Landscapes are considered as layers of different activity and ways of generalizing, smoothing, interpolating both real and abstract properties of these landscapes have been evolved. In particular, interpolation of point patterns into gridded surfaces are now widely available in GIS. Methods of doing this are still being researched intensively as spatial properties such as autocorrelation, local variance, and such like are important in producing good generalizations. In a sense, the surfaces produced are geographic rather than geometric. The biggest development in GI Science which has generated VR-like systems is the idea that different windows on such surfaces can be produced within the viewing environment. The notion of taking different renditions of the map or surface represented in both 2D and 3D and displaying its properties using more abstract spaces – graphs, networks, flow charts, and so on – displayed in windows linked to the basic surface are now widely used as part of spatial statistical software. There are few proprietary systems which take these display mechanisms as far as the narrower form of VR but there are examples in which such windowing has been reported into artificial environment such as CAVES and VR theaters. The work of the *GeoVista* group at Penn State has pioneered developments in this area (for example, MacEachren 2001; see Chapter 16 by Gahegan in this volume for additional details as well), but it is important to stress that such developments are hard to classify and generalize as applications in VR generally are eclectic.

What is beginning to happen, however, is the mixing of geographic with geometric realities. In one sense, this might be seen as simply putting geographic and geometric layers together to search for coincidences, correlations and patterns but the fact that 3D GIS is now driving so much of this field means that users of geometric environments can be treated to glimpses of their geographic equivalents and vice versa. We are a long way from truly embedding geographic attributes into the real geometry of the 3D world and in one sense, our perspectives from each of these directions are not very well-developed conceptually (Batty 2000). But the ability to switch layers on and off in virtual environments give us the power to associate radically different conceptions and in one sense that is what VR is all about – evoking unusual and perceptive insights into real environments through the virtual. In Figure 17.2, we show an example of how a rather abstract surface of the intensity of retailing activity is overlaid and viewed translucently on the 3D geometry of central London with the higher buildings poking through this sea of retailing. Our abilities to deal with these intersections and layering are far from perfect but we consider this to be the way forward in VR for GI Science.

Before we move to VR media, we should say a little about how to navigate in such 3D environments. Navigation is always problematic as there is a steep learning curve for most software. 3D GIS still tends to offer bird's eye rather than detailed

Fig. 17.2 A geographic "sea" showing the intensity of retail activity layered onto the 3D geometry of central London

ground level viewing in contrast to CAD where ground level movement is easier. This is changing but in delivering most 3D environments, navigation needs to be restricted and thus predetermined flight paths are the rule rather than the exception. Much depends of course on the complexity of the environment but as the geometry becomes more intricate, navigation skills become more important. This is in contrast to the wider user interface which relates to why the user is immersed in the environment in the first place but this too can interact with our ability to navigate. Currently our knowledge of best practice in this area is primitive and this does represent a constraint on applications of such technologies. In Figure 17.3, we show a typical scene in the block model of Virtual London from predetermined fly-through along the river.

Media for Delivering VR: Headsets, CAVE, Theaters, Nets

The oldest VR simulations were based on immersing the user in the virtual environment, usually a 3D scene, in such way that the user is part of the machine which generates the "whole body experience". The real world is excluded as the user is hooked to the machine and its software by a helmet or similar device which is configured so the user is part of the visual scene being simulated. Sometimes there are other sensory input and output, particular touch the simulated through data gloves, and hearing transmitted within the helmet. Part and parcel of the scene is the ability

Fig. 17.3 A typical fly-through in virtual London, west along the River Thames

to engage in two-way interaction, not passively but actively invoking changes in the elements comprising the scene. By the early 1990s, the idea that two or more users could be connected together, adding interactions between themselves as well as within the environment in which they were moving, became a powerful driver to new kind of hardware-software. The notion of the VR theater in which a collective of users could interact with each other online and offline (for such theaters are real physical settings) as well as within the simulated environment, became cutting edge. These extensions opened the virtual world to external interactions. At the same time, the notion of something closer to full immersion – the CAVE – was developed. CAVES are based on approximately 10 m^3 cubes on whose faces different elements of the virtual environment are projected in synchronized fashion so that users standing in the CAVE are literally within a closed physical world made virtual. Normal conversation is possible as users see each other within the virtual scene.

Since then a variety of other VR devices have been produced, in particular desktop-like tables in which the scene is projected, holographic-like and users wearing glasses or helmets are synchronized with one other and with the scene. However what has blown VR completely out of its original niche has been the net. As computing has drifted onto the net and as the dominant paradigm has veered towards collective interactive computing over nets of various kinds, VR has adapted to let many users take part in the same kinds of "out-of-body experience" that traditional immersive handset technologies allow. In fact, the sense of total exclusion can never be completely simulated on the desktop but the fact than many remote users linked through the net can by synchronized is shared environments has produced new reactions and insights into the virtual world. The best examples of these environments are multi-user worlds. These initially began as gaming environments, multi-users domains/dungeons (MUDs), based on adventure style games, but involving more considered "chat" rooms and bulletin boards. Their visual equivalents, of which the most well-known is *AlphaWorld*, enable players or users to construct and colonize, appearing as avatars and being able to chat with those also present in the world (Dodge 2002). These are quintessential examples of virtual users in virtual space. In part, the objects in those worlds can also have a degree of intelligence in that they can be molded to enable different functions to be generated, particularly with respect to design (Smith 2002).

There have been some limited experiments with such worlds in GI Science but mainly for purposes of controlled design rather than entertainment. In fact in our Virtual London model, one way in which we are making the model available to users is to take its contents and manipulate it in such a world in which is set up as a virtual planning forum, virtual design studio, or virtual exhibition hall. To give a sense of what is possible, Figure 17.4 shows how digital panoramas as well as the digital block model of central London developed using 3D GIS can be used to populate such a virtual world. Figure 17.4(a) shows the virtual exhibition hall with Figure 17.4(b) showing how one can walk from this into a more realistic rendition of a well-known galleria on London's South Bank. A little further along the bank is City Hall and in Figure 17.4(c), we show the avatar walking onto the balcony of the Hall which overlooks the financial quarter of the City across the river.

This is the kind of media that is fast developing in the network world but so far, the nature of the user interaction has been somewhat informal and it remains to be seen as to how GI Science might make use of such interactivity. In contrast but with equally rapid development at present, wireless VR is being to make an impact. Of course whatever can be achieved over wires can in principle be developed wirelessly and the whole range of media from headset to CAVE to theater can be treated in this manner. In Figure 17.5, we show an example of how 3D GIS which forms once again a virtual image of central London can be projected wirelessly onto elaborate headsets which contain their own screens and which are linked wirelessly to other users. The user can see the virtual model and any other users who might be interacting with it as in Figure 17.5(a) while in Figure 17.5(b) we see the entire segment of the block model that each user sees projected into their headset.

These last two examples show how VR is still stretching the ways in which we can interact with 3D environment but so far, the technology continues to dominate. Scientific applications are beginning but a lot more needs to be done on the process of user interaction as well as tailoring these methods to specifics applications (Schroeder, Huxor, and Smith 2001). We have remarked before that 3D GIS is a technology in search of a problem in that we have very little theory of the third dimension on which we are able to build the requisite functionality which is the workhorse of 2D GIS (Batty 2000). In fact, 3D GIS is leading the way but the limits here are small in comparison with VR technologies where applications are still largely focuses on demonstrating what is possible.

(a)

(b)

(c)

Fig. 17.4 Porting the digital model and various panoramas into a virtual world: (a) the "fictional" exhibition space which has doors and/or windows into/onto, (b) a digital panorama of a "real" galleria which in (c) leads the avatar to City Hall which overlooks the "real" virtual city

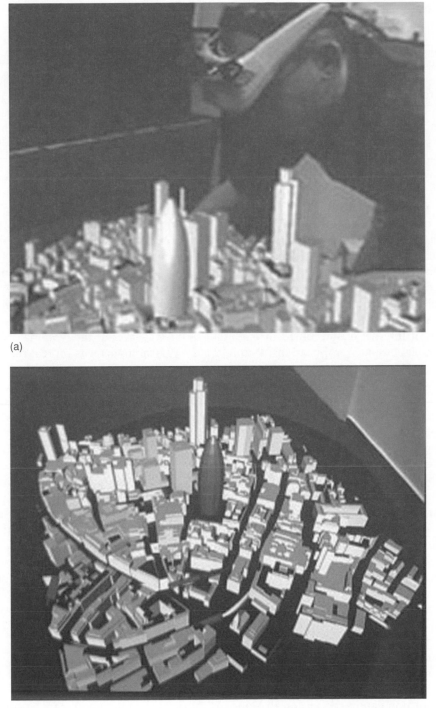

(a)

(b)

Fig. 17.5 The Arthur interface – viewing the virtual city in a multi-user wireless headset fully immersive environment: (a) another user, (b) the projected model within the headset from http://www.fit.fraunhofer.de/projeckte/arthur/index_en.xml

VR as an Interface to GIS

In one sense, VR is simply an interface, albeit one of the richest available, to systems in which human users are intimately involved, and which enable diverse perspectives on the system in question to be accessed in parallel or simultaneously. There is a temptation to think of all science and entertainment as being enabled by VR but the number of possible applications where the power of VR is not contestable is quite small. In GI Science, VR has not been widely applied and much of its power is in potential as yet unrealized. A glance at a recent survey of VR in Geography (Fisher and Unwin 2002) shows that current applications tend to stress VR with respect to visualization which is intrinsically geographic, rather than VR in GI Science or GI Science within VR. So far, although there are some useful examples of how data structures, geographic representation, and spatial analysis more generally can be enriched using VR, by far the majority of applications are like those we have illustrate here – geographic landscapes of various kinds which enable users to search and visualize patterns in the broadest sense.

One of the best examples of VR as an interface dates back almost ten years to a geographic visualization of traffic on the Internet, constructed in real time and visualized using the various screens defined by the walls of a CAVE. Lamm, Reed, and Scullin (1996) pioneered such a real time display as an interface to the space-time series of Internet traffic routed through the MOSAIC server at the National Center for Supercomputer Applications. Within the CAVE, the focus was on real time display of the pulsating nature of traffic on the globe but other windows within the CAVE displayed the data in numerical form as we show in the interface in Figure 17.6. The focus of the effort was an analysis of load balancing but the potential of systems like this for visualizing any kind of electronic transactions are still enormous.

Since the mid-1990s, VR systems based on elaborate immersive hardware have become more usual generating much less hype than before. In fact as VR has spread out of these environments and onto the net, many of interfaces to desktop- and net-based software originally pioneered in headsets and theaters, are being adapted to stand-alone software. An excellent example of this which is one of the newest additions to ESRI's *ArcGIS* family of software, is *ArcGlobe* which lets users map any geographic file, suitably projected and tagged to the globe, thus enabling a clear analysis of scale to be imposed on any problem. *ArcGlobe* provides a "revolutionary way to support multi-resolution global data visualization in 3D" (ESRI 2003) and integrates with a variety of files formats which are common in VR such as VRML, 3D Studio Max, and Open Flight. In essence, the user can zoom, pan and analyze surfaces on the globe at any scale. Although the global cannot be spun in the kind of holographic environment available in a CAVE as in NSCA's web traffic application, the user can spin the global on the desktop and thus view different parts of the world at different levels of resolution in 3D. We show a typical application of this technology in Figure 17.7 where we have mapped the degree of world urbanization onto three globes from three different perspectives and growth rates of population onto a fourth globe. All of these data are shown at the country level but of course, visualization can be accomplished at any scale if the requisite data is available.

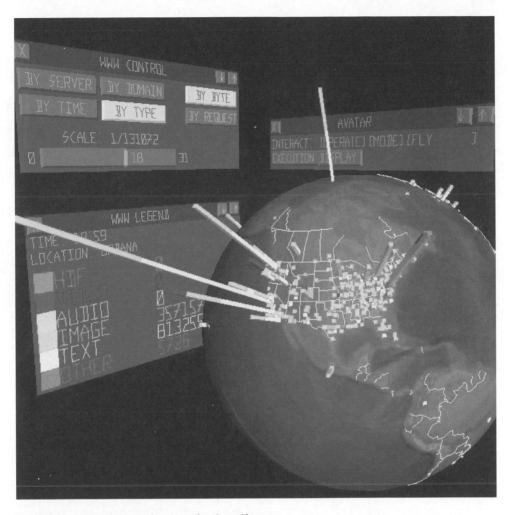

Fig. 17.6 Geographic visualization of web traffic

This kind of interface, which by the time this book is published may well be commonplace within GIS, raises on of the most important issues involved in VR – the way time, navigation and scale interact. We will illustrate these issues with respect to the way we can move around the globes shown in Figure 17.7 for many kinds of time are associated with these visualizations. Digital navigation is quasi-independent of scale and time but at least three varieties are implicit. As the user moves around the globe, the time taken is that associated with a rocket ship but as the user zooms, speeds reduce. Once down at the 3D level of a city, navigation can be slowed to walking page. This is a massive contrast in terms of the initial speed and time implied by panning at the world level and it defines two ends of a time spectrum that is fixed by scale. Moreover different layers of spatial data can be dissolved on the globe. At the world level again, we can move from one time period to another by dissolving layers associated with different times and this implies

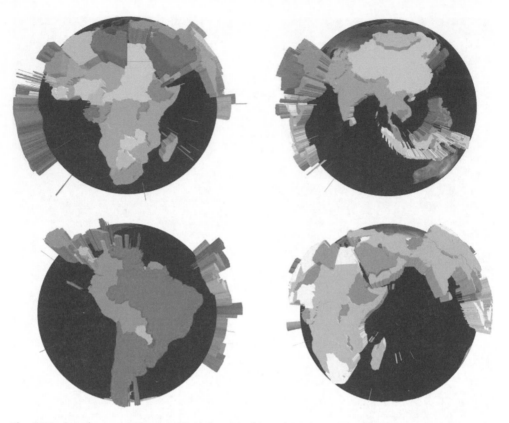

Fig. 17.7 Interfaces to GIS using 3D globes: (a), (b), and (c) three views of urbanization 2005, and (d) growth rates 1995–2005

a different time line where the speed of dissolution is related to the datum point of the data layer in question. In this way, one might visualize the evolution of activities on the planet. The intersection of space and time in this way has barely been thought though in mainstream geography or GI Science with this being one of the best examples of how VR is stretching our imagination about how structures evolve in space and time.

To complete the confusion that VR brings, readers are referred to visualizations of city growth in 3D which are being developed for historical and heritage studies. One of the best examples is Bologna where the growth of the city from the eleventh century to the modern day can be visualized by pulling up different historical visualizations configured from the location where the user is rooted in the city. The idea of time travel as a user walks along a street, watching the city changing through its actual temporal development, is implicit in these interfaces although to date, they remain experimental (Bocchi, Calori, Fraticelli, Guidazzoli, and Mariani 2004). Another example under active development involves a history of the city of London from the twelfth century based on the merging of detailed historical archives and archaeology with more contemporary data (see http://www.casa.ucl.ac.uk/people/Melina.html for additional details).

Beyond VR: Back to Reality, of Sorts

Once geographic information is represented in 3D, then there is the prospect that this media can be translated not only into VR but also back into more material renditions. There is little doubt that users are ambivalent about the digital world, knowing implicitly that the material world from which the digital might be derived offers qualitatively different insights into their understanding and feeling. A good example of this involves the use of physical models of the city to aid in our understanding of design proposals. The City of London, for example, has a large conventional model built of traditional materials. They consider this invaluable to economic development and related visualizations, and developers are encouraged to add their proposals to the model in the form of detailed renderings of buildings. All this could be replaced by a digital model of the kind illustrated above but the City's Officers are reluctant (Batty, Chapman, Evans, et al. 2001). However what is now in prospect is the transformation of our digital model into a form which is nearer the material but still digital. We can place the model in a virtual design studio and users as avatars can visit the room and discuss design proposals virtually. In Figure 17.8, we extend the virtual worlds that we illustrated above to show how avatars might view the model in such world, as a simulation of the traditional model into which they are able to inject their design proposals, virtually. In some sense, this represents a recursion that has no end: a "simulacra" or "simulation of a simulation" as Baudrillard (1994) terms it. Imagine entering a digital 3D model, panning and zooming into a room which housed a virtual world in which the digital model was placed, the very digital model that was entered in the first instance. The recursion might continue indefinitely, Escher-like, while at any point the user might jump through a window onto some other part of the world, to a world of analysis, to a theoretical world, whatever and wherever.

It is entirely possible to move in another way. Once the model has been developed, it is possible to print it as hard copy. Photographs and maps are not routinely manufactured in a material form from their digital equivalents and it is only a matter of time before geographical space in all three dimensions can be so generated. In Figure 17.9, we show how one section of Virtual London can be printed using CAD/CAM device. The model is tiny compared to its traditional form notwithstanding that it took two days to print but this shows the future. How long will it be before such models can be produced routinely, thus giving back to reality, their traditional equivalents? In the future, VR in GI Systems and GI Science may be as much about moving from the virtual to the real as from the real to virtual. The future is both ways.

ENDNOTE

1 "CAVE the name selected for the virtual reality theatre, is both a recursive acronym (Cave Automatic Virtual Environment) and a reference to 'The Simile of the Cave' found in Plato's Republic, in which the philosopher explores the ideas of perception, reality and illusion. Plato used the analogy of a person facing the back of a cave alive with shadows that are his/her only basis for ideas of what real objects are" (from http://www.sv.vt.edu/future/vt-cave/whatis/ accessed July 20, 2004).

Fig. 17.8 Constructing the Simulacra: The digital model as a "material artifact" in a virtual exhibition space

ACKNOWLEDGEMENTS

The author wishes to thank Elena Besussi, Steve Evans, Andy Hudson-Smith, and Sinesio Alves Junior for their help with the various examples reproduced here.

REFERENCES

Batty, M., Dodge, M., Doyle, S., and Smith, A. 1998. Modeling virtual environments. In Longley, P. A., Brooks, S., McDonnell, R., and Macmillan, B. (eds) *Geocomputation: A Primer*. Chichester, John Willey and Sons: 139–61.

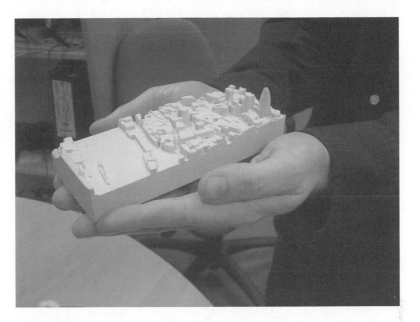

Fig. 17.9 Printing the virtual city using CAD/CAM technology

Batty, M. 2000. The new geography of the third dimension. *Environment and Planning B* 27: 483–4.

Batty, M., Chapman, D., Evans, S., Haklay, M., Kueppers, S., Shiode, N., Smithm, A., and Torrens, P. 2001. Visualizing the city: Communicating urban design to planners and decision-makers. In Brail, R. and Klosterman, R. (eds) *Planning Support Systems: Integrating Geographic Information Systems, Models, and Visualization Tools*. Redlands, CA, ESRI Press: 405–43.

Baudrillard, J. 1994. *Simulacra and Simulation*. Ann Arbor, MI, University of Michigan Press.

Bocchi, F., Calori, L., Fraticelli, L., Guidazzoli, A., and Mariani, M. 2004. The Four-Dimensional City. WWW document, http://www.cineca.it/editions/ssc97/html/guidazzo.htm.

Dodge, M. 2002. Explorations in Alpha World: The geography of 3D virtual worlds on the Internet. In Fisher, P. F. and Unwin, D. J. (eds) *Virtual Reality in Geography*. London, Taylor and Francis: 305–31.

ESRI 2003. Introducing ArcGlobe: An ArcGIS 3D Analyst application. *ArcNews Online* (Summer 2003; available at http://www.esri.com/news/arcnews/summer03articles/introducing-arcglobe.html).

Fisher, P. F. and Unwin, D. J. 2002. *Virtual Reality in Geography*. London, Taylor and Francis.

Lamm, S. E., Reed, D. A., and Scullin, W. H. 1996. Real-time geographic visualization of World Wide Web traffic. In *Proceedings of the Fifth International World Wide Web Conference*, Paris, France (available at http://www5conf.inria.fr/fich_html/papers/P49/Overview.html).

MacEachren, A. M. 2001. Cartography and GIS: Extending collaborative tools to support virtual teams. *Progress in Human Geography* 25: 431–44.

Rheingold, H. 1991. *Virtual Reality*. New York, Touchstone Books.

Schroeder, R., Huxor, A., and Smith, A. 2001. ActiveWorlds: Geography and social interaction in Virtual Reality. *Futures* 33: 569–87.

Shiffer, M. J. 2001. Spatial multimedia for planning support. In Brail, K. and Klosterman, R. (eds) *Planning Support Systems: Integrating Geographic Information Systems, Models, and Visualization Tools*. Redlands, CA, ESRI Press: 361–85.

Smith, A. 2002. 30 Days in ActiveWorlds: Community, design and terrorism in a virtual world. In Shroeder, R. (ed.) *The Social Life of Avatars: Presence and Interaction in Shared Virtual Environments*. Berlin, Springer: 77–89.

Part IV   Knowledge Elicitation

The fourth section of the book consists of three chapters looking at the increasingly important task of knowledge elicitation. The first chapter (Chapter 18 by Chris Brunsdon) considers some fundamental ideas about inference and the difficulties of applying these ideas to spatial processes. The chapter starts with a brief overview of the classical and Bayesian statistical inferential frameworks and moves from there to explore the particular nature of statistical inference when spatial processes are considered and the ways in which these two sets of inferential tasks are related. The chapter concludes with a brief discussion of computational approaches and broader issues – including other forms and types of inference and the availability of software to implement the types of applications noted in this particular chapter.

In the second chapter in this group (Chapter 19), Harvey J. Miller discusses the process of geographic knowledge discovery (GKD) and one of its central components, geographic data mining. The rapid growth in number and importance of these techniques derives from the large volumes of geographically referenced data that are being collected, archived, and shared by researchers, public agencies and the private sector, and the realization that traditional statistical techniques may not help with knowledge discovery in these instances. The chapter starts out with a review of the more general problem of knowledge discovery from databases and related data mining techniques. It moves on from there to a discussion of GKD and geographic data mining techniques, and concludes by noting some of the important research frontiers in GKD at the time of writing.

In Chapter 20, the final chapter in this set, Frederico Fonseca examines the prospects for building the geospatial semantic web. This enterprise seeks to translate the description of all of the available data on the Web – existing geospatial data as well as less structured informal information sources that may or may not contain geographic information – into a formal language that computers can understand and process. The central roles played by trust and meaning in the semantic web and why the semantic web and geospatial semantic web in particular, must be viewed and understood as works in progress are explained. This chapter concludes by noting why we need to create ontologies and to match them to data on the Web in order to implement in computers something similar to the human use of space and time metaphors.

Chapter 18

Inference and Spatial Data

Chris Brunsdon

It is often necessary to make informed statements about something that cannot be observed or verified directly. It is equally useful to assess how reliable these statements are likely to be. A great deal of research is based on the collection of data, both qualitative and quantitative, in order to make such statements. For this reason, inference in science is a fundamental topic, and the development of theories of *statistical inference* should be seen as a cornerstone of any field of study claiming to be based on scientific method.

However, despite this clear recognition of the importance of statistical inference, many commercial Geographic Information System (GIS) packages claiming to offer "spatial analysis" facilities offer no tools for statistical inference. One might ask why this is. The answer is complex, but one thing to note is that it was the Chi-Squared test, and not statistical inference in general that was cited by the American Association for the Advancement of Science (AAAS) as a key development. Chi-Squared tests are relatively simple computationally, and make a number of assumptions about the simplicity of the underlying processes about which inferences are to be made. In particular, they assume that each observation is probabilistically independent and is drawn from the same distribution. For spatial data this is unlikely to be the case – observations may well not be independent. In addition, the distributions of observations may well be conditional on their geographic location. This violates the "drawn from the same distribution" assumption. Thus, although tools of inference are just as important for geographic data as for any other kind of data, there are potential problems when "borrowing" standard statistical methods and applying them to spatial phenomena. Thus, the aim of this chapter is to consider some fundamental ideas about inference, and then to discuss some of the difficulties of applying these ideas to spatial processes – and hopefully offer a few constructive suggestions. It is also important to note that although for some areas a degree of consensus has been reached, the subject of statistical inference is not without its controversies – see Fotheringham and Brunsdon (2004) for example – and in particular there are unresolved issues in inference applied to geographic data.

Basic Ideas of Inference

To begin with, it is important to identify – and distinguish between – some key concepts of statistical inference. These are:

- **The inferential framework:** This is essentially the model of how inferences are made. Examples of these are *Bayesian inference* (Bayes 1763) and *classical inference*. Each model provides a set of general principles describing how some kind of decision related to a model (or set of models) can be arrived at, given a set of observations.
- **The process model:** This is a model, with a number of unknown parameters, describing the process that generated the observations. This will take a mathematical form, describing the probability distribution of the observations.
- **The inferential task:** The task that the analyst wishes to perform having obtained their observations. Typical tasks will be testing whether a hypothesis about a given model is true, estimating the value of a parameter in a given model, or deciding which model out of a set of candidates is the most appropriate.
- **The computational approach:** Having chosen a process model, the inferential framework should determine what mathematical procedure is necessary to carry out the inferential task. In many cases, the procedure is the relatively simple application of a simple formula (for example, a Chi-Squared test). However, sometimes it is not. In such cases alternative strategies are needed. Sometimes they involve numerical solution of equations or optimizations. In other cases Monte-Carlo simulation-based approaches are used, where characteristics of statistical distributions are determined by simulating variables drawn from those distributions. The strategy used to carry out the task is what will be termed the "computational approach" here.

Probably the most fundamental of these concepts is the inferential framework. This is also the most invariant across different kinds of statistical applications – even if geographers have special process models or computational approaches, most of the time they are still appealing to the same fundamental principles when they draw inferences from their data. For example, one frequently sees geographers declare parameters in models to be "significantly different from zero," or quote confidence intervals. When they do so, they are making use of two key ideas from classical inference that may be applied to geographic and non-geographic problems alike.

The most geographically specific of the concepts is the process model. As stated earlier, many inferential tests are based on the assumption that observations are independent of one another – in many geographic processes (such as those influencing house prices) this is clearly not the case. In some cases, the geographic model is a generalization of a simpler aspatial model – perhaps the situation where geography plays no role is a special case where some parameter equals zero. In these situations, one highly intuitive inferential task is to determine this parameter *does* equal zero. In other cases, the task is to estimate the parameters (and find confidence intervals) that appear in both spatial and aspatial cases of the models (for example regression coefficients). In these cases, the spatial part of the model is essentially a nuisance, making the inferential task related to another aspect of the model more difficult.

The previous examples are relatively simple from a geographic viewpoint, but more sophisticated geographic inferential tasks can be undertaken. In particular, the tasks above are related to what Openshaw (1984) terms "whole-map statistics." That is, they consider single parameters (or sets of parameters) that define the nature of spatial interaction at all locations, but supply no information about any specific locations. To the geographer, or GIS user, it is often more important to identify *which* locations are in some way different or anomalous. Arguably, this is a uniquely geographic inferential task. Although this inferential task can be approached with standard inferential frameworks, some careful thought is required.

Thus, to address the issue of statistical inference for geographic data one must consider the nature of statistical inference *in general*, the particular nature of statistical inference when spatial processes are considered, and the way in which these two are related. This provides a broad framework for the chapter. First, a (very) brief overview of the key statistical inferential frameworks will be outlined. Next, spatial process models and related inferential tasks will be considered, together with a discussion of how the inferential approaches may be applied in this context. Finally, a set of suggested computational approaches will be considered.

An Overview of Formal Inferential Frameworks

The two most commonly encountered inferential frameworks are Classical and Bayesian. Suppose we assume a model M with some unobserved parameters θ, and some data x. Two kinds of tasks commonly encountered are the following:

1   Given M and x, to infer whether some statement about θ is likely to be true;
2   Given M and x, to estimate the value of θ.

Classical inference

The classical framework is most commonly used, and will be defined first. The classical framework generally addresses two kinds of inferential tasks. The first task is dealt with using the *significance* test.

Hypothesis testing

The statement about θ mentioned above is termed the *null hypothesis*. Next a *test statistic* is defined. Of interest here is the distribution of the test statistic if the null hypothesis is true. The *significance* (or p-value) of the test statistic is the probability of obtaining a value at least as extreme as the observed value of the test statistic if the null hypothesis is true. When the significance is very low, this suggests that the null hypothesis is unlikely to be true. To perform an $\alpha\%$ significance test one calculates the value of the test statistic with a significance of $\alpha/100$ – this is called the *critical value*. Typical values of α are 0.05 and 0.01. If the observed value is more extreme than the critical value, then the null hypothesis is rejected. Note that adopting the above procedure has a probability of α of rejecting the null hypothesis when it is actually true.

This may seem rather abstract without an example. One commonly used technique based on these principles is the two-sample t-test. Here $\theta = (\mu_1, \mu_2)$ where μ_1 and μ_2 are means of two normally distributed samples having the same variance. The null hypothesis here is that $\mu_1 = \mu_2$. Here the test statistic is the well-known t-statistic

$$t = \frac{\bar{x}_1 - \bar{x}_2}{\sqrt{s^2\left(\frac{1}{n_1} + \frac{1}{n_2}\right)}} \qquad (18.1)$$

where $\bar{x}_1$ and $\bar{x}_2$ are the sample means from the two samples, n_1 and n_2 are the respective sample sizes, and s^2 is defined by

$$s^2 = \frac{(n_1 - 1)s_1^2 + (n_2 - 1)s_2^2}{n_1 + n_2 - 2} \qquad (18.2)$$

where s_1^2 and s_2^2 are the respective sample variances for the two samples.

The above outlines the procedure of a significance test, one of the two inferential tasks performed using classical inference. Of course, such inference is probabilistic – one cannot be certain if we reject the null hypothesis that it really is untrue. However, we do know what the probability of incorrectly rejecting the null hypothesis is. This kind of error is referred to as the *type I error*. Another form of error results when we incorrectly accept the null hypothesis – this is called a *type II error*. It is generally harder to compute the probability of committing a type II error – usually denoted as $1 - \beta$. The relationship between α and β is given in Table 18.1.

Estimating parameters

The other inferential task is that of estimating θ. As with hypothesis testing, we cannot be sure that our estimate is exact – indeed given the fact that it is derived from a sample we can be almost certain that it is not. Thus, in classical inference the key method provides upper and lower bounds – the so-called *confidence interval*. Note that this assumes that θ is a scalar quantity. The situation when they are not will be discussed later. A confidence interval is a pair of numbers a and b computed from the sample data, such that the probability that the interval (a, b) contains θ is $1 - \alpha$. This probability is computed on the assumption that the model is known in advance, up to the specification of θ. A very important distinguishing

Table 18.1 Relationship between α and β

Probability	Reject Null Hypothesis	
	Yes	No
Null Hypothesis True	$1 - \alpha$	α
Null Hypothesis False	$1 - \beta$	β

characteristic of this approach is that the probability quoted for a confidence interval is NOT the probability that θ lies within the interval (a,b). θ is not a random variable – under classical inference θ is a fixed but unobservable quantity. The variables a and b are the random variables, since they are computed from the random sample of observations.

In situations where θ is not a scalar, one may specify confidence *regions* from the data. For example, in the two-dimensional case we could represent it as a point in the plane. A confidence region is some sub-region of the plane determined from the sample data that has a $1 - \alpha$ probability of containing the true θ.

Other issues for classical inference

Earlier (see "Hypothesis testing" above) it was assumed that the quantity α could be easily calculated. In some situations this is not the case and a *Monte-Carlo* (Metroplois and Ulam 1949) approach may be more helpful. In this approach, a large number of random numbers are drawn from the probability distribution of the test statistic that would apply under the null hypothesis, and the observed value of the statistic is compared against this list (see Manly 1991 for some examples). It may be checked that the percentage rank of the observed test statistic when it is merged with the list of randomly generated test statistics is itself a significance level. Thus, provided we may generate random numbers from the distribution of the test statistic, this provides an alternative approach to the classical significance test – albeit one with a very different computational approach. This approach may also be used to generate confidence intervals.

Another important observation is that the derivation of the test statistics hinges on the model for the distribution of the observational data being known – however, it is possible to draw inference from data when such a model is unknown. This is known as a *non-parametric* approach (see, for example, Siegel 1957). The advantage of this approach is that it allows tests to be made when one has no strong evidence of the distribution generating the data. A price paid for this is that the computational overhead is much higher – and typically non-parametric tests are not as powerful as the simpler parametric equivalents, provided the assumptions underlying the parametric tests hold.

Simple classical inference in action

To illustrate some of the above ideas a simple example is given. Here, the data consists of a number of sale prices of houses from two adjacent districts in the greater London area in 1991. The location of the districts in the context of greater London as a whole is shown in Figure 18.1, as are the locations of the houses in the sample. There are 220 houses in district 1 and 249 in district 2 (the district to the west).

If we assume that house prices in both districts have independent normal distributions with equal variances, we may test the hypothesis that the mean house price is the same in each district. Together, this null hypothesis along with the assumptions set out above lead to the use of the *t*-test as set out in equation (18.1). The values of the relevant quantities are set out in Table 18.2.

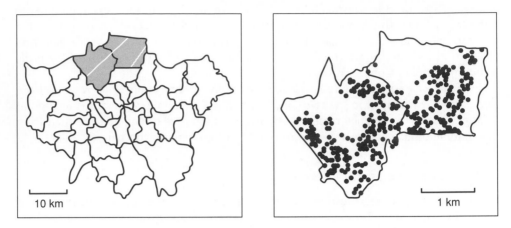

Fig. 18.1 The location of study area (LHS) and the houses in the samples (RHS)

Table 18.2   Two sample *t*-test

District 1		District 2	
n_1	220	n_2	249
x_1	77.7	x_2	86.4
s_1	37.3	s_2	41.5
s_2	39.6		
v	467		
t	−2.37		

Since we are interested in detecting differences in the mean value of either sign, we use the absolute value (that is, 2.37). However, from tables, the critical value of *t* for (two-tailed) $a = 0.05$ is 1.96 – suggesting we should reject the null hypothesis at the 5% level. Thus, with a 5% chance of making an incorrect statement if the null hypothesis is true, we reject the null hypothesis – and state that there is a difference in average house price between two zones.

Bayesian inference

The Bayesian approach views θ in a very different way. Whereas classical inference regarded θ as a deterministic but unknown quantity, Bayesian inference regards it as a random variable. The idea is that the probability distribution of θ represents the analyst's knowledge about θ – so that, for example, a distribution with very little variance suggests a great deal of confidence in knowing the value of θ. If we accept that θ is a random quantity, as is x, the observed data, we can consider the joint probability density of the two items given model M, say $f(x,\theta|M)$. Standard probability theory tells us that

$$f(\theta \mid x) = f(x \mid \theta)f(\theta)/f(x) \tag{18.3}$$

where $f(x \mid .)$ and $f(\theta \mid .)$ denote marginal distributions of x and θ respectively. Assuming we have a given observed data set x, we may regard $f(x)^{-1}$ as a normalizing constant and write

$$f(x \mid \theta) \propto f(x \mid \theta)f(\theta) \tag{18.4}$$

This is essentially Bayes' theorem, and is the key to the inferential model here. If we regard $f(\theta)$ as the analysts knowledge about θ regardless of $\mathbf{x}$, then multiplying this by the probability of observing x given theta (that is, $f(x \mid \theta)$) gives an expression proportional to $f(\theta \mid x)$. Note that in this framework, $f(x \mid \theta)$ is our process model, as set out in the initial section, "Basic Ideas of Inference." We can interpret this last expression as the analyst's knowledge about θ *given* the observational data x. Thus, we have updated knowledge about θ in the light of the observations x – this is essentially the inferential step.

In standard Bayesian terminology $f(\theta)$ is referred to as the *prior* or *prior distribution* for θ and $f(x \mid \theta)$ is referred to as the *posterior* or *posterior distribution* for θ. Thus, starting out with a prior belief in the value of θ, the analyst obtains observational data x and modifies his or her belief in the light of these data to obtain the posterior distribution. The approach has a number of elegant properties – for example, if individual data items are uncorrelated and if data is collected sequentially, one can use the posterior obtained from an earlier subset of the data as a prior to be input to a later set of data. However, the approach does require a major change in world view. The requirement of a prior distribution for θ from an analyst could be regarded as removing objectivity from the study. Where does the knowledge to derive this prior come from?

One way of overcoming this is the use of *non-informative* priors which represent no knowledge of the value of θ prior to analysis. For example, if θ were a parameter between 0 and 1, then $f(\theta) = 1$ – a uniform distribution – would be a non-informative prior since no value of θ has a greater prior probability density than any other. Sometimes this leads to problems – for example if θ is a variable taking any real value. In this case, $f(\theta)$ = constant is not a well defined probability density function. However, this shortcoming is usually ignored provided the posterior probability thus created is valid (typically the posterior in this case could be regarded as a limiting value of an infinite sequence of posteriors derived from well-defined priors – for example, if a sequence of priors with variances increasing without bound were supplied). A prior such as this is termed an *improper prior*.

Having arrived at a posterior distribution $f(x \mid \theta)$ we may begin to address the two key inferential questions:

1 **Estimate the value of θ:** Since we have a posterior distribution for θ we can obtain point estimates of θ using estimates of location for the distribution – such as the mean or median. Alternatively, we can obtain interval estimates such as the inter-quartile range derived from this distribution. Typically, one would compute an interval $[\theta_1, \theta_2]$ between which θ has a 0.95 probability of lying. Note that this is subtly different from the confidence interval of classical inference.

The 95% in a confidence interval refers to the probability that the randomly sampled data provides a number pair that contains the unobserved, but non-random θ. Here we treat θ as a random variable distributed according to the posterior distribution obtained from equation (4). To emphasize that these Bayesian intervals differ from confidence intervals, they are referred to as *credibility intervals*.

2   **Infer whether some statement about θ is likely to be true:** If our statement is of the form $a < \theta < b$ where either a or b are infinite, then this may be answered by computing areas under the posterior probability density function. However, questions of the form addressed by classical inference – such as "is θ zero?" where typically one is concerned with *point* values of θ present more difficulties. With a posterior probability density, the probability attached to any point value is zero. One approach is to decide how far from zero θ could be for the difference to be unimportant and to term this e. If this is done, we may then test the statement $-e < \theta < e$ using the above approach. Other approaches do attempt to tackle the exact value test directly – see Lee (1997) for further discussion.

Bayesian inference in action

In this section, we revisit the house price example, this time applying a Bayesian inferential framework to the problem. As before, we assume that house prices are independently normally distributed in each of the two districts. If we regard our list of house prices as x, then $\theta = (\mu_1, \mu_2)$ the respective means of the house price distribution for districts 1 and 2, and $f(x|\theta)$ is just the product of the house price probability densities for each observed price. Here we are interested in the quantity $\mu_1 - \mu_2$. In this case we have a non-informative prior in μ_1 and μ_2 and also in $\log \sigma$ where σ is the standard deviation of house prices in both districts. The choice of the prior for σ may seem strange, but essentially stems from the fact that this is a scale parameter, rather than one of location – see Lee (1997), for example. In this case, it can be shown that the posterior distribution for the quantity $\delta = \mu_1 - \mu_2$ is that of the expression

$$(\bar{x}_1 - \bar{x}_2) + s \left(\frac{1}{n_1} + \frac{1}{n_2} \right)^{\frac{1}{2}} t \tag{18.5}$$

where all variables are as defined in equation (18.1) except for t, which is a random variable with a t distribution with v degrees of freedom (again v is as defined earlier). The posterior distribution for δ is shown in Figure 18.2.

Here, the hypothesis under test differs from that of the classical test. Rather than a simple test of whether $\delta = 0$ – which makes little sense given the posterior curve above, we test whether $|\delta| < G$ where G is defined as some quantity below which a difference in means would be of little consequence. This is very different from the standard classical approach. In that framework, if a test were sufficiently powerful, differences in mean house prices of pennies could be detected. However, in terms of housing markets such a difference is of no practical importance. For this example we choose G to be £1,000 (UK). If this is the case, the probability

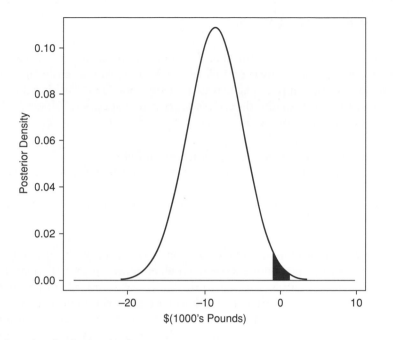

Fig. 18.2 Posterior distribution for $\delta = \mu_1 - \mu_2$

that $|\delta| < G$ corresponds to the shaded area in Figure 18.2. This is equal to 0.014 – alternatively one could state the probability that $|\delta|$ exceeds £1,000 is $1 - 0.014 = 0.986$. Thus, from a Bayesian perspective, it seems very likely that there is a non-trivial difference between the mean house prices for the two districts. Another possibility is to compute the probability that district 2 has a higher mean than district 1. This is just the posterior probability that $\delta > 0$, which, from the curve is equal to 0.99 – again suggesting this is highly likely.

Bayesian approaches – some closing comments

The Bayesian approach is regarded by some as very elegant. Certainly the simplicity of the underpinning equation (18.4) and the natural way that hypotheses may be assessed, and parameters estimated from the posterior distribution do have a directness of appeal. However, there is a sting in the tail. Equation (18.4) gives the posterior distribution *up to a constant* – implying that the expression for the probability distribution can only be obtained by integrating its un-normalized form. Herein lies the problem: in many cases the integral is not analytically tractable. At the time of writing, this presents fewer problems than in the past as numerical quadrature techniques may be used to estimate the integrals. Alternatively, techniques based on Monte-Carlo simulation and the Metropolis algorithm allow random values of θ to be generated according to the posterior distribution. In this case, hypotheses about θ are investigated by generating large numbers of random values and investigating their properties.

What is Special about Spatial?

In the above section, two of the most common approaches to formal statistical inference were discussed. However, this was done in a general sense – nothing stated in the previous section applied exclusively to geographic data. As hinted in the introduction, working with spatial data introduces a few specific problems.

This raises a number of issues:

1 What are the consequences of ignoring spatial effects?
2 Does one need to modify the above ideas of inference when working with spatial data?
3 If some spatial effects are present, can they be represented as geographic patterns or images?

All of these issues lead to important questions – questions without unique answers. If there are no serious problems encountered when ignoring spatial effects then there is little that spatial analysis can add to the canon of standard statistical methods. Perhaps unsurprisingly, it is argued here that there are serious consequences of ignoring such effects. There are many examples of the consequences of ignoring space – a striking analysis is that by Fotheringham, Brunsdon, and Charlton (1998), which follows the work of Rees (1995) in modeling the relationship between limiting long-term illness (LLTI) as defined in the 1991 UK census of population and a number of predictor variables. The study area consists of the four counties Tyne and Wear, Durham, Cleveland, and North Yorkshire, in the north-east of England. Of particular interest here is the population density variable. An ordinary least squares regression model was fitted to the data, giving a coefficient of -5.6. A t-test based on principles of classical inference showed this to be significantly different from zero. In general, this suggests that an increase in population density leads to a decrease in LLTI. This is perhaps counterintuitive. Normally one associates higher morbidity rates with urban areas, which have higher population densities. However, the study went on to consider geographically weighted regression (GWR) (Brunsdon, Fotheringham, and Charlton 1996) – a technique in which regression parameters vary over space. It was found that the regression parameter for population density was at its most negative in areas in the region around the coalfields of east Durham. Here, it is likely that LLTI is linked to employment in the coalfields, and that most people in such employment lived in settlements near to the coalfields, where population density is low. However, those people living in urbanized areas in that part of the region are less likely to be employed in occupations associated with high LLTI. Thus, in that locality a negative relationship between population density and LLTI holds. However this is unusual in general, and in other parts of the study area (west Durham, North Yorkshire), there is a positive relationship. Here, low population density corresponds to a more typical rural environment, and in these places a more conventional urban/rural trend occurs. The key point here is that the global model told only one story, while the spatially-oriented GWR identified two different processes occurring in different parts of the study area. Ignoring geography can lead to misinterpretation!

How can this difficulty be overcome? This leads on to the second issue listed above. To address this, we return once again to the four aspects of statistical inference listed in the first section of the chapter: both Bayesian and classical *inferential frameworks* can handle the key *inferential tasks* of hypothesis evaluation and parameter estimation for spatial processes. However, for spatial data the *process model* must allow for geographic effects. Finally, it is also the case that the *computational approach* must also be altered on some occasions. These two key issues will be considered in turn.

Process models for spatial data

The process models for spatial data can differ from more commonly used ones in a number of ways. The two most common ones are that they exhibit *spatial non-stationarity* and *spatial autocorrelation*. Spatial non-stationarity is essentially the characteristic of the LLTI example above. The unknown parameter θ is not a constant, but in fact a function of spatial location. In this case, a technique like GWR may be used to estimate θ at a set of given localities. Using this approach, one can apply the classical inferential framework to obtain estimates of θ, and test hypotheses such as "is θ a global fixed value?" A classical inferential framework for GWR is detailed in Fotheringham, Brunsdon, and Charlton (2002).

The phenomenon of spatial autocorrelation occurs when each of the observed x values are not drawn from statistically independent probability distributions, but are in fact correlated. In the geographic context, the correlation is generally related to proximity – nearby x values are more correlated than values located far apart. Typical examples are the SAR (spatial autoregression) and CAR (conditional auto-regression) models. Unlike GWR, these regression models do not assume that the regression parameters vary over space – however they do assume that the dependent variables are correlated. Typically here, each record of variables is associated with a spatial unit, such as a census tract, and the spatial dependence occurs between adjacent spatial units. As well as the regression coefficients and the variance of the error term, CAR and SAR models have an extra parameter controlling the degree to which adjacent dependent variables are related. In the classical inference case, parameter estimation is typically based on maximum likelihood, with the parameter vector θ containing the extra parameter described above as well as the usual regression parameters. There is much work on the classical inferential treatment of such models, see, for example, Cressie (1991). LeSage (1997) offers a Bayesian perspective.

The computational approach

Computational issues for geographic data are generally complex. The whole field of geocomputation has grown to address this. As well as problems of data storage, data retrieval, and data mining, there are many computational overheads attributable to inference in spatial data, for a number of reasons. In some cases, the issue is related to Monte-Carlo or randomization methods – this is particularly true of the Monte-Carlo Markov Chain approach to Bayesian analysis. In others, it is linked to developing efficient algorithms to access large geographic datasets – this can be an issue in localized methods such as GWR. In each case, it is true that specific algorithms may need to be created to handle the geographic situation. A very good example is found in Diggle, Tawn, and Moyeed (1998).

The final question in the earlier list also raises some interesting problems. The formal (Bayesian or classical) approach to hypothesis testing is essentially founded on the notion of testing a single hypothesis. However, many geographers would like answers to more complex hypotheses. In the spatial context, one of the key questions is "Is there an unusually high or low value of some quantity in region R?" Typically this quantity might be the average price of a house, or an incidence rate of some disease. This phenomenon is often termed *clustering* (see Chapter 22 by Jacquez in this volume for more discussion on this topic). In some situations R is known in advance – for instance it may represent the catchment area of a particular school in the house price example. If it is known in advance the approach is relatively simple. One creates a proximity measure to reflect how close to R each observation is, or creates a "membership function" of R for each observation, and then builds this into a model, using a parameter that may vary the influence of this new variable. Then one goes on to test the hypothesis that this parameter is zero (or whatever value of the parameter implies that proximity to R has no influence on the quantity of interest).

This approach fits in well with conventional theory – there is one single hypothesis to test, and it may be tested as set out above. However, on many occasions we have no prior knowledge of R, possibly even on whether R is a single region or a number of disjoint areas. On such occasions, a typical approach would be to carry out a test such as that described above on every possible region and map the ones that have a significant result. This is essentially the approach of the Geographical Analysis Machine (GAM; Openshaw 1987) – here the Rs are circular regions of several radii centered on grid points covering the study area. However, there is a difficulty with this approach. Suppose we carry out a significance test on each of the Rs. There could be a large number of tests, possibly hundreds. Even if no clustering were present, the chance of obtaining a false positive is α, the significance level of the test. If $\alpha = 0.05$ as is common practice, we would expect to find $N\alpha$ significant results even when no clustering occurs, where N is the number of regions to be tested. For example, if $N = 200$ and $\alpha = 0.05$, we would expect to find 10 significant regions *even when in reality no clustering occurs*. Thus, in an unadjusted form, this procedure is very prone to false positive findings. Essentially this is a problem of *multiple hypothesis testing*. Because the test has a positive probability of incorrectly rejecting the null hypothesis, carrying out enough tests will give some positive results even if in reality there are no effects to detect. A typical way of tackling the problem is to apply the *Bonferroni* adjustment to the significance levels of the test. For example, this is done by Ord and Getis (1995) for assessing local autocorrelation statistics.

The correction is derived by arguing that to test for clustering, we wish to test that *none* of the regions R have a significant cluster centered on them. Thus, the probability of a false positive overall is the probability that any one of the regions has a false positive result. If it is assumed that each test is independent, then it can be shown that this probability (which will be called α') is given by

$$\alpha' = 1 - (1 - \alpha')^N \tag{18.6}$$

Now, if we wish to develop an overall test for clustering, with say $\alpha' = 0.05$ then equation (18.6) may be solved for α – giving a significance level for the individual tests needed in order to achieve the overall level of significance. For example, if $N = 200$ and we require $\alpha' = 0.05$ then $\alpha = 0.000256$. This is a fairly typical result. To counter the risk of false positives, the individual tests must have very low values of α.

However, one thing of note about the above approach is that the assumption that the tests are independent is often incorrect for geographic studies. Typically, a large number of regions R are used, and many overlap, sharing part of the sample data used for the local tests – and clearly the results of these tests cannot be independent. It is usually argued that the Bonferroni procedure provides conservative tests – in the case where the tests are correlated the estimate of α' in equation (18.6) is an underestimate. In the attempt to avoid false positives, we insist on very strong evidence of clustering around each of the test regions. In the presence of correlated testing, we will be insisting on stronger evidence than is actually necessary. As a result of this, there is some chance that genuine clustering is overlooked. In a nutshell this is a typical dilemma when looking for clusters – ignoring multiple hypothesis testing leads to false positives, but overcompensation for this could lead to false negatives.

Other Types of Inference

Although classical and Bayesian methods are both covered in this chapter, these are not the only possible approaches. For example, Burnhan and Anderson (1998) outline ways in which Akaike's Information Criterion (AIC; Akaike 1973) may be used to compare models. This approach is quite different in terms of its inferential task – rather than testing whether a statement about a particular model is true – or assuming a specific model holds and then attempting to estimate a parameter of that model, this approach takes several models and attempts to identify which one is "best" in the sense that it best approximates reality. The AIC is an attempt to measure the "nearness" of the model to reality – obviously the true model is not known, but the observations have arisen from that model, and this is where the "clues" about the true model come from. This is very different from the other approaches because it regards all potential models as compromises – none is assumed to be perfect – and attempts to identify the best compromise. This area may prove fruitful in the future – for example, Fotheringham, Brunsdon, and Charlton (2002) use a method based on this idea to calibrate GWR models. The idea of finding a "best approximation" also sits comfortably with the idea of approximating a large finite sample with a continuous distribution put forward in the previous section.

Of course, exploratory data analysis can be thought of as yet another inferential framework, albeit a less formal one. Although this can provide a very powerful framework for discovering patterns in data, it could be argued that this is an entire subject in its own right, and that there will be many examples elsewhere in this book, where the production of maps and associated graphics by various software packages provide excellent examples exhibiting the power and utility of graphical data exploration.

Software

No chapter about inference would be complete without some discussion of software. Having argued that making inferences about data is central to knowledge discovery in spatial analysis, one has every right to expect that software for inferential procedures will be readily available. However, as mentioned in the introduction, most readily available GIS packages do not contain code for many of the procedures outlined here. Unfortunately, although several commercial statistics packages do contain code for carrying out general inferential procedures, such as the t-test example discussed earlier in the chapter, they offer less support for more specific inferential tasks developed for spatial data. Until recently, for a number of spatial inferential tasks one was forced to write one's own code. However this situation is now improving. A number of packages that are either dedicated to the analysis of spatial data or sufficiently flexible that they may be extended to provide spatial data analysis now exist. Although by no means the only option, the statistical programming language R provides good spatial analysis options – all of the examples (most notably the spatial one) in this chapter were based on calculations done in R. The package is also "Open Source" so it provides an easy entry option for anyone wishing to experiment more with inferential approaches for geographic data.

ACKNOWLEDGEMENTS

I am grateful to the Nationwide Building Society for providing the house price data first introduced in the section entitled "Simple Classical Inference in Action."

REFERENCES

Akaike, H. 1973. Information theory and an extension of the maximum likelihood principle. In B. Petrov and F. Csaki (eds) *Proceedings of the Second International Symposium on Information Theory*. Budapest: Akademiai Kiado, pp. 267–78.

Bayes, T. [1763] 1958. Studies in the history of probability and statistics: IX, Thomas Bayes's essay towards solving a problem in the doctrine of chances. *Biometrika* 45: 296–315. (Bayes's essay in modernized notation.)

Brunsdon, C., Fotheringham, A. S., and Charlton, M. 1996. Geographically weighted regression: A method for exploring spatial non-stationarity. *Geographical Analysis* 28: 281–9.

Burnhan, K. P. and Anderson, D. R. 1998. *Model Selection and Inference: A Practical Information-Theoretic Approach*. New York: Springer.

Cressie, N. A. C. 1991. *Statistics for Spatial Data*. New York: John Wiley and Sons.

Diggle, P. J., Tawn, J. A., and Moyeed, R. A. 1998. Model-based geostatistics. *Applied Statistics* 47: 299–350.

Fotheringham, A. S. and Brunsdon, C. 2004. Some thought on inference in the analysis of spatial data. *International Journal of Geographical Information Science* 18: 447–57.

Fotheringham, A. S., Brunsdon, C., and Charlton, M. 1998. Scale issues and geographically weighted regression. In N. Tate (ed.) *Scale Issues and GIS*. Chichester: John Wiley and Sons: 123–40.

Fotheringham, A. S., Brunsdon, C., and Charlton, M. 2002. *Geographically Weighted Regression: The Analysis of Spatially Varying Relationships*. Chichester: John Wiley and Sons.

Lee, P. M. 1997. *Bayesian Statistics: An Introduction*. London: Arnold.

LeSage, J. 1997. Bayesian estimation of spatial autoregressive models. *International Regional Science Review* 20: 113–29.

Manly, B. 1991. *Randomization and Monte Carlo Methods in Biology*. London: Chapman and Hall.

Metropolis, N. and Ulam, S. 1949. The Monte-Carlo method. *Journal of the American Statistical Association* 44: 335–41.

Openshaw, S. 1984. *The Modifiable Areal Unit Problem*. Norwich: Quantitative Methods Research Group, Royal Geographical Society and Institute of British Geographers, Concepts and Techniques in Modern Geography Publication No. 38.

Openshaw, S. 1987. A mark 1 geographical analysis machine for the automated analysis of point data sets. *International Journal of Geographical Information Systems* 1: 335–58.

Ord, J. K. and Getis, A. 1995. Local spatial autocorrelation statistics: Distributional issues and an application. *Geographical Analysis* 27: 286–306.

Rees, P. 1995. Putting the census on the researcher's desk. In S. Openshaw (ed.) *Census Users' Handbook*. Cambridge: GeoInformation International: 27–82.

Siegel, S. 1957. *Nonparametric Methods for the Behavioral Sciences*. New York: McGraw-Hill.

Chapter 19

Geographic Data Mining and Knowledge Discovery

Harvey J. Miller

Geographic information science exists in an increasingly data- and computation-rich environment. The coverage and volume of digital geographic data sets are extensive and growing. High spatial, temporal, and spectral resolution remote sensing systems and other environmental monitoring devices gather vast amounts of geo-referenced digital imagery, video, and sound (see Chapter 3 by Lees in this volume for more details on this topic). Geographic data collection devices linked to location-aware technologies (LATs) such as the global positioning system allow field researchers to collect unprecedented amounts of data. Other LATs, such as cell phones, in-vehicle navigation systems, and wireless Internet clients, can capture data on individual movement patterns. Information infrastructure initiatives such as the US National Spatial Data Infrastructure are facilitating data sharing and interoperability. The growth of computing power is widely expected to continue the exponential rate implied by Moore's Law for at least two or three more decades.

Traditional spatial analytical methods were developed when data collection was expensive and computational power was weak. The increasing volume and diverse nature of digital geographic data easily overwhelm techniques that are designed to tease information from small, scientifically sampled, and homogenous data sets. Traditional statistical methods, particularly spatial statistics, have high computational burdens. They are also confirmatory and require the researcher to have a priori hypotheses, meaning that they cannot discover unexpected or surprising information (Miller and Han 2001).

This chapter discusses the process of *geographic knowledge discovery* (GKD) and one of its central components, namely, *geographic data mining*. GKD is based on a belief that there is novel and useful geographic knowledge hidden in the unprecedented amount and scope of digital georeferenced data being collected, archived, and shared by researchers, public agencies, and the private sector. This knowledge cannot be revealed using traditional methods that require a priori hypotheses or cannot be scaled to handle massive data. Since GKD is an extension of a broader trend in computer science, we will first review the more general problem of knowledge discovery from databases (KDD) and related data mining techniques. We will then discuss why GKD is a meaningful extension of KDD, as well as identify major geographic data mining techniques. This chapter concludes with a discussion of research frontiers in GKD.

It is important to make a distinction between geographic data, GKD, and geographic data mining on the one hand, and the closely related but broader field of spatial databases, knowledge discovery, and data mining on the other. "Spatial" concerns any phenomena for which the data objects can be embedded within some formal space that generates implicit relationships among the objects. Examples include genetics and astronomy (Shekhar and Chawla 2003). "Geographic" refers to the specific case in which the data objects are georeferenced and the embedding space relates (at least conceptually) to locations on or near the Earth's surface. Although many spatial databases and related techniques can be applied to the specific problem of GKD, these techniques can also be applied more widely.

Knowledge Discovery from Databases

The knowledge discovery process

Data mining is only one step of the KDD process. Data mining involves the application of techniques for distilling data into *information* or facts implied by the data. KDD is the higher level process of obtaining facts through data mining and distilling this information into *knowledge* or ideas and beliefs about the mini-world described by the data. This generally requires a human-level intelligence to guide the process and interpret the results based on pre-existing knowledge (Miller and Han 2001). The data miner is the critical interface between the syntactic knowledge or *patterns* generated by machines and the semantic knowledge required by humans for reasoning about the real world (Gahegan, Wachowicz, Harrower, and Rhyne 2001).

The KDD process does not seek any arbitrary pattern from a database; rather, data mining seeks only those that are *interesting*. These patterns are *valid* (a generalizable pattern, not simply a data anomaly), *novel* (unexpected), *useful* (relevant), and *understandable* (can be interpreted and distilled into knowledge) (Fayyad, Piatetsky-Shapiro, and Smyth 1996). In addition to the scale of the data involved, the requirement for novelty distinguishes data mining from traditional statistics oriented towards hypothesis confirmation rather than generation. From a KDD perspective, anything that can be hypothesized *a priori* is not novel and therefore not interesting.

The KDD process typically involves the following major steps grouped into larger activity categories (Fayyad, Piatetsky-Shapiro, and Smyth 1996, Han and Kamber 2001, Qi and Zhu 2003):

1 **Background**
   (i) Developing an understanding of the application domain; this is often referred to as *background knowledge*;

2   **Data pre-processing**
  (i)   *Data selection*, or determining a subset of the records or variables in the database for focusing the search for interesting patterns;
  (ii)   *Data cleaning*, including removal of noise and outliers;
  (iii) *Data reduction*, including transformations, projections and aggregations to find useful representations for the data;
3   **Data mining**
  (i)   *Choosing the data mining task*, involving the selection of the generic type of pattern sought through data mining, this is the language for expressing facts in the database; generic pattern types include classes, associations, rules, clusters, outliers and trends (discussed in more detail below);
  (ii)   *Choosing the data mining technique* for discovering patterns of the generic type selected in the previous step; since data mining algorithms are often heuristics (due to scalability requirements), there are typically several techniques available for a given pattern type, with different techniques concentrating on different properties or possible relationships among the data objects;
  (iii) *Data mining*: applying the data mining technique to search for interesting patterns;
4   **Knowledge construction**
  (i)   *Interpreting the mined patterns*, often through visualization;
  (ii)   *Consolidating the discovered knowledge*, either by incorporating the knowledge into a computational system (such as a knowledge-based database) or through documenting and reporting the knowledge to interested parties.

The KDD process is not necessarily sequential: it is likely that the analyst will re-sequence and even revisit steps based on the knowledge sought and the nature of the information uncovered within the process. The data pre-processing steps of selection, cleaning, and reduction can be applied in different sequences and iteratively. The three data mining steps are also highly flexible and often iterative. The analyst can also jump back and forth between major tasks such as background knowledge, pre-processing, mining, and knowledge construction.

Although human-level intelligence is required to guide the complex KDD process, it is possible to support the process through computational representations of background knowledge and the interestingness measures. *Concept hierarchies* are simple but powerful representations of multilevel background knowledge. Each node represents a concept at some level of abstraction, and a tree arranges these levels of abstraction from the highest level (at the root) to its lowest levels (at the leaves). Concept hierarchies support *rolling up* (generalization) and *drilling down* (specialization) of data for the mining process; this allows the user to explore and interpret patterns at different semantic levels. Figure 19.1 illustrates a concept hierarchy for the geographic concept *location* within a particular knowledge domain (Han and Kamber 2001).

A data mining technique can generate an overwhelmingly large number of patterns. *Interestingness measures* attempt to quantify the concept of interesting to limit the candidate patterns presented to the analyst. Types of interestingness measures include *simplicity* (such as rule length), *certainty* (confidence measures), *utility* (database

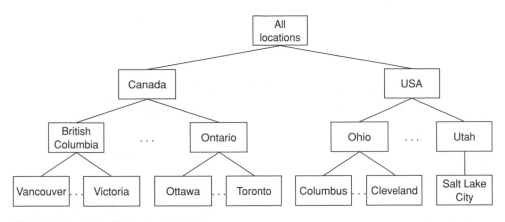

Fig. 19.1 A concept hierarchy for *location*
Based on Han and Kamber 2001, Figure 2.7

support; the number of objects for which it is true) and *novelty* (redundancy with patterns already stored in the database). Most measures require specified relevancy threshold values (Han and Kamber 2001).

Data Warehousing and Related Technologies

Data warehouses are critical, enabling technologies underlying the data mining and knowledge discovery process. A data warehouse is a non-transactional (non-editable) database, comprising read-only historical copies of one or more of the transactional databases used in an organization or enterprise. Data warehouses integrate and represent these data in a manner that supports very efficient querying and processing. Consequently, the design principles for data warehouses are different from transactional databases. Transactional databases should be *normalized* or converted into the simplest logical representation in order to avoid inconsistencies associated with multiple users editing replicated data within the same database. In contrast, data warehouses need to be as connected as possible to support efficient query processing; this implies redundant data. Data warehouse design schemes include the *star design*, the *snowflake design*, and *fact constellations* (Han and Kamber 2001). Data warehouses are not necessarily centralized; they can be distributed, multi-tiered, and federated. For example, some systems can include *data marts* or smaller scale data warehouses specific to particular departments or divisions within a larger enterprise (Bédard, Merrett, and Han 2001).

Data warehouses include tools for quick multi-dimensional and multi-level data summaries. *Online analytical processing* (OLAP) tools allow users to manipulate simple database summaries and explore the data associated with these views. This can support data mining and other stages of the KDD process by allowing a synoptic sense of the database before applying more computationally intensive techniques. A common OLAP technique is the *data cube*; this reports all possible cross-tabulations of database attributes in the format of an *m* dimensional hypercube,

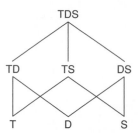

Fig. 19.2 A data cube for traffic data attributed by date, time of day, and station
Shekhar, Lu, Zhang, and Liu 2002

where m is the number of attributes (Gray, Chaudhuri, Bosworth 1997). Figure 19.2 illustrates a simple data cube for traffic data attributed according to the date (D), time of day (T), and traffic counter station (S).

Seeking interesting patterns through data mining

Most data mining techniques are heuristics tailored to discover patterns of a generic type. Generic patterns include classes, associations, rules, clusters, and outliers (Han and Kamber 2001). Since these techniques are heuristics, there is no single optimal algorithm for discovering patterns of a given type; different techniques highlight different aspects of the information space implied by the database at the expense of other characteristics. The following discussion defines these patterns and some associated techniques:

1 **Classification** involves mapping data into specified *classes* or meaningful categories whose cardinality is much less than the number of data objects. The mapping can be achieved through *data characterization* (classification based on shared characteristics) or *data discrimination* (methods that highlight differences among records). Characterization techniques include *attribute-oriented induction*: this compresses data into increasingly general relations through aggregating attributes based on the generalization operators or concept hierarchies (Han and Fu 1996). Discrimination techniques generate statistical and visual summaries intended to highlight differences among user-specified classes such as those based on data cubes.

2 **Association analysis** involves the study of dependency relationships between attributes in a database to determine their *association rules*. These are typically expressed in the form $X \Rightarrow Y(c,r)$ where X, Y are disjoint sets of database attributes, c is the *confidence* or the conditional probability $P(Y|X)$ and r is *support* or the probability $P(X \cup Y)$. Algorithms for association analysis include breadth-first and depth-first search heuristics based on concept hierarchies (Hipp, Güntzer, and Nakhaeizadeh 2000).

3 **Classification and prediction** involves finding simple functions or models that can distinguish between data classes or concepts. Classification and prediction are also common in traditional statistics, but data mining demands highly scalable techniques that can be applied in an exploratory rather than a confirmatory manner.

Techniques include decision tree induction, fitted linear or curvilinear relationships, Bayesian classification, and artificial neural networks (Han and Kamber 2001).

4 **Cluster analysis** includes techniques for classifying data objects into similar groups. Unlike concept description techniques, clusters are not pre-specified but rather emerge from the inherent similarity and dissimilarity among objects. Clustering is a computationally intensive problem and it is only since the 1990s that efficient techniques for massive databases have emerged in the literature. Clustering techniques have different objectives that affect the resulting cluster morphology. Important data characteristics include the type of data (nominal, ordinal, numeric), data dimensionality (since some techniques perform better in low-dimensional spaces), and error (since some techniques are sensitive to noise) (Han, Kamber, and Tung 2001).

5 **Outlier analysis** refers to the examination of data objects (*outliers*) that appear inconsistent with respect to the remainder of the database (Barnett and Lewis 1994). While in many cases these can be anomalies or noise, they sometimes represent rare or unusual events to be investigated further. For example, outlier analysis has been used in detecting credit fraud, determining voting irregularities, and in severe weather prediction (Shekhar, Lu, and Zhang 2003a). Most tools for data mining can also be used in outlier detection since they typically also generate exceptions to the discovered patterns. However, using standard data mining techniques in outlier detection can be restrictive (since outliers are defined indirectly as observations that do not meet some specified pattern) and computationally inefficient (since finding patterns can require more effort than finding outliers). Direct methods for outlier detection include *distribution-*, *depth-*, and *distance-based approaches*. Distribution-based approaches use standard statistical distributions, depth-based techniques map data objects into an *m*-dimensional information space (where *m* is the number of attributes), and distance-based approaches calculate the proportion of database objects that are a specified distance from a target object (Ng 2001).

Visualization and knowledge discovery

Visualization is a powerful strategy for leveraging the visual orientation of sighted human beings. Sighted humans are extraordinarily good at recognizing visual patterns, trends, and anomalies; these skills are valuable at all stages of the knowledge discovery. Visualization can be used in conjunction with OLAP to aid the user's synoptic sense of the database. Visualization can also be used to support data preprocessing, the selection of data mining tasks and techniques, interpretation, and integration with existing knowledge (Keim and Kriegel 1994). Visualization creates an opportunity for machines and humans to cooperate in ways that exploit the best abilities of both (fast but dumb calculation and record-keeping versus slow but smart recognition and interpretation, respectively) (Gahegan, Wachowicz, Harrower, and Rhyne 2001).

Methods for visual data mining and exploratory analysis include map-based, chart-based, projection, pixel, iconographic, and network techniques. *Map-based techniques* allow the user to interactively represent georeferenced data in cartographic form. *Chart-based techniques* plot data using graphs and charts such as scatterplots and pie charts. *Projection techniques* use statistical transformations to represent data in alternative (non-Euclidean) spaces. *Pixel techniques* map data values to individual pixels that are in some meaningful order or position on the screen (such as temporal ordering or similarity-based clusters). *Iconographic techniques* use complex symbols (such as stick figures) to give the viewer a sense of the whole while highlighting some differentiation in the data. *Network methods* organize visual representation based on specified logical structures such as trees (Gahegan 2000).

Geographic Knowledge Discovery

Why geographic knowledge discovery?

Geographic knowledge discovery (GKD) is the process of extracting information and knowledge from massive georeferenced databases. The nature of geographic entities, relationships, and data means that standard KDD techniques are not sufficient (Shekhar, Zhang, Huang, and Vatsavai 2003b). Specific reasons include the nature of geographic space, the complexity of spatial objects and relationships as well as their transformations over time, the heterogeneous and sometimes ill-structured nature of georeferenced data, and the nature of geographic knowledge.

Objects in aspatial databases are typically discrete with explicitly defined relationships codified into the database. In contrast, spatial objects by definition are embedded in a continuous space that serves as a measurement framework for all other attributes. This framework generates a wide spectrum of implicit distance, directional, and topological relationships, particularly if the objects are greater than one dimension (such as lines, polygons, and volumes). Also, although most mapping techniques and spatial analyses assume Euclidean space for this framework, there are many physical and human geographic processes that exhibit non-Euclidean spatial properties (examples include migration, disease propagation, and travel times in congested urban areas). Exploring alternative geo-spaces for representing geographic data is a form of data pre-processing that could substantially enhance the GKD process. It is also possible to calculate relationships through attributed geographic space that represents terrain, land cover, velocity fields, or other cost fields that condition movement (Miller and Wentz 2003).

Geographic data often exhibits the properties of spatial dependency and heterogeneity. Spatial dependency is the tendency of observations that are more proximal in geographic space to exhibit greater degrees of similarity or dissimilarity (depending on the phenomena). Proximity can be defined in highly general terms, including distance, direction, and/or topology. Spatial heterogeneity or the non-stationarity of the process with respect to location is often evident since many geographic processes are local. Spatial dependency and heterogeneity can be evidence of misspecification (such as missing variables) but can also reflect the inherent nature of the geographic process. Either way, these relationships are information-bearing (Miller and Wentz 2003, Shekhar, Zhang, Huang, and Vatsavai 2003b).

Data objects in typical KDD applications can be reduced to points in some multidimensional space without information loss. In contrast, many geographic entities cannot be reduced to point objects without significant information loss. Characteristics such as the size and morphology of geographic entities can have non-trivial influences on geographic processes. Geographic objects can also be measurement artifacts; aggregate spatial units such as census districts are often chosen for administrative reasons or convenience rather than reality. The sensitivity of model results to the spatial measurement units is a well-known quandary often referred to as the *modifiable areal unit problem* (MAUP) in spatial analysis (see Fotheringham and Wong 1991 for additional details). The implications of arbitrary spatial zoning and aggregation should be explored within the GKD process to determine if a discovered pattern is robust or simply an artifact of the spatial measurement units.

Including time introduces additional complexity to the GKD process. A simple strategy that treats time as an additional spatial dimension is not sufficient. Time has different semantics than space: time is directional, has unique scaling and granularity properties, and can be cyclical and even branching with parallel local time streams (Roddick and Lees 2001). Spatial transformations are also complex: for example, a formal model of possible transformations in spatial objects with respect to time includes the operators *create, destroy, kill, reincarnate, evolve, spawn, identity, aggregate, disaggregate, fusion,* and *fission* (see Frank 2001 for additional details). Roddick and Lees (2001) argue that the potential complexity of spatio-temporal patterns may require *meta-mining* techniques that search for higher-level patterns among the large number of patterns generated from spatio-temporal mining.

Digital geographic databases are also expanding to include more heterogeneous data types, including ill-structured data. Traditional logical data models for spatial data include vector (spatial objects in an embedding space) and raster (discrete representations of continuous spatial fields using a grid-cell tessellation). Real-time environmental monitoring systems such as intelligent transportation systems and location-based services are generating georeferenced data in the form of dynamic flows and space-time trajectories. *Georeferenced multimedia* includes audio, imagery, video, and text that can be related unambiguously to a location on the Earth's surface based on the location of the data collection or its content (Câmara and Raper 1999). Despite the ill-structured nature of these data, they contain a potential wealth of information about particular places and times, including secondary (interpreted) information.

The complexity of spatial objects and relationships in georeferenced data, as well as the computational intensity of many spatial algorithms, means that geographic background knowledge can play an important role in managing the GKD process. Geographic concept hierarchies are particularly useful for guiding knowledge discovery at different levels of spatial aggregation (and therefore levels of computational complexity due to the spatial representations and data volumes). Figure 19.1 illustrated a simple geographic concept hierarchy. More sophisticated hierarchies are available from formal theories such as central place theory as well as the qualitative knowledge in traditional regional geography (as discussed in more detail in the final section).

Spatial Data Warehousing

Spatial data warehouses (SDWs) are data warehouses that include both spatial and aspatial data. SDWs often include georeferenced data, but other non-geographic data (such as medical imagery) can also be archived using SDW techniques. Examples of geographic SDWs include the US Census database, Sequoia 2000, and archives from transportation operations centers (Shekhar and Chawla 2003).

Functional differences between SDW and standard data warehouses include capabilities for visualization and spatial aggregation. Conventional OLAP methods such as the data cube generate summary cross-tabs in tables; spatial data requires capabilities for data summaries in cartographic form. Conventional OLAP tools also have clear standards for aggregation and cross-tabulation, namely, the one-dimensional attributes associated with each data object. Conversely, spatial aggregation is more complex, and standards for aggregation operators on geometric data types have not yet emerged (Shekhar and Chawla 2003). In addition to purely spatial aggregations, SDW must also support *non-spatial aggregations* (such as those involving administrative or political units treated as nominal categories), *spatial-to-non-spatial aggregations* (where data are spatial at low aggregation levels but aspatial at some higher level and beyond; an example is aggregating polygons representing states or provinces that eventually become nominal categories such as regions or countries). These spatial and hybrid aggregation operators can require background knowledge in the form of concept hierarchies such as the one illustrated in Figure 19.1. Spatial aggregation and related measures can also be computationally-demanding; the SDW designer can choose whether to compute them on-the-fly, to selectively pre-compute some measures, or use filter and refine methods (Bédard, Merrett, and Han 2001).

Shekhar, Lu, Tan, Chawla, and Vatsavai (2001) develop the *map cube* as a spatial analog of the data cube. A map cube includes standard summaries and cross-tabulations as well as spatial summaries at different levels of aggregation with pointers to the corresponding spatial objects. The map cube includes geographic visualizations of these summaries and cross-tabulations: it generates an album of maps corresponding to all possible aspatial and spatial summaries of the data based on a specified spatial aggregation hierarchy. Figures 19.3 and 19.4 illustrate map cube visualizations based on the traffic data cube in Figure 19.2. Figure 19.3

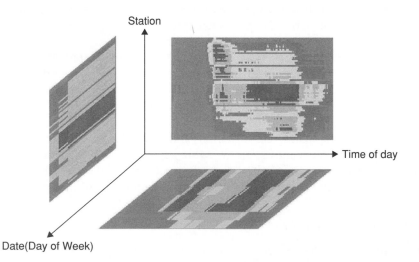

Fig. 19.3 Traffic map cube visualization by date, time of day, and station
Shekhar, Lu, Zhang, and Liu 2002

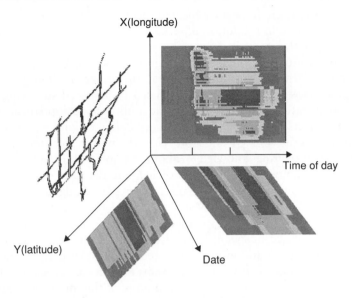

X(longitude)

Time of day

Y(latitude)

Date

Fig. 19.4 Traffic map cube visualization by date, time of day, and geographic location

illustrates a visualization of cross-tabulation by date, time of day, and station, and Figure 19.4 illustrates visualization by date, time of day, and geographic location.

Spatial data mining

Pattern types such as classes, associations, rules, clusters, outliers, and trends all have spatial expressions since these patterns can be conditioned by the morphology as well as spatial relationships among these objects. This section reviews major techniques and applications of spatial data mining.

1   **Spatial classification** These techniques map spatial objects into meaningful categories that consider the distance, direction, or connectivity relationships and/or the morphology of these objects. Koperski, Han, and Stefanovic (1998) use spatial buffers to classify objects based on attribute similarity and distance-based proximity. Ester, Kriegel, and Sander (1997) generalize this approach through a spatial classification learning algorithm that considers spatial relationships defined as path relationships among objects in a defined neighborhood of a target object. These paths are highly general and can be defined using any spatial relationship (see Ester, Kriegel, and Sander 2001 for a generalization of this approach to other data mining tasks).

2   **Spatial association** *Spatial association rules* are association rules as defined above that also contain spatial predicates in their precedent or antecedent. Koperski, Han, and Stefanovic (1998) pioneered this concept, providing detailed descriptions of their formal properties as well as a top-down tree search technique that exploits background knowledge in the form of a geographic concept hierarchy. A specific type of association rule is a *co-location pattern*: these are subsets of

spatial objects that are frequently located together. Huang, Shekhar, and Xiong (2000) develop a multi-resolution filtering algorithm for discovering co-location patterns in spatial data.

3 **Spatial classification and prediction** Malebra, Esposito, Lanza, and Lisi (2001) use inductive learning algorithms to extract information from general purpose topographic maps such as the type produced by national surveying and cartographic organizations. A search heuristic builds logical predicates based on the spatial objects, background knowledge, defined higher-level concepts and a performance criterion. Qi and Zhu (2003) apply a decision tree induction algorithm to extracting knowledge about complex soil-landscape processes. Their system combines background knowledge in the form of a soil survey map with other environmental data to extract the expert's judgments underlying the subjective map. Gopal, Liu, and Woodcock (2001) use a type of artificial neural network known as adaptive resonance theory networks to extract knowledge from remotely sensed imagery. They also illustrate the use of visualization to support interpretation and insights into neural network performance.

4 **Spatial clustering** *Spatial clustering* algorithms exploit spatial relationships among data objects in determining inherent groupings of the input data. Since finding the optimal set of k clusters is intractable (where k is some integer much smaller than the cardinality of the database), a large number of heuristic methods for clustering exist in the literature. Many of these can be adapted to or are specially tailored for spatial data (Han, Kamber, and Tung 2001). Traditional *partitioning* methods such as k-means and the *expectation-maximization* (EM) method can capture simple distance relationships and are therefore available for massive spatial databases. *Hierarchical methods* build clusters through top-down (by splitting) or bottom-up (through aggregation) methods. *Density-based methods* define clusters as regions of space with a relatively large number of spatial objects; unlike other methods, these can find arbitrarily-shaped clusters. *Grid-based methods* divide space into a raster tessellation and clusters objects based on this structure. *Model-based methods* find the best fit of the data relative to specific functional forms. *Constraints-based methods* can capture spatial restrictions on clusters or the relationships that define these clusters. An example is the *clustering with obstructed distance* algorithm that can account for geographic obstacles such as rivers, borders, and mountains.

5 **Spatial outlier analysis** Shekhar, Lu, and Zhang (2003a) define a spatial outlier as a spatially-referenced object whose non-spatial attributes appear inconsistent with other objects within some spatial neighborhood. Note that, other than is the case with aspatial outliers, this definition does not imply that the object is significantly different than the overall database as a whole: it is possible for a spatial object to appear consistent with the other objects in the entire database but nevertheless appear unusual within a local neighborhood. They develop a unified modeling framework and identify efficient computational structures and strategies for detecting these types of spatial outliers based on a single (non-spatial) attribute. More generally, geographic objects can also exhibit unusual spatial properties such as size and shape. Ng (2001) uses distance-based measures to detect unusual paths in two-dimensional space traced by individuals through a monitored environment. These measures allow the identification of unusual

trajectories based on entry/exit points, speed and geometry; these trajectories may correspond to unwanted behaviors such as theft.

Geographic visualization and knowledge discovery

Geographic visualization (GVis), or the integration of scientific visualization with traditional cartography, is highly complementary to the GKD process and can be exploited at all stages, including data pre-processing, data mining, and knowledge construction (see Gahegan, Wachowicz, Harrower, and Rhyne 2001 and Chapter 16 by Gahegan in this volume for additional details). In addition to the data volumes involved, a unique challenge is the extraordinary richness of geographic data with respect to the number and types of attributes that can be associated with geographic locations, particularly when diverse data sets are integrated based on place (Gahegan, Wachowicz, Harrower, and Rhyne 2001). The problem is how to preserve the richness of this information space when restricted to the low dimensional information spaces that can be easily related to geographic space by the user (these being two or three spatial dimensions, and possibly time through animation). For example, dense and complex symbols and colors within low dimensionality spaces can create visual interaction effects that are poorly understood and can confound knowledge discovery and communication (Gahegan 1999).

GKD Research Frontiers

GKD is a dynamic field that is only just beginning at the time of writing. GKD will continue to grow as the scope and volume of digital georeferenced data expands and GI scientists develop new techniques to exploit these data as well as the increasing power of computational platforms. The following points identify some particularly important research frontiers in GKD.

1 **Representation and integration of background geographic knowledge** There is a rich source of existing geographic models, theories, laws, and knowledge that can be codified to serve as background knowledge for the GKD process. For example, central place theory offers a hierarchy of market centers that can be applied at scales from local to global. Other theories such as spatial interaction theory suggest general principles regarding the role of spatial separation and complementary attributes of origins and destinations in conditioning the movement of materials or information, as well as related spatial processes. Qualitative and subjective knowledge about regional geography at all geo-scales from local to global could also be represented as concept hierarchies or other semantic networks. Physical geography also contains sophisticated geographic concept hierarchies based on geomorphology, river networks, biotic regimes, and so on. A critical research frontier is extracting and representing this rich geographic background knowledge to guide GKD.

2 **Knowledge discovery from georeferenced multimedia** A multimedia database system stores and manages large collections of multimedia objects such as audio, image, video, and text, including metadata about where and when the media was

collected, or the locations and times described by the media. Multimedia data mining is challenging since these data are implicit and the media must be processed to extract even its most basic features and structures (Han and Kamber 2001). Mining georeferenced multimedia encompasses these challenges plus the difficulties associated with geographic data mining and knowledge discovery.

3 **Parallel algorithms and distributed infrastructures for geographic data mining** Parallel processing and distributed computational platforms such as grid computing environments can be exploited in GKD. Spatial data mining techniques can sometimes be decomposed into parallel tasks. Even if task parallelism is impossible, georeferenced data can often be divided into spatial subsets for parallel processing (see Healy, Dowers, Gittings, and Mineter 1998, Wang and Armstrong 2003). Developing and testing parallel and distributed algorithms and architectures for GKD is an important research frontier.

4 **Spatio-temporal knowledge discovery** As mentioned earlier in this chapter, including time in GKD greatly increased its logical complexity. There is a critical need to develop formal representations, database designs, data mining techniques, and visualization methods that can extract meaningful information from these data. A particularly challenging frontier is developing representation and methods for data on mobile objects. LATs and wireless network-based *location-based services* (LBS) are a potentially rich source of knowledge about human activities in space and time, as well as emergent spatio-temporal phenomena such as traffic jams, urban dynamics, and migration patterns (Smyth 2001, Wolfson 2002). Physical phenomena such as weather, predator–prey dynamics, invasive species, disease propagation, and environmental change also have emergent spatio-temporal properties that are notoriously difficult to analyze using traditional methods (Mesorobian, Muntz, Santos 1994, 1996).

5 **New questions for geographic research** A critical and broad research question for GI scientists and geographers is "What are the questions that we could not ask before?" Geographic and other domain scientists should articulate fundamental and daunting research questions and challenges to guide the development of new GKD techniques. There are also needs for benchmarking and proof-of-concepts to demonstrate that these techniques can discover interesting geographic knowledge.

6 **Integrating discovered knowledge into spatial analysis and Geographic Information Systems (GIS)** Most existing spatial analysis techniques and GIS databases use simple representations of geographic knowledge such as primitive distance and topological relationships. Discovered geographic knowledge should be used to develop knowledge-based GIS as well as intelligent spatial analytical techniques. There is also a need to develop user-friendly interfaces for these techniques and software so they can be exploited by domain scientists.

ACKNOWLEDGEMENTS

Figure 19.4 was reproduced courtesy of Professor Shashi Shekhar and Mr. Pusheng Zhang, Spatial Databases Group, University of Minnesota (http://www.cs.umn.edu/research/shashi-group/).

REFERENCES

Barnett, V. and Lewis, T. 1994. *Outliers in Statistical Data* (3rd edn). Chichester: John Wiley and Sons.

Bédard, Y., Merrett, T., and Han, J. 2001. Fundamentals of spatial data warehousing for geographic knowledge discovery. In H. J. Miller and J. Han (eds) *Geographic Data Mining and Knowledge Discovery*. London: Taylor and Francis, pp. 53–73.

Câmara, A. S. and Raper, J. (eds). 1999. *Spatial Multimedia and Virtual Reality*. London: Taylor and Francis.

Ester, M., Kriegel, H.-P., and Sander, J. 1997. Spatial data mining: A database approach. In M. Scholl and A. Voisard (eds) *Advances in Spatial Databases*. Berlin: Springer Lecture Notes in Computer Science No. 1262: 47–66.

Ester, M., Kriegel, H.-P., and Sander, J. 2001. Algorithms and applications for spatial data mining. In H. J. Miller and J. Han (eds) *Geographic Data Mining and Knowledge Discovery*. London: Taylor and Francis, pp. 160–87.

Fayyad, U. M., Piatetsky-Shapiro, G., and Smyth, P. 1996. From data mining to knowledge discovery: An overview. In U. M. Fayyad, G. Piatetsky-Shapiro, P. Smyth, and R. Ulthurusamy (eds) *Advances in Knowledge Discovery and Data Mining*. Cambridge, MA: MIT Press, pp. 1–34.

Fotheringham, A. S. and Wong, D. W. S. 1991. The modifiable areal unit problem in multivariate statistical analysis. *Environment and Planning A* 23: 1025–44.

Frank, A. 2001. Socio-economic units: Their life and motion. In A. Frank, J. Raper, and J.-P. Cheylan (eds) *Life and Motion of Socio-economic Units*. London: Taylor and Francis, pp. 21–34.

Gahegan, M. 1999. Four barriers to the development of effective exploratory visualisation tools for the geosciences. *International Journal of Geographical Information Science* 13: 289–309.

Gahegan, M. 2000. On the application of inductive machine learning tools to geographical analysis. *Geographical Analysis* 32: 113–39.

Gahegan, M., Wachowicz, M., Harrower, M., and Rhyne, T. M. 2001. The integration of geographic visualization with knowledge discovery in databases and geocomputation. *Cartography and Geographic Information Systems* 28: 29–44.

Gopal, S., Liu, W., and Woodcock, X. 2001. Visualization based on fuzzy ARTMAP neural network for mining remotely sensed data. In H. J. Miller and J. Han (eds) *Geographic Data Mining and Knowledge Discovery*. London: Taylor and Francis, pp. 315–36.

Gray, J., Chaudhuri, S., Bosworth, A., Layman, A., Reichart, D., Venkatrao, M., Pellow, F., and Pirahesh, H. 1997. Data cube: A relational aggregation operator generalizing group-by, cross-tab and sub-totals. *Data Mining and Knowledge Discovery* 1: 29–53.

Han, J. and Fu, Y. 1996. Exploration of the power of attribute-oriented induction in data mining. In U. M. Fayyad, G. Piatetsky-Shapiro, P. Smyth, and R. Ulthurusamy (eds) *Advances in Knowledge Discovery and Data Mining*. Cambridge, MA: MIT Press, pp. 399–421.

Han, J. and Kamber, M. 2001. *Data Mining: Concepts and Techniques*. San Francisco, CA: Morgan Kaufman.

Han, J., Kamber, M., and Tung, A. K. H. 2001. Spatial clustering methods in data mining: A survey. In H. J. Miller and J. Han (eds) *Geographic Data Mining and Knowledge Discovery*. London: Taylor and Francis, pp. 188–217.

Healy, R., Dowers, S., Gittings, B., and Mineter, M. (eds). 1998. *Parallel Processing Algorithms for GIS*. London: Taylor and Francis.

Hipp, J., Güntzer, U., and Nakhaeizadeh, G. 2000. Algorithms for association rule mining: A general survey and comparison. *SIGKDD Explorations* 2: 58–64.

Huang, Y., Shekhar, S., and Xiong, H. 2000. Discovering Co-location Patterns from Spatial Data sets: A General Approach. WWW document, http://www.cs.umn.edu/research/hashi-group/.

Keim, D. A. and Kriegel, H.-P. 1994. Using visualization to support data mining of large existing databases. In J. P. Lee and G. G. Grinstein (eds) *Database Issues for Data Visualization*. Berlin: Springer Lecture Notes in Computer Science No. 871: 210–29.

Koperski, K., Han, J., and Stefanovic, N. 1998. An efficient two-step method for classification of spatial data. *In Proceedings of the International Symposium on Spatial Data Handling (SDH '98)*, Vancouver, Canada. Columbus, OH: International Geographical Union, pp. 45–54.

Malerba, D., Esposito, F., Lanza, A., and Lisi, F. A. 2001. Machine learning for information extraction from topographic maps. In H. J. Miller and J. Han (eds) *Geographic Data Mining and Knowledge Discovery*. London: Taylor and Francis, pp. 291–314.

Mesrobian, E., Muntz, R., Santos, J. R., Shek, E., Mechoso, C. R., Farrara, J. D., and Stolorz, P. 1994. Extracting spatio-temporal patterns from geoscience data sets. In *Proceedings of the IEEE Workshop on Machine Vision*, Seattle, WA, USA. Los Alamitos, CA: Institute of Electrical and Electronic Engineers, pp. 92–103.

Mesrobian, E., Muntz, R., Shek, E., Nittel, S., La Rouche, M., Kriguer, M., Mechoso, C. R., Farrara, J. D., Stolorz, P., and Nakamura, H. 1996. Mining geophysical data for knowledge. *IEEE Expert* 11: 34–44.

Miller, H. J. and Han, J. 2001. Geographic data mining and knowledge discovery: An overview. In H. J. Miller and J. Han (eds) *Geographic Data Mining and Knowledge Discovery*. London: Taylor and Francis, pp. 3–32.

Miller, H. J. and Wentz, E. A. 2003. Representation and spatial analysis in geographic information systems. *Annals of the Association of American Geographers* 93: 574–94.

Ng, R. 2001. Detecting outliers from large data sets. In H. J. Miller and J. Han (eds) *Geographic Data Mining and Knowledge Discovery*. London: Taylor and Francis, pp. 218–35.

Qi, F. and Zhu, A.-X. 2003. Knowledge discovery from soil maps using inductive learning. *International Journal of Geographical Information Science* 17: 771–95.

Roddick, J. F. and Lees, B. 2001. Paradigms for spatial and spatio-temporal data mining. In H. J. Miller and J. Han (eds) *Geographic Data Mining and Knowledge Discovery*. London: Taylor and Francis, pp. 33–49.

Shekhar, S. and Chawla, S. 2003. *Spatial Databases: A Tour*. Upper Saddle River, NJ: Prentice-Hall.

Shekhar, S., Lu, C. T., Tan, X., Chawla, S., and Vatsavai, R. R. 2001. Map cube: A visualization tool for spatial data warehouses. In H. J. Miller and J. Han (eds) *Geographic Data Mining and Knowledge Discovery*. London: Taylor and Francis, pp. 74–109.

Shekhar, S., Lu, C. T., Zhang, P., and Liu, R. 2002. Data mining for selective visualization of large spatial data sets. *In Proceedings of Fourteenth IEEE International Conference on Tools with Artificial Intelligence (ICTAI '02)*, Washington DC, USA. Los Alamitos, CA: Institute of Electrical and Electronic Engineers, pp. 41–8.

Shekhar, S., Lu, C. T., and Zhang, P. 2003a. A unified approach to detecting spatial outliers. *GeoInformatica* 7: 139–66.

Shekhar, S., Zhang, P., Huang, Y., and Vatsavai, R. R. 2003b. Trends in spatial data mining. In H. Kargupta, A. Joshi, K. Sivakumar, and Y. Yesha (eds) *Data Mining: Next Generation: Challenges and Future Directions*. Menlo Park, CA: AAAI/MIT Press, pp. 357–79.

Smyth, C. S. 2001. Mining mobile trajectories. In H. J. Miller and J. Han (eds) *Geographic Data Mining and Knowledge Discovery*. London: Taylor and Francis, pp. 337–61.

Wang, S. and Armstrong, M. P. 2003. A quadtree approach to domain decomposition for spatial interpolation in grid computing environments. *Parallel Computing* 29: 1481–504.

Wolfson, O. 2002. Moving objects information management: The database challenge. *In Proceedings of the Fifth Workshop on Next Generation Information Technologies and Systems*, Caesarea, Israel. Los Alamitos, CA: Institute of Electrical and Electronic Engineers.

Chapter 20

The Geospatial Semantic Web

Frederico Fonseca

The Web's continuous growth has facilitated the increased availability of information in general and of geospatial information in particular. At the same time it has become clear that such a wealth of data sources is not useful if we are not able to efficiently index, retrieve, and integrate all this information. The situation regarding geospatial data is even worse because of the new means for collecting spatial data such as the widespread use of Global Positioning System (GPS) technology and the availability of new and more sophisticated satellites (see Chapters 3 and 28 by Lees and Dana respectively, in this volume, for more information on both of these options). The only way to deal efficiently with such an amount of information is to delegate to computers some of the tasks of indexing, organizing, and retrieving information. In order to do this it is necessary for computers to be able to understand the metadata, the data that describes the data. Therefore the main challenge of the enterprise called the Semantic Web is to translate the description of all the available data on the Web into a formal language that computers can understand and process.

The idea of the Semantic Web can be compared to what you do when you are looking for some hard-to-find product. It might be, for instance, a vintage computer game. You cannot find it in large chain stores. You have to go there and talk with the sales people. They will give you a phone number or an address of a small store specializing in vintage games. Now you call the small store and are forwarded to another out-of-town store. You call the new store and the phone number is disconnected. You then go to the Web and do a search and find the store's new phone number. Now you are able to call and finally reach them. They do not have the game you are looking for but they write down your name and number and promise to call you back whenever they are able to get the game. Finally, they call you two months later with the game. In the Semantic Web vision you can delegate all the tasks that you have performed in this example to a software agent. You can think of an agent as a piece of software that you trust and to which you can delegate some functions such as finding and buying you a vintage game cartridge. In order to do this the agent has to know your preferences and constraints. It also has to know the intended meanings of terms that you and the parts it will be negotiating with use. So at the heart of the Semantic Web we have two main issues: trust and meaning. Since the work is being accomplished by computers another crucial issue is that whatever way this meaning is expressed it has to be machine-readable.

Since we do not have yet these resources available on the Web we can say that today's agents that look for the best price for you to buy a book or the best deal for you vacation use brute force. The semantics and the knowledge are hardwired into the software agent itself. If one of the travel websites changes their tags for price or location, the software agent has to be reprogrammed. Someone has to go through the code and make the changes there or this website will not be taken into account in the next search for an affordable dream vacation. Berners-Lee, Hendlers, and Lassila (2001) considers that "the challenge of the Semantic Web, therefore, is to provide a language that expresses both data and rules for reasoning about the data and that allows rules from any existing knowledge-representation system to be exported onto the Web." The Semantic Web has to be viewed and understood as a work in progress. From the World Wide Web Consortium (W3C 2001) Introduction we can read "The Semantic Web is a vision: the idea of having data on the web defined and linked in a way that it can be used by machines not just for display purposes, but for automation, integration and reuse of data across various applications."

Semantics

But what is semantics? The term "semantic" is related to signification or meaning. What is the meaning of the database with data about deforestation in the Amazon that I just found on the Web? What did the vendor of a game cartridge mean by price on its web page? All the (human) negotiation in the search for meaning that we gave in our introductory example has to be performed by computers in the Semantic Web vision. In order to do this we need standard vocabularies, formal definitions, and formal languages. Uschold (2003) suggests a semantic continuum starting from shared human consensus which has little formalism and ranging to the completely formal languages to be used in computers. Human beings communicate and share meaning through the use of language. The process of communication implies the acceptance of the meaning embedded in words used by one agent in the communication process by a second agent in the same process. When there are questions the agents engage in a discussion trying to clarify the precise meaning of certain terms. This process of interaction is considered to be the basis of being human and it is addressed in the hermeneutical literature, especially by Heidegger (1962) and Gadamer (1975). One of the main challenges of the Semantic Web is how to close the gap between the shared human consensus of our everyday lives (the one that the user brings with him/her when he/she uses a search engine) and a language with formally specified semantics, a language that computers can manipulate and use to make inferences.

Main Components of the Semantic Web

We are interested in the information resources that are available on the Internet. For us to have access to these resources we have to name them with identifiers so that: (1) we are able to locate them; and (2) they are uniquely identified. The unique name of each resource on the Web is called the Uniform Resource Identifier (URI). Once we have a name tag for each resource we need to be able to describe what is in each resource. The description should be understandable to different communities and to computers as well. The proposed solution for a language that can be used to describe the contents of web pages is the Resource Description Framework (RDF). The RDF language was created by consensus among the members of the World-Wide Web Consortium (W3C). What RDF does is to provide syntax specification and a schema specification. Thus RDF allows the advertisement of the contents of web resources in a machine-readable (and understandable) form. Once web pages have their RDF-ready descriptions, multiple uses will arrive such as improved searches, more effective and secure web commerce, collaboration, and customizations. In RDF each resource is seen as having properties. Each of these properties has a Property Type and Value. Therefore, using RDF we can make assertions that any information resource with a URI has properties which in their turn have values. RDF uses the eXtensible Markup Language (XML) that lets users create their own tags referring to the content of a document. Although RDF gives users a tool to express simple statements about resources, it is still necessary to create descriptions of the vocabularies used in the statements. What is the meaning of the terms being described? An RDF Schema, usually referred to as RDF-S, provides the resources to describe a hierarchy of classes and the attached properties of the entities. Therefore, RDF-S can be used to describe taxonomies. More complex descriptions may be necessary in the Semantic Web scenario and new languages that can be considered as extensions of RDF-S are being created and will be discussed below.

Now that we have a way to describe the content of Internet resources the next question is how can we share meaning? For information sharing to be efficient and to deliver the kind of data that the users are expecting it is necessary to have an agreement on the meaning of the data. In broader terms, it is necessary to reach an agreement about the meaning of the entities representing the content of the web information resources. These entities are parts of a mental model that represents concepts of the real world. A concept such as body of water carries with it a definition and the mental image that users have of it. But what kinds of agreement can be reached among users? The question of whether it is possible to reach such an agreement among all users regarding the basic entities of the world is a subject under discussion by researchers. We can see this issue under two different perspectives. In the first, there is one Ontology and we can reach a consensus about it through the refinement of the concepts step by step over time. The other perspective does not accept this single Ontology and says that it is necessary to live with possibly incompatible views of reality and try to map concepts from one ontology into another whenever possible. A solution that can reach out to both perspectives supports the notion that small agreements can be made within small communities. Later, these agreements can be expanded to reach larger communities. When this larger agreement occurs, part of the original meaning is lost, or at least some level of detail is lost. For instance, inside a community of biology scholars, a specific body of water can be a lake that serves as the habitat for a specific species and, therefore, it can have a special concept or name to refer to it. Nonetheless, it is still a body of water, and when a biologist is working at a more general level it is considered as a body of water and not as a lake. At this higher level it is more likely that this real-world entity – body of water – can find a match with the same concept in another community. So the biologist and a member of another community can exchange information about bodies of water. The information will be more general than when the body of water is seen as the habitat of a specific fish species.

Before information sharing happens among different communities it is necessary first to have explicit formalizations of the mental concepts that users have about the real world. Furthermore, these concepts need to be grouped by communities representing the basic agreements that exist within each community. Once these mental models are explicitly formalized, mechanisms must be created for generalizing a specific type of lake into a body of water or for adding sufficient specification to the concept of body of water so that it becomes a specific lake. People perform such operations in their minds all the time. The requirement to formalize them comes from the need to have these operations available as computer implementations. The explicit formalization of the mental models of a certain community is called an ontology. The basic description of the real things in the world, the description of what would be the truth, is called Ontology (with an upper-case O). The result of making explicit the agreement within communities is what the Artificial Intelligence community calls ontology (with a lower-case o). Therefore, there is only one Ontology, but many ontologies.

For agents to be able to carry on a successful negotiation it is necessary that not only the ontologies are formally expressed but also that they are in a computer-readable language. Therefore, the Semantic Web needs ontology languages (Fensel 2002). The most recent recommendation from W3C is OWL, the Web Ontology Language. OWL is a semantic markup language for sharing ontologies on the Web; it is a vocabulary extension of RDF and is derived from the DAML+OIL. DAML, the DARPA Agent Markup Language, was created with the objective of surpassing the limitations of RDF, such as the lack of resources for specifying data types and a consistent expression for enumerations. OIL, the Ontology Inference Layer, was a proposal for a web-based representation and inference layer for ontologies. OIL gives the user a precise semantics for describing term meanings and also, therefore, information resources on the Web. These are the fundamentals of the Semantic Web. But what are the implications for geospatial data on the Web? In the next section, the specifics of spatial information that make the challenges of indexing, retrieving, and using geographic information available on the Web so daunting are discussed.

The Geospatial Semantic Web

In their vision paper about the Semantic Web, Berners-Lee, Hendlers, and Lassila (2001) used concepts of space and time:

> At the doctor's office, Lucy instructed her Semantic Web agent through her handheld Web browser. The agent promptly retrieved information about Mom's prescribed treatment from the doctor's agent, looked up several lists of providers, and checked for the ones in-plan for Mom's insurance within a 20-mile radius of her home and with a rating of excellent or very good on trusted rating services.

Egenhofer (2002) also emphasizes that the future of the Semantic Web includes not only the geographic component but the spatial component. In his perspective, Egenhofer addresses the combination of geospatial data with other kinds of data. Another perspective for the Web user is to look for exclusively geospatial data. This second type of user is looking for images, maps, spatial databases, tables in general. In this case, the geospatial data was previously classified as such. But a third and widely available geospatial information resource on the Web is the geographic descriptions present in personal web pages. Historically, humans in every civilization and all over the world have gathered information about their environment for utilitarian purposes. These can vary from directions of how to get to a favorite restaurant to descriptions of landscapes and reports of bird watching activities. People write about their spatial surroundings and post it on the Web. Therefore, one of the research questions for the Geospatial Semantic Web is to come up with methods to recognize and distinguish the different geospatial information resources.

For example, when you are looking for a place to rent or buy you want your (web) agent to collect:

- Information from your online realtor according to your criteria for location, price, or other conveniences;
- Information from a governmental environmental agency giving you data about potential health hazards in the region;
- Informal reports from people who live in the neighborhood;
- Aerial photos from an Internet map service; and
- Photos from the neighborhood taken by the people that live there.

When your software agent gives you all the data then you can make an informed decision. In the next sections we discuss the nature of the different geospatial web information resources from the point of view of their structure, the queries that can be applied to them, and their nature. The work underway in the Semantic Web is related to the current research effort in interoperability. We can see the Semantic Web as another layer in the interoperability arena. All the results achieved in interoperability research are going to be used in the Semantic Web. What was missing in the previous interoperability research efforts was a common semantic understanding. A growing interest in the development of a common data model led to new lines of research in geographic information integration. One of the largest initiatives following this line of research is the OpenGIS Consortium (McKee and Buehler 1996). This association of software developers, government agencies, and systems integrators aims at defining a set of requirements, standards, and specifications to support geographic information interoperability. The OpenGIS data model deals primarily with representations of geographic information.

The OpenGIS Consortium has also created a language based on XML to store and exchange geographic data. The Geography Markup Language (GML) (see http://www.opengis.org/docs/02-023r4.pdf for additional details), is able to encode OpenGIS geographic features. New approaches are needed to step up to a higher level of abstraction where the more valuable information about the meaning of the data can be handled. Neither a standard data format nor a common data model allows for the transfer of the meaning of information. The more complex issue of what is represented instead of how it is represented needs to be addressed. For instance, the user looking for water in New Mexico can obtain this information from the files of the Environmental Protection Agency or from data stored by the New Mexico Parks and Recreation Department. The important thing here is if and how these two agencies share the idea of what a body of water is. An active agent that uses the concept of body of water can actively look for this information, retrieve it, and make it available for the user. Addressing the issues of how to share meaning in a computational environment is the main objective of the geospatial semantic web.

The structure of geospatial information resources

Geospatial information on the web is available in different forms ranging from highly structured resources such as databases and web services to loosely structured resources such as personal web pages (Table 20.1). Highly structured information resources are closer to what is necessary for the geospatial semantic web to come into being. These resources are organized according to some explicit structure be it either a data model or a taxonomy.

What is necessary in terms of the geospatial semantic web for these information resources to be available is for them to describe their contents in RDF and commit to a known public ontology expressed in a semantic web language such as OWL. Since they are already in a structured format all that is necessary is that the ontologies become widely available as well as translation mechanisms between their structure and a language such as RDF. Structured resources, such as a spatial data portal, have passed through the first step in organizing data. Usually portals have taxonomies leading to data sources. These taxonomies need to be translated into a semantic web compatible format. Services that negotiate between the portal and the information resources are also necessary. Therefore, there is a need for one level of translation here that will transform a structured information resource into a highly structured one. Scientific papers on Geographic Information Science (GISc) are also a structured source of geo-information. Scientific papers usually have a common ontology with title, abstract, keywords, introduction, findings, and conclusion among other things. Journals and other sources of papers can commit a general ontology of papers expressed in OWL and communicate their contents in RDF. Unstructured documents such as textual web pages are a rich resource but also the most difficult one to be converted into a semantic web ready form. How can an agent tell that this is a "geospatial" web page? What is it that makes a web page spatial? The

Table 20.1 Organization of different geospatial web resources

Organization	Resource	Examples
Highly structured	Web services	Finding the closest restaurant to your hotel
Highly structured	Spatial databases	US Census Bureau boundary files
Structured	Portals	US Geological Survey portal for geographic data
Structured	Scientific papers	CiteSeer with a GISc query
Unstructured	Textual web pages	Best restaurants in Pittsburgh

ability to find these kinds of web pages and extract the correct resources and format them to be used by the semantic web is a challenging task. All these tasks of formalization of web content need support from geo-ontologies. Previous work in GISc will help with the creation of ontologies of different geographic kinds to enable the advent of the geospatial semantic web.

Queries

There are two important aspects to the queries created by users of the geospatial semantic web: the form and the presentation of the results. Egenhofer (2002) suggests a canonical form to pose geospatial queries:

```
<geospatial request> ::= <geospatial constraint> [<logical connective> <geospatial request>]
<geospatial constraint> ::= <geospatial term> <geospatial comparator> <geospatial term>
<geospatial comparator> ::= ! based on the geospatial-relation ontology used
<geospatial term> ::= <geospatial class> | <geospatial label>
<geospatial class> ::= ! based on a geospatial feature ontology
<geospatial label> ::= ! based on a geospatial gazetteer
```

The second aspect in a geospatial query is how to represent the results. Depending on the nature of the query the result may be only an address such as an URI or real spatial data. What kind of spatial data and how to represent it is a question that an ontology of spatial data can help to answer. Again ontologies come into the picture. The solution for most semantic web problems involves the user committing to an ontology. This means that a user relies on some previous compilation of some explanation of facts that he/she thinks is truthful. In our query problem, the user will accept as a representation of the answer the representation of geographic data as stated in a specific ontology.

Nature of geospatial information resources

The Geospatial Semantic Web is intended to handle information that is meaningful both for humans and computers. Therefore, the way potential users understand and use information is very important. In particular, we see three basic dimensions for users of geographic information in the Geospatial Semantic Web context:

1 **Professional** – highly structured geographic information stored in geographic databases which are indexed, stored, or described on the Web;
2 **Naive** – the retrieval of unstructured, subjacent, informal geographic information in web pages;
3 **Scientific** – geographic information science papers, models, and theories available on the Web.

Professional

In order to improve the results of queries looking for information stored in geographic databases it is necessary to support better definition for spatial concepts and terms used across different disciplines and the development of multiple spatial and terminological ontologies. Here we also have web services – the most common examples are called location-based services – that are able to locate restaurants, hotels, and other facilities depending on the user's location through the use of spatial databases (see Chapter 32 by Brimicombe later in this volume, for more a detailed discussion of the linkages between location-based services and GIS databases). The OpenGIS consortium has a standard for web services. This shows again the importance of the integration between OpenGIS and the W3C. The geospatial semantic web will need to build a semantic layer over the standards that the OpenGIS is developing or have already deployed.

Naive

In this case we are looking for geographic information in the text of web pages. It can be pages with descriptions of geographic features such as cities, geographic phenomena, or facilities. The user may be looking for complementary information or the Web may even be the main source of data. A case in which we can mix the content of textual web pages with more structured data such as the geographic distribution of machines on the Web is the ability to answer queries such as "I found this interesting web page, where is its geographic location?" or "Find other websites that contain information about places close to places mentioned in this website" or "List (or even graphically display) all the location information on the IBM web site (offices, research centers, etc.)". The user may also want to find pages that describe a special region of San Francisco and uses as a constraint that the web server in which the page is stored is located in San Francisco because he or she thinks that locals will give better naive information.

Scientific

Here the situation is similar to what Citeseer (Giles, Bollacker, and Lawrence 1998) does today for computer science. It provides a specialized search engine for scientific papers in a specific domain. Besides being able to index GISc documents it is also necessary to create ontologies that express the different theories expressed in the papers.

Challenges of the Geospatial Semantic Web

There are many the challenges we face to make these types of queries feasible. Most of them are related to ontologies because ontologies are the main structure in the semantic web to represent semantics.

Creation and management of geo-ontologies

Activities involved in ontology management include designing, developing, storing, registering, discovering, visualizing, maintaining, and querying ontologies. One aspect that makes ontology management particularly challenging is that ontology is based on agreements among domain experts that can be geographically distributed. Ultimately, the endurance of an ontology is based on users' acceptance. GIS communities can support an initiative in ontology management that can include developing or adapting effective methodologies and tools for ontology management, and applying them to develop domain specific ontologies with broad community acceptance.

Matching geographic concepts in web pages to geo-ontologies

It is necessary to apply a geospatial characteristic to the interpretation of texts (hermeneutics). It is also necessary to research in innovative methods of building ontologies from maps, images, and sketches available on the web.

Ontology integration

In order to provide better results for queries it is necessary to integrate different ontologies not only in the geographic dimension (scientific, professional, naive) but also in the non-geographic domain. Future research needs to address the necessity of developing and testing the theory of the integration of multi-disciplinary ontologies by: (1) performing an empirical study of how different communities categorize the relationship between the different geographic entities; (2) creating relevant geo-ontologies; and (3) designing, prototyping, and assessing computational models to specify, represent, access, and share multiple ontologies of geographic information.

Trust

It is necessary to implement mechanisms that enable users to assign different levels of trustfulness to different information resources. The necessary level of confidence in the information is different if the user is looking for information about the restaurants around his/her hotel or if the user is downloading geospatial information to decide if he/she is going to open a million-dollar business at this location. Onsrud (1999) provides a fine review of the potential liability problems for producers and consumers of spatial information. He later reminds us of even further implications because of the international aspect of the Web (Onsrud 2004). Different laws and perspectives come into play in a global scenario such as the Internet (see Chapter 29 by Cho in this volume for an in-depth discussion of the various ways in which GIS and personal privacy issues are dealt with in different legal jurisdictions).

CONCLUSIONS

The Geospatial Semantic Web is about bringing more semantics to the web. It is a way to make computers work closer to the way people do. The Web is here to stay and more and more content will be available online. There is no use for this massive amount of information if we cannot have at least reasonably easy access to it. Furthermore, spatial information is present everywhere. Some people estimate that 80 percent of government and business information is spatial in some way (Eichelberger 1993, McKee and Buechler 1996). This proportion should be valid for data on the Web as well and means that there is a vast amount of spatial data on the Web with the potential to advance our understanding of the world around us. The semantic web vision includes the creation of descriptions of the data sources on the Web. These descriptions should be expressed both in machine-readable format and in natural language. We need to create ontologies (expressing our world views) and match them to data on the Web. This is the great geospatial semantic web challenge. The second challenge is the building of what Egenhofer (2002) calls the Semantic Spatial Web. Then we will be able to have implemented in computers something similar to the human use of metaphors of space and time.

REFERENCES

Berners-Lee, T., Hendler, J., and Lassila, O. 2001. The Semantic Web: A new form of Web content that is meaningful to computers will unleash a revolution of new possibilities. *Scientific American* 284(5): 34–43.

Egenhofer, M. J. 2002. Toward the Semantic Geospatial Web. In K. Makki and N. Pissinou (eds) *The Tenth ACM International Symposium on Advances in Geographic Information Systems.* New York: ACM Press, pp. 1–4.

Eichelberger, P. 1993. The importance of addresses: The locus of GIS. In *Proceedings of the Annual Conference of the Urban and Regional Information Systems Association (URISA '93)*, Atlanta, GA, USA. Park Ridge, IL: Urban and Regional Information Systems Association, pp. 212–22.

Fensel, D. 2002. Language standardization for the Semantic Web: The long way from OIL to OWL. In J. Plaice, P. G. Kropf, P. Schulthess, and J. Slonim (eds) *Distributed Communities on the Web: Proceedings of the Fourth International Workshop (DCW 2002)*. Berlin: Springer Lecture Notes in Computer Science No. 2468: 215–27.

Gadamer, H.-G. 1975. *Truth and Method.* New York: Seabury Press.

Giles, C. L., Bollacker, K. D., and Lawrence, S. 1998. CiteSeer: An automatic citation indexing system. In *Proceedings of the Third ACM International Conference on Digital Libraries*, Pittsburgh, PA, USA, pp. 89–98.

Heidegger, M. 1962. *Being and Time.* New York: Harper.

McKee, L. and Buehler, K. (eds). 1996. *The Open GIS Guide: An Introduction to Interoperable Geoprocessing and the OpenGIS Specification.* Wayland, MA: Open GIS Consortium, Inc.

Onsrud, H. 1999. Liability in the use of GIS and geographical data sets. In P. A. Longley, M. F. Goodchild, D. J. Maguire, and D. W. Rhind (eds) *Geographical Information Systems: Management Issues and Applications.* New York: John Wiley and Sons, pp. 643–52.

Onsrud, H. 2004. Geographic information legal issues. In C. Medeiros (ed.) *The Encyclopedia of Life Support Systems (EOLSS).* Developed under the auspices of the UNESCO, EOLSS Publishers, Oxford, UK, [http://www.eolss.net].

Uschold, M. 2003. Where are the semantics in the semantic web? *AI Magazine* 24(3): 25–36.

W3C. 2001. The Semantic Web Activity Statement. WWW document, http://www.w3.org/2001/sw/.

Part V Spatial Analysis

The next group of four chapters examines selected topics in spatial analysis. Chapter 21 by Martin E. Charlton, the first of these chapters, offers a straightforward but insightful account of the links between quantitative analysis and Geographic Information Systems (GIS). The various strands indicate the ways in which GIS and statistical packages complement one another and the circumstances in which GIS analysts might consider using the latter and when statisticians might make use of GIS. From this vantage point, it provides an important backdrop to the three subsequent chapters dealing with specific techniques for analyzing spatial data.

In the next chapter in this set (Chapter 22) by Geoffrey M. Jacquez examines spatial cluster analysis. The first part of this chapter provides a working definition of a cluster, the role of cluster analysis in exploratory spatial data analysis, and the five components that contribute to statistical pattern recognition. Global, local, and focused clustering methods, cluster morphology descriptors, approaches for quantifying cluster change and persistence, and the value of multiple testing are discussed next. These details are used by Jacquez to demonstrate why the "one size fits all" approach to cluster analysis is likely to yield an incomplete picture of cluster morphology and why integrated approaches are likely to yield substantial benefits in documenting cluster change and persistence and for identifying disparities in two or more geographically referenced variables. The chapter concludes with an overview of software resources for conducting cluster analysis.

Yongxin Deng, John P. Wilson, and John C. Gallant review several key characteristics of terrain analysis and the importance of fuzzy logic, equifinality, shape-based data quality evaluations, and multi-scale terrain analysis in Chapter 23. The chapter begins by describing the most common terrain attributes as scale- and algorithm-dependent descriptions of the terrain surface and related biophysical processes. The contributions of terrain analysis to soil erosion/deposition modeling, soil mapping/ landform delineation, and TOPMODEL model applications are discussed next and these examples are used to illustrate several enduring challenges in terrain analysis. The final part of the chapter examines the effects of spatial scale and data quality on computed terrain attributes and identifies some of the key questions that will need to be answered to advance terrain analysis applications in the years ahead.

In the final chapter in this set on spatial analysis (Chapter 24), Jochen Albrecht recasts some recent work usually subsumed under the geocomputation banner by focusing on the temporal dimension and uncovering the low-level structural problems that make it so difficult to merge the spatial and temporal aspects of GIS data models. The chapter starts out with a review of current approaches to dynamic GIS that covers several important innovations, such as the development of fully-fledged dynamic modeling languages, cellular automata, and agent-based modeling, before moving to brief descriptions of examples related to air traffic control and regional storm prediction. The chapter concludes with the observation that further progress in building and implementing dynamic GIS will depend on the development and implementation of a language of "change."

Chapter 21

Quantative Methods and Geographic Information Systems

Martin E. Charlton

This chapter is concerned with the relationship between quantitative methods and Geographic Information Systems (GIS). It does not explore the use of particular quantitative methods – some of the following chapters deal in greater detail with particular techniques which may be applied to spatial data. Rather, it considers when and how we might use GIS and appropriate quantitative techniques together to deal with some larger problem. In practice this relationship is often based on moving information between applications programs on a computer. Anyone who has had to do this will understand that this is a sometimes rocky path. It is worth, therefore, spending some time considering what is involved and when. Readers interested in a rather more detailed treatment of quantitative methods viewed from a geographical perspective might usefully refer to Fotheringham, Brunsdon, and Charlton (2000) – this chapter draws on material from chapters 2 and 3 of that text.

The reader of this book might well feel that by the time this chapter has been reached he or she has an understanding of the nature and purpose of GIS. In practice an analyst is usually faced with a computer program – among the more popular ones we would include Arc/Info, ArcMap, ArcView, and MapInfo: this list is by no means exhaustive. The analyst needs to understand the nature of data which is appropriate for storage and manipulation in a GIS, whether it be vector with a locational component and some associated attributes or raster in which the phenomenon of interest is spatially discretized in a regular fashion. It is helpful to have some understanding of coordinate systems and map projections, and it is useful to know what sort of operations can be undertaken in the program that is being used. In short it is also useful to be conversant with the manner in which data may be stored, interrogated, manipulated, mixed, and displayed in the particular program. A good understanding of the principles which underlie GIS will help greatly in dealing with its practice.

If the data of interest have a spatial component, what quantitative methods are appropriate for such data? A related question is: if the data have a spatial component do we have to use that component explicitly in its analysis? A glance at the 59 or so titles in the Concepts and Techniques in Modern Geography series which were published under the auspices of the Quantitative Methods Study Group of the Institute of British Geographers, mainly during the 1970s and 1980s, would suggest that the spatial component was not particularly important. While introductory pedagogic guides to cartograms (Dorling 1996), Voronoi polygons (Boots 1986), spatial autocorrelation (Goodchild 1986), the modifiable areal unit problem (Openshaw 1984), centrographic measures (Kellerman 1981), and directional statistics (Gaile and Burt 1980) suggested that we should take notice of some of the properties of spatial data in our analyses, other author's volumes on logit models, linear regression, factor analysis, and contingency table analysis implied that we need not. However, while many of the techniques in the former group do not exist within some of the mainstream statistical applications programs, all of the latter group do and are frequently and widely applied to data which relates to locations on the surface of the Earth.

One of the notable developments since the 1980s has been a set of techniques which are applicable to spatial data – the analyst might turn to Cressie (1993) for help on geostatistical methods, and a wide range of models for lattice and point data. None of the techniques that he describes can be found in mainstream statistical software such as SPSS. However, we find in volumes such as Davis (2002) treatment of a wide range of techniques which might be thought of as aspatial (t-test, analysis of variance, correlation, linear regression, spectral analysis, cluster analysis, factor analysis, correspondence analysis, and canonical correlation), as well as the analysis of point, line and spherical data, shape analysis, contouring, trend surfaces, and elementary geostatistics. Again the aspatial techniques are in SPSS and the spatial ones are not.

Where does GIS fit into this picture? Assuming that the appropriate quantitative method has been chosen, and suitable software is available, the analyst has an interest in four areas:

1   Creating the data for the analysis;
2   The exploration of the data prior to the analysis;
3   The statistical analysis of the data; and
4   The interpretation of the results.

Each of these stages may involve several sub-stages and several pieces of software. There may also be feedback loops where an examination of the results of the analysis may lead to the realization that an important item of data is missing, in which case a return to an earlier step may be required. In this sense we may view GIS as a tool which assists in the analysis.

Despite the best efforts of GIS software developers, no single GIS applications program will provide all that an analyst is likely to need. Indeed GIS software which incorporated all of the techniques in Cressie (1993) and Davis (2002) would be unwieldy in use. This does not stop manufacturers releasing add-ons for their software to carry out geostatistical operations, for example. Sensibly many software developers provide a macro language to permit analysts to incorporate their own techniques into software; good examples are the Arc Macro Language used in Arc/Info and Avenue which was used with ArcView. A search of ESRI's own website and elsewhere on the Internet will reveal a rich variety of software written in these macro languages which extends the capabilities of the base product to deal with various problems which are of interest. It would be pointless to catalog such software here, as the body of user-produced software increases daily.

It is not always necessary to resort to coding one's own software. Let us consider a simple example as a prelude to thinking about the issues involved. An analyst is interested in the relationship between the price of housing and the characteristics of that housing. There is a rich body of theory and practice in economics which deals with hedonic models which are appropriate for such an analysis. Hedonic models require individual data on each house. Areas in cities are sometimes characterized as being "cheap" or "expensive" in terms of housing cost, perhaps for their proximity to an industrial neighborhood or a well kept park. Location may well be important. A real-estate or mortgage company may be able to provide anonymous details of housing data on an individual level, although they are unlikely to be able to provide any spatial reference for it beyond a zip code or postcode. As well as the housing data, the analyst may be able to obtain data about the social structure of the immediate neighborhood of each house. If proximity to some facility is thought to be important, then the analyst may need to measure the distance to that facility, or perhaps its direction. A mail company may be able to provide an index which gives spatial references for the zipcodes or postcodes.

Such a problem is not atypical. Long before any hedonic modeling can take place, the analyst has to combine these data sources to extract the appropriate data. Any data exploration may well include mapping the data (are all the detached properties in the same location in the city?). While an initial analysis may involve nothing more and a linear regression model fitted using ordinary least squares, the interpretation of the results would be partial if the analyst did not consider the possibility of the existence of spatial patterns in the results, so another map will be necessary. What issues are there in dealing with such a problem?

An important initial question is that of how many houses are involved in the study. One might be tempted to use Excel to deal with some of the initial data integration. There is currently a limit of 65,536 rows in Excel, so if the data set is larger, then an alternative has to be found. To add spatial information to the housing data, the zipcodes or postcodes will need to be merged with the index. What is the granularity of the spatial references in the index? In the UK postcodes have a variety of formats which may cause problems in matching the data sets. Indeed the analyst might use the file matching facilities available in SPSS to link the coordinate information from the index to the housing records.

Having gone to the extent of creating an SPSS data file, can this be transferred easily to another package? Linking in the census data will probably require the GIS software. If the housing records are stored with point references, how are they to be linked to the census data – the latter may have a point reference or centroid, or may be represented as an area. If the census data have point references then association of the census and housing data might be the result of an operation to find the nearest census centroid to each house, and assign its attributes to the house record. If the census data are area data, then a geometric operation is required that will identify in which census area each house lies – once this is done, the census area attributes can be transferred to the house data. Other proximity or containment computations are best left to the GIS software. At this point, some exploration of the spatial patterns in the data can take place – what are appropriate symbols, what classification is to be used with the continuous data (are any patterns in the prices a reflection of some spatial phenomenon or merely an artifact of the choices the software has made for its symbols)?

It is unlikely that the hedonic modeling would be carried out in the GIS software. A transfer to another package is required. In the case of ArcView the dBase files in which it stores attribute data can be relatively easily transferred to SPSS, although some attention may be required to the characteristics of each attribute, such that, say, the number of decimal places is preserved. Depending on which statistical software is being used – the choice is large – this transfer may be more or less reliable.

The nature of the modeling will be driven by the analyst's expertise and experience. Perhaps stepwise methods might be employed to remove variables which do not appear to influence the variation in the price. It would be wise to calculate some residuals for each of the models that are used. Among the assumptions of OLS regression are that the observations are independent of one another and that the residuals are random and heteroskedastic. A map of the residuals may help to determine whether they have these properties. Spatial pattern in the residuals suggests that something has been omitted from the model, or perhaps that OLS regression is not the most appropriate technique. Either way, the residuals will require transfer from the statistical software to the GIS program. This may be a simple as cutting and pasting from one desktop window to another, or it may require the creation and linking of tables. In the latter case, it helps if the data in the statistical program and the GIS bear some corresponding identifier (even if this is just a sequence number).

Coupling

It is worth noting here that the GIS software has been used in three of the four steps outlined above and the statistical program in two. There is another issue which is sometimes referred to as "coupling." Outlined above is an example of "loose coupling." Two software programs have been used for tasks which appear appropriate to their strengths, and the data has been transferred between them in one or more files. This is at once flexible yet error prone, and does demand some facility with at least two software programs. The converse is "tight coupling" in which the analytical method is incorporated in the GIS, either by initial design, or by using an extension or add-in (such as ESRI's "Spatial Analyst" which permits simple geostatistical analysis). The advantage is that the problems of data transfer in the loose-coupled approach do not occur, but the analyst is now restricted to whatever is programmed into the extension.

A third approach is to consider adding simple functions to other statistical environments, such as R (Ihaka and Gentleman 1996). One such extension allows R to read data from ArcView shapefiles and then use these for modeling. The analyst benefits from the ease of data transfer, and the full and rich analytical functionality of R that is available. However, transfer to a GIS for mapping may be less easy, although cut/paste operations are available. While GIS does provide some facilities for interactive exploration of data, a highly interactive environment of the sort provided by XLispStat (Tierney 1990), S-Plus (Venables and Ripley 2002), or R (Ihaka and Gentleman 1996) might provide greater flexibility. It is perhaps worthwhile noting that if the automatic class-interval selection provided with many GIS display software is relied on, a number of alternative classifications should be explored to ensure that the pattern seen on the screen is not an artifact of the classification algorithm.

Anselin and Bao (1997) have provided a SpaceStat extension for ArcView 3.x. SpaceStat itself is a program for undertaking various spatial analyses (Anselin 1988, 1992). As a stand-alone program it has facilities for the input of spatial data and a variety of analyses including creation and manipulation of spatial weights, descriptive spatial statistics and spatial regression. The ArcView extension allows the user to transfer locational and associated attribute data from a shapefile to SpaceStat, and to transfer results of the analyses back into ArcView for display via a menu which is added to the ArcView menubar.

Aldstat and Getis (2002) have programmed Visual Basic (VBA) macros for use with ArcGIS to carry out a range of point pattern analyses. Visual Basic is embedded within ArcGIS as a development tool to allow extensions to be coded to carry out analyses not included within ArcGIS. VBA has the disadvantage that it is an interpreted rather than compiled language, so the applications will run somewhat slower than a standalone program. However, dynamic link libraries may be written for "core" spatial analysis tasks and VBA used to link them to ArcGIS in a rather more tightly coupled fashion.

Unix users may use Xgobi (Cook, Majure, Symanzik, and Cressie 1996, Symanzik, Majure, and Cressie 1997) which links with ArcView to provide a range of tools for visually exploring multivariate data. Haining (1998) has been responsible for the generation of a set of procedures for use with Arc/Info which considerably extend the capabilities of that software to include relative risk models for disease data and spatial regression models. MacEachren, Brewer, and Pickle (1998) detail an extension to ArcView to provide a set of additional tools for the exploration of spatial data including linked brushing and outlier detection. ESRI's replacement of ArcView 3.x with the integrated ArcGIS system where the scripting language is Visual Basic appears to have created a market for tools to convert existing Avenue scripts into Visual Basic (an Internet search reveals a number of organizations offering such services, among them: Compass Informatics in Dublin, Ireland; CEDRA Corporation in New York; and Data East LLC in Novosibirsk, Russia).

Data Integration and Management

In many studies involving spatial data analysis, the role of GIS is that of providing an integrating environment. There are a number of related issues that require a little exploration so that the result of the data manipulation and integration is reliable. Goodchild (2000) attributes the observation that "GIS technology lets us produce rubbish faster, more cheaply, and in greater volume than ever before" to David Rhind. It is probably worth bearing in mind these words each time you undertake a GIS operation on your spatial data. To give an example, a recent study in the UK used the number of motorway interchanges per head of population in an area as one measure of accessibility to that area. A point shapefile was created for the motorway interchanges in the UK, this was then intersected with a polygon shapefile of the areas of interest, and a count produced of the number of motorway interchanges in each area. Such an operation is a relatively common GIS task. In assessing the results of the study a problem came to light which had potentially serious implications for the modeling that had taken place. In several areas, one of the boundaries ran along a section of motorway which happened to include an interchange. What should be done about such interchanges – are they to be allocated to one area and one only, or are they to be allocated to the two neighboring areas (or perhaps half an interchange assigned to each area)?

The software used assumed that the point which represented the interchange could be assigned to one area only. Blakemore (1984) identified the status of a point with respect to a polygon as being: (1) wholly inside; (2) probably inside; (3) on the boundary; (4) probably outside; or (5) wholly outside. In the situation of the interchanges on the motorways, cases (2), (3), or (4) presented themselves. In cases (2) and (4), depending on where the point representing the interchange had been chosen, the interchange would be assigned to one polygon and not its neighbor on the other side of the motorway. In case (3) the software made a choice that may have been random or simply the result of the manner in which the point-in-polygon algorithm underlying the assignment had been coded by its programmers. Prudent analysts would have at least mapped the assignments following the intersection operation, rather than making the discovery after a great deal of modeling effort.

Underlying this are issues of accuracy and appropriateness. Doing the wrong thing with GIS is no more beneficial than doing the wrong thing without GIS, and GIS will not rescue bad initial decisions.

Location

The data used in GIS are given some form of spatial reference. However, not all the data might have compatible spatial references. A commonly asked first question is whether the data being gathered together for the study have the same coordinate system, and whether they have the same level of resolution. Many of the algorithms in GIS software assume that the data being manipulated is in some Cartesian coordinate system, where the coordinates of each location area are obtained with reference to an origin and a pair of axes at right angles to each other (usually referred to as the x- and y-axis). The coordinates are usually referred to as the "x-coordinate" and the "y-coordinate," or "eastings" and "northings." For instance, in Great Britain, the national mapping agency, the Ordnance Survey, has its own projection based on a Transverse Mercator projection – published maps and digital data are provided with coordinates measured in meters relative to the origin for the projection which is to the south west of the Scilly Isles. Other data-providing agencies use this projection in supplying mappable data, notably the Office of National Statistics which publishes data from Britain's decennial censuses. The national mapping agencies of most countries have their own projections. The Universal Transverse Mercator projection divides the globe into a set of segments running north–south which occupy six degrees of arc at the equator. The availability of a consistent set of coordinates removes one set of problems.

How do we deal with the case where data to be integrated is in a series of different projections? For example, data in UTM coordinates might be being linked with data taken from a hand-held Global Positioning System (GPS) receiver. The measurements of location used in the GPS are in latitude and longitude based on a model of the Earth or datum known as WGS84. It is likely that the GPS receiver will have built into its firmware a series of commonly used projections such that the position display on the screen is in the coordinate system of the locally used projection. The formulae used for these on-the-fly conversions are generally approximate, and the user is best advised to download the stored locations (waypoints) in latitude and longitude and use the projection conversion facilities of the GIS program at hand to change from geographic to projected coordinates. This may require knowledge of the projection name, datum, and spheroid, although for commonly used projections, this is provided in the software. Using an incorrect datum will usually result in coordinates being shifted some distance from their "true" position. An accessible text which deals with some of the detail of projections and datum in GIS is Iliffe (2000). The goal of creating the spatial database for manipulation is to have all the layers with a consistent spatial reference.

A second consideration is that of the spatial resolution of the coordinates. It may appear to be obvious that the coordinates in each layer should have the same level of resolution (for example that all should be in 1 m coordinates) but it is easy to overlook this if the layers are from a variety of sources. For example, the Ordnance Survey grid reference to 1 m of a point somewhere in the Geography Department at Newcastle University is 424787, 565115. The 100 m resolution reference would be 4247, 5651. Grid references from earlier versions of the Central Postcode Directory in the UK are to 100 m resolution. Many GIS programs have a facility which allows the transformation from one Cartesian coordinate system to another. This allows data from digitizing tablets to be converted into the coordinate system being employed in the spatial database and apparently makes the problem of resolution one that is easily solved – the coordinate values need to be multiplied by 100. Unfortunately this introduces a consistent south-west bias in the grid references. This may have repercussions in data integration – for example, points may be assigned to the wrong polygon in an intersection operation. One solution might be to *jitter* the locations after the transformation by the addition of a random number between zero and 99 to each of the coordinates. While this may be reasonable for point data (what is "reasonable" depends on the application) it may cause problems in preserving topological relationships in polygon data.

Allied to the problem of the resolution of the coordinate data is a programming problem. In the days when computer storage was expensive, the choice of data precision for the storage of layers was important. The Arc/Info system, for example, allows users to choose whether the locational components of a layer are stored in single or double precision. What does this mean and what are the implications? A floating point number in a computer may be stored in a 32- or a 64-bit location, single or double precision respectively. Seven bits of each location are reserved for the abscissa, and one bit for the sign. That leaves 24 bits for the mantissa in single precision and 56 bits in double precision. The advantage of single precision is that the locational data only requires half the storage that would be required if the coordinates were stored in double precision. The disadvantage is that typically seven digits of precision are available in single precision compared with 14 for double precision. With coordinate values that are in the hundred thousands – which includes most of those at 1 km resolution in UK data – roundoff errors become a problem in GIS operations carried out in single precision. I did not worry too much about this until I was examining data from repeated differential GPS surveys of a river in northern England – estimates were needed of the movement of gravel in a particular reach of the river over a period of time. The coordinates were to two places of decimals – effectively eight significant figures. Quite different answers were obtained when the computations were carried out in double precision rather than single precision. With the relative cheapness of storage, this is less of a problem. The disadvantage of being cavalier with storage is the cost of transfer across a network – however, technological improvements in networks will bring these costs down in future years.

Spatial Data and Quantitative Methods

Spatial data does pose some interesting challenges to the analyst. Since the early 1970s geographers and others working with spatial data have realized that some care needs to be taken with its analysis. While recent examples in the geographic literature (Bailey and Gatrell 1995, Fotheringham and Rogerson 1994, Fischer and Getis 1997, Longley and Batty 1996) suggest that geographers are beginning to concern themselves with the problem of analysis with spatial data, there have been developments in other disciplines notably statistics and computer science (Cressie 1993).

There has been a diffusion of interest in GIS outside geography and this has led to some interesting developments. Openshaw's Geographical Analysis Machine (Openshaw, Charlton, Wymer, and Craft 1987) was the result of inquiry on the part of a group of pediatric oncologists, which engendered further work by others, notably Besag and Newell (1991) and Fotheringham and Zhan (1996), as well as a range of scan statistics for dealing with similar point-based problems, for example Kulldorff's spatial scan statistics (Kulldorff 1997).

However, there are a number of problems posed by spatial data which should be borne in mind. These include, but are not limited to: spatial autocorrelation, spatial outliers, edge effects, spatial dependency, and the modifiable areal unit problem. While geographers in the 1960s were quite happy to punch up the FORTRAN codes from texts such as Cooley and Lohnes (1962) and run the programs on their data ignoring any geographical solecisms, the benefit of hindsight brings circumspection.

One of the underlying assumptions of classical inference is that the objects forming the basis of the analysis are independent of each other. However, Waldo Tobler observed in 1970 that "everything is related to everything else, but near things are more related than distant things" (Tobler 1970); for a recent discussion on this see Phillips (2004) Smith (2004), Goodchild (2004), and Tobler (2004). Stephan (1934) made the observation 70 years ago that

Data of geographic units are tied together, like bunches of grapes, not separate, like balls in an urn. Of course, mere contiguity in time and space does not itself indicate independence between units in a relevant variable or attribute, but in dealing with social data, we know that by the very virtue of their social character, persons, groups, and their characteristics are interrelated and not independent. Sampling error formulas may yet be developed which are applicable to these data, but until then, the older formulas must be used with great caution. Likewise, other statistical measures must be carefully scrutinized when applied to these data.

Cressie (1993, pp. 14–15) presents an example of the influence of autocorrelation on the estimate of the confidence interval for the population mean. If there is positive autocorrelation, and the usual formula is used, the confidence interval will be too narrow. If the data are negatively autocorrelated and we ignore it, the confidence interval will be too wide. This does have implications if we are comparing means for equality. If the data for two samples are positively spatially autocorrelated and we choose to ignore this, then we may decide that the sample means are not sufficiently different for us to be confident that there is an observable difference in the population means. Thus, if autocorrelation is ignored, then we run the risk of making the wrong decision. The subject of autocorrelation and its associated problems has interested geographers since the early 1970s (for example, Cliff and Ord 1973, 1981; Griffith 1987).

Aggregation

Aggregation of data from either individual observations to a set of zones, or from one system of areas to another is a further source of potential problems. A common operation in GIS is to take some data which are the attributes of a point coverage and count or aggregate these in some fashion to a set of larger areal units. There might be a number of very good reasons for doing this. For example, the areal units being used might be those of some administrative areas which form the basis of the study, or the areal units might be those at which a large corpus of data produced by a third party is available. The GIS operations of intersection or union, followed by some summation and perhaps creation of ratios is typical. As an example, a study of illness among a particular cohort of children might have the location of the children given as a point reference. If data about the "at-risk" population are only known at an aggregated scale, one option is to aggregate the individual records for the children to the boundaries of the areas. In Openshaw, Charlton, Wymer, and Craft's (1987) Geographical Analysis Machine GAM, the aggregation was to circular regions, combining data from two layers – point references of the location of the residences of children with acute lymphoblastic leukaemia and point references of the centroids of enumeration districts (small areas used in data collection in the UK Census of Population).

Gehlke and Biehl (1934) observed that if data for census tracts in Cleveland were aggregated to larger units, the correlation coefficient between male juvenile delinquency and median rental income increased as the size of the areal units increased. They observed a similar phenomenon on the correlation between numbers of farmers with the value of farm products taking data from 1,000 rural counties in 1910. Robinson (1950) showed that correlations based on data aggregated to some spatial units were different from those which were based on data about the individuals in those units. Thomas and Anderson (1965) came to the conclusion that the findings of Gehlke and Biehl could not be substantiated. However, Openshaw and Taylor (1979) revisited the problem in bivariate analysis, and Fotheringham and others have investigated the problem in multivariate analysis and spatial modeling (Fotheringham and Wong 1991, Fotheringham, Densham, and Curtis 1995). The problem has two components, one due to scale (results may be conditioned on unit size) and the other due to zoning (different results can be obtained with different regroupings of zones). If suitable individual level data are available then the methodology proposed by Holt, Steel Tranmer, and Wrigley (1996) and Tranmer and Steel (1998) shows some promise. King (1997) uses an approach based on local forms of regression to deal with the problem.

This does have implications for the use of GIS in data manipulation, particularly in the choice of the areal units used. One GIS operation, often referred to as cross-area aggregation, involves a set of procedures used to estimate data available for one set of areal units to another set of areal units. Typically, the areal units have entirely different geographies, without the nesting which characterizes administrative spatial hierarchies. In the UK, areas used for the delivery of local government services (districts and unitary authorities) are different in number and size and are not conterminous with those used for the organization of postal deliveries (postal districts). If data are available for a phenomenon at the postal district level (perhaps readership of a particular newspaper) and data on some other phenomenon, such as the social characteristics of the population, are available for administrative districts, then it is tempting to consider how the relationship between the two might be modeled. However, to convert the census counts from the administrative geography to the postal geography requires cross-area aggregation. This is effectively an interpolation procedure.

There are several ways of achieving this. A union operation between the two sets of areal units will produce another set of areal units in which each new area has the attributes of its parent areas. Aggregation may take place using weighting by area. If we have two sets of areas, the first numbered 1, 2, 3, . . . and the second identified by letter, A, B, C, . . . then the result of the intersection might produce three new areas that together comprise area A, but are composed of 50 percent of area 1, 30 percent of area 2, and 20 percent of area 3. If the variable in question is, say, p population, then the population of area A is the sum of 50 percent of the population of area 1, 30 percent of the population of area 2, and 20 percent of the population of area 3. Other attributes would then be created using a similar weighting. The underlying assumption is that the population is distributed uniformly around each area, which may or may not be the case. If some third variable is available whose distribution is known in both the source and target zones (say the locations of dwellings), then weighting may take this into account.

Greater levels of sophistication are available for cross-area aggregation using Poisson regression for count data (Flowerdew 1988) and the Dempster, Laird, and Rubin (1977) EM algorithm (Flowerdew and Green 1991). Kehris (1989) has created a suitable GLIM macro together with code to interchange data with Arc/Info as an implementation of this method. Tobler (1991) has suggested pycnophylactic inter-polation as another means of interpolating data from one set of areal units to another. Dummer, Dickinson, Charlton, and Parker (2000) use conditional probabilities in the estimation of the population of young people by ethnic group for one region in the UK based on aggregated data – one advantage of this method is that it pro-duces confidence intervals on the resulting counts. Although the authors suggest it could be applied to other regions in the UK, its data requirements might preclude its adoption in other parts of the world. However, in most cases, aggregation will be based on whatever has been provided by the software manufacturer and this usually means areal weighting or weighting by another variable.

The Post-Analysis Role of GIS

GIS has also a role after any analysis in the exploration and interpretation of the results. The extraction of the mean of a distribution or the median might lead to mapping those locations with values above or below the statistic – is there a spatial pattern, and is it regional or local? With techniques such as regression, among the outputs are a set of predicted values and the residuals (the difference between the observed and predicted values). Here, GIS has a more obvious role. The analyst should certainly plot the spatial distribution of the residuals. Useful diagnostics are the standardized or studentized residuals as these values have a mean of zero and a standard deviation of unity. This allows identification of pattern in the residuals – the class intervals for the symbolism might be in units of 1 standard deviation either side of the mean. It also allows the identification of unusually high or low values – the class intervals might be chosen to identify those in the tails of the dis-tribution of values (that is, ± 1.96 and ± 2.58 standard deviations from the mean). Again, the questions which suggest themselves include the identification of patterns in the residuals. If a linear regression has been fitted using the OLS procedure in SPSS, for example, and pattern is observed in the plot of the residuals, this would suggest either that one or more variables have been omitted, or that OLS regres-sion is inappropriate and a method needs to be used that will deal with the spatial dependency in the data. Models to think about here include drift analysis (Casetti and Can 1999), the expansion method (Casetti 1972, 1992), or geographically weighted regression (Fotheringham, Brunsdon, and Charlton 2002).

Other multivariate techniques also have mappable outputs. For example, one out-put from a discriminant analysis is the prediction that an observation will have mem-bership of a particular group. Mis-classification occurs when the predicted group membership differs from the actual group membership. A plot of the misclassified and correctly classified locations may suggest some further steps in the analysis. As well as a predicted group membership, another output is the probability of member-ship of each group – plotting either the probability of membership of the observed or the predicted class again may suggest further steps to take in the analysis.

Principal components analysis is a frequently used data reduction technique in which an attempt is made to create a set of new variables ("principal components") a few of which will account for most of variation in the data set. One output is a set of values of the new variables ("component scores") which again may be mapped to give some insights into the problem being addressed – where are the outliers, for example, and why might those locations have such unusual values?

Cluster analysis is a set of classification techniques which attempt to partition a set of objects into some (generally) distinct groups. There are some techniques which attempt to find a hierarchy in the data (agglomerative techniques such as Ward's method, or divisive techniques such as monothetic and polythetic division) and some which are non-hierarchical, such as k-means or partitioning around medoids. If a cluster analysis is being undertaken on spatial data, then a plot of the locations of the predicted members of each cluster may be a helpful tool in the interpretation of the cluster results. K-means provides a measure of the distance from the location of an object in k-space to its cluster centroid; again, a plot of the values will show whether the objects which are "distant" from a cluster center exhibit any spatial pattern. Other cluster methods, for example those based on density methods, also provide a value which can be mapped.

Geographically weighted regression (GWR) (Fotheringham, Brunsdon, and Charlton 2002) is an example of a technique where the outputs are intended for mapping. GWR attempts to deal with spatial dependency in the data set being modeled by incorporating this dependency into a weighting scheme in a kernel regression. Among the outputs are parameter estimates of the local model structure at the locations at which the data have been collected. A number of different weighting schemes are possible, but they share the common feature that data locations near the point where the parameters are being estimated are given a greater weight in the estimation procedure than locations further away. In one sense this embodies the view expressed by Tobler (1991, 2004). If a set of parameters is desired for some location $\mathbf{u}$, then these are obtained with the estimator

$$\hat{\beta}(\mathbf{u}) = (\mathbf{X}^T\mathbf{W}(\mathbf{u})\mathbf{X})^{-1}\mathbf{X}^T\mathbf{W}(\mathbf{u})\mathbf{Y} \tag{21.1}$$

where $\mathbf{W}(\mathbf{u})$ is the weight matrix for the location $\mathbf{u}$. It should be noted that the location $\mathbf{u}$ need not be one at which data have been collected. GWR can be thus used to interpolate a surface of parameter estimates. As well as a parameter vector for each $\mathbf{u}$, there is a vector of location-specific standard errors. If the regression points (that is, those at which parameter estimates are being provided) coincide with the data points, further diagnostics are obtained, including a set based around the hat matrix (the matrix which converts the observed Ys into the predicted Ys). The diagnostics include: measures of influence and leverage, and Cook's D values. A local goodness of fit measure can also be computed (see an early suggestion in Fotheringham and Rogerson 1994). In addition to providing a model with a Gaussian error term for fitting data from a continuous distribution, GWR models also exist for dichotomous response variables (logistic) and response variables which are positive non-zero integers (Poisson) (Fotheringham, Brunsdon, and Charlton 2002).

The plethora of parameter estimates and associated diagnostics from GWR brings in turn its own set of visualization problems. If the data relate to point objects, say, houses or people, then symbolism of varying sizes and colors can be used to show the variation in the estimates. Should the range include zero, then the symbolism can be a conjunction of size and color, with say, blue for negative values and red for positive values. There may be cases in which the number of symbols to be displayed on a map to the spatial variation in a parameter estimate is so great that it confuses any pattern. One solution is to convert the point output into a surface – this can be done by a further process of interpolation (using some form of distance weighted smoothing, for example). A post-GWR interpolation stage does, however, introduce further choices by the analyst, such as a smoothing parameter, which may obscure rather than reveal any spatial variation if it is injudiciously chosen. An alternative approach is to make use of the ability of GWR to create parameter estimates at locations other than those at which the data were selected. If these latter locations are the mesh points of a regular grid, then these can be plotted as a smoothly varying surface. The disadvantage with the second approach is that diagnostics based on the hat matrix are not available.

If the geographically weighted regression is based on data for areas, the area values can themselves be mapped. If the software developed by Fotheringham, Brunsdon, and Charlton (2002) is being used, the output has no ID variable to link it through a tabular-merge with an existing areal database, so an intersect operation is required to tag the records in the GWR output with the IDs of the areas to which they belong. After that, choropleth mapping is one approach to displaying the results – again with different colors and intensities of color to model the spatial variation.

One advantage of GIS with a macro language, such as ESRI's Arc Macro Language, is that plotting a range from maps from a GWR run can be automated to the extent of creating a linked series of macros, with a single "driver" macro to generate and plot the necessary maps. Many of the maps in Fotheringham, Brunsdon, and Charlton (2002) were generated in this fashion. The advantage is that new results from a change of, say, bandwidth, can be mapped with the minimum of effort leaving maximum time for their interpretation. One perhaps ought to bear in mind Rhind's strictures mentioned earlier in this chapter when adopting such an approach.

CONCLUSIONS

The upsurge of interest since the 1980s in the analysis of spatial data has been paralleled by the wide diffusion geographic information software and expertise. It is notable that this diffusion has taken place in many disciplines outside of geography. This has been accompanied by a welcome interest in the problems of spatial data analysis from the statistical community. None of these three strands of activity shows any sign of abating.

The link between quantitative analysis and GIS is bi-directional. Inevitably, many of those who have an interest in the quantitative analysis of spatial data will encounter GIS software in its role of providing an integrating environment and perhaps a tool to assist in the interpretation of the results. On the other hand, there will be those who are frequent users of GIS who encounter a problem that requires some statistical analysis. The former group need to be aware of the strengths and weaknesses of GIS; whether they use a particular piece of software or employ somebody to operate that software is a separate question, but the necessity for appreciation remains. Those in the latter group should perhaps be aware of the problems and opportunities offered by the statistical analysis of spatial data.

That the software is easy to use does not absolve users from thinking about the solutions and whether they are appropriate.

REFERENCES

Aldstadt, J. and Getis, A. 2002. Point pattern analysis in an ArcGIS environment. In *Proceedings of the CSISS Specialist Meeting on New Tools for Spatial Data Analysis*, Santa Barbara, CA, USA. Santa Barbara, CA: Center for Spatially Integrated Social Science.

Anselin, L. 1988. *Spatial Econometrics: Methods and Models*. Dordrecht, Kluwer.

Anselin, L. 1992. *SpaceStat Tutorial: A Workbook for Using SpaceStat in the Analysis of Spatial Data*. Santa Barbara, CA: National Center for Geographic Information and Analysis Technical Software Series S-92-1.

Anselin, L. and Bao, S. 1997. Exploratory spatial data analysis linking SpaceStat and ArcView. In M. M. Fischer and A. Getis (eds) *Recent Developments in Spatial Analysis*. Berlin: Springer-Verlag, pp. 35–59.

Bailey, T. C. and Gatrell, A. C. 1995. *Interactive Spatial Data Analysis*. Harlow: Longman.

Besag, J. and Newell, D. 1991. The detection of clusters in rare diseases. *Journal of the Royal Statistical Society, Series A* 154: 143–55.

Blakemore, M. J. 1984. Generalisation and error in spatial databases. *Cartographica* 21: 131–9.

Boots, B. 1986. *Voronoi (Thiessen) Polygons*. Norwich: GeoBooks.

Casetti, E. 1972. Generating models by the expansion method: Applications to geographic research. *Geographical Analysis* 4: 81–91.

Casetti, E. 1992. Bayesian regression and the expansion method. *Geographical Analysis* 24: 58–74.

Casetti, E. and Can, A. 1999. The econometric estimation and testing of DARP models. *Journal of Geographical Systems* 1: 91–106.

Cliff, A. D. and Ord, J. K. 1973. *Spatial Autocorrelation*. London: Pion.

Cliff, A. D. and Ord, J. K. 1981. *Spatial Processes: Models and Applications*. London: Pion.

Cook, D., Majure, J. J., Symanzik, J., and Cressie, N. 1996. Dynamic graphics in a GIS: Exploring and analyzing multivariate spatial data using linked software. *Computational Statistics* 11: 467–80.

Cook, D., Symanzik, J., Majure, J. J., and Cressie, N. 1997. Dynamic graphics in a GIS: More examples using linked software. *Computers and Geosciences* 23: 371–85.

Cooley, W. W. and Lohnes, P. R. 1962. *Multivariate Procedures for the Behavioural Sciences*. New York: John Wiley and Sons.

Cressie, N. A. C. 1993. *Statistics for Spatial Data*. New York: John Wiley and Sons.

Davis, J. C. 2002. *Statistics and Data Analysis in Geology*. New York: John Wiley and Sons.

Dempster, A., Laird, N., and Rubin, D. B. 1977. Maximum likelihood from incomplete data via the EM algorithm (with discussion). *Journal of the Royal Statistical Society B* 39: 1–38.

Dorling, D. 1996. *Area Cartograms: Their Use and Creation*. Norwich: University of East Anglia.

Dummer, T. B. J., Dickinson, H. O., Charlton, M. E., and Parker, L. 2000. Estimating the population of young people by ethnic group in the Northern Region of England 1971–1991. *Environment and Planning A* 32: 1935–58.

Fischer, M. M. and Getis, A. (eds). 1997. *Recent Developments in Spatial Analysis*. Berlin: Springer-Verlag.

Flowerdew, R. 1988. *Statistical Methods for Areal Interpolation: Predicting Count Data from a Binary Variable*. Lancaster: Universities of Lancaster and Newcastle, Northern Regional Research Laboratory Research Report 16.

Flowerdew, R. and Green, M. 1991. Data integration: Statistical methods for transferring data between zonal systems. In I. Masser and M. Blakemore (eds) *Handling Geographic Information*. Harlow: Longman, pp. 38–54.

Fotheringham, A. S., Brunsdon, C., and Charlton, M. E. 2000. *Quantitative Geography*. London: Sage.

Fotheringham, A. S., Brunsdon, C., and Charlton, M. E. 2002. *Geographical Weighted Regression: The Analysis of Spatially Varying Relationships*. Chichester: John Wiley and Sons.

Fotheringham, A. S., Densham, P. J., and Curtis, A. 1995. The zone definition problem in location-allocation modelling. *Geographical Analysis* 27: 60–77.

Fotheringham, A. S. and Rogerson, P. A. (eds). 1994. *Spatial Analysis and GIS*. London: Taylor and Francis.

Fotheringham, A. S. and Wong, D. 1991. The modifiable areal unit problem in multivariate statistical analysis. *Environment and Planning A* 23: 1025–44.

Fotheringham, A. S. and Zhan, F. 1996. A comparison of three exploratory methods for cluster detection in spatial point patterns. *Geographical Analysis* 28: 200–18.

Gaile, G. L. and Burt, G. E. 1980. *Directional Statistics*. Norwich: GeoBooks.

Gehlke, C. D. and Biehl, K. 1934. Certain effects of grouping on the size of the correlation coefficient in census tract material. *Journal of the American Statistical Association, Supplement* 29: 169–70.

Goodchild, M. F. 1986. *Spatial Autocorrelation*. Norwich: GeoBooks.

Goodchild, M. F. 2000. Cartographic futures on a digital earth. *Cartographic Perspectives* 36: 3–11.

Goodchild, M. F. 2004. The validity and usefulness of laws in geographic information science and geography. *Annals of the Association of American Geographers* 94: 300–3.

Griffith, D. A. 1987. *Spatial Autocorrelation: A Primer*. Washington, DC: Association of American Geographers.

Haining, R. P. 1998. Spatial statistics and the analysis of health data. In A. C. Gatrell and M. Loytonen (eds) *GIS and Health*. London: Taylor and Francis, pp. 29–47.

Holt, D., Steel, D. G., Tranmer, M., and Wrigley, N. 1996. Aggregation and ecological effects in geographically based data. *Geographical Analysis* 28: 244–61.

Ihaka, R. and Gentleman, R. 1996. R: A language for data analysis and graphics. *Journal of Computational and Graphical Statistics* 5: 299–314.

Iliffe, J. C. 2000. *Datums and Map Projections for Remote Sensing GIS and Surveying*. Boca Raton, FL: CRC Press.

Kehris, E. 1989. *Interfacing Arc/Info and GLIM: A Progress Report*. Lancaster: University of Lancaster, Northwest Regional Research Laboratory Research Report No. 5.

Kellerman, A. 1981. *Centrographic Measures in Geography*. Norwich: GeoBooks.

King, G. 1997. *A Solution to the Ecological Inference Problem: Reconstructing Individual Behaviour from Aggregate Data*. Princeton, NJ: Princeton University Press.

Kulldorf, M. 1997. A spatial scan statistic. *Communications in Statistics: Theory and Methods* 26: 1481–96.

Longley, P. A. and Batty, M. 1996. *Spatial Analysis: Modelling in a GIS Environment*. Cambridge: GeoInformation International.

MacEachren, A. M., Brewer, C. A., and Pickle, L. 1998. Visualizing georeferenced data: Representing reliability of health statistics. *Environment and Planning A* 30: 1547–61.

Openshaw, S. 1984. *The Modifiable Areal Unit Problem*. Norwich: GeoBooks.

Openshaw, S., Charlton, M. E., Wymer, C., and Craft, A. W. 1987. A mark 1 geographical analysis machine for the automated analysis of point data sets. *International Journal of Geographical Information Systems* 1: 359–77.

Openshaw, S. and Taylor, P. J. 1979. A million or so correlation coefficients: Three experiments on the modifiable areal unit problem. In N. Wrigley (ed.) *Statistical Applications in the Spatial Sciences*. London: Pion, pp. 127–44.

Phillips, J. D. 2004. Doing justice to the law. *Annals of the Association of American Geographers* 94: 290–3.

Robinson, W. R. 1950. Ecological correlation and the behaviour of individuals. *American Sociological Review* 15: 351–7.

Smith, J. M. 2004. Unlawful relations and verbal inflation. *Annals of the Association of American Geographers* 94: 294–99.

Stephan, F. F. 1934. Sampling errors and interpretations of social data ordered in time and space. *Journal of the American Statistical Association* 29: 165–6.

Thomas, E. N. and Anderson, D. L. 1965. Additional comments on weighting values in correlation analysis of areal data. *Annals of the Association of American Geographers* 55: 492–505.

Tierney, L. 1990. *LISP-STAT: An Object-oriented Environment for Statistical Computing and Dynamic Graphics.* New York: John Wiley and Sons.

Tobler, W. R. 1970. A computer movie simulating urban growth in the Detroit region. *Economic Geography* 46: 234–40.

Tobler, W. R. 1991. Frame-independent spatial analysis. In M. F. Goodchild and S. Gopal (eds) *Accuracy of Spatial Databases.* London: Taylor and Francis, pp. 115–22.

Tobler, W. R. 2004. On the first law of geography: A reply. *Annals of the Association of American Geographers* 94: 304–10.

Tranmer, M. and Steel, D. G. 1998. Using census data to investigate the causes of the ecological fallacy. *Environment and Planning A* 30: 817–31.

Venables, W. N. and Ripley, B. D. 2002. *Modern Applied Statistics with S* (4th edn). New York: Springer.

Chapter 22

Spatial Cluster Analysis

Geoffrey M. Jacquez

We may at once admit that any inference from the particular to the general must be attended with some degree of uncertainty, but this is not the same as to admit that such inference cannot be absolutely rigorous, for the nature and degree of the uncertainty may itself be capable of rigorous expression. In the theory of probability, as developed in its application to games of chance, we have the classic example proving this possibility. If the gambler's apparatus are really true or unbiased, the probabilities of the different possible events, or combinations of events, can be inferred by a rigorous deductive argument, although the outcome of any particular game is recognized to be uncertain. The mere fact that inductive inferences are uncertain cannot, therefore, be accepted as precluding perfectly rigorous and unequivocal inference. (Fisher 1935)

Humility is indeed wise for the spatial analyst! (Bailey and Gatrell 1995)

Spatial cluster analysis plays an important role in quantifying geographic variation patterns. It is commonly used in disease surveillance, spatial epidemiology, population genetics, landscape ecology, crime analysis, and many other fields, but the underlying principles are the same. This chapter provides an overview of a probabilistic approach that is the foundation of spatial cluster analysis. It first provides a working definition of a cluster, founded on the type of data to be analyzed. The role of cluster analysis in Exploratory Spatial Data Analysis (ESDA) is discussed and provides an entrée into five components that underlie statistical pattern recognition. The clustering typology of global, local, and focused methods is then defined, followed by an overview of descriptors of cluster morphology. Approaches to quantifying cluster change and persistence are summarized, and issues of multiple testing are addressed. The chapter concludes with an overview of some software resources for undertaking cluster analyses.

What is a Cluster?

In order to define a spatial cluster we first must consider the kinds of data that are being studied. The information to be clustered may be event-based, population-based, field-based, or feature-based. Event-based data include point locations (such as the places of residence and time of diagnosis of cases of disease in people, or the locations of a species of tree in a forest) and counts (accidents at particular road intersections). Population-based data incorporate information on the population from which the events arose, and include disease rates with case counts in the numerator and size of the at-risk population in the denominator. Field-based data are observations that are continuously distributed over space, and include concentrations and temperatures. Feature-based data include boundaries and polygons that may be derived from field-based data, such as zones of rapid change in an attribute's value.

A spatial cluster might then be defined as an excess of events (for event- and population-based data, such as a cancer cluster) or of values (for field-based data, such as a grouping of excessively high concentrations of cadmium in soils) in geographic space. For feature-based data, a cluster might be a spatial aggregation of boundaries. But in practice, the term "cluster" is too generic and does not convey information on *cluster morphology*, such as descriptors of magnitude of the excess or deficit, geographic size and shape of the cluster and the locations of spatial outliers, and descriptors of boundary shape, as described in more detail later in this chapter. For now, it is useful to think of a "cluster" as a spatial pattern that differs in important respects from the geographic variation expected in the absence of the spatial processes that are being investigated. This is a key concept – that "clustering" is always measured relative to a null expectation.

Why Search for Clusters?

Cluster analysis plays important roles in the construction of spatial models and in ESDA. Model construction requires an understanding of the patterns of spatial variation as one often wants to incorporate relevant features of attribute variability into the model. ESDA involves the identification and description of spatial patterns (such as outliers, clusters, hotspots, cold spots, trends, and boundaries) and has two primary objectives:

- **Objective 1**: Pattern recognition using visualization, spatial statistics and geostatistics to identify the locations, magnitudes, and shapes of statistically significant pattern descriptors;
- **Objective 2**: Hypothesis generation to specify realistic and testable explanations for the geographic patterns found under Objective 1.

Statistical Pattern Recognition

Spatial patterns are of interest because they are the trace of space-time processes that are the focus of geographic studies. For example, in cancer research spatial patterns contain the geographic trace of processes, covariates, and factors (such as exposures to environmental carcinogens; access to cancer screening facilities; behaviors mediating cancer risks, and so on) that determine how cancer risk varies across and is expressed within human populations.

There are several approaches to pattern recognition, including visualization techniques that rely on the human visual cortex (for example, "eye-balling"), kernel-based methods that accentuate differences on a surface (for example, median smoothing techniques such as headbanging – see Kafidar 1996, Gelman, Price, and Lin 2000); artificial intelligence approaches (for instance, neural network and genetic algorithms – Pandya and Macey 1996), and methods of statistical pattern recognition that are the foundation of ESDA (Bailey and Gattrell 1995; Moore and Carpenter 1999; Jacquez 1998, 2000; Rushton and Elmes 2000). Most commonly used spatial cluster analysis methods are founded on statistical pattern recognition.

Statistical pattern recognition proceeds by calculating a statistic (for instance, spatial cluster statistic, autocorrelation statistic, boundary statistic, etc.) that quantifies a relevant aspect of spatial pattern in event-, population-, field- or feature-based data. For human disease this might be a health outcome (for example, case/control locations, incidence, or mortality rate), putative cause (for example, exposure to agricultural pesticides), risk factor (smoking prevalence, for example) or access measure (for instance, availability of prostate screening). The numerical value of this statistic is then compared to the distribution of that statistic's value under a null spatial model. This provides a probabilistic assessment of how unlikely an observed spatial pattern is under the null hypothesis (Gustafson 1998). Waller and Jacquez (1995) formalized this approach by identifying five components of a test for spatial pattern:

- The *test statistic* quantifies a relevant aspect of spatial pattern (for example, Moran's I, Geary's c, LISA, a spatial clustering metric, etc.);
- The *alternative hypothesis* describes the spatial pattern that the test is designed to detect. This may be a specific alternative, such as clustering near a focus, or it may be the omnibus "not the null hypothesis";
- The *null hypothesis* describes the spatial pattern expected when the alternative hypothesis is false (for instance, uniform cancer risk);
- The *null spatial model* is a mechanism for generating the reference distribution; this may be based on distribution theory, or it may use randomization (for instance, Monte Carlo) techniques – for example, many disease cluster tests employ heterogeneous Poisson and Bernoulli models for specifying null hypotheses (see Lawson and Kulldorff 1999);
- The *reference distribution* is the distribution of the test statistic when the null hypothesis is true.

Comparison of the test statistic to the reference distribution allows calculation of the probability of observing that value of the test statistic under the null hypothesis of no clustering. This five-component mechanism underpins most tests commonly used in spatial cluster analysis.

Null or Neutral Hypothesis? There still is some debate as to how the null hypothesis and null spatial model should be specified. Many implementations of spatial cluster tests employ the null hypothesis of Complete Spatial Randomness or CSR. In the real world, CSR is an appropriate model for pure noise processes such as the static on a television screen when a channel with no signal is selected. But most geographic systems are highly complex and a null hypothesis of CSR is rarely, if ever, appropriate. While CSR is useful in some situations, it is not a relevant null hypothesis for highly complex and organized systems such as those encountered in the physical, environmental, and health sciences including fields such as geography, spatial epidemiology, and exposure assessment (Liebisch, Jacquez, Goovaerts, and Kaufmann 2002). CSR is not relevant because spatial randomness rarely, if ever, occurs – some spatial pattern is almost always present. Hence, rejecting CSR has little scientific value because it does not describe any plausible state of the system. The term "Neutral Model" captures the notion of a plausible system state that can be used as a reasonable null hypothesis (for example, "background variation"). A typology of neutral models that account for differences in underlying population sizes and for regional and local variation in mean values is one mechanism for constructing null hypotheses that are more plausible than CSR (Goovaerts and Jacquez 2004, 2005). The problem then is to identify spatial patterns above and beyond that incorporated into the neutral model. In this chapter we will continue to use the terms "null model" and "null hypothesis," with the proviso that they denote appropriate levels of spatial pattern expected in the absence of the hypothesized alternative spatial process.

Types of Tests

There are dozens of cluster statistics (see Jacquez, Grimson, Waller, and Wartenberg 1996, Jacquez, Waller, Grimson, and Wartenberg 1996; Lawson and Kulldorff 1999, among others, for reviews), and presentation of these statistics would fill most of this book. Instead, we now present the characteristics of *global*, *local*, and *focused* tests with a few commonly used statistics as examples.

Global cluster statistics are sensitive to spatial clustering, or departures from the null hypothesis, that occur anywhere in the study area. Many early tests for spatial pattern, such as Moran's *I* (Moran 1950) were global in nature and provided one statistic, such as a global autocorrelation coefficient, that summarized spatial pattern over the entire study area. While global statistics can identify whether spatial structure (for instance, clustering, autocorrelation, uniformity) exists, they do not identify where the clusters are, nor do they quantify how spatial dependency varies from one place to another.

Local statistics such as Local Indicators of Spatial Autocorrelation (LISA) (Anselin 1995, Ord and Getis 1995) quantify spatial autocorrelation and clustering within the small areas that together comprise the study area. Many local statistics have global counterparts that often are calculated as functions of local statistics. For example, Moran's global autocorrelation statistic is the scaled sum of the LISA statistics that are calculated as:

$$L_I = z_i \sum w_{ij} z_j \qquad\qquad (22.1)$$

Here L_I is the LISA statistic for area i, z_i is the observation at location i, scaled to have a mean of 0 and unit standard deviation (a z-score), and the term in the summation is the average within those areas immediately adjacent to the i th area. Local statistics thus can tell you the nature of spatial dependency (for example, not significantly different from the null expectation, cluster of high values, cluster of low values, and high or low spatial outlier) in a given locality, while also providing a global test.

Focused statistics quantify clustering around a specific location called a *focus*. These tests are particularly useful for exploring possible clusters of disease near potential sources of environmental pollutants. For example, Lawson (1989) and Waller, Turnbull, Clark, and Nasca (1992) proposed tests that score each area for the difference between observed and expected disease counts, weighted by degree of exposure to the focus:

$$U = \sum_{i=1}^{N} g_i(o_i - e_i) \tag{22.2}$$

Here there are N areas, g_i is a function defining the exposure to the focus, o_i is the observed number of cases in area i, and e_i is the expected number of cases in that area. A commonly used exposure function is the inverse distance to the focus (1/d). The null hypothesis is no clustering relative to the focus, and the expected disease count thus is calculated as the Poisson expectation using the population at risk in each area and the assumption that disease risk is uniform over the study area.

Waller, Turnbull, Clark, and Nasca (1992) used this test to explore whether cases of leukemia clustered near 12 hazardous waste sites in upstate New York that were injecting trichloroethylene (TCE) into the groundwater. The Score test found some of the foci to be associated with high leukemia risk, and was significant after adjusting for the 12 repeated tests.

Tests for Cluster Association

Once clusters are identified they define feature sets that can be compared to the configuration of other feature sets. This is an exercise in pattern matching, rather than statistical pattern recognition. So, for example, one can ask whether the edges of disease clusters are near the edges of pollutant plumes. An ecologist might ask whether the spatial distribution of species abundance is associated with habitat patches, and so on. This kind of pattern matching task has been called the map comparison problem (Jacquez 1995), and has been addressed using methods of geographic boundary analysis (Jacquez, Maruca, and Fortin 2000). Two approaches will be discussed. The first quantifies boundary overlap, the second quantifies area overlap.

Boundary overlap

Jacquez (1995) proposed four tests for the overlap of geographic boundaries. For ease of reference, we will term one set of boundaries boundary **G** and the other Boundary **H**. For example, **H** might correspond to the edges of clusters in a health related variable and **G** might be the cluster edges for a pollutant plume. The statistics are based on the nearest neighbor distances between *boundary elements*

(BEs), which are those geographic coordinates that define the cluster boundary. The first statistic, O_S, is the count of the number of BEs that are included in both sets of boundaries, and is a measure of exact boundary overlap. The second statistic, O_G, is the mean minimum distance from the BEs in $\mathbf{G}$ to the nearest BE in $\mathbf{H}$. O_H is the mean minimum distance from the BEs in $\mathbf{H}$ to the nearest BE in $\mathbf{G}$. O_{GH} is the mean distance from a BE in either boundary set to the nearest BE in the other boundary set. These statistics are useful for evaluating whether the edges of geographic features, such as zones of rapid change and clusters, are significantly near one another. These statistics thus evaluate spatial association and are a tool that can be used in conjunction with non-spatial tests for association such as correlation and regression.

Boundary overlap examples

Jacquez (1995) explored boundary overlap in respiratory illness and environmental ozone in southern Ontario. Exposure to high ozone can cause acute respiratory distress leading to pulmonary edema or even emphysema. Jacquez asked whether zones of rapid change in environmental ozone induced concomitant zones of rapid change in respiratory health. Ozone boundaries appeared by visualization to coincide with boundaries in hospital respiratory admissions; however, the overlap statistics were not significant. Most likely other factors were involved that may have obscured the relationship between ozone and respiratory health, and these results demonstrate the need to statistically evaluate apparent associations in order to avoid the "Gee Whiz" effect (Jacquez 1998).

Fortin, Drapeau, and Jacquez (1996) used boundary overlap to assess the relationships between edaphic factors (soil types and moisture) and vegetation boundaries. They found that vegetation boundaries based on species' stem density and species' presence/absence overlapped boundaries in edaphic factors, but vegetation boundaries based on species diversity and richness did not. This pattern suggests a hierarchy of effects, with edaphic factors predicting species presence but not plant community structure.

To determine how much the variable examined influences boundary delineation, Fortin (1997) evaluated overlap among vegetation boundaries calculated from different data sets. She found that density, percent coverage, and presence/absence for trees, shrubs, and trees and shrubs together significantly overlapped. While most variables concurred, the tree-only and the shrub-only data did not. This study demonstrated the use of boundary overlap analysis to distinguish variables that are spatially associated from those that are not.

Hall and Maruca (2001) compared vegetation boundaries to those in songbird abundance in a 45 ha swamp in Michigan. They found that bird abundance boundaries were significantly associated with vegetation boundaries, but not vice versa. Upon investigating the composition of the eight vegetation clusters, they found that the variable driving the vegetation clusters, and hence their boundaries, was the density of coniferous trees, a potentially important factor influencing the selection of songbird nesting and foraging areas. The authors suggested boundary analysis may aid in the development of monitoring and recovery plans for threatened bird species that use mosaic landscapes.

Cluster overlap

Maruca and Jacquez (2002) developed tests for identifying the amount of overlap between two spatial patterns. These tests differ from boundary overlap tests in that they focus on overlap of the areas enclosed within cluster boundaries. Recognizing the ubiquity of edge effects in natural systems and that spatial heterogeneity typically occurs on several spatial scales, they developed a test for association between two spatial patterns (for example, sets of clusters calculated on two different variables) that is not biased by edge effects and is based on null spatial models that can incorporate spatial heterogeneity found in real-world systems. These methods can be used to determine whether landscape classification maps have geographic partitions that overlap to a significant extent, and to determine whether spatial clusters defined on different variables (for instance, health outcomes and putative exposures) are significantly close to one another, existence of which is consistent with (but does not prove) a causal relationship.

Assume two sets of clusters, I and J, each comprised of N_I and N_J clusters and obtained as a spatial cluster analysis of different variables in the same geographic space. For cluster i in set I and cluster j in set J, *relative cluster overlap* is calculated as:

$$a_{ij} = \frac{a_{(i \cap j)}}{a_{(i \cup j)}} \tag{22.3}$$

Here $a_{(i \cap j)}$ is the area of intersection and $a_{(i \cup j)}$ is the area of union for clusters i and j. This statistic is zero for non-overlapping clusters, and increasing values represent better overlap, with a maximum value of 1 for perfectly overlapping clusters (where $a_{(i \cap j)} = a_{(i \cup j)}$). For each cluster i in I we can then find the cluster in J that i overlaps best with, by finding the maximum value of a_{ij} over all clusters in J (called the *maximum relative cluster overlap*):

$$A_i = \max(a_{i \bullet}) \tag{22.4}$$

A cluster overlap statistic from the perspective of set I is the *average maximum relative cluster overlap* (Equation 22.5). This also can be calculated for set J (Equation 22.6), and a simultaneous area overlap statistic is shown in Equation 22.7:

$$A_I = \frac{\sum_{i=1}^{N_I} \max(A_{i \bullet})}{N_I} \tag{22.5}$$

$$A_J = \frac{\sum_{j=1}^{N_J} \max(A_{\bullet j})}{N_J} \tag{22.6}$$

$$A_{IJ} = \frac{\sum_{i=1}^{N_I} \max(A_{i \bullet}) + \sum_{j=1}^{N_J} \max(A_{\bullet j})}{N_I + N_J} \tag{22.7}$$

The statistic A_I is a measure of how well the clusters in **I** overlap the clusters in **J**, and A_J is a measure of how well clusters in **J** overlap with those in **I**. A_{IJ} is a general (or bi-directional) measure of overlap between the two sets of clusters. These statistics are best suited to instances where the two sets of clusters contain roughly the same number of clusters, and with clusters of about the same size. However, in the real world we expect to find cluster sets with very different numbers of clusters. Further, a given cluster set may contain clusters of drastically different sizes, as occurs when spatial variation is heterogeneous. The constraint on cluster number and size distribution is relaxed by calculating the *average maximum relative overlap* (A_I') as a weighted average, where the weight is the area of the focus cluster (a_i for cluster i in set **I**; a_j for cluster j in set **J**). In this scenario, the statistics would be calculated as follows:

$$A_I' = \frac{\sum_{i=1}^{N_I} [a_i \max(A_{i\bullet})]}{\sum_{i=1}^{N_I} a_i} \tag{22.8}$$

$$A_J' = \frac{\sum_{j=1}^{N_J} [a_j \max(A_{\bullet j})]}{\sum_{j=1}^{N_J} a_j} \tag{22.9}$$

$$A_{IJ}' = \frac{\sum_{i=1}^{N_I} [a_i \max(A_{i\bullet})] + \sum_{j=1}^{N_J} [a_j \max(A_{\bullet j})]}{\sum_{i=1}^{N_I} a_i + \sum_{j=1}^{N_J} a_j} \tag{22.10}$$

These tests have local versions, just as the global cluster tests discussed earlier have corresponding local tests. The local version identifies those locations on the map where statistically significant cluster overlap and overlap avoidance are found. Implementation of this involves decomposing the summation for either the raw (equations 22.5–22.7) or weighted versions (from equations 22.8–22.10) into local contributions for each cluster. Thus, for the global unweighted overlap statistic in Equation 22.5 the local version is:

$$A_{ll} = \frac{\max(A_{l\bullet})}{N_I} \tag{22.11}$$

Here the "index" cluster is cluster l, and A_{ll} is the contribution of cluster l to the global overlap statistic A_I. Local counterparts can also be constructed for the other global statistics in equations (22.6–22.10). The value of each local overlap statistic can indicate overlap, no association, or overlap avoidance. A probability for overlap of cluster of the size of cluster l is calculated from the distribution of A_{ll} under a Monte Carlo simulation that conditions on both the number and size distribution of the clusters, but assumes no association between the geographies of the clusters in sets **I** and **J**.

Method Selection Advisors

So far the reader has been introduced to global, local, and focused cluster statistics, as well as spatial association tests for boundary and area overlap. Researchers often ask "which spatial cluster test should I use?" and online cluster analysis advisors, such as those found at http://www.terraseer.com/bsr/boundaryseer_advisor.html and http://www.terraseer.com/csr/clusterseer_advisor.html can aid in this regard.

But looking for the "one" suitable cluster test is appropriate only when one has prior knowledge of cluster shape. For example, if one knows cancer clusters are circular then it would be appropriate to use a spatial scan statistic with a circular scanning window. However, this reasoning is also circular since one usually undertakes a cluster analysis in order to locate and describe the clusters – prior knowledge of cluster shape therefore is lacking.

Cluster Morphology

There is a growing awareness that clusters can take on a variety of different shapes, but cluster tests are usually sensitive to only one profile (Sun 2002, Smith 2003, Jacquez 2004, Tango and Takahashi 2005). Different techniques are sensitive to different aspects of *cluster morphology* – some detect boundaries, some detect outliers, some detect circular clusters, some detect elliptical clusters, and so on. Areas of high value can take many shapes, yet most cluster-detection techniques employ geographic "templates" such as circular scanning windows. These include the GAM (Openshaw, Charlton, Craft, and Birch 1988), the scan statistic (Kulldorff and Nagarwalla 1995), Turnbull's test (Turnbull, Iwano, Burnett, Howe, and Clark 1990), Besag and Newell's (1991) test, the score test of Lawson and Waller (1996) that uses a circular risk function, first-order adjacencies such as LISA statistics, and nearest-neighbors relationships such as used in Cuzick and Edward's (1990) test. These tests are most sensitive to cluster shapes that correspond to the geographic templates they employ. But spatial variation and hence cluster morphology in geographic systems is highly complex, and cannot be well described by a single geographic template or clustering approach.

This observation recently motivated the development of an integrative approach that provides a more complete description of cluster morphology (Jacquez and Greiling 2003a, b). This integrative framework employs a battery of ESDA techniques including geographic boundary analysis, spatial agglomerative clustering, local Moran tests, and scan statistics (Table 22.1). Jacquez and Greiling employed it to more fully describe multi-scalar and multivariate patterns in the incidence of breast, colorectal, and lung cancers on Long Island, New York. They used global, local, and focused tests to explore the spatial scale of clustering, LISA statistics to identify spatial outliers, hot spots and cool spots, boundary analysis to find zones of rapid change, boundary overlap to evaluate possible associations between lung cancer and airborne

Table 22.1 Cluster Morphology Descriptors

Descriptor	Example
Amount of excess or deficit	Relative Risk in a disease cluster, number of cases in the cluster
Extent	Geographic area, number of sub-areas in the cluster
Length	Length of major and minor axes in an elliptical cluster
Boundary	
Length	Length of cluster boundary
Crenellation	Boundary fractal dimension
Fuzziness	Alpha core and surrounding zone of uncertainty
Shape	Ratio of boundary length / cluster area
Bivariate spatial association	Boundary overlap analysis; Cluster overlap analysis
Multivariate spatial structure	Clusters from multivariate spatially agglomerative clustering; boundaries from multivariate boundary analysis

carcinogens, and spatial agglomerative clustering to identify multivariate clusters that were homogeneous in lung, breast, and colorectal cancer incidence (Figure 22.1). This integrative approach yielded a detailed description of the morphology of statistically significant geographic variation patterns for breast, lung, and colorectal cancer incidence on Long Island.

An alternative approach is to use techniques for which the geographic template is flexible and can assume any shape. The first method, called the Upper Level Set scan statistic (Patil and Taillie 2004), involves estimation of cluster morphology (for example, shape, extent, and configuration) from the data itself. The second method involves the "growth" of clusters by grouping adjacent areas that have similar (high or low) rates (see Urban 2004). While techniques for spatially agglomerative clustering have been available for some time (Legendre 1987, for example), they often do not assign probabilities to the resulting clusters. A new approach called B-statistics was recently proposed that simultaneously detects agglomerative clusters of arbitrary shape as well as edges (borders where two adjacent areas of significantly different rates abut), and provide cluster probabilities under realistic null hypotheses (Jacquez, Kaufmann, and Goovaerts 2006). Finally, kernel density estimation methods result in spatially continuous maps of the probability of a disease outcome (Rushton 1997, Rushton, Peleg, Banerjee, Smith, and West 2004) and appear capable of circumscribing clusters of variable shape, but the impact of kernel-based smoothing on the type I and type II error has yet to be fully quantified.

Cluster Change and Persistence

With the advent of routine remote sensing and improved environmental monitoring and health surveillance it now is possible to analyze data that are spatially and temporally referenced. In particular, there now are Space-Time Intelligence Systems (STIS) designed specifically to deal with georeferenced data through time. Analysis of how spatial patterns change through time is quite straightforward in such systems.

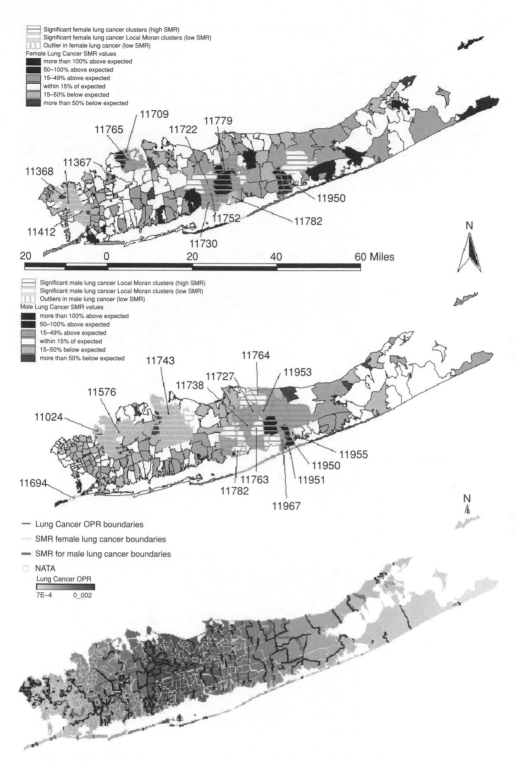

Fig. 22.1 Cluster morphology of cancer incidence on Long Island. Top: LISA clusters in male lung cancer incidence; Middle: LISA clusters in female lung cancer incidence; Bottom: Boundaries in male and female lung cancer incidence and air toxics from EPA's National Air Toxics Assessment program

From Jacquez and Greiling 2003 a, b

We now consider two aspects of cluster change and persistence: temporal change in the spatial distribution of clusters and clustering of attributes from two different time periods. In this discussion we use the LISA statistic, but the approach is general and can apply to most cluster tests.

- **Temporal change in the spatial distribution of clusters** An obvious first-step in the exploration of cluster change and persistence is to cluster an attribute at time t and compare the spatial distribution of those clusters to those obtained for that attribute at time $t + 1$ (Figures 22.2 and 22.3). Boundary and area overlap

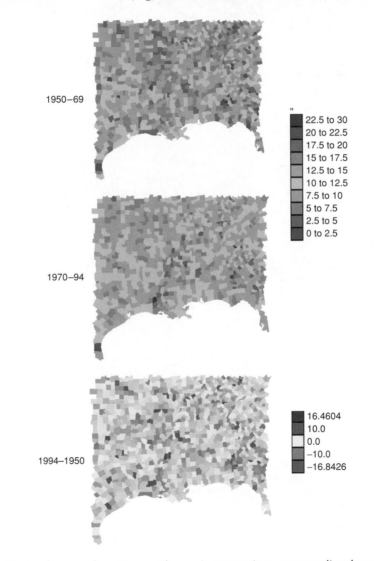

Fig. 22.2 Cluster change and persistence. Changes in pancreatic cancer mortality along the lower Mississippi River for white males for all ages from 1950–69 to 1970–94. Top: Mortality 1950–69; Middle: Mortality 1970–94; Bottom: Difference in mortality rates for the two time periods Cancer data from CancerAtlas Viewer; http://www.terraseer.com/products/atlasviewer.html

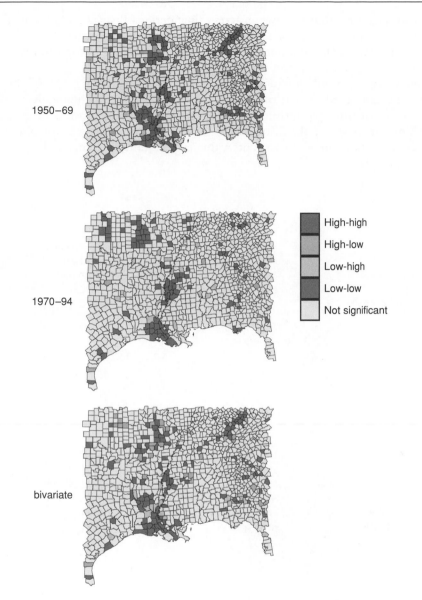

1950–69

1970–94

bivariate

High-high

High-low

Low-high

Low-low

Not significant

Fig. 22.3 Cluster change and persistence continued. LISA clusters in pancreatic cancer for 1950–69 (top) and 1970–94 (middle). High pancreatic cancer mortality is spreading north in counties along the Mississippi River. Bivariate LISA clusters identify counties high in pancreatic cancer in 1950–69 that are surrounded by counties with high mortality in 1970–94. Analyses conducted in TerraSeer's STIS software

statistics such as those summarized earlier may be used to determine the amount of association between the clusters at the different time periods. The local overlap statistics are used to distinguish those clusters that significantly overlap from those that do not. This approach is useful for identifying where cluster existence changes through time and where clusters are persistent.

- **Clustering of attributes at two different time periods** Bivariate LISA statistics are useful for identifying those areas with high values at time t that are surrounded by areas with high values at time $t + 1$. This tool is useful for gaining insights into cluster persistence and spread (Figure 22.3).
- **Clustering temporal difference** Rather than working with maps of the attribute at times t and $t + 1$, one can first calculate difference maps that subtract the value at time $t + 1$ from the value at time t. These clusters (Figure 22.4) identify areas where the difference is high, and thus are useful for pinpointing those localities where the attribute value is uncertain, unstable and/or changing dramatically.

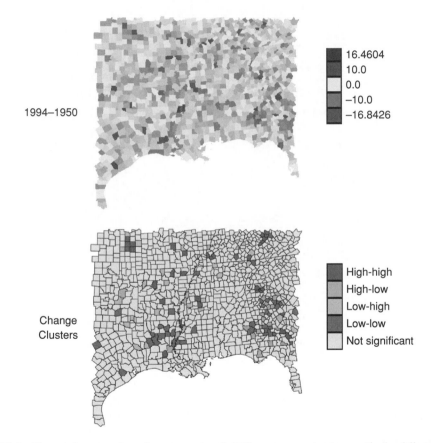

Fig. 22.4 Cluster change and persistence continued. Difference maps (top) quantify the difference in mortality rates between 1950–69 and 1970–94. LISA clusters of the difference maps (bottom) identify those localities where the change in morality is high. Analyses conducted in TerraSeer's STIS software

Disparity clusters

When faced with different classes of an item (say males and females) the question often arises as to whether spatial clusters of disparity exist. This is an important problem in the health sciences, as substantial disparities in disease incidence and mortality are observed for different race, gender, and ethnic groups. Consider cancer. According to the National Cancer Institute's (NCI) planning and budget proposal for 2004

> The unequal burden of cancer in our society is more than a scientific and medical challenge. It is a moral and ethical dilemma for our Nation. Certain populations experience significant disparities in cancer incidence, the care they receive, and the outcomes of their disease. These differences have been recognized, or at least suspected, for some time. They now are being documented with increasing frequency and clarity. (NCI 2003)

The identification of locations of high disparity in a health outcome – a disparity cluster – is an important step that allows one to target interventions and to address inequities in access to health care and provision of screening services. How can health disparities be identified?

The approach employed involves three steps similar to those employed for evaluating cluster change and persistence. Comparison of cluster maps, bivariate cluster analysis, and clustering of difference maps.

Consider an example. Pancreatic cancer incidence and mortality have changed little over the last three decades. Mortality rates in this period have been relatively stable for black and white males, have decreased for white females, and increased for black females. In each racial/ethnic group males have higher incidence and mortality than women. Blacks have incidence and mortality rates nearly 50 percent higher than whites. Rates for native Hawaiians are higher than for whites, while those for Hispanic and Asian-American rates are lower (Miller, Kolonel, Bernstein, et al. 1996). Risk is highest in the older population, and pancreatic cancer is rare among those 30–54 years old. Incidence for blacks 55–69 years of age is 60 percent higher than for whites of the same age, although this difference diminishes in ages 70 years and older. Age-based racial mortality patterns are similar to those observed in the incidence rates. Does the disparity in cancer mortality between white and black males cluster geographically?

The difference in standardized rates can be used to create cancer mortality disparity maps (for example, for BM–WM), and clusters of high values on these maps (hot spots) then identify locations of elevated mortality for black males. In addition to univariate clustering of the difference, bivariate LISAs for detecting clusters and anomalies of disparities in cancer mortality can be constructed of the form:

$$L_{i,BMxWM} = z_{i,BM} \sum_j w_{ij} z_{j,WM} \qquad (22.12)$$

Here $L_{i,BMxWM}$ is the bivariate LISA at location i for the disparity in pancreatic cancer mortality between black and white males at the local spatial scale. $z_{j,WM}$ is the standardized mortality rate for black males at that location and $z_{j,WM}$ is the local component for white males at location j adjacent to i. The term $\sum_j w_{ij} z_{j,\text{WM}}$ is the average white male pancreatic cancer mortality for locations (for instance, counties) adjacent to i. The disparity statistic $L_{i,\text{BM}x\text{WM}}$ is positive when pancreatic cancer mortality at neighboring locations is similar for both races, and negative if there is a disparity. Significance of the statistic is evaluated under conditional randomization, and the Moran scatter plot identifies clusters and hotspots of racial disparity in pancreatic cancer mortality. P-values under randomization are then used to construct pancreas cancer mortality disparity maps (Figure 22.5).

The two approaches (univariate on difference maps and bivariate on standardized rates) inform two different aspects of the geography of disparity. The univariate difference clusters detect significant spatial clusters and outliers in the difference between standardized rates (for example, BM–WM). The bivariate LISA statistic identifies spatial clusters and outliers in the standardized rates for one race–gender combination (for example, BM) relative to the average of the standardized rates for the second race–gender combination (for example, WM) in surrounding areas. Spatial cluster analysis has obvious utility for geographically pinpointing locations of statistically significant disparities in pancreatic cancer mortality.

An alternative approach to evaluating statistical significance of health disparities was recently proposed by Goovaerts (2005), who recognized that tests for differences in means, such as that based on the students t-distribution, would be useful for detecting health disparities provided differences in population sizes could be accounted for. His disparity statistic is an adaptation of the classical test for inference of two population proportions (Devore 2000) to the comparison of rates measured

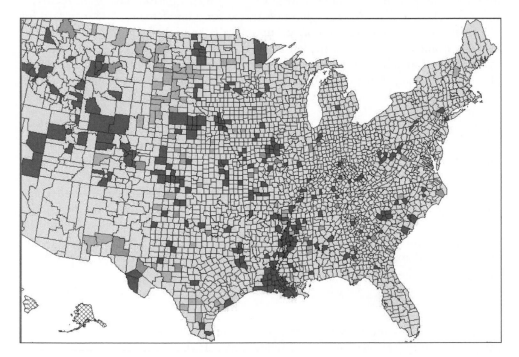

Fig. 22.5 Disparity clusters

in two sub-populations, labeled as reference and target populations. For a given region, the disparity statistic is calculated as the standardized difference between the target and reference rates, weighted by the population proportions, and has been demonstrated to detect regions with statistically significant health disparities that account for the population sizes of the reference and target populations (see Goovaerts 2005).

Multiple Testing

ESDA and the description of cluster morphology may involve iterations of visualization and statistical analyses to elucidate different aspects of spatial patterns and to successively refine the alternative hypotheses explored in pattern recognition procedures. Hence a typical analysis, say of prostate cancer incidence, may begin with the creation of maps using appropriately adjusted rates (for example, to stabilize the rates and to adjust for covariates such as age), and may then involve the use of global, local, and focused tests to determine whether the rates are spatially autocorrelated (using global tests such as those of Moran 1950, Oden 1995, Tango 1995 and others), to identify the locations of cold-spots, hot-spots, and outliers (using local tests such as Anselin 1995, Getis and Ord 1992, Turnbull, Iwano, Burnett, Howe, and Clark 1990, Besag and Newell 1991, among others), and to assess whether cases tend to cluster near the locations of putative environmental exposures (using focused tests such as Lawson and Waller 1996). To maintain statistical rigor the impact of repeated statistical procedures must be accounted for (Jacquez, Waller, Grimson, and Wartenberg 1996b), and this may be accomplished within the structure of the test itself (as in Kulldorff's 1997 scan test) or by adjusting P-values or Type I errors using the methods of Bonferroni (Sidak 1967, Simes 1986), Holms (Holland and Copenhaver 1987) or Hochberg (1988). Recently, Tango (2007) proposed using a "min-P" approach in which the test statistic itself is the minimum p-value observed from a group of tests. Because of the exploratory nature of the analyses there is some question as to whether a formal approach to inferential statistics (for example, comparing a P-value to the alpha level to determine whether the null hypothesis is rejected) is applicable. Most experts now advocate interpretation of P-values within the context of other information, such as the biological plausibility of the cluster, the quality of the data, and the costs associated with false positives and negatives (Waller and Jacquez 1995; Jacquez, Grimson, Waller, and Wartenberg 1996, Jaquez, Waller, Grimson, and Wartenberg 1996). Some software packages such as ClusterSeer (Jacquez, Maruca, Greiling, et al. 2001, Jacquez, Greiling, Durbeck 2002) account for multiple tests automatically and provide appropriately adjusted probability values.

Tools

Information Frames. There are literally dozens of spatial cluster tests, and cogent summaries of the different tests are needed to support method selection and to remind one of the properties of a given test. In a recent study funded by the NCI, researchers at BioMedware, Inc. and the University of Michigan School of Public

k-Nearest Neighbor test	• spatial/temporal
	• global
	• region-based
		• Similar Tests:
		    ◦ Mantel
		    ◦ Knox
		    ◦ Direction
		    ◦ Grimson
		• All available tests

Indications/Recommendations for use: When space-time interaction is present nearby cases will occur at about the same time, (space nearest neighbors tend to be time nearest neighbors), and the test statistic will be large (more).

Description: A k-nearest neighbor method used to detect space-time interaction.

Test statistic: The number of pairs of cases that are k-nearest neighbors in both space and time.	$J_k = \sum\limits_{i=1}^{N} \sum\limits_{j=1}^{N} s_{ijk} t_{ijk}$

Null Hypothesis: Whether cases are nearest neighbors in space is independent of whether they are nearest neighbors in time.

Alternative Hypothesis: Nearest neighbors in space tend to be nearest neighbors in time.

GeoMed Inputs: Space and time distances between pairs of cases.

GeoMed Outputs: Results table includes:

- ◦ k, the number of nearest neighbors being considered
- ◦ J_k, the number of space-time k nearest neighbors
- ◦ significance of J_k
- ◦ ΔJ_k, the number of space-time nearest neighbors added when k increased from k-1
- ◦ significance of ΔJ_k
- ◦ p-values for the probability, under H_0, of observing a statistic as large or larger than that observed and combined for the k tests in 3 ways:
    - ▪ Bonferroni
    - ▪ Simes
    - ▪ Statistical Distance
- ◦ A linkage map
    - ▪ case locations are mapped
    - ▪ k-order linkages displayed as selected by user

Example Analysis	Reference:

Fig. 22.6 Information frames provide cogent summaries of the properties of spatial cluster statistics and are available through the cluster advisor found at http://zappa.nku.edu/~longa/geomed/stathelp/advisor.html

Health developed one-page information frames that give a quick overview of the properties of a test (Figure 22.6).

Software. Cluster analysis software includes Satscan, which implements a scan-type statistic employing circular scanning windows. The commercial software ClusterSeer (Jacquez, Greiling, Estberg, et al. 2001, Jacquez Greiling, Durbeck, et al. 2002) has dozens of statistical techniques that employ a variety of geographic templates (for instance, circular scanning windows, nearest neighbor relationships, LISA tests, global tests, and focused tests). It comes with a cluster advisor and adjusts for multiple testing. BoundarySeer (Jacquez, Maruca, Greiling, et al. 2001) employs techniques to detect the edges of clusters, and these clusters can be of any shape. Both univariate and multivariate methods are included. The CancerAtlas viewer (http://www.terraseer.com/atlasviewer.html) comes with county, State Economic Area, and State level mortality data for 43 site-specific cancers, and employs LISA statistics. TerraSeer's STIS (http://www.terraseer.com/products/stis.html) supports linked windows, statistical brushing, spatial and space-time cluster statistics, animation as well as spatio-temporal georeferencing.

CONCLUSIONS

This chapter provided a quick overview of some of the issues and approaches in spatial cluster analysis. The reader should now have some appreciation of the role spatial cluster analysis plays in ESDA, and of global, local, and focused techniques. It should now be apparent that the "one size fits all" approach to cluster analysis yields an incomplete picture of cluster morphology. The integrated approach conveys a far more complete quantification of cluster morphology descriptors and ultimately leads to a better understanding of spatial variation. Spatial cluster analysis can yield substantial benefits in documenting cluster change and persistence, and for identifying disparities in two spatially referenced variables. The field is evolving rapidly, and as the volume of spatially and temporally referenced data increases cluster analysis will play an increasing role in pattern recognition and data reduction.

ACKNOWLEDGEMENTS

Some of the findings reported in this publication were developed under grants CA92669 from the National Cancer Institute and ES10749 from the National Institute of Environmental and Health Sciences. The perspectives of this publication are solely those of the author and do not necessarily represent the official views of the funding organizations.

REFERENCES

Anselin, L. 1995. Local indicators of spatial association: LISA. *Geographical Analysis* 27: 93–115.

Bailey, T. C. and Gatrell, A. C. 1995. *Interactive Spatial Data Analysis*. Harlow: Addison Wesley Longman.

Besag, J. and Newell, J. 1991. The detection of clusters in rare diseases. *Journal of the Royal Statistical Society Series A* 154: 143–55.

Cuzick, J. and Edwards, R. 1990. Spatial clustering for inhomogeneous populations. *Journal of the Royal Statistical Society Series B* 52: 73–104.

Devore, J. L. 2000. *Probability and Statistics for Engineering and the Sciences*. Belmont, CA: Duxbury Press.

Fisher, R. A. 1935. *Design of Experiment*. (First Edition). London: Oliver and Boyd.

Fortin, M. J. 1997. Effects of data types on vegetation boundary delineation. *Canadian Journal of Forest Research* 27: 1851–8.

Fortin, M. J., Drapeau, P., and Jacquez, G. M. 1996. Statistics to assess spatial relationships between ecological boundaries. *Oikos* 77: 51–60.

Gelman, A., Price, P. N., and Lin, C. 2000. A method for quantifying artefacts in mapping methods illustrated by application to headbanging. *Statistics in Medicine* 19: 2309–20.

Getis, A. and Ord, J. K. 1992. The analysis of spatial assosciation by use of distance statistics. *Geographical Analysis* 24: 189–206.

Goovaerts, P. 2005. Analysis and detection of health disparities using geostatistics and a space-time information system: The case of prostate cancer mortality in the United States, 1970–1994. In *Proceedings of GIS Planet 2005*, Estoril, Portugal. Lisboa: Instituto Geografico Portugues.

Goovaerts, P. and Jacquez, G. 2004. Accounting for regional background and population size in the detection of spatial clusters and outliers using geostatistical filtering and spatial neutral models: The case of lung cancer in Long Island, New York. *International Journal of Health Geographics* 3:14.

Goovaerts, P. and Jacquez, G. M. 2005. Detection of temporal changes in the spatial distribution of cancer rates using LISA statistics and geostatistically simulated spatial neutral models. *Journal of Geographical Systems* 7: 137–59.

Gustafson, E. J. 1998. Quantifying landscape spatial pattern: What is the state of the art? *Ecosystems* 1: 143–56.

Hall, K. R. and Maruca, S. L. 2001. Mapping a forest mosaic: A comparison of vegetation and bird distributions using geographic boundary analysis. *Plant Ecology* 156: 105–20.

Hochberg, Y. 1988. A sharper Bonferroni procedure for multiple tests of significance. *Biometrica* 75: 800–2.

Holland, B. S. and Copenhaver, M. D. 1987. An improved sequentially rejective Bonferroni test procedure. *Biometrics* 43: 417–23.

Jacquez, G. M. 1995. The map comparison problem: Tests for the overlap of geographic boundaries. *Statistics in Medicine* 14: 2343–61.

Jacquez, G. M. 1998. GIS as an enabling technology. In A. Gatrell and M. Loytonen (eds) *GIS and Health*. London: Taylor and Francis, pp. 17–28.

Jacquez, G. M. 2000. Spatial epidemiology: Nascent science or a failure of GIS? *Journal of Geographical Systems* 2: 91–7.

Jacquez, G. M. 2004. Current practices in the spatial analysis of cancer: Flies in the ointment. *International Journal of Health Geographics* 3:22.

Jacquez, G. M. and Greiling, D. A. 2003a. Local clustering in breast, lung, and colorectal cancers in Long Island, New York. *International Journal of Health Geographics* 2: 3 (available at http://www.ij-healthgeographics.com/content/2/1/3).

Jacquez, G. M. and Greiling, D. A. 2003b. Geographic boundaries in breast, lung, and colorectal cancers in relation to exposure to air toxics in Long Island, New York. *International Journal of Health Geographics* 2: 4 (available at http://www.ij-healthgeographics.com/content/2/1/4).

Jacquez, G. M., Greiling, D. A., Durbeck, H., Estberg, L., Do, E., Long, A., and Rommel, B. 2002. *ClusterSeer User Guide 2: Software for Identifying Disease Clusters*. Ann Arbor, MI: TerraSeer Press.

Jacquez, G. M., Greiling, D., Estberg, L., Do, E., Long, A., and Rommel, B. 2001. *ClusterSeer User Guide: Software for Identifying Disease Clusters*. Ann Arbor, MI: TerraSeer Press.

Jacquez, G. M., Grimson, R., Waller, L., and Wartenberg, D. 1996. The analysis of disease clusters: Part 2, Introduction to techniques. *Infection Control and Hospital Epidemiology* 17: 385–97.

Jacquez, G. M., Kaufmann, A., and Goovaerts, P. 2006. Boundaries, ladders and clusters: A new paradigm in spatial analysis? *Environmental and Ecological Statistics* 13: 31–48.

Jacquez, G. M., Maruca, S. L., and Fortin, M. J. 2000. From fields to objects: A review of geographic boundary analysis. *Journal of Geographical Systems* 2: 221–41.

Jacquez, G. M., Maruca, S. L., Greiling, D. A., Kaufmann, A., Muller, L., Rommel, B., Sengupta, S., Agarwal, P., and Hall, K. 2001. *BoundarySeer User Guide: Software for Geographic Boundary Analysis*. Ann Arbor, MI: TerraSeer Press.

Jacquez, G. M., Waller, L., Grimson, R., and Wartenberg, D. 1996. The analysis of disease clusters: Part I, State of the art. *Infection Control and Hospital Epidemiology* 17: 319–27.

Kafidar, K. 1996. Smoothing geographical data, particularly rates of disease. *Statistics in Medicine* 15: 2539–60.

Kulldorff, M. 1997. A spatial scan statistic. *Communications in Statistics: Theory and Methods* 26: 1481–96.

Kulldorff, M. and Nagarwalla, N. 1995. Spatial disease clusters: Detection and inference. *Statistics in Medicine* 14: 799–810.

Lawson, A. B. 1989. *Score Tests for Detection of Spatial Trend in Morbidity Data*. Dundee: Dundee Institute of Technology.

Lawson, A. B. and Kulldorff, M. 1999. A review of cluster detection methods. In A. B. Lawson, A. Biggeri, D. Böhning, E. Lesaffre, J.-F. Viel, and R. Bertollin (eds) *Advanced Methods of Disease Mapping and Risk Assessment for Public Health Decision Making*. London: John Wiley and Sons, pp. 99–110.

Lawson, A. B. and Waller, L. A. 1996. A review of point pattern methods for spatial modelling of events around sources of pollution. *Environmetrics* 7: 471–87.

Legendre, P. 1987. Constrained clustering. In P. Legendre and L. Legendre (eds) *Developments in Numerical Ecology*. Berlin: Springer-Verlag, pp. 289–307.

Liebisch, N., Jacquez, G. M., Goovaerts, P., and Kaufmann, A. 2002. New methods to generate neutral images for spatial pattern recognition. In M. J. Egenhofer and D. M. Mark (eds) *GIScience2002: The Second International Conference on Geographic Information Science*. Berlin: Springer-Verlag Lecture Notes in Computer Science No. 2478: 181–95.

Maruca, S. L. and Jacquez, G. M. 2002. Area-based tests for association between spatial patterns. *Journal of Geographical Systems* 4: 69–84.

Miller, B. A., Kolonel, L. N., Bernstein, L., Young Jr., J. L., Swanson, G. M., West, D., Key, C. R., Liff, J. M., Glover, C. S., Alexander, G. A., Coyle, L., Hankey, B. F., Gloeckler Ries, L. A., Kosary, C. L., Harras, A., Percy, C., and Edwards, B. K. 1996. *Racial/Ethnic Patterns of Cancer in the United States 1988–1992*. Bethesda, MD: National Cancer Institute Publication No. 96–4104.

Moore, D. A. and Carpenter, T. E. 1999. Spatial analytical methods and Geographic Information Systems: Use in health research and epidemiology. *Epidemiologic Reviews* 21: 143–61.

Moran, P. A. P. 1950. Notes on continuous stochastic phenomena. *Biometrika* 37: 17–23.

National Cancer Institute. 2003. *The National Cancer Institute's (NCI) Planning and Budget Proposal for Fiscal Year 2004: The Nation's Investment in Cancer Research*. WWW document, http://plan.cancer.gov/discovery/index.html.

Oden, N. 1995. Adjusting Moran's I for population density. *Statistics in Medicine* 14: 17–26.

Openshaw, S., Charlton, M., Craft, A. W., and Birch, J. M. 1988. Investigation of leukaemia clusters by use of a geographical analysis machine. *Lancet* 1: 272–3.

Ord, J. K. and Getis, A. 1995. Local spatial autocorrelation statistics: Distributional issues and an application. *Geographical Analysis* 27: 286–306.

Pandya, A. S. and Macey, R. B. 1996. *Pattern Recognition with Neural Networks in C++*. Boca Raton, FL: CRC Press.

Patil, G. P. and Taillie, C. 2004. Upper level set scan statistic for detecting arbitrarily shaped hotspots. *Environmental and Ecological Statistics* 11: 183–97.

Rushton, G. 1997. Improving public health through geographical information systems: An instructional guide to major concepts and their implementation (CD-ROM; Version 2.0). Iowa City, IA: Department of Geography, University of Iowa (available at http://www.uiowa.edu/~geog/).

Rushton, G. and Elmes, G. 2000. Considerations for improving Geographic Information System research in public health. *Journal of the Urban and Regional Information Systems Association* 12: 31–49.

Rushton, G., Peleg, I., Banerjee, A., Smith, G., and West, M. 2004. Analyzing geographic patterns of disease incidence: Rates of late-stage colorectal cancer in Iowa. *Journal of Medical Systems* 28: 223–36.

Sidak, Z. 1967. Rectangular confidence regions for the means of multivariate normal distributions. *Journal of the American Statistical Association* 62: 626–33.

Simes, R. J. 1986. An improved Bonferroni procedure for multiple tests of significance. *Biometrika* 73: 751–4.

Smith, G. H. 2003. Disease cluster detection methods: The impact of choice of shape on the power of statistical tests. Unpublished report, Department of Geography University of Iowa.

Sun, Y. 2002. Determining the size of spatial clusters in focused tests: Comparing two methods by means of simulation in a GIS. *Journal of Geographical Systems* 4: 359–70.

Tango, T. 1995. A class of tests for detecting "general" and "focused" clustering of rare diseases. *Statistics in Medicine* 14: 2323–34.

Tango, T. 2007. A class of multiplicity-adjusted tests for spatial clustering based on case-control point data. *Biometrics* 63: in press.

Tango, T. and Takahashi, K. 2005. A flexibly shaped spatial scan statistic for detecting clusters. *International Journal of Health Geographics* 4:1.

Turnbull, B. W., Iwano, E. J., Burnett, W. S., Howe, H. L., and Clark, L. C. 1990. Monitoring for clusters of disease: Application to leukemia incidence in upstate New York. *American Journal of Epidemiology* 132: S136–43.

Urban, D. L. 2004. Multivariate Analysis: Nonhierarchical Agglomeration, Spatially Constrained Classification. WWW document, http://www.env.duke.edu/landscape/classes/env358/mv_pooling.pdf.

Waller, L. A. and Jacquez, G. M. 1995. Disease models implicit in statistical tests of disease clustering. *Epidemiology* 6: 584–90.

Waller, L. A., Turnbull, B. W., Clark, L. C., and Nasca, P. 1992. Chronic disease surveillance and testing of clustering of disease and exposure: Application to leukemia incidence and TCE-contaminated dumpsites in upstate New York. *Environmetrics* 3: 281–300.

Chapter 23

Terrain Analysis

Yongxin Deng, John P. Wilson, and John C. Gallant

Digital terrain analysis seeks to construct mathematical abstractions of the terrain surface (for example, Moore, Grayson, and Landson 1991, Florinsky 1998a) to delineate or stratify landscapes (for example, Hammond 1964, Dikau 1989, Burrough, Wilson, van Gaans, and Hanson 2001) and to examine or define the relationships between the terrain surface and various biophysical processes/patterns (for example, Moore, Gessler, Nielsen, and Peterson 1993, Franklin 1995, Beven 1997). The essential scientific value of these three tasks relies on three simple facts: (1) terrain poses significant control over other biophysical elements, (2) the former is much easier to measure than the latter; and (3) both tend to vary continuously over space in a correlated fashion (Burrough and McDonnell 1998). The computed terrain attributes often provide important, if not the only, clues indicating key biophysical patterns and processes, and sometimes serve as a spatial prediction tool directly (for example, Moore, Gessler, Nielsen, and Peterson 1993, Bell, Grigal, and Bates 2000). These roles of terrain analysis represent a bridge from the known to the unknown, and are often vital for resource inventory and environmental modeling, especially at topo- (that is, hillslopes of 50–200 m in length) to meso-scales (that is, watersheds of 10–100 km^2 in extent). They also point to the strong multi-disciplinary character of many terrain analysis applications.

Terrain analysis is nonetheless different from most scientific approaches to the study of the biophysical environment in that it is enabled by GIS and related computer technologies, and is supported more by digital terrain data (mostly gridded DEMs – digital elevation models) than by direct field observations or laboratory measurements of biophysical properties. Terrain analysis is quantitative, implying high precision in terms of outputs, but these results may simultaneously be plagued with uncertainties in terms of the relationship between terrain and biophysical attributes – implying the possibility of low accuracy. The common approaches employed in soil-landscape analyses (Park, McSweeney, and Lowery 2001) – statistical correlation (for example, Moore, Gessler, Nielsen, and Peterson 1993), classification of terrain attributes based on pre-defined criteria (for example, Zhu 1997), and statistical clustering of terrain indices (for example, Irvin, Ventura, and

Slater 1997, McBratney and Odeh 1997) – provide typical examples of how terrain analysis results tend to be used. They all deal primarily with the probability, instead of the certainty, that we can: (1) use knowledge of soil–landscape relationships to infer soil conditions from terrain properties, and (2) extrapolate the relationships to other places or interpolate them to other scales. This is fundamentally different from conventional sciences such as chemistry in which chemical reactions can both be reproduced and explained with certainty.

The relative ease with which terrain analysis can be performed points to numerous opportunities, but implies tremendous challenges because of these inherent uncertainties. In other words, terrain analysis is more a science dealing with uncertainty than with certainty. These uncertainties are often linked to issues such as terrain data quality (Adkins and Merry 1994, Bolstad and Stowe 1994, Hunter and Goodchild 1997, Krupnik 2000, Deng, Wilson, and Goodchild 2006), algorithm reliability (Skidmore 1989, Desmet and Govers 1996a, Florinsky 1998b, Quinn, Beven, Chevallier, and Planchon 1991), spatial scale effects (Chang and Tsai 1991, Zhang and Montgomery 1994, Florinsky and Kuryakova 2000, Gertner, Wang, Fang, and Angerson 2002), objects with indeterminate boundaries (Burrough and Frank 1996), and ontological discrepancies regarding landform definitions (Hudson 1992, Zhu 1997, Burrough, Wilson, van Gaans, and Hanson 2001, Mark and Smith 2003). They signify the intrinsic complexity of terrain-environment relationships, as well as our relative lack of appreciation and understanding of these issues. The growth of new terrain data sources, terrain analysis programs, and terrain-based environmental models provide many new opportunities for biophysical study, although they should all be assessed carefully in terms of their scientific basis prior to widespread deployment. They may directly lead to an increase in precision, but do not necessarily imply a corresponding improvement in accuracy.

This chapter examines several essential aspects of terrain analysis based on the aforementioned general vision. Under the heading "Terrain Attributes – State of the Art" we describe terrain attributes as scale- and algorithm-dependent descriptions of the terrain surface and related biophysical processes. Three examples are used to identify the strengths and weaknesses of terrain-based environmental models and landscape stratifications in the second section, "Modeling and Synthesis." The third section, "Enduring Challenges," examines the effects of spatial scale and data quality on terrain analysis, and the final section highlights some of the developments that are likely to occur by the 2020s.

Terrain Attributes – State of the Art

All three of the terrain analysis tasks listed at the start of this chapter rely on calculated terrain attributes. A distinction is generally drawn between primary attributes that are computed directly from the DEM and composite attributes that involve combinations of primary attributes (Moore, Lewis, and Gallant 1993, Florinsky 1998a, Wilson and Gallant 2000a). Elevation is unique because its computation does not rely on other points; however, we often make assumptions about the character of the land surface – in terms of its continuity and smoothness – to estimate elevation in a DEM using sparse source data in practice (for example, Hutchinson 1989).

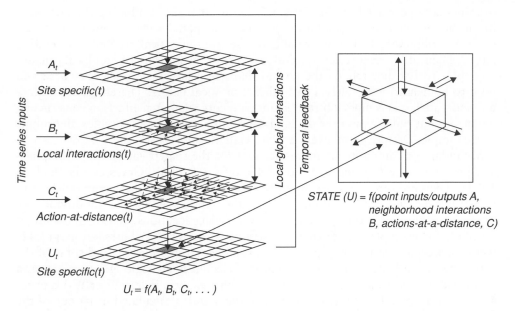

Fig. 23.1 Schematic diagram showing site-specific, local, and regional interaction as a function of time

From Wilson and Burrough 1999

Florinsky (1998a) also distinguished local primary attributes that are calculated as a function of their surroundings and non-local primary attributes that require the analysis of a larger, non-local land surface area from a computational perspective. Wilson and Burrough (1999) later explained this distinction between local versus non-local terrain attributes in terms of the existence of local interactions between neighboring points and "action-at-distance" forces (see Figure 23.1 for details).

Most primary attributes are calculated from the geometric derivatives of the terrain surface using either a second-order finite difference scheme (for example, Skidmore 1989, Moore, Lewis, and Gallant 1993b, Florinsky 1998b) or a bivariate interpolation function $z = f(x,y)$ that has been fitted to the DEM (Mitasova, Hofierka, Zlocha, and Iverson 1996). Typical examples of local primary attributes include slope, aspect, and plan and profile curvatures; non-local primary attributes include flow path length, proximity to nearest ridgeline, dispersal area, and upslope contributing area. More complete lists can be found in Moore, Grayson, and Ladson (1991), Moore, Lewis, and Gallant (1993b), Florinsky (1998a), and Gallant and Wilson (2000).

By definition, the local terrain shape – which is usually thought of as the continuous variation of elevation values over the terrain surface from point to point – has an enormous impact on local terrain attributes, but this role is influenced by data and computational factors. Florinsky (1998b) suggested that local attributes, such as slope gradient, aspect, and curvatures, are mathematical variables rather than real-world values. This statement may be extended to all local terrain attributes for two reasons. First, local terrain shape can have different mathematical descriptions, so that the calculated local attributes depend on algorithm selection. Second, the terrain shape portrayed by DEMs is a function of scale, combining the complexity of the terrain, scale or resolution of data, and spatial scale at which the terrain surface is observed. Thus it is possible to use the same local attribute to describe terrain shape at different scales (resolutions). The special feature of non-local primary attributes is that they rely on the terrain shape of a larger, non-neighbor area and need to be defined with reference to other, non-local points. Therefore, calculating non-local attributes is more difficult because it incurs additional efforts in constructing point-to-point connections over the landscape and involves more complex algorithms (for example, Desmet and Govers 1996a, Gallant and Wilson 2000).

Secondary or composite attributes account for the spatial variability of biophysical processes as a function of topographic effects (Moore, Grayson, and Ladson 1991). They are often used to quantify the role played by the terrain surface in redistributing water and sediments over the landscape and in modifying the amount of solar radiation received at various surface locations. Wilson and Gallant (2000b), for example, described three sets of composite attributes – topographic wetness, sediment transport capacity, and solar radiation indices – and some of the ways they have been deployed to interpret selected hydrologic, geomorphic, and ecological processes and patterns. The topographic wetness index (W or W_T) is probably the most popular of these composite indices and is calculated using one of the following equations depending on whether uniform soil transmissivity (T) under saturation is assumed or not:

$$W = \ln\left(\frac{a}{\tan\beta}\right)$$
(23.1)

or

$$W_T = \ln\left(\frac{a}{T\tan\beta}\right)$$
(23.2)

where a ($m^2\ m^{-1}$) is the specific catchment area and β (°) is the slope gradient (Kirkby 1976, Beven and Kirkby 1979, Moore, Grayson, and Ladson 1991). Provided certain conditions are met (Beven and Kirkby 1979), the wetness index describes the pattern of depth to water table in a catchment and hence the pattern of hydrologic response. It has frequently been used as an index of position in the landscape and of accumulation of materials for predicting soil properties.

Modeling and Synthesis

Both primary and composite attributes are frequently used to provide input data for various environmental models or landscape delineations based on attribute distributions. The outputs of these applications take the form of identified terrain–environment relationships, spatio-temporal predictions of environmental properties, and the delineation of meaningful spatial units over the landscape. Three examples are utilized below to demonstrate how new knowledge may be garnered through these modeling and synthesis activities. They also illustrate three enduring issues in terrain analysis: (1) the impact of the choice of spatial scale or landscape unit on model predictions; (2) the difficulties encountered representing spatial continuity; and (3) the problem of "equifinality."

Soil erosion/deposition modeling

The large number and complexity of factors influencing soil erosion rates coupled with the relative paucity of data at fine scales (that is, 5–10 meter grid cells) have slowed soil erosion model development since the 1950s. The first proposals for combining soil erosion models and Geographic Information Systemns (GIS) were published nearly two decades ago (for example, Ventura, Chrisman, Connors, Gurda, and Martin 1988, Warren, Diersing, Thompson, and Goran 1989) and may be contrasted with several of the more recent models that have been implemented in GIS environments from the outset (for example, Mitasova, Hofierka, Zlocha, and Iverson 1996, Mitas and Mitasova 1998).

The most popular, and in many ways most important, soil erosion model up to this point in time is the Universal Soil Loss Equation (USLE), an empirical equation derived from observations of more than 10,000 plot-years on farmlands (Wischmeier and Smith 1978; see Wilson and Lorang 1999 for a detailed review). It incorporates six factors – rainfall–runoff, soil erodibility, slope length, slope gradient, crop management, and conservation practices – and calculates the mean erosion rate (t ha^{-1} yr^{-1}) by comparing the conditions of the target slope (slope length, gradient, erodibility, management, etc.) with a standard soil-loss plot that is 22.13 m long and has a uniform width and slope gradient (9 percent).

This model is intrinsically limited to: (1) landscapes in which erosion is detachment limited; (2) planar slopes (except where the special rules for irregular slopes proposed by Foster and Wischmeier (1974), which divided irregular slopes into a series of planar slope facets, are implemented); and (3) those parts of the landscape that experience net erosion over the long term (this requirement will often exclude foot-slopes and valley bottoms in semi-arid and humid areas for example; see Wilson and Lorang 1999 for additional discussion of this limitation). Two reasons explain these limits. First, the model uses the entire slope as the basic spatial unit, and does not incorporate within-slope change in runoff direction and speed (that is, convergence, divergence, acceleration, and deceleration). These changes are a function of surface shape (that is, curvatures along or perpendicular to the steepest slope direction) and they are key to the successful prediction of within-slope variation of sediment transport processes (detachment, transport, deposition, and detainment). Second, slopes are conceptualized as isolated spatial units, so that the impact of input sediment flow and the possibility of net deposition are not taken into account.

The topographic or length-slope factor (LS) for the USLE, originally computed as a function of overall slope length and average slope gradient (Wischmeier and Smith 1978), is primarily responsible for the aforementioned limits of the USLE (Wilson and Lorang 1999). Wilson (1986) proposed a way around these limits that involved sampling slopes in watersheds and using topographic map information along with the irregular slope estimation method of Foster and Wischmeier (1974) to generate frequency distributions of LS for specific watersheds (that is, catchments). This method facilitated watershed-level comparisons (for example, Wilson and Ryan 1988, Wilson

1989) but was not able to characterize the erosion hazard at finer scales. Griffin, Beasley, Fletcher, and Foster (1988) later generalized the topographic or length-slope factor so that the USLE was able to estimate the soil erosion potential at specific places (that is, points) in the landscape (so long as the original model assumptions noted earlier held true). However, this approach greatly increased the time and effort needed to implement the USLE and, as a consequence, it attracted little attention prior to the widespread adoption and use of GIS for natural resource assessment.

It was therefore not surprising when Desmet and Govers (1996b) proposed a GIS-based method to calculate the topographic factor over a two-dimensional landscape that automated the approach of Griffin, Beasley, Fletcher, and Foster (1988), although with one important modification. They utilized the upslope contributing area in place of upslope flowpath length and then concluded that the original (that is to say, manual) method leads to an underestimation of the erosion risk because the effect of flow convergence is not taken into account.

Moore and Wilson (1992) had several years earlier proposed a dimensionless, unit stream power-based sediment transport capacity index T to replace LS in certain landscape conditions:

$$T = \left(\frac{a}{22.13}\right)^m \left(\frac{\sin \beta}{0.0896}\right)^n \tag{23.3}$$

where a is the specific catchment area (m^2 m^{-1}) and β (°) is the slope gradient. Foster (1994) criticized this approach because it relied on different assumptions (most notably that the erosion rate is transport- rather than detachment-limited) and attempted to modify just one of several components in this empirical model (the USLE is fundamentally a series of nested regression models and changing one component may necessitate changes to one or more other components). It is clear that these same criticisms would apply to the approach of Desmet and Govers (1996b) given the similarities between the two methods. Moore and Wilson (1994) subsequently acknowledged these shortcomings and went on to show that their equation produced similar results to the original USLE for certain slopes (that is, planar slopes with lengths <100 m and gradients <14°) despite the fact their approach relied on different assumptions to the original USLE.

Moore and Wilson (1992, 1994) also suggested calculating a second index:

$$\Delta T_{cj} = \Phi[a_{sj-}^m(\sin \beta_{j-})^n - a_{sj}^m(\sin \beta_j)^n] \tag{23.4}$$

where ΔT_{cj} is the change in T along the flow direction over a grid cell, Φ is a constant, a_s is the specific catchment area (m^2 m^{-1}), subscript j signifies the outlet of cell j and j- signifies the inlet to cell j, and β (°) is the slope gradient (as in Equation 23.3). They proposed using this equation to distinguish those parts of the landscape likely to experience net erosion ($\Delta T > 0$) from those parts likely to experience net deposition ($\Delta T \leq 0$), although this would clearly only work for landscapes in which soil erosion is transport limited.

Mitasova, Hofierka, Zlocha, and Iverson (1996) and Mitas and Mitasova (1998) later incorporated some of these same ideas in a soil erosion model that relied on the solution of bivariate first principles water and sediment flow equations. These equations can be used to characterize the relationship between erosion/deposition rates and terrain curvatures on slopes with varying soil and cover properties. Their models, which incorporated detachment- as well as transport-limited conditions and both profile and tangential curvatures, provide a sound theoretical explanation for the results of field experiments reported by Sutherland (1991), Busacca, Cook, and Mulla (1993), Quine, Desmet, Govers, Vandaele, and Walling (1994), and Heimsath, Deitrich, Nishiizumi, and Finkel (1997). The highest erosion rates were observed on divergent shoulder elements and deposition on convergent footslope elements in the first pair of studies, whereas the maximum soil loss was observed from the slope convexities and maximum gain in both the slope concavities and the main thalwegs in the final two studies. These results illustrate how small variations in terrain shape and soil and land cover can have a dramatic impact on the location and rates of soil erosion and deposition.

This discussion of terrain analysis and soil erosion models would not be complete without some mention of the Water Erosion Prediction Project (WEPP) model (Flanagan and Nearing 1995). The tremendous progress towards physically-based erosion models achieved within this project since at least the mid 1990s means that the USLE in its various forms is best suited to preliminary assessments and/or situations where data is limited nowadays. The WEPP model can be implemented at various levels and can predict erosion and deposition. The WEPP watershed model (Ascough, Baffaut, Nearing, and Liu 1997, Baffaut, Nearing, Ascough, and Liu 1997), for example, is an extension of the WEPP hillslope model and can be used for estimating watershed erosion and sediment yield. However, the application of WEPP to watersheds requires that hillslopes be delineated and channels identified (Figure 23.2). Each hillslope, represented as a rectangle in WEPP, must be assigned a representative length, width, and slope profile (as illustrated in the third part of Figure 23.2). Cochrane and Flanagan (1999) noted that GIS analysis using DEMs provides a useful tool for parameterization of hillslopes, channels, and representative slope profiles for WEPP simulations and set out to describe and evaluate three methods for integrating GIS and WEPP to facilitate watershed level applications that are often of interest to resource managers and policy analysts. This integration is relatively straightforward but it cannot overcome the fact that the WEPP family of models is based on a one-dimensional sediment routing over planar hillslopes (Foster, Flanagan, Nearing, et al. 1995) and that in many instances this approach will only partially explain the impact of terrain shape and the spatial variability of soil and land cover at the watershed scale (Mitas and Mitasova 1998).

Soil mapping and landform classification

Conventional soil–landscape analysis and soil mapping (Hudson 1992) are based on a crisp conceptual model that allows a data point to belong to only one class, and thus a sample location (such as a grid point) to fall in only one map unit. Soil variation only occurs across boundaries in geographic space and between the central concepts of prescribed soil classes in attribute space (Zhu 1997). From a utilitarian point of view, this leads to loss of soil information because: (1) minor deviations of local soil from the prescribed soil central concept may be known by local soil experts but cannot be included in a crisp soil map; and (2) soil bodies

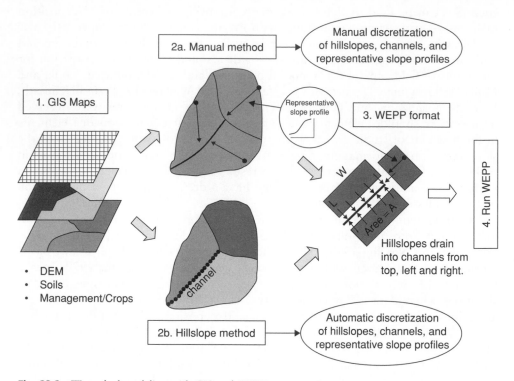

Fig. 23.2 Watershed modeling with GIS and WEPP
From Cochrane and Flanagan 1999

smaller than the minimum map unit either will be ignored or combined into another soil class (Zhu, Hudson, Burt, Lubich, and Simonson 2001).

The problems with this soil–landscape model follow from the fact that natural soils often exist as spatial continua and natural soil boundaries only exist under special circumstances (Burrough 1993, Burrough and Frank 1996, McBratney and Odeh 1997, Zhu 1997). It is also debatable whether soil should always be viewed as spatial objects or as the surrogate of aggregated soil properties, because there is "no agreement on what a basic or fundamental unit of soil is" (Arnold 1983). Hence, there could be endless combinations of dynamic soil conditions (properties), although individual soil properties are usually the major concern in applications such as soil and water conservation, non-point source pollution control, and precision agriculture (for example, Burrough 1993, Indorante, et al. 1996, Berry, Delgado, Khosla, and Pierce 2003). These two "discontinuity" and "object" problems in conventional soil mapping often occur simultaneously and help to explain the limitations of knowledge-based interpretations of soil–landscape relationships (Zhu, Hudson, Burt, Lubich, and Simonson 2001), which can be more precisely captured as correlations between soil properties and landscape attributes (for example, Burrough 1993, Moore, Gessler, Nielsen, and Peterson 1993, Florinsky and Kuryakova 2000).

Based on fuzzy logic, Zhu (1997, 2000), Zhu, Band, Dutton, and Nimlos (1996), Zhu, Band, Vertessy, and Dutton (1997), and Zhu, Hudson, Burt, Lubich, and

Simonson (2001) provided a solution for both sets of above-mentioned problems using a soil similarity model. This approach describes local soil as a similarity vector to prescribed soil classes (central concepts) through three steps: (1) identification of a set of central concepts according to existing soil classifications or expert knowledge; (2) definition of the linkages between these central concepts (or soil classes) and higher resolution landscape properties; and (3) calculation of the similarity (for example, values from 0 to 1) of each data point to each central concept through a comparison of their landscape properties. The soil properties of a point can then be derived by combining its similarity vector with the soil properties of the central concepts (Zhu, Band, Vertessy, and Dutton 1997, Zhu, Hudson, Burt, Lubich, and Simonson 2001). Terrain analysis offers several useful inputs for describing environmental conditions and constructing the soil–landscape models in this approach (Zhu, Hudson, Burt, Lubich, and Simonson 2001). The net effect of using terrain attributes is to incorporate continuous spatial variations of the biophysical environment into the output fuzzy soil classes at a higher resolution than conventional soil maps.

Soil mapping in the US, nevertheless, is a special case of natural resource inventory because it has a long history and has involved large human investments to provide rich resources for the identification of central soil concepts. When sufficient mapping resources and/or expert knowledge do not exist to support the central concepts (for example, Franklin 1995) the fuzzy k-means landform classification method will usually provide a better solution for the continuous delineation of the biophysical environment. This approach uses terrain attributes as input data to define the most representative clusters of the data following an iterative, unsupervised clustering procedure that can differentiate the data to a maximum extent (McBratney and Odeh 1997, Irvin, Ventura, and Slater 1997, Burrough and McDonnell 1998, Burrough, Wilson, van Gaans, and Hanson 2001). Class centers so defined are similar to central concepts in soil classification. The membership (similarity) of each data point to each class center is defined as the attribute distance between the point and the class center in the attribute space, which is calculated using a selected distance function (Irvin, Ventura, and Slater 1997). The biophysical meanings of fuzzy k-means landform class centers must be post-interpreted, rather than pre-defined, based on the terrain attributes that were used. Furthermore, some additional work will usually be required to select the terrain attributes (and weights) that will be used to identify the specific biophysical pattern(s) of interest.

Terrain analysis might also be used to create explicit environmental stratifications for survey design and to provide quantitative spatial predictions of individual soil properties. McKenzie, Gessler, Ryan, and O'Connell (2000), for example, described a two-step stratified random sampling strategy for the Bago-Maragle study area in New South Wales, Australia that combined geologic information (that is, published geologic map units supported by airborne gamma radiometric remote sensing) with the Prescott Index (which is a function of mean monthly precipitation and potential evapotranspiration), and the topographic wetness index. A field survey of 144 selected sample sites was then conducted and the data used for quantitative spatial prediction of key land qualities, including soil erodibility, nutrient status, and the soil–water regime (McKenzie and Ryan 1999). Terrain analysis variables can be used in conjunction with these other variables (climate geology, remote sensing, etc.) to extend point observations of individual soil properties using statistical models so long as the explanatory variables are easier to obtain than soil variables. Hence, Gessler (1996) used a regression tree approach to predict solum depth in the above-mentioned study area, and Bell, Grigal, and Bates (2000) used a series of linear and exponential statistical models to predict soil organic carbon in the Cedar Creek Natural History Area in Minnesota. Interested readers can learn more about the challenges and subtleties of the statistical methods employed in these types of modeling applications in McKenzie and Austin (1993), Gessler, Moore, McKenzie, and Ryan (1995), Gessler (1996), and McKenzie and Ryan (1999).

TOPMODEL

The original TOPMODEL introduced by Beven and Kirkby (1979) estimates overland runoff by integrating a spatially variable contributing area model with a simple, lumped soil water response (storage) model (Kirkby 1976). It defines the saturated area on the landscape A_c as the area where:

$$W > S_T/m - S_3/m + \lambda \tag{23.5}$$

where W is the topographic wetness index (Equation 23.1), S_T is the local maximum water storage (sum of surface interception, depression, near surface infiltration, and subsurface storage), and m and λ are constants. Given the assumption of a time-independent steady state rainfall rate, overland flow (q_{of}) is estimated as:

$$q_{of} = iA_c \tag{23.6}$$

where i is an instantaneous rainfall intensity and A_c is the saturated area calculated with Equation 23.5.

W and A_c were the only spatial variables used in the original model. All of the other variables – such as flow velocity, interception, infiltration, subsurface storage, and channel routing – were treated as lumped parameters and had to be measured or estimated. Beven (1997) discussed the equifinality problem in his critique of TOPMODEL and noted that many different sets of parameter values can simulate observed data (that is, the hydrograph) almost equally well in terms of some quantitative goodness-to-fit measure. Beven and Binley (1992) attributed the difficulty of finding a global optimum parameter set to the complexity of the multi-dimensional attribute space involved in hydrological modeling that is a function of the use of threshold parameters, intercorrelation between parameters, autocorrelation and heteroscedascity in the residuals, and inclusion of insensitive parameters. Beven (1993) and Savenije (2001) also linked this equifinality problem to scales of hydrological processes and "laws," as well as to the effects of aggregation and averaging (lumping) across scales during modeling.

Beven and Binley (1992), Beven (1997) and Beven and Freer (2001) have advocated using the GLUE (generalized likelihood uncertainty estimation) method to manage the equifinality problem. This method is based on Monte Carlo simulations in which the predictions of each parameter set realization are given a likelihood weighting according to how well that model fits the observed data (Beven 1997). The likelihood weights of the parameter sets can be updated when more data (that is, observations) become available. With the use of the GLUE method, the uncertainty of models can be defined more precisely and TOPMODEL applications can potentially become an iterative process of model selection, rejection, and optimization. Approaches similar to GLUE may see potential use in terrain analysis because the selection of terrain attributes and terrain analysis scales is currently based on data availability and it is likely that different combinations of scales (resolutions) and attributes (as well as their weights) may achieve the same terrain analysis goal. The advent of finer resolution DEMs may improve the definition of uncertainty in models using these types of uncertainty estimation methods.

Enduring Challenges

Data quality

The key role of terrain shape in terrain analysis indicates that the distribution, instead of the magnitude, of elevation errors should be the primary focus of DEM quality assessments (Hunter and Goodchild 1997; Burrough and McDonnell 1998, pp. 244–7; Heuvelink 1998; see Hutchinson, Chapter 8 of this volume, for additional discussion of the role of shape in DEM quality assessments). However, most DEM producers only provide aggregated error indicators (for example, RMSE or root mean squared error) that are calculated based on more accurate elevations of a few control points to report the mean magnitude of elevation errors over one tile (or map area) of the DEM. Some DEM users have utilized mutually independent sample points to link point elevation errors to the accuracy of calculated point terrain attributes (for example, Isaacson and Ripple 1990, Adkins and Merry 1994, Bolstad and Stowe 1994). The reliance on isolated points in both instances precludes the assessment of local distribution of error and the impact of these errors on terrain shape (see Wise 2000 for an extensive review of some of the key issues here). Table 23.1 shows that terrain shape may be severely distorted by local errors in DEMs that have a small RMSE and vice versa, and that the local standard deviation of errors calculated with a moving window is a better statistic to describe local distortions of terrain shape, given the availability of an error surface.

The focus on point or average errors dominated error assessments provided with DEMs of coarser resolutions (for example, 30 m or 3 arc second US Geological Survey DEMs), which are aggregated representations of toposcale surface conditions (that is, they are capable of identifying and characterizing 50–150 m long slopes). This focus presents even larger problems when applied to new DEM data sources of higher resolutions, because 10 m or finer DEMs can delineate within-slope variations much more precisely and accurately than previous 30 m or 3 arc second DEMs. For example, the subtle variations of landform structures on a slope (topographic hollows or local convexities) may be contained, or hidden, in the width of one 30 × 30 m DEM cell, but be manifest on a 10 m or finer resolution DEM. In the meantime, a 0.2 m elevation error of one point on a 10 percent uniform slope may cause the estimate of slope gradient to change from 10 percent to 10.7 percent on a 30 m DEM, but from 10 percent to 12 percent on a 10 m DEM, and from 10 percent to 30 percent on a 1 m DEM. These two examples indicate that the emergence

Table 23.1   DEM errors, shape representation, and appropriate/inappropriate statistics (adapted from Deng, Wilson, and Goodchild n.d.)

Scenarios	1			2			3			4		
Distribution of point	2	2	2	8	8	8	0	2	4	0	8	16
elevation errors	2	2	2	8	8	8	2	0	4	8	0	16
(imagined units)	2	2	2	8	8	8	4	0	2	16	0	8
Local mean error	2			8			2			8		
Local RMSE	2			8			$\sqrt{(20/3)}$			$\sqrt{(320/3)}$		
Error description based on RMSE	low			high			low			high		
Local standard deviation of errors	0			0			$2/\sqrt{3}$			$8/\sqrt{3}$		
Distortion of terrain shape	none			none			low			high		

Note: The word "local" refers to the 3 × 3 window shown in the four scenarios. It also implies that a 3 × 3 moving window can be applied to the entire area of interest to generate distributed error statistics.

of 10 m, 5 m, or even 1 m DEMs implies more than increasing detail in terrain surface modeling. Instead, these higher resolutions, compared to previous 30 m or 3 arc second ones, may represent a major shift of spatial scales incorporated in the DEMs, thus a more imminent need to evaluate the effects of DEM error distributions. In other words, we may be simultaneously facing more precise (in terms of horizontal resolution) and more erroneous (in terms of terrain shape) terrain surface depictions, given the same magnitude of RMSE of elevations reported for these high resolution DEMs as for the previous coarser resolution ones.

New methods are needed to circumvent the unavailability of a true error surface. Hunter and Goodchild (1997) proposed that, instead of dealing with error itself, we could define data or model uncertainty, or the extent to which we are unconfident in the obtained results. Specifically, they suggested that a worst-case scenario could be identified by introducing into DEMs a set of error fields that incorporate different degrees of spatial autocorrelation. All possible uncertainties caused by DEM errors could only occur within a range defined by this worst case scenario. This approach thus directs the research focus from the DEM errors themselves to the possible effects of DEM errors. A similar approach was adopted by Ehlschlaeger, Shortridge, and Goodchild (1997), who added a series of error surfaces with various autocorrelation and disturbance variables to an interpolation process and generated animated visualizations of data uncertainty with 250 realizations of interpolated 30 m DEMs (from a 3 arc second source DEM). Deng, Wilson, and Goodchild (n.d.) argue that it is also necessary to adopt a spatially explicit view to define the differences between various DEM sources and suggest that a DEM difference surface calculated from two DEMs can be used to develop spatial tools to estimate either the DEM errors themselves or the effects of terrain shape distortion on calculated terrain attributes.

Spatial scale

Moore, Lewis, and Gallant (1993) identified four scale-related issues in terrain analysis – basic element size, choice of attribute algorithms, merging of data sources, and scale differences between model and data – that still resonate today. Indeed, two additional issues should be added to this list given the recent emergence of numerous high-resolution DEMs; the need to: (1) define geomorphic units at various scales; and (2) calculate terrain attributes at appropriate scales with high-resolution data. The DEM spatial resolution was linked to issues of data quality and modeling uncertainty in the previous section – a fact that has been long observed (for example, Chang and Tsai 1991, Wolock and Price 1994, Zhang and Montgomery 1994, Mitasova, Hofierka, Zlocha, and Iverson 1996, Bian 1997, Wilson, Spangrud, Nielsen, Jacobsen, and Tyler 1998, Hutchinson and Gallant 2000, Gertner, Wang, Fang, and Angerson 2002) but has increased in importance in recent years. All these perspectives indicate the need to interpret the spatial scale in terrain analysis as an independent dimension that is related to all terrain analysis practices in either an explicit or implicit manner.

The basic (spatial) element size has attracted the most attention in the study or treatment of scale-dependencies of terrain analysis. For example, the topographic analysis for the first applications of TOPMODEL were based on a set of uniform areal elements that were approximately 5,500 m^2 in extent and delineated by dividing the basin according to flow lines, contour lines, and steepest slope lines (Beven and Kirkby 1979). Wood, Sivapalan, Beven, and Band (1988) adopted a similar approach based on the concept of representative elementary areas to define the scale effects of hydrological modeling. The consideration of landscape features (for example, flow lines, uniform slopes, etc.) in these applications produced irregular subdivisions that may provide a higher "actual" resolution than regular grid cells of the same size (Florinsky 1998a).

Most of these studies have examined the effects of the selected DEM resolution on calculated terrain attributes and modeling results given the widespread use of gridded DEMs. Isaacson and Ripple (1990), for example, observed very low correspondence between grid point slope gradient and aspect values calculated from 30 m and 3 arc second (roughly 65 × 92 m) US Geological Survey DEMs respectively. Chang and Tsai (1991) concluded that the accuracy of the same two attributes decreased with the increase of DEM cell size from 8 m to 80 m. Zhang and Montgomery (1994), Mitasova, Hofierka, Zlocha, and Iverson (1996), and Florinsky and Kuryakova (2000) identified threshold DEM resolutions for the modeling of soil moisture, overland flow, and erosion processes. Florinsky and Kuryakova (2000) interpreted the regular grid cell size, or DEM resolution, in terms of its adequacy for the description of specific landscape properties. Various statistical measures (for example, means, standard deviations, terrain–environment correlation coefficients, fractal dimensions, etc.) have been used in these types of studies to characterize the effects of spatial scale on computed terrian attributes (for example, Florinsky and Kuryakova 2000; see Moore, Lewis, and Gallant 1993 for a comprehensive review).

Figure 23.3 shows that spatially aggregated statistical analysis may not sufficiently capture the impact of DEM resolution on calculated terrain attributes. Hence, the effect of resolution variation varies from place to place and a simple assessment

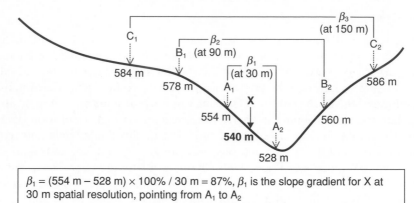

β_1 = (554 m − 528 m) × 100% / 30 m = 87%, β_1 is the slope gradient for X at 30 m spatial resolution, pointing from A_1 to A_2

β_2 = (578 m − 560 m) × 100% / 90 m = 20%, β_2 is the slope gradient for X at 90 m spatial resolution, pointing from B_1 to B_2

β_3 = (586 m − 584 m) × 100% / 150 m = 1%, β_3 is the slope gradient for X at 150 m spatial resolution, pointing from C_2 to C_1

Fig. 23.3   Scale effects of terrain analysis. Slope gradients (β_1, β_2, and β_3) for the same point B (or *d*) are defined in different ways due to the change of spatial resolution. The resultant slope gradients are different not only in magnitudes, but also in terms of topographic meanings

of attribute value change in magnitude may hide the fact that β_1, β_2, and β_3 in Figure 23.3 have different topographic, as well as biophysical, meanings. A more dramatic impact could be reasonably expected with greater change of spatial resolutions and/or in more complex terrain. Therefore, a spatially explicit approach that incorporates more complete interpretations of terrain attributes (for example, combining slope gradient with aspect) may need to be developed to account for the scale effects of terrain analysis.

Several other scale issues warrant further investigation as well. One is the potential problem of using a single-sized neighborhood window to estimate terrain characteristics. Multi-scale terrain analysis – the use of expanding neighborhood windows to calculate and compare terrain attributes – can potentially identify threshold window sizes across which the attribute values change abruptly to help clarify the meaning of different attributes and delineate natural landform boundaries. Gallant and Dowling (2003) demonstrate a method that combines terrain attributes at different scales into a single multi-scale attribute to represent geomorphic objects (valley bottoms) that occur at a range of scales.

Other key questions regarding spatial scales in terrain analysis that also require answers include:

- How do the model and attribute scale interact with one another?
- How can the various spatial resolutions of different attributes be incorporated into the same model according to their different process scales (or scales at work, see Bian 1997)?
- How do different terrain analysis algorithms behave with the change of scales?

- What modeling effects should be expected (that is to say anticipated) when combining data sources that incorporate different spatial scales?

SUMMARY

We briefly reviewed several key characteristics of terrain analysis and discussed several emerging perspectives, including the role of fuzzy logic, equifinality, shape-based data quality evaluations, and multi-scale terrain analysis. Several soil erosion models were reviewed to demonstrate the importance and great difficulties that are encountered delineating landscape units in these types of modeling applications. The traditional approach to soil–landscape analysis was described to portray how digital terrain analysis, together with new methodologies such as fuzzy logic, can improve soil classification and support a shift from crisp to continuous paradigms in various environmental modeling domains. The guiding principles of TOPMODEL were briefly discussed and used as an example to explain the equifinality problem and its potential significance to terrain analysis. The limitations encountered in managing spatial scale and data quality problems were identified as enduring challenges to terrain analysis and our ability to use computed terrain attributes to describe environmental patterns and processes of interest.

REFERENCES

Adkins, K. F. and Merry, C. J. 1994. Accuracy assessment of elevation data sets using the global positioning system. *Photogrammetric Engineering and Remote Sensing* 60: 195–202.

Arnold, R. W. 1983. Concepts of soils and pedology. In L. P. Wilding, N. E. Smeck, and G. F. Hall (eds) *Pedogenesis and Soil Taxonomy: Concepts and Interactions*. New York: Elsevier, pp. 1–22.

Ascough, J. C., Baffaut, C., Nearing, M. A., and Liu, B. Y. 1997. The WEPP watershed model: I. Hydrology and erosion. *Transactions of the American Society of Agricultural Engineers* 40: 921–33.

Baffaut, C., Nearing, M. A., Ascough, J. C., and Liu, B. Y. 1997. The WEPP watershed model: II. Sensitivity and discretization. *Transactions of the American Society of Agricultural Engineers* 40: 935–43.

Bell, J. C., Grigal, D. F., and Bates, P. C. 2000. A soil-terrain model for estimating spatial pattern of soil organic carbon. In J. P. Wilson and J. C. Gallant (eds) *Terrain Analysis: Principles and Applications*. New York: John Wiley and Sons, pp. 295–310.

Berry, J. K., Delgado, J. A., Khosla, R., and Pierce, F. J. 2003. Precision conservation for environmental sustainability. *Journal of Soil and Water Conservation* 58: 332–9.

Beven, K. J. 1993. Prophecy, reality and uncertainty in distributed hydrological modeling. *Advances in Water Resources* 16: 41–51.

Beven, K. J. 1997. TOPMODEL: A critique. *Hydrological Processes* 11: 1069–85.

Beven, K. J. and Binley, A. 1992. The future of distributed models: Model calibration and uncertainty prediction. *Hydrological Processes* 6: 279–98.

Beven, K. J. and Freer, J. 2001. Equifinality, data assimilation, and uncertainty estimation in mechanistic modeling of complex environmental systems using the GLUE methodology. *Journal of Hydrology* 249: 11–29.

Beven, K. J. and Kirkby, M. J. 1979. A physically based variable contributing area model of basin hydrology. *Hydrology Science Bulletin* 24: 43–69.

Bian, L. 1997. Multiscale nature of spatial data in scaling up environmental models. In D. A. Quattrochi and M. F. Goodchild (eds) *Scale in Remote Sensing and GIS*. New York: Lewis Publishers, pp. 13–26.

Bolstad, P. V. and Stowe, T. 1994. An evaluation of DEM accuracy: Elevation, slope and aspect. *Photogrammetric Engineering and Remote Sensing* 23: 387–95.

Burrough, P. A. 1993. Soil variability: A late 20th century view. *Soils and Fertilizers* 56: 531–61.

Burrough, P. A. and Frank, A. U. (eds). 1996. *Geographic Objects with Indeterminate Boundaries*. London: Taylor and Francis.

Burrough, P. A. and McDonnell, R. A. 1998. *Principles of Geographical Information Systems*. Oxford: Oxford University Press.

Burrough, P. A., Wilson, J. P., van Gaans, P. F. M., and Hanson, A. J. 2001. Fuzzy *k*-means classification of topo-climatic data as an aid to forest mapping in the Greater Yellowstone Area, USA. *Landscape Ecology* 16: 523–46.

Busacca, A. J., Cook, C. A., and Mulla, D. J. 1993. Comparing landscape scale estimation of soil erosion in the Palouse using Cs-137 and RULSE. *Journal of Soil and Water Conservation* 48: 361–7.

Chang, K. and Tsai, B. 1991. The effect of DEM resolution on slope and aspect mapping. *Cartography and Geographic Information Systems* 18: 69–77.

Cochrane, T. A. and Flanagan, D. C. 1999. Assessing water erosion in small watersheds using WEPP with GIS and digital elevation models. *Journal of Soil and Water Conservation* 54: 678–85.

Deng, Y. X., Wilson, J. P., and Goodchild, M. F. n.d. New DEM data sources and the portrayal of terrain shape. *International Journal of Geographic Information Science*: In preparation.

Desmet, P. J. J. and Govers, G. 1996a. Comparison of routing algorithms for digital elevation models and their implications for predicting ephemeral gullies. *International Journal of Geographical Information Systems* 10: 311–31.

Desmet, P. J. J. and Govers, G. 1996b. A GIS procedure for automatically calculating the USLE LS factor on topographically complex landscape units. *Journal of Soil and Water Conservation* 51: 427–33.

Dikau. R. 1989. The application of a digital relief model to landform analysis in geomorphology. In J. Raper (ed.) *Three Dimensional Applications in Geographic Information Systems*. London: Taylor and Francis, pp. 1–77.

Ehlschlaeger, C. R., Shortridge, A. M., and Goodchild, M. F. 1997. Visualizing spatial data uncertainty using animation. *Computers and Geosciences* 23: 387–95.

Flanagan, D. C. and Nearing, M. A. (eds). 1995. *Water Erosion Prediction Project: Hillslope Profile and Watershed Model Documentation*. West Lafayette, IN: USDA-ARS National Soil Erosion Research Laboratory Report No. 10.

Florinsky, I. V. 1998a. Combined analysis of digital terrain models and remotely sensed data in landscape investigations. *Progress in Physical Geography* 22: 33–60.

Florinsky, I. V. 1998b. Accuracy of local topographic variables derived from digital elevation models. *International Journal of Geographic Information Science* 12: 47–61.

Florinsky, I. V. and Kuryakova, G. A. 2000. Determination of grid size for digital terrain modeling in landscape investigations: Exemplified by soil moisture distribution at a micro-scale. *International Journal of Geographical Information Science* 14: 815–32.

Foster, G. R. 1994. Comment on "Length-slope factors for the Revised Universal Soil Loss Equation: Simplified method of estimation." *Journal of Soil and Water Conservation* 49: 171–3.

Foster, G. R., Flanagan, D. C., Nearing, M. A., Lane, L. J., Risse, L. M., and Finkner, S. C. 1995. Hillslope erosion component. In D. C. Flanagan and M. A. Nearing (eds) *Water Erosion Prediction Project: Hillslope Profile and Watershed Model Documentation*. West Lafayette, IN: USDA-ARS National Soil Erosion Research Laboratory Report No. 10: 11.1–11.12.

Foster, G. R. and Wischmeier, W. H. 1974. Evaluating irregular slopes for soil loss prediction. *Transactions of the American Society of Agricultural Engineers* 17: 305–9.

Franklin, J. 1995. Predictive vegetation mapping: Geographic modeling of biospatial patterns in relation to environmental gradients. *Progress in Physical Geography* 19: 474–99.

Gallant, J. C. and Dowling, T. I. 2003. A multi-resolution index of valley bottom flatness for mapping depositional areas. *Water Resources Research* 39: 1347–60.

Gertner, G., Wang, G., Fang, S., and Angerson, A. B. 2002. Effect and uncertainty of digital elevation model spatial resolutions on predicting the topographical factor for soil loss estimation. *Journal of Soil and Water Conservation* 57: 164–74.

Gallant, J. C. and Wilson, J. P. 2000. Primary topographic attributes. In J. P. Wilson and J. C. Gallant (eds) *Terrain Analysis: Principles and Applications*. New York: John Wiley and Sons, pp. 51–86.

Gessler, P. E. 1996. Statistical Soil-landscape Modeling for Environmental Management. Unpublished Ph.D. Dissertation, Centre for Resource and Environmental Studies, Australian National University.

Gessler, P. E., Moore, I. D., McKenzie, N. J., and Ryan, P. J. 1995. Soil-landscape modeling and spatial prediction of soil attributes. *International Journal of Geographical Information Science* 9: 421–32.

Griffin, M. L., Beasley, D. B., Fletcher, J. J., and Foster, G. R. 1988. Estimating soil loss on topographically non-uniform field and farm units. *Journal of Soil and Water Conservation* 43: 326–31.

Hammond, E. H. 1964. Analysis of properties in landform geography: An application to broad-scale landform mapping. *Annals of the Association of American Geographers* 54: 11–9.

Heimsath, A. M., Dietrich, W. E., Nishiizumi, K., and Finkel, R. C. 1997. The soil production function and landscape equilibrium. *Nature* 388: 358–61.

Heuvelink, G. B. M. 1998. *Error Propagation in Environmental Modelling with GIS*. London: Taylor and Francis.

Hudson, B. D. 1992. Soil genesis, morphology and classification: The soil survey as paradigm-based science. *Soil Science Society of America Journal* 56: 836–41.

Hunter, G. J. and Goodchild, M. F. 1997. Modelling the uncertainty of slope and aspect estimates derived from spatial databases. *Geographical Analysis* 29: 35–49.

Hutchinson, M. F. 1989. A new procedure for gridding elevation and streamline data with automatic removal of pits. *Journal of Hydrology* 106: 211–32.

Hutchinson, M. F. and Gallant, J. C. 2000. Digital Elevation Models and representations of terrain shape. In J. P. Wilson and J. C. Gallant (eds) *Terrain Analysis: Principles and Applications*. New York: John Wiley and Sons, pp. 29–50.

Indorante, S. J., Hammer, R. D., Thompson, B. W., and Alexander, D. L. 1996. Positioning soil survey for the 21st century. *Journal of Soil and Water Conservation* 51: 21–8.

Irvin, B. J., Ventura, S. J., and Slater, B. K. 1997. Fuzzy and isodata classification of landform elements from digital elevation data in Pleasant Valley, Wisconsin. *Geoderma* 77: 137–54.

Isaacson, D. L. and Ripple, W. J. 1990. Comparison of 7.5-minute and 1-degree digital elevation models. *Photogrammetric Engineering and Remote Sensing* 56: 1523–27.

Kirkby, M. J. 1976. Hydrograph modelling strategies. In R. Peel, M. Chrisholm, and P. Haggett (eds) *Process in Physical and Human Geography*. London: Heinemann, pp. 69–90.

Krupnik, A. 2000. Accuracy assessment of automatically derived digital elevation models from SPOT images. *Photogrammetric Engineering and Remote Sensing* 66: 1017–23.

McBratney, A. B. and Odeh, I. O. A. 1997. Application of fuzzy sets in soil science: Fuzzy Logic, fuzzy measurements, and fuzzy decisions. *Geoderma* 77: 85–113.

McKenzie, N. J. and Austin, M. P. 1993. A quantitative Australian approach to medium- and small-scale surveys based on soil stratigraphy and environmental correlation. *Geoderma* 57: 329–55.

McKenzie, N. J., Gessler, P. E., Ryan, P. J., and O'Connell, D. A. 2000. The role of terrain analysis in soil mapping. In J. P. Wilson and J. C. Gallant (eds) *Terrain Analysis: Principles and Applications*. New York: John Wiley and Sons, pp. 245–65.

McKenzie, N. J. and Ryan, P. J. 1999. Spatial prediction of soil properties using environmental correlation. *Geoderma* 89: 67–94.

Mark, D. M. and Smith, B. 2003. A science of topography: Bridging the qualitative–quantitative divide. In M. P. Bishop and J. Shroder (eds) *Geographic Information Science and Mountain Geomorphology*. Chichester: Springer-Praxis, pp. 1–23.

Mitas, L. and Mitasova, H. 1998. Distributed soil erosion simulation for effective erosion prevention. *Water Resources Research* 34: 505–16.

Mitasova, H., Hofierka, J., Zlocha, M., and Iverson, L. R. 1996. Modeling topographic potential for erosion and deposition using GIS. *International Journal of Geographical Information Systems* 10: 629–41.

Moore, I. D., Gessler, P. E., Nielsen, G. A., and Peterson, G. A. 1993. Soil attribute prediction using terrain analysis. *Soil Science Society of America Journal* 57: 443–52.

Moore, I. D., Grayson, R. B., and Ladson, A. R. 1991. Digital terrain modeling: A review of hydrological, geomorphological, and biological applications. *Hydrological Processes* 5: 3–30.

Moore, I. D., Lewis, A., and Gallant, J. C. 1993. Terrain attributes: Estimation methods and scale effects. In A. J. Jakeman, M. B. Beck, and M. J. McAleer (eds) *Modelling Change in Environmental Systems*. New York: John Wiley and Sons, pp. 189–214.

Moore, I. D. and Wilson, J. P. 1992. Length-slope factors for the revised Universal Soil Loss Equation: Simplified method of estimation. *Journal of Soil and Water Conservation* 47: 423–8.

Moore, I. D. and Wilson, J. P. 1994. Reply to comments by Foster. *Journal of Soil and Water Conservation* 49: 174–80.

Park, S. J., McSweeney, K., and Lowery, B. 2001. Identification of the spatial distribution of soils using process-based terrain characterization. *Geoderma* 103: 249–72.

Quine, T. A., Desmet, P. J. J., Govers, G., Vandaele, K., and Walling, D. E. 1994. A comparison of the roles of tillage and soil erosion in landform development and sediment export on agricultural land near Leuven, Belgium. In L. J. Olive (ed.) *Variability in Stream Erosion and Sediment Transport*. Gentbrugge, Belgium: International Association of Hydrological Sciences Publication No. 224: 77–86.

Quinn, P. F., Beven, K. J., Chevallier, P., and Planchon, O. 1991. The prediction of hillslope flow paths for distributed hydrological modeling using digital terrain models. *Hydrological Processes* 5: 59–79.

Savenije, H. H. G. 2001. Equifinality, a blessing in disguise? *Hydrological Processes* 15: 283–8.

Skidmore, A. K. 1989. A comparison of techniques for calculating gradient and aspect from a gridded digital elevation model. *International Journal of Geographical Information Systems* 3: 323–34.

Sutherland, R. A. 1991. Caesium-137 and sediment budgeting within a partially closed drainage basin. *Zeitschrift für Geomorphologie* 35: 47–63.

Ventura, S. J., Chrisman, N. R., Connors, K., Gurda, R. F., and Martin, R. W. 1988. A land information system for soil erosion control planning. *Journal of Soil and Water Conservation* 43: 230–3.

Warren, S. D., Diersing, V. E., Thompson, P. J., and Goran, W. D. 1989. An erosion-based land classification system for military installations. *Environmental Management* 13: 251–7.

Wilson, J. P. 1986. Estimating the topographic factor in the Universal Soil Loss Equation in watersheds. *Journal of Soil and Water Conservation* 41: 179–84.

Wilson, J. P. 1989. Soil erosion from agricultural land in the Lake Simcoe-Couchiching Basin, 1800–1981. *Canadian Journal of Soil Science* 69: 137–51.

Wilson, J. P. and Burrough, P. A. 1999. Dynamic modeling, geostatistics, and fuzzy classifications: New sneakers for a new geography? *Annals of the Association of American Geographers* 89: 736–46.

Wilson, J. P. and Gallant, J. C. 2000a. Digital terrain analysis. In J. P. Wilson and J. C. Gallant (eds) *Terrain Analysis: Principles and Applications*. New York: John Wiley and Sons, pp. 1–27.

Wilson, J. P. and Gallant, J. C. 2000b. Secondary topographic attributes. In J. P. Wilson and J. C. Gallant (eds) *Terrain Analysis: Principles and Applications*. New York: John Wiley and Sons, pp. 87–132.

Wilson, J. P. and Lorang, M. S. 1999. Spatial models of soil erosion and GIS. In A. S. Fotheringham and M. Wegener (eds) *Spatial Models and GIS: New Potentials and New Models*. London: Taylor and Francis, pp. 83–108.

Wilson, J. P. and Ryan, C. M. 1988. Landscape change in the Lake Simcoe-Couchiching Basin, 1800–1983. *Canadian Geographer* 206–22.

Wilson, J. P., Spangrud, D. J., Nielsen, G. A., Jacobsen, J. S., and Tyler, D. A. 1998. Global positioning system sampling intensity and pattern effects on computed topographic attributes. *Soil Science Society of America Journal* 62: 1410–7.

Wischmeier, W. H. and Smith, D. D. 1978. *Predicting Rainfall Erosion Losses: A Guide to Conservation Planning*. Washington DC: US Department of Agriculture Handbook No. 537.

Wise, S. 2000. Assessing the quality for hydrological applications of digital elevation models derived from contours. *Hydrological Processes* 14: 1909–29.

Wolock, D. M. and Price, C. B. 1994. Effects of digital elevation model and map scale and data resolution on a topography-based watershed model. *Water Resources Research* 30: 3041–52.

Wood, E. C., Sivapalan, M., Beven, K. J., and Band, L. E. 1988. Effects of spatial variability and scale with implications to hydrologic modeling. *Journal of Hydrology* 102: 29–47.

Zhang, W. H. and Montgomery, D. R. 1994. Digital elevation model grid size, landscape representation, and hydrologic simulations. *Water Resources Research* 30: 1019–28.

Zhu, A.-X. 1997. A similarity model for representing soil spatial information. *Geoderma* 77: 217–42.

Zhu, A.-X. 2000. Mapping soil landscape as spatial continua: The neural network approach. *Water Resources Research* 36: 663–77.

Zhu, A.-X., Band, L. E., Dutton, B., and Nimlos, T. J. 1996. Automated soil inference under fuzzy logic. *Ecological Modeling* 90: 123–45.

Zhu, A.-X., Band, L. E., Vertessy, R., and Dutton, B. 1997. Derivation of soil properties using a soil land inference model (SoLIM). *Soil Science Society of America Journal* 61: 523–33.

Zhu, A.-X., Hudson, B., Burt, J., Lubich, K., and Simonson, D. 2001. Soil mapping using GIS, expert knowledge, and fuzzy logic. *Soil Science Society of America Journal* 65: 1463–72.

Chapter 24

Dynamic Modeling

Jochen Albrecht

Most Geographic Information Systems (GIS) are an epitome of static, which is why to many people the word "dynamic GIS" is an oxymoron. GIS usually is about data and to a much lesser degree about what can be accomplished with the data. As such, GIS can be compared to early twentieth-century geography – an Aristotelian description of "where is what" and "what is where" (Barnes 1984). Process, and the notion of change, are acknowledged on the sidelines of Geographic Information Science (GISc) research but have not yet become the mainstream in terms of everyday GIS use. This is in spite of some 15 years since Tomlin's (1990) "cartographic modeling," the success of the GIS and Environmental Modeling conference series (for example, Goodchild, Steyaert, Parks, et al. 1996, Clarke and Parks 2001), and well-established pockets of spatial process modeling research such as Peter Burrough's group at the University of Utrecht in the Netherlands. This chapter will reframe some of the work usually subsumed under the header of geocomputation by focusing on a single but new dimension – time – and uncovering low-level structural problems (Burrough 1992) that make it so hard to merge the spatial and temporal aspects of GIS data models (see also Chapter 9 by Yuan in this volume, for further discussion of this topic).

Similar to the diversity of disciplinary origins of GIS in the 1970s (US Census Bureau 1969, Tomlinson 1973, Dutton 1978), current approaches to dynamic GIS come from a variety of backgrounds. IT-related approaches include cellular automata and agent-based modeling systems (Itami 1988, Clarke and Gaydos 1998), classical physical geography software development focuses on the marriage between traditional fluid dynamics modeling and GIS (Maidment 1993, Wesseling, Karssenberg, Burrough, and van Deursen 1996), civil engineers expand the realm of computer aided design (CAD) by adding sophisticated schedulers (SCADA 1991, Miller and Shaw 2001, Zlatanova, Holweg, and Coors 2003), and 2004 marked the first year that a major vendor released a scripting tool that allows end users rather than developers to create their own dynamic spatial models (ESRI 2003). So far, there have been few efforts to combine all these approaches and to give them a coherent scientific foundation. As such, this chapter is as much defining a GISc research agenda as it is reporting on the early successes of such research.

Among the earliest of such endeavors were, quite appropriately, a series of dissertations in the early 1990s that independently aimed to summarize different approaches to the incorporation of temporal elements into GIS analysis (Hazelton 1991, Kelmelis 1991, Al-Taha 1992, Langran 1992, Hamre 1994). Accompanied by academic prototypes, they all sought to add temporal querying capabilities to existing GIS structures. These temporal extensions to GIS have predominantly been modeled either as snapshots, where each layer represents an instance in time, or by amendment vectors, where each entity is associated with a list that contains information regarding each change in the entity (Langran 1992, Peuquet 1994).

The "snapshot" data model, the earliest representation of time in GIS, organizes space over time, where each raster layer is used to represent a state of the world at a point in time (Wachowicz 1999). A collection of those spatio-temporal snapshots is used to represent a 4D space-time cube, where at each time step there is a tuple of object id, space, and time (Peuquet 2001). This snapshot model is conceptually intuitive, convenient, and easily adapts to available data sources such as satellite imagery, hence it remains prevalent due to its simplicity (for example, Chen and Zaniolo 2000). Problems of large-scale data redundancy, where over time phenomena do not change everywhere, produced an alternative, the base-state with amendment model. This model updates states from the initially complete snapshot for only those objects that undergo change (Langran 1992). For both these approaches, change is interpolated, whether it be between system states or object states.

Incorporating time into the raster and vector data models is seen as the obvious solution to representing dynamics. However, as argued by Peuquet (1994), time and space exhibit important differences that do not comply with the neat addition of dimensions. Recognition that "simply extending a spatial data model to include temporal data, or vice versa, will result in inflexible and inefficient representations for space-time data" (Peuquet 2001, p. 15) has produced a slew of spatio-temporal alternatives. Then again, time can be represented by space, as has been developed in time geography, which implements Hägerstrand's (1967) classic model of temporal phenomena. Computational implementations of time geography represent the potential path of an individual as a spatial extent which changes over time as the individual moves through space over time (Huisman and Forer 1998, Miller 2003).

A different approach, first described by Kirby and Pazner (1990) and expanded by Smith, Su, El-Abbadi, et al. (1995), Pullar (2001, 2003), and Pedrosa, Fonseca, Câmara, Carneiro, and Souza (2002) is the idea to concatenate GIS procedures to modeling scripts. While these are add-ons to existing GIS structures, Wesseling, Karssenberg, Burrough, and van Deursen (1996) went one step further with the integration of a full-fledged dynamic modeling language into their PCRaster system. Both ESRI's geoprocessing framework (ESRI 2003) and PCRaster's modeling language integration have proven the validity of this approach by being marketable. However, there are many limitations to this coupling of inherently discrete and continuous modeling approaches, which have been well documented (Waters 2002). As Kemp (1997, p. 232) notes, "In order to fully integrate the two we need to add dynamics and continuity to our understanding of spatial data and spatial interaction and functionality to the environmental models."

Object-orientation has been hailed as a new basis for representing environmental processes (for example, Raper and Livingstone 1995, Wachowicz 1999, Bian 2000).

Object-orientated approaches typically handle time by time-stamping objects or time-stamping their attributes (Stefanakis 2003). Yuan (1996, 2001; Chapter 9 in this volume) developed typologies of modeling change, where change is represented as a new state with a new time stamp. This expression draws apart the temporal, spatial and attribute dimensions, reducing change to a variety of distinct forms. For capturing change in spatial objects, various temporal interpolation methods have been proposed for geometric changes of spatial objects (Zhang and Hunter 2000).

While all of the above-mentioned work can be seen as a natural extension of GIS technology, other, usually more social-science oriented researchers started to look at cellular automata (CA) and agent-based modeling (ABM) systems in an attempt to capture individual spatial behaviors. The majority of CA-based research (Couclelis 1985, 1997; Batty and Xie 1994; Wu 1998; Batty, Xie, and Sun 1999; Shi and Pang 2000; O'Sullivan 2001a, b; Benenson, Omer, and Portugali 2002) uses this simulation environment to determine which spatial configurations and what set of rules (behavior) lead to a desired or observed urban form. University College of London's Centre for Advanced Spatial Analysis has been the source for an impressive array of software tools that mix and match CA and ABM to mimic urban landscapes. Similar to Batty and Longley's (1994) fractal cities, the emphasis is on prediction and the exploration of new techniques.

More theoretically oriented is the use of ABM by a group of European regional scientists (Bura, Guerin-Pace, Mathian, Pumain, and Sanders 1996, Nijkamp and Reggiani 1998, Sembolini 2000, Benenson and Torrens 2004), who seek to develop software environments that help to create explanatory spatiotemporal models. These models are formal specifications of conceptual models in settlement geography, regional economics, political geography, and in one rare case coastal geomorphology (Raper and Livingstone 1995, Raper, Livingstone, Bristow, and Horn 1999) and as such are aimed at confirming existing or developing new theories. This is in stark contrast to their US counterparts (for example, Smith, Beckman, Baggerly, Anson, and Williams 1993, Westervelt and Hopkins 1999, Villa 2000, Agarwal, Green, Grove, Evans, and Schweik 2001, Jenerette and Wu 2001, Gimblett 2002, Waddell 2002), who are very application oriented.

These developments have lead to what has been termed dynamic GIS (De Vasconcelos, Gonçalves, Catry, Paúl, and Barros 2002). Here the lines between the traditional fields of GIS and computational simulation are rapidly blurring, with both the increasing integration of GIS data structures into computational simulation tools and the converse of the import of simulation tools into a GIS environment. For example, De Vasconcelos, Gonçalves, Catry, Paúl, and Barros (2002) present a dynamic GIS which is based on a "geounit", a CA-like data structure which extends that simple formalism to any form of spatial structure and is combined with scheduled and event-based events. A further example of the integration of computational simulation and GIS is the development of CA within a GIS, for example, Wesseling, et al. (1996) developed a spatially distributed hydrological model in PCRaster (an open source GIS developed at the University of Utrecht), which is essentially a CA.

In either case, CA and ABM systems do little to address the fundamental shortcomings of GIS (Chrisman 1987, Burrough 1992, Raper 2000). One of the first to basically start from scratch with the development of a new four-dimensional data model was Peuquet (1992, 1994), whose work resulted in a series of research projects

(MacEachren, Wachowicz, Edsall, and Haug 1999, Peuquet and Guo 2000) and sparked a new generation of truly geographic (as opposed to computer-science as discussed later) data modeling literature (Yuan and Lin 1992, Peuquet and Wentz 1994, Peuquet and Duan 1995, Tryfona and Jensen 1999, Wachowicz 1999, Yuan 1999 and Chapter 9 in this volume, Mennis, Peuquet, and Qian 2000, Renolen 2000).

Although the discipline of geography adopted the notion of process around the turn of the twentieth century, it failed almost completely to scrutinize the fundamental role of time. Non-computational exceptions were Blaut (1961) Hägerstrand (1967), Carlstein, Parkes, and Thrifts (1978), and Pred (1981). Now, we experience a renaissance of "time geography" and the works of Egenhofer and Golledge (1998), Kwan (1998), Frank, et al. (2001), Bian (2000), Raper (2000), Frihida, Marceau, and Thériault (2002), and Pereira (2002) provide the first usable geographic conceptualizations that were compiled with an implementation on a computer in mind. The crucial difference to FORTRAN hacks of previous generations is that these are genuine geographic models and not mere adaptations of physics. Miller (Miller and Wentz 2003; Miller 2005) summarizes the current state of geographic conceptualizations of space and time.

The next step in implementing these new geographic models on a computer is to develop formal spatio-temporal specifications or ontologies. Starting with Casati, Smith, and Varsi (1998), this has become an important new thread of GISc research (for example, Hornsby and Egenhofer 2002, Mota, Bento, and Botelho 2002, Bittner and Smith 2003, Reitsma and Bittner 2003). Philosophers, computer scientists, and geographers are now concentrating on two phenomena and their conceptualization, representation, and analysis (Erwig, Güting, Schneider, and Vazirgiannis 1999): one is the notion of "process" (Claramunt and Thériault 1996, Chen and Molenaar 1998, Pang and Shi 2002) and the other that of "change" (Frank et al. 2001, Galton 2000, Hornsby and Egenhofer 2000, Worboys 2001).

Traditional representations of geographic phenomena (Berry 1964, Goodchild 1990) are object centered, where an object has a location (x,y,z), some attributes (a,b,c), and rarely also some time stamp (t). This is usually represented as $G = f(x,y,z,[t],a,b,c, \dots)$. The geographic object G can be one or more raster cells or some vector geometry, in which case a, b, c, t would be held constant over whatever area G covers.

Complementary to this object-centered perspective is a process-based one, where the primitive is a tuple of the form $(x_1,y_1,z_1,t_1,a_1,b_1,c_1,\Delta,x_2,y_2,z_2,t_2,a_2,b_2,c_2)$. Vector Δ represents some form of transformation, including the null transformation, where nothing changes, in which case $x_1 = x_2$, $a_1 = a_2$, etc. Process as primitive provides leverage for querying and analyzing the geographic processes that have for so long eluded analysis. Crucial to the identity of a process is its pattern of change, vector Δ from above.

Example 1: Air Traffic Control (ATC)

The ultimate goal of ATC is to get airplanes safely from one place to another. Safely means that there should be a minimum distance between planes, measured in minutes (between takeoffs) or miles (3 miles horizontal, 1000' vertical near the airport, 5 miles

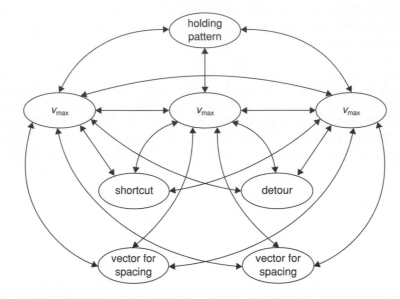

Fig. 24.1 Eight possible ATC commands for a single aircraft
From Bayen, Grieder, and Tomlin 2005

and 2000′ vertical further away). The main process is the flight itself, a path determined by waypoints and time stamps. Both aircraft separation and flight plan are obviously easily represented in the tuple structure given above. This main process can be divided into eight sub-processes, which form the eight different classes of flight directives that an air traffic controller may communicate to the pilot (Figure 24.1).

Auxiliary processes govern the communication between air traffic controllers and management of up to 30 airplanes per individual controller; they are well codified (Wickens, Mavor, Purusurum, and McGee 1998) and include all the contextual and relational information that we would place into our blackboard structure such as: the hierarchy of the system command center, route traffic control centers, sectors, TRACONs (Terminal Radar Approach Control – typically, the TRACON controls aircraft approaching and departing between 5 and 50 miles of the airport), and terminal air traffic control towers; the system of airspace classes, and especially the protocol for transferring responsibility from one controller to the next one along the flight path. This all works fairly well until bad weather or human error moves a complicated system into a state of complexity. Example 2 deals with the weather.

Example 2: Advanced Regional Prediction System (ARPS)

ARPS is a development of the Center for Analysis and Prediction of Storms at the University of Oklahoma. In their 700+ page user manual (APRS 2006) the Center describes a model for storm prediction.

A storm is the movement of air masses, the prediction of which is dependent on momentum, heat, mass, water transfer, and turbulent kinetic energy. The model is essentially an equation of state that is initialized using prescribed analytical functions. The basic model variables are defined as:

$$u(x,y,z,t) = \mathbf{u}(z) + u'(x,y,z,t)$$
$$v(x,y,z,t) = v(z) + v'(x,y,z,t)$$
$$w(x,y,z,t) = w'(x,y,z,t)$$
$$\theta(x,y,z,t) = \mathbf{\theta}(z) + \theta'(x,y,z,t)$$
$$p(x,y,z,t) = p(z) + p'(x,y,z,t)$$
$$\varrho(x,y,z,t) = \varrho(z) + \varrho'(x,y,z,t)$$
$$qv(x,y,z,t) = qv(z) + qv'(x,y,z,t)$$
$$qli(x,y,z,t) = qli'(x,y,z,t)$$

where u, v and w are the Cartesian components of velocity (momentum), θ the potential temperature, p the pressure, ϱ the density, qv the water vapor mixing ratio, and qli one of the hydrometeorological categories (vapor, rain, snow, hail). The bold-faced variables represent the base state and the primed variables are the deviations (APRS 2006, p. 6:5).

ARPS solves prognostic equations for u, v, w, θ', p' and q, ψ, which are, respectively, the x, y and z components of the Cartesian velocity, the perturbation potential temperature and perturbation pressure, and the six categories of water substance (water vapor, cloud water, rainwater, cloud ice, snow, and hail). The whole ARPS model is hence one huge tuple of the form $(x_1,y_1,z_1,t_1,a_1,b_1,c_1,\Delta,x_2,y_2,z_2,t_2,a_2,b_2,c_2)$.

The challenge (and opportunity) in linking the air traffic control model with the storm prediction model is to determine when what characteristic of Δ of the ARPS model is likely to have a negative impact on Δ in the ATC model. The beauty of this example is that everything is fairly straightforward because we use compatible mathematical notations. Wesseling's success with PCRaster is based on its design for application areas whose conceptual models lend themselves to be formalized in algebra. The advantage built into the above example is a shared conceptualization of space and time.

This is unfortunately not true across all geographically relevant domains. What we need is a "language of change." This language needs to be: (1) high-level to let domain scientists rather than highly specialized information scientists encode their models; and (2) extensible and open enough to allow for different notions of space and time at a range of scales. Conceptually, Reitsma and Albrecht's (2005) *nen* model has this kind of openness but it has yet to be supported by a process ontology that can express rules defining the thresholds of process change, and operations expressing the behavior of the process. The semantic web rule language (W3C 2005) is currently the most promising candidate for such tools that would allow us to build truly interoperable libraries of geospatial process models.

REFERENCES

Agarwal, C., Green, G., Grove, M., Evans, T., and Schweik, C. 2001. *A Review and Assessment of Land-use Change Models: Dynamics of Space, Time, and Human Choice.* Bloomington, IN: Indiana University, Center for the Study of Institutions, Population, and Environmental Change Collaborative Report No. 1.

Al-Taha, K. 1992. Temporal Reasoning in Cadastral Systems. Unpublished PhD Dissertation, Department of Spatial Information Science and Engineering, University of Maine.

ARPS. 2006. Advanced Regional Prediction System User's Guide (Version 5). Norman, OK: University of Oklahoma, Center for Analysis and Prediction of Storms (available at http://www.caps.ou.edu/ARPS/arpsdoc.html).

Barnes, J. (ed.). 1984. *The Complete Works of Aristotle*. Oxford: Oxford University Press.

Batty, M. and Longley, P. A. 1994. *Fractal Cities*. London: Academic Press.

Batty, M. and Xie, Y. 1994. From cells to cities. *Environment and Planning B* 21: 31–48.

Batty, M., Xie, Y., and Sun, Z. 1999. Modeling urban dynamics through GIS based cellular automata. *Computers Environment and Urban Systems* 23: 205–33.

Bayen, A., Grieder, P., and Tomlin, C. 2005. Lagrangian delay predictive model for sector based air traffic flow. *AIAA Journal on Guidance, Control and Dynamics* 28: 1015–26.

Benenson, I., Omer, I., and Portugali, J. 2002. Entity-based modeling of urban residential dynamics: the case of Yaffo, Tel Aviv. *Environment and Planning B* 29: 491–512.

Benenson, I. and Torrens, P. 2004. *Geosimulation: Automata-based Modeling of Urban Phenomena*. London: John Wiley and Sons.

Berry, B. 1964. Approaches to regional analysis: A synthesis. *Annals of the Association of American Geographers* 54: 2–11.

Bian, L. 2000. Object-oriented representation of modeling mobile objects in an aquatic environment. *International Journal of Geographical Information Science* 14: 603–23.

Bittner, T. and Smith, B. 2003. Granular spatio-temporal ontologies. In *Proceedings of the AAAI Spring Symposium on Foundations and Applications of Spatio-temporal Reasoning* (FASTR). Palo Alto, CA, USA. Menlo Park, CA: American Association for Artificial Intelligence.

Blaut, J. 1961. Space and process. *Professional Geographer* 13: 1–7.

Bura, S., Guerin-Pace, F., Mathian, H., Pumain, D., and Sanders, L. 1996. Multiagent systems and the dynamics of a settlement system. *Geographical Analysis* 28: 161–78.

Burrough, P. A. 1992. Are GIS data structures too simple minded? *Computers and Geosciences* 19: 395–400.

Carlstein, T., Parkes, D., and Thrift, N. (eds). 1978. *Timing Space and Spacing Time*. London: Edward Arnold.

Casati, R., Smith, B., and Varzi, A. 1998. Ontological tools for geographic representation. In N. Guarino (ed.) *Formal Ontology in Information Systems*. Amsterdam: IOS Press, pp. 77–85.

Chen, C. and Zaniolo, C. 2000. SQLST: A spatio-temporal data model and query language. In *Proceedings of the Nineteenth International Conference on Conceptual Modeling (ER'00)*, Salt Lake City, UT, USA. New York: Association of Computing Machinery, pp. 96–111.

Chen, T. and Molenaar, M. 1998. A process-oriented spatio-temporal data model to support physical environmental modeling. In *Proceedings of the Eighth International Symposium on Spatial Data Handling*, Vancouver, British Columbia, pp. 418–30.

Chrisman, N. 1987. Fundamental principles of geographic information systems. In *Proceedings of AutoCarto 8*, Baltimore, MD: 32–41.

Claramunt, C. and Thériault, M. 1996. Toward semantics for modeling spatio-temporal processes within GIS. In M. Kraak and M. Molenaar (eds) *Proceedings of the Seventh International Symposium on Spatial Data Handling*. London: Taylor and Francis, pp. 47–63.

Clarke, K. and Gaydos, L. 1998. Loose coupling a cellular automaton model and GIS: Long term urban growth prediction for San Francisco and Washington/Baltimore. *International Journal of Geographical Information Systems* 12: 699–714.

Clarke, K. and Parks, B. O. (eds). 2001. *Geographic Information Systems and Environmental Modeling*. Upper Saddle River, NJ: Prentice-Hall.

Couclelis, H. 1985. Cellular worlds: A framework for modeling micro-macro dynamics. *Environment and Planning A* 17: 585–96.

Couclelis, H. 1997. From cellular automata to urban models: New principles for model development and implementation. *Environment and Planning B* 24: 165–74.

De Vasconcelos, M., Gonçalves, A., Catry, F., Paúl, J., and Barros, F. 2002. A working prototype of a dynamic geographical information system. *International Journal of Geographic Information Science* 16: 69–91.

Dutton, G. (ed.). 1978. *Harvard Papers on Geographic Information Systems: Proceedings of the First International Symposium on Topological Data Structures for Geographic Information Systems*. Reading, MA: Addison-Wesley.

Egenhofer, M. and Golledge, R. (eds). 1998. *Spatial and Temporal Reasoning in Geographic Information Systems*. Oxford: Oxford University Press.

Erwig, M., Güting, R., Schneider, M., and Vazirgiannis, M. 1999. Spatio-temporal data types: An approach to modeling and querying moving objects in databases. *GeoInformatica* 3: 269–96.

ESRI. 2003. ArcGIS 9: Extending the ArcGIS platform. *ArcNews Online Fall 2003* (available at http://www.esri.com/news/arcnews/fall03articles/arcgis9.html).

Frank, A., Raper, J., and Cheylan, J. (eds). 2001. *Life and Motion of Socio-economic Units*. London: Taylor and Francis.

Frihida, A., Marceau, D., and Thériault, M. 2002. Spatio-temporal object-oriented data model for disaggregate travel behavior. *Transactions in GIS* 6: 277–94.

Galton, A. 2000. *Qualitative Spatial Change*. Oxford: Oxford University Press.

Gimblett, H. (ed.). 2002. *Integrating Geographic Information Systems and Agent-based Modeling Techniques for Simulating Social and Ecological Processes*. Oxford: Oxford University Press.

Goodchild, M. F. 1990. Geographical data modeling. In A. Frank and M. F. Goodchild (eds) *Two Perspectives on Geographical Data Modeling*. Santa Barbara, CA: University of California National Center for Geographic Information and Analysis Technical Report No. 90–11.

Goodchild, M. F., Steyaert, L., Parks, B., Johnston, C., Maidment, D., Crane, M., and Glendenning, S. (eds). 1996. *GIS and Environmental Modeling: Progress and Research Issues*. Fort Collins, CO: GIS World Books.

Hägerstrand, T. 1967. *Innovation Diffusion as a Spatial Process*. Chicago: University of Chicago Press.

Hamre, T. 1994. An object-oriented conceptual model for measured and derived data in 3D space and time. In T. Waugh and R. Healey (eds) *Proceedings of the Sixth International Symposium on Spatial Data Handling*: London: Taylor and Francis, pp. 868–81.

Hazelton, N. 1991. Integrating Time, Dynamic Modeling and Geographical Information Systems: Development of Four-dimensional GIS. Unpublished PhD Dissertation, Department of Geospatial Sciences, Royal Melbourne Institute of Technology.

Hornsby, K. and Egenhofer, M. 2000. Identity-based change: A foundation for spatio-temporal knowledge representation. *International Journal of Geographical Information Science* 14: 207–24.

Hornsby, K. and Egenhofer, M. 2002. Modeling moving objects over multiple granularities. *Annals of Mathematics and Artificial Intelligence* 36: 177–94.

Huisman, O. and Forer, P. 1998. Computational agents and urban life-spaces: A preliminary realization of the time geography of student lifestyles. In *Proceedings of the Third International Conference on GeoComputation*. Bristol, United Kingdom (available at http://www.geocomputation.org/1998/68/gc_68a.htm).

Itami, R. 1988. Cellular automatons as a framework for dynamic simulations in geographic information systems. In *Proceedings of GIS/LIS '88*, Phoenix, AZ, USA. Bethesda, MD: American Congress on Surveying and Mapping and American Society for Photogrammetry and Remote Sensing, pp. 590–7.

Jenerette, G. and Wu, J. 2001. Analysis and simulation of land-use change in the central Arizona–Phoenix region, USA. *Landscape Ecology* 16: 611–26.

Kelmelis, J. 1991. Time and Space in Geographic Information: Towards a Four-dimensional Spatio-temporal Data Model. Unpublished PhD Dissertation, Department of Geography, Pennsylvania State University.

Kemp, K. 1997. Fields as a framework for integrating GIS and environmental process models: Part 1, Representing spatial continuity. *Transactions in GIS* 1: 219–34.

Kirby, K. and Pazner, M. 1990. Graphic map algebra. In *Proceedings of the Fourth International Symposium on Spatial Data Handling*, Zürich, Switzerland. Columbus, OH: International Geographical Union.

Kwan, M.-P. 1998. Space-time and integral measures of individual accessibility: A comparative analysis using a point-based network. *Geographical Analysis* 30: 191–216.

Langran, G. 1992. *Time in Geographic Information Systems*. London: Taylor and Francis.

MacEachren, A., Wachowicz, M., Edsall, R., and Haug, D. 1999. Constructing knowledge from multivariate spatio-temporal data. *International Journal of Geographic Information Science* 13: 311–34.

Maidment, D. 1993. GIS and hydrologic modeling. In M. Goodchild, B. Parks, and L. Steyaert (eds) *Environmental Modeling with GIS*. New York: Oxford University Press, pp. 147–67.

Mennis, J., Peuquet, D., and Qian, L. 2000. A conceptual framework for representing complex and dynamic phenomena in geographic information systems. *International Journal of Geographical Information Science* 14: 501–20.

Miller, H. 2003. Activities in space and time. In D. Hensher (ed.) *Handbook in Transport: 5, Transport Geography and Spatial Systems*. New York: Pergamon, pp. 647–60.

Miller, H. 2005. A measurement theory for time geography. *Geographical Analysis* 37: 17–45.

Miller, H. and Shaw, S. 2001. *Geographic Information Systems for Transportation: Principles and Applications*. New York: Oxford University Press.

Miller, H. and Wentz, E. 2003. Representation and spatial analysis in geographic information systems. *Annals of the Association of American Geographers* 93: 574–94.

Mota, L., Bento, J., and Botelho, L. 2002. Ontology definition languages for multi-agent systems: The geographical information ontology case study. In *Proceedings of the Workshop on Ontologies in Agent Systems (OAS '02)*, Bologna, Italy. New York: Association of Computing Machinery.

Nijkamp, P. and Reggiani, A. 1998. *The Economics of Complex Spatial Systems*. Amsterdam: Elsevier.

O'Sullivan, D. 2001a. Exploring spatial process dynamics using irregular graph-based cellular automaton models. *Geographical Analysis* 33: 1–18.

O'Sullivan, D. 2001b. Graph-cellular automata: A generalized discrete urban and regional model. *Environment and Planning B* 28: 687–705.

Pang, M. and Shi, W. 2002. Development of a process-based model for dynamic interaction in spatio-temporal GIS. *GeoInformatica* 6: 323–44.

Pedrosa, B., Fonseca, F., Câmara, G., Carneiro, T., and Souza, R. 2002. TerraML: A language to support spatial dynamic modeling. In *Proceedings of the Second International Conference on Geographic Information Science*, Boulder, CO, USA. Santa Barbara, CA: National Center for Geographic Information and Analysis.

Pereira, G. 2002. A typology of spatial and temporal scale relations. *Geographical Analysis* 34: 21–33.

Peuquet, D. 1992. *Toward the Representation and Analysis of Spatiotemporal Processes in Geographic Information Systems*. University Park, PA: Pennsylvania State University Technical Report.

Peuquet, D. 1994. It's about time: A conceptual framework for the representation of temporal dynamics in geographical information systems. *Annals of the Association of American Geographers* 84: 441–61.

Peuquet, D. 2001. Making space for time: Issues in space-time data representation. *GeoInformatica* 5: 11–32.

Peuquet, D. and Duan, N. 1995. An event-based spatiotemporal data model (ESTDM) for temporal analysis of geographic data. *International Journal for Geographic Information Systems* 9: 7–24.

Peuquet, D. and Guo, D. 2000. Mining spatial data using an interactive rule-based approach. In *Proceedings of the First International Conference on Geographic Information Science*, Savannah, GA, USA. Santa Barbara, CA: National Center for Geographic Information and Analysis.

Peuquet, D. and Wentz, E. 1994. An approach for time-based analysis of spatio-temporal data. In *Proceedings of the Sixth International Symposium on Spatial Data Handling*, Edinburgh, Scotland, pp. 489–504.

Pred, A. (ed.). 1981. *Space and Time in Geography: Essays Dedicated to Torsten Hägerstrand*. Lund, Sweden, Gleerup.

Pullar, D. 2001. MapScript: A map algebra programming language incorporating neighborhood analysis. *GeoInformatica* 5: 145–63.

Pullar, D. 2003. Simulation modeling applied to runoff modeling using MapScript. *Transactions in GIS* 7: 267–84.

Raper, J. 2000. *Multidimensional Geographic Information Science*. London: Taylor and Francis.

Raper, J. and Livingstone, D. 1995. Development of a geomorphological spatial model using object-oriented design. *International Journal of Geographical Information Systems* 9: 359–83.

Raper, J., Livingstone, D., Bristow, C., and Horn, D. 1999. Developing process response models for spits. In D. Kraus (ed.) *Proceedings of Coastal Sediments*. Long Island, NY: American Society of Civil Engineers, pp. 1755–69.

Reitsma, F. and Albrecht, J. 2005. Implementing a new data model for simulating processes. *International Journal of Geographic Information Science* 19: 1073–90.

Reitsma, F. and Bittner, T. 2003. Scale in object and process ontologies. In W. Kuhn, M. Worboys, and S. Timpf (eds) *Spatial Information Theory: Foundations of Geographic Information Science*. Berlin, Springer Lecture Notes in Computer Science No. 2825: 13–27.

Renolen, A. 2000. Modeling the real world: Conceptual modeling in spatiotemporal information system design. *Transactions in GIS* 4: 23–42.

SCADA. 1991. *SCADA: A Special Collection of Past Conference Proceedings on AM/FMSCADA Integration*. Englewood, CO: AM/FM International.

Semboloni, F. 2000. The growth of an urban cluster into a dynamic self-modifying spatial pattern. *Environment and Planning B* 27: 549–64.

Shi, W. and Pang, M. 2000. Development of Voronoi-based cellular automata: An integrated dynamic model for geographical information systems. *International Journal of Geographical Information Science* 14: 455–74.

Smith, L., Beckman, R., Baggerly, K., Anson, D., and Williams, M. 1993. *Overview of TRANSIMS*. Los Alamos, NM, Los Alamos National Laboratories.

Smith, T., Su, J., El-Abbadi, A., Agrawal, D., Alonso, G., and Amitabah, S. 1995. Computational modeling systems. *Information Systems* 20: 127–53.

Stefanakis, E. 2003. Modeling the history of semi-structured geographical entities. *International Journal of Geographic Information Science* 17: 517–46.

Tomlin, D. 1990. *Geographic Information Systems and Cartographic Modeling*. Englewood Cliffs, NJ: Prentice-Hall.

Tomlinson, R. 1973. *A Technical Description of the Canada Geographic Information System.* Ottawa, Lands Directorate, Environment Canada.

Tryfona, N. and Jensen, C. 1999. Conceptual data modeling for spatio-temporal applications. *GeoInformatica* 3: 245–68.

US Census Bureau. 1969. *The DIME Encoding System.* Washington, DC: US Census Bureau Census Study Report No. 4.

Villa, F. 2000. Design of multi-paradigm integrating modeling tools for ecological research. *Environmental Modeling Software* 15: 169–77.

Wachowicz, M. 1999. *Object-Oriented Design for Temporal GIS.* London: Taylor and Francis.

Waddell, P. 2002. UrbanSim: Modeling urban development for land use, transportation and environmental planning. *Journal of the American Planning Association* 68: 297–314.

Waters, N. 2002. Modeling the environment with GIS: A historical perspective. In K. Clarke, B. Parks, and M. Crane (eds) *Geographic Information Systems and Environmental Modeling.* Upper Saddle River, NJ: Prentice-Hall, pp. 1–35.

Wesseling, C. G., Karssenberg, D., Burrough, P. A., and van Deursen, W. P. A. 1996. Integrating dynamic environmental models in GIS: The development of a dynamic modeling language. *Transactions in GIS* 1: 40–8.

Westervelt, J. and Hopkins, L. 1999. Modeling mobile individuals in dynamic landscapes. *International Journal for Geographical Information Science* 13: 191–208.

Wickens, C., Mavor, A., Purusurum, R., and McGee, B. 1998. *The Future of Traffic Control.* Washington, DC: National Academy of Sciences.

Worboys, M. 2001. Modeling changes and events in dynamic spatial systems with reference to socio-economic units. In A. Frank, J. Raper, and J. Cheylan (eds) *Life and Motion of Socio-Economic Units.* London: Taylor and Francis, pp. 129–37.

Wu, F. 1998. SimLand: A prototype to simulate land conversion through the integrated GIS and CA with AHP derived transition rules. *International Journal of Geographical Information Science* 12: 63–82.

W3C. 2005. Semantic Web Rule Language First Order Logic Submission to W3C. WWW document, http://www.w3.org/Submission/2005/SUBM-SWRL-FOL-20050411/.

Yuan, M. 1996. Temporal GIS and spatio-temporal modeling. In NCGIA (eds) *Proceedings of the Third International Conference on Integrating GIS and Environmental Modeling,* Santa Barbara, CA: University of California, National Center for Geographic Information and Analysis: CD-ROM.

Yuan, M. 1999. Use of three-domain representation to enhance GIS support for complex spatio-temporal queries. *Transactions in GIS* 3: 137–59.

Yuan, M. 2001. Representing complex geographic phenomena in GIS. *Cartography and Geographic Information Science* 28: 83–96.

Yuan, M. and Lin, H. 1992. Spatio-temporal modeling of wildfire in geographic information systems. In *Proceedings of the Symposium of Chinese Professionals in Geographic Information Systems,* pp. 194–200.

Zhang, W. and Hunter, G. 2000. Temporal interpolation of spatially dynamic objects. *GeoInformatica* 4: 403–18.

Zlatanova, S., Holweg, D., and Coors, V. 2003. Geometrical and topological model for real-time GIS. In *Proceedings of the Next Generation Geospatial Information Conference,* Cambridge, MA (available at http://dipa.spatial.maine.edu/NG2I03/CD_Contents/EA/latanovaa_Siyka_01.pdf).

Part VI Geographic Information Systems and Society

The next set of six chapters examines a series of broader issues that influence the development, conduct, and impacts of geographic information technologies. The first of these chapters (Chapter 25 by David L. Tulloch) examines institutional GIS and GI partnering. This chapter starts out with a description of institutional GIS, building a working model of the processes at hand and describing recent research on monitoring the status of institutional GIS, including the models, barriers, and benefits that characterize GIS implementation, and the incorporation of GIS into decision-making and public participation. The chapter then examines GI partnering in some detail – discussing the role of consortia, collaboration, and cooperation (the 3 Cs) and the character and importance of national and global spatial data infrastructures. The chapter concludes with a brief discussion of future trends and directions.

In Chapter 26, the next chapter in this group, Daniel Weiner and Trevor M. Harris examine public participation GIS (PPGIS). It shows how the GIS and Society debates concerning the diffusion of geographic technologies shaped early participatory GIS projects and explores how the term has evolved through several variant forms that reflect important developments. The various examples introduced in this chapter explore PPGIS methods, the nature of participation in PPGIS, and the regional and domain areas in which these systems have been developed – taken as a whole, these examples show how the focus has remained centered on systems and projects that link communities with GIS and related geospatial technologies.

Chapter 27, the third chapter in this group, by Piotr Jankowski and Timothy L. Nyerges, examines GIS and participatory decision-making. The chapter commences with a review of empirical research on collaborative and group decision-making – exploring group support systems, spatial decision support systems, and frameworks for using information technologies to support participatory decision-making – and then takes the best of the concepts and ideas to specify methods for GIS-supported participatory decision-making. The chapter concludes with a plea for systematic evaluation of these technologies to ensure that they efficiently, effectively, and equitably support both large and small groups of decision makers.

The fourth chapter of the group (Chapter 28 by Peter H. Dana) uses examples from several participatory mapping projects in Central America to illustrate the dynamic interplay between conceptions of people and place and the methods used to survey them. Dana reviews many different mapping methods (including GPS technologies in some detail) and explains why surveys of people and place are a part of a process through which ideas of ethnicity and landscape are formulated, explored, altered, and articulated. The examples used throughout this chapter show why the mapping of people and place is a complex process involving numerous methodological choices that may impact the ways in which territory is perceived as well as the ways in which territory is eventually portrayed and/or used as the basis for spatial analysis.

George C. H. Cho examines the relationship between GIS, personal privacy, and the law across a variety of jurisdictions in the next chapter in the group, Chapter 29. This chapter is divided into three major parts. The first examines the philosophical and doctrinal issues that define the structure and character of the problem of privacy and geographic information technologies. The legal, regulatory, and policy framework that underlies the source of this interest is evaluated in the context of considering privacy as a "right" in the middle part. The European Union Data Directive is used as a case study and the "Safe Harbor" framework is described together with a series of alternatives adopted in other countries to respond to this data directive here. The final section of this chapter reviews some of the different geographic information technologies that promote intrusiveness, enhance privacy protection, and are sympathetic to privacy protection.

The final chapter in this set is Chapter 30 by Joseph J. Kerski, which examines the history and character of GIS in education, including the major developments and organizations involved and the opportunities for educating oneself in GIS. GIS in education is conceptualized in four distinct ways – professional development, research about GIS, teaching about GIS, and teaching and learning with GIS – and the history, goals, and needs in each of these diverse settings are described. Kerski concludes this chapter by noting that the diversity of the field of Geographic Information Science (GISc) is reflected in the diversity of educational issues, and why the steadily increasing need for educated practitioners will continue to drive the development of new educational opportunities worldwide.

Chapter 25

Institutional Geographic Information Systems and Geographic Information Partnering

David L. Tulloch

Institutional Geographic Information Systems (GIS) has gained recognition as a critical element in the ongoing global establishment and expansion of Geographic Information (GI) practices, particularly in the public sector. When GIS first emerged they were often used for short-term applications. The adoption of this innovative technology by agencies for use in addressing their day-to-day needs brought new problems ranging from funding to data maintenance to access to interoperability. Quickly, system developers found themselves less involved in developing new code and more involved in establishing relationships, rules, standards, and budgets. The result is something often described as institutional GIS.

While institutional GIS includes some of the more traditional Geographic Information Science (GISc) research regarding status and implementation, it goes much further. Institutionalized GIS is necessary to provide sufficient support for many public resource decisions (planning, public works, etc.) with the incorporation of public participation into many of those decisions. Similarly, institutional GIS drives much of the spatial data infrastructure research, increasingly referred to as GI partnering. Institutional GIS research also includes the investigation of the benefits and costs of system implementation. Underlying much of this work is an understanding that relationships or partnerships are the glue that makes institutional GIS function. Understanding these GI partnerships can help us with many of the remaining questions about institutional GIS.

As organizations have undertaken the process of implementing GIS, two major categories of GIS activities have emerged shaping community systems development. Institutional GIS and GI partnering are two closely related concepts that reflect the evolving ways in which systems are implemented and GIS is applied by organizations with long-term interests. Institutional GIS refers to permanent organizational and technological structures that have evolved to provide and support GI over extended periods of time. Within the larger conception of institutional GIS is GI partnering which generally refers to the active relationships that individuals and institutions formed for cooperation on projects and data exchange, but it places a great emphasis on the complex spatial data infrastructures (SDIs) that rely heavily on partnering. Both represent ways GIS is incorporated into the working structure of countless institutions, impacting the uses and understandings of the technology in some of its most common and accessible forms.

Like other emerging topics in GIS, institutional GIS and GI partnering have both surfaced supported by a significant amount of relevant existing research but very little defining scholarship. This chapter endeavors to present these topics as they are generally treated presently in the literature and describe the directions in which each is advancing. However, these concepts have often been used and treated somewhat inconsistently (particularly institutional GIS) and multiple interpretations should be expected for the time being. The professional nature of many of the related problems makes this a difficult area to track; much of the derived information does not come through traditional research, and much of the information is shared through various forms of "grey literature." This chapter is broken into two distinct sections – first institutional GIS and then GI partnering – with unavoidable overlap between the two. The institutional GIS section establishes a working understanding of the processes at hand and describes three major areas of related research. The GI partnering section provides an overview of two major categories of GI partnering activities. Finally, a discussion is provided exploring implications and directions for this rich research area.

Institutional GIS

Permanence was not always a common attribute or goal of GIS developers; instead, many of the early examples were project-oriented systems designed and implemented to serve on a short-term basis. Many GIS histories (Sinton 1991; Steinitz 1993a, b, c; Foresman 1997) describe individual projects as exemplars of foundational work in the field. These early projects often assembled data and used it for the duration of a single project, sometimes lasting only a semester or a year. While the data may have been preserved for future parallel projects, practical limitations like funding often forced the data sets to remain static, not reflecting the constantly changing landscape that they purported to represent. Certainly, there was little expectation that new projects could benefit significantly from pre-existing data produced elsewhere. Without a reliance on external data, partnering and data sharing relationships were not common attributes of systems.

In pace with the technological changes in GIS, changes occurred in the integration of GIS data as a central element in the processes of government. Organizations began to update their digital data as regularly as their analog records. The institutionalization of these systems came with the development of appropriate arrangements to support and maintain the systems including: sustained funding, appropriate staffing, development of rules and practices for the operation of the systems and the treatment of data, the incorporation of the system into a larger organizational context, and the development of relationships for data sharing, coordination, and partnering. This change in perspective is reflected in the consensus derived definition of GIS that emerged in 1989, describing it as "a system of hardware, software, data, people, organizations, and institutional arrangements for collecting, storing, analyzing, and disseminating information about areas of the Earth" (Dueker and Kjerne 1989, pp. 7–8). This broad definition clearly reflects the sort of long-term perspective commonly associated with institutionalized systems.

Institutionalizing GIS requires significant commitment. Institutional GIS needs sustained, reliable funding to ensure that system maintenance is sufficient and to prevent the system from being co-opted by one-shot project funding. The institutionalization of GIS requires leadership that recognizes the future value of the system as well as the current costs. Sometimes institutionalizing GIS requires organizational restructuring and occasionally institutionalization requires a significant change in the very culture of an institution. For example, an agency that has grown comfortable with obscure land use decisions made without public input may find that a transparent and publicly accessible database makes these decisions much harder.

Institutional GIS can be described and perceived according to a variety of perspectives. Campbell and Masser (1992) present three distinct perspectives of implementation (and ultimately institutionalization): technological determinism, managerial rationalism, and social interactionism. The importance of understanding differing perspectives has been driven home in the intersecting (or colliding) dialogs of the early GIS and Society literature (for example, Dobson 1983, 1993, Taylor and Overton 1991, Openshaw 1991, Smith 1992, Sheppard 1995, Pickles 1995, Curry 1998). This is one area where the multi-disciplinary nature of GIS can be problematic as it promotes miscommunication and misunderstandings. The human element and the roles it can play including concerns about power structures, discomfort with uncertainty, and broad cultural biases make far-reaching GI applications complex (de Man 2003; see Chapter 28 by Dana in this volume for additional discussion of this topic).

Only recently have there been some clear efforts to define and contextualize institutional GIS as an explicit developing sub-field within GIS (Campbell 1999, de Man 2003). Since institutional GIS is only recently coming into its own, a theoretical framework for understanding institutional GIS must be assembled relying on research from a variety of related areas. The GIS literature includes several areas from which this research area draws:

- Monitoring status and modeling implementation;
- Barriers and benefits; and
- Decision making and public participation.

Each of these is described in the following sections. Additionally, important theoretical foundations come from other disciplines. Much of the literature relating to institutional GIS is founded on understandings of economics and institutions (North 1990), organizations and organizational change (Handy 1993), and the roles of computers, technology, and innovations in changing institutions and organizations (Zaltman, Duncan, and Holbek 1973, Keen 1981, Eason 1988, Morton 1991) from outside the traditional GI literature.

Monitoring status and modeling implementation

A significant area of early research about the development of institutional GIS was monitoring the adoption and diffusion of the technology. Since many researchers quickly recognized that GIS adoption comes in many flavors (for example, both within an agency and among different agencies), the research sometimes evolved into a process of monitoring the status of system development. Much of the interest in this area focused on larger, public systems making these investigations ultimately part of an ongoing study of the institutionalization of GIS.

A number of the early efforts at monitoring system status came as part of agency efforts to identify potential or future data sources within their jurisdiction. In the US, many states used surveys of local governments as a means of identifying potential partners or participants in statewide GI initiatives (for example, Arizona Geographic Information Council 1992, Sandberg 1992, Wisconsin Land Information Board 1995, Renner and Waldon 1995, Minnesota Governor's Council on Geographic Information 1994, New Jersey State Mapping Advisory Committee 1997). Some states, like Wisconsin, have assessed status longitudinally and fairly consistently in ways that become quite informative in regards to the overall development patterns of institutional GIS (Hart, Koch, Moyer, and Niemann 2001). These monitoring surveys continue for purposes of governmental accountability, but have the additional consequence of monitoring the slow and subtle changes that occur as these systems evolve from project-oriented to permanent, institutional systems.

Similar, although generally more rigorous or consistent work has been conducted at regional and national levels. British local governments were surveyed by Campbell and Masser (1995) showing that GIS had been widely adopted while raising questions about the ways it was being used. In the US surveys included military bases (Cullis 1995), planning agencies (Budic 1994), and local governments (Budic 1993, Tulloch and Niemann 1996, Tulloch, Niemann, Ventura, and Epstein 1996). In what might have been the largest such survey, the US Federal Geographic Data Committee (FGDC) and the National States Geographic Information Consortium (NSGIC) collaborated in the late 1990s to distribute over 15,000 questionnaires of potential data producers (Tulloch and Robinson 2000). Of the 5,200 surveys returned, over half indicated they were producing geospatial data that might contribute to the US National Spatial Data (NSDI) – 836 of these were county governments (Tulloch and Fuld 2001). More recently, similar efforts have expanded to catalog, examine, and compare multiple nations. In a European Science Foundation (ESF) funded effort, Masser, Campbell, and Craglia (1996) compiled a series of assessments of the status of local government GIS within a number of European countries. Most recently, a variety of efforts have loosely followed this tradition in developing information for SDI – and particularly Global Spatial Data Infrastructure (GSDI) – assessments.

These survey efforts led to the development of theoretical models and descriptions of the ways in which these systems evolve. While unique to GIS, this research often built on the larger tradition of research on the adoption and diffusion of innovations. At the center of this tradition is the work of Everett Rogers (summarized well in Rogers 1995) whose achievements in the field include the widely used s-shaped adoption curve (see an example of its use in Goodchild 1997). While Rogers' writing was mostly meant as an overarching evaluation of all types of innovations, he did contribute to the GIS literature (Rogers 1993). The theoretical models of GIS adoption and development reflect greatly on this early innovation and technology research, a form of GIS institutionalization (Masser and Onsrud 1993).

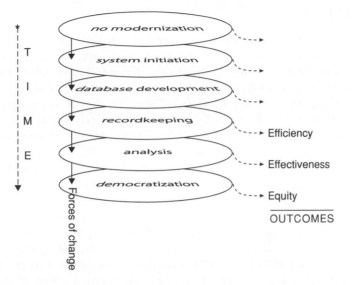

Fig. 25.1 A conceptual diagram of multipurpose land information system development. The diagram illustrates six primary stages and the associated outcomes – efficiency, effectiveness, and equity
From Tulloch 1999

While not often explicitly stated, the GIS adoption and development models often describe the institutionalization of GIS in public agencies. While this adoption and diffusion literature has ranged from conceptual frameworks (for example, Anderson 1996, Nedovic-Budic and Pinto 1999) to theoretical models (for example, Crain and MacDonald 1984, Ezigbalike, Coleman, Cooper, and McLaughlin 1988, Azad and Wiggins 1995, Tulloch 1999, Chan and Williamson 1999), it has developed an important basis for describing the ways that GIS has moved from being discussed as a temporary, project-based technology plagued by instability to a complex, permanent, enterprise-wide, institution. The author has previously published a six-stage theoretical model of multi-purpose land information systems (MPLIS) development (Tulloch 1999) which culminates in a final stage of democratization. The MPLIS development process (Figure 25.1) is described with a specific understanding that public (or community-based) systems are often not fully established until they are utilized by a user community that extends far beyond the internal group of GIS users contributing to the system's development. This carries the understanding of system development far past implementation and allows an examination of what happens after GIS is inside the institution.

Monitoring research could be typified by traditional surveys of GIS institutions seeking to measure the nature and extent of GIS implementation and partnerships. This research could include improved monitoring of clearinghouses and data holdings as a means of creating comprehensive assessments of data availability. Certainly, this should be conducted in ways that contribute to a longitudinal understanding of a GIS program. A key step is the establishment of the best possible baseline data about as many systems as possible.

Barriers and benefits

As part of the exploration of adoption and diffusion research, researchers have examined the factors that either facilitate or limit GIS implementation. An improved understanding of the reasons for system development and investment would be very beneficial to the broader GIS-user community. This has been an area negatively impacted by the struggle to keep GIS from being treated as simply an appendage of IT work and research. Within GIS this vein of research has already emerged as a strong area of study but it remains an understudied area in need of continued attention.

Early research exploring both the barriers and benefits of institutionalizing GIS emerged from anecdotal reports of GIS successes and failures (for instance, Huxhold 1982, Croswell 1989, 1991, Coffeen 1994). While technical barriers received a great deal of attention (as described in Obermeyer and Pinto 1994), organizational and institutional issues were clearly identified as playing a key role in the process (Budic 1994, Ventura 1995, Cullis 1995). With an emphasis on measuring the impact of distinct barriers, Onsrud and Pinto (1993) undertook a thorough survey of GIS users. The results demonstrated the importance of overcoming a series of non-technical barriers in order for a system to become established. While this area of research has not fully captured the change, the gradual reduction in cost of technology and increase of IT support within most public agencies has changed the nature and extent to which cost and technical issues serve as barriers to implementation and institutionalization.

With similar roots in the professional literature, descriptions of the benefits of system development have played an important role in the contemporary understanding of successful system development. Ranging from early conjecture (Cook and Kennedy 1966) to case studies (Onsrud, Pinto, and Azad 1992, 1993, Gillespie 1994, Craig and Johnson 1997) to spreadsheet accounting (Antenucci, Brown, Croswell, Kevany, and Archer 1991), extensive work has been invested in accounting for the beneficial outcomes of GIS and its institutionalization. Traditional economics serve as a useful basis for measuring outcomes (Larsen, Clapp, Miller, Niemann, and Ziegler 1978, Wunderlich and Moyer 1984, Moyer 1993, Gillespie 1994), however many find this to be either daunting or inappropriate for their needs. These economic studies have often placed a premium on measuring efficiency and/or effectiveness. A broad perspective was brought to the topic by the National Center for Geographic Information and Analysis (NCGIA) Initiative 4 studying the Use and Value of Geographic Information (Onsrud, Calkins, and Obermeyer 1989). More socially-oriented research has incorporated concerns about benefits (and associated costs) of systems that are passed along to the communities they are intended to serve (Chrisman 1987, Kishor, Niemann, Moyer, et al. 1990, Cowen 1994, Sheppard 1995, Epstein and Marble 2002). These system benefits have been summarized by Tulloch and Epstein (2002) in three broad categories: efficiency, effectiveness, and equity/empowerment/engagement. Efficiency is related to improvements on traditional recordkeeping, usually internal within the office in which the automation has occurred. Effectiveness describes potential benefits from improved analysis and related techniques, often with the larger agency within which the GIS has been institutionalized. Equity, empowerment, and engagement describe the ways that members of the larger community can benefit (sometimes unknowingly) from the use and applications of an institutionalized system. These system benefits (referred to colloquially as the 3 Es) represent an important element in institutional GIS because they create the support needed to preserve and maintain permanent and accessible public systems.

Decision making and public participation

Much of the significance of GIS in institutions involves its incorporation into formal decision-making processes. For public institutions this often includes public participation in determining the outcome of spatially-based policies and individual decisions. In the USA, a variety of federal and state laws mandate the incorporation of public participation into proposed government plans, decisions and actions before final decisions are made. The incorporation of the technology into existing decision processes has required developments in areas such as multiple criteria decision making which are slowly become institutionalized like the systems used to apply them. Since other chapters of this book include significant information on both decision making (DM) (see Chapter 27 by Jankowski and Nyerges) and public participation GIS (PPGIS) (see Chapter 26 by Weiner and Harris) they will only be discussed briefly here.

Institutionally, DM becomes critical because many larger systems are developed with a direct interest in supporting policy and decision making. An insufficient effort in the development of the supporting system can render the resulting decisions and policy weakened or indefensible. An ongoing and repetitive DM process (like wetlands permitting or long-term management of a timber property) requires data to be accessible, consistent, reliable, and appropriate to the legally required standards for the permit or process. For frequently changing data, this creates a demand for a systematic institutionalization providing a rigorous system of rules and procedures as well as assurances regarding data quality. As major institutions and agencies move towards incorporating GIS data directly into decisions or policy processes, it will become increasingly important for a measure of data quality to be a part of an institutionalized GIS. While DM and PPGIS may seem like entirely distinct topics, Jankowski and Nyerges (2001, 2003) have done an elegant job of demonstrating their intersection.

Technological advances have played a significant role in altering access to spatial data as GIS has moved from mainframes and magnetic tape to networked PCs and internet map services. Increased access to GIS, especially through Internet mapping and downloadable data, have created a burgeoning sub-field of PPGIS (a strong set of examples are presented clearly in Craig, Harris, and Weiner 2002). While access and participation are distinct characteristics (Tulloch and Shapiro 2003) they are both embedded with institutional issues (de Man 2003) with some of the more dramatic PPGIS examples relying partially on an existing institutionalized system. This represents a larger trend within institutional GIS with movement from an enterprise-wide model to a larger societal model. As institutional GIS develops, formal and informal policies develop addressing format, access, and dissemination, all impacting the ability of groups to participate in policy processes or view the public processes with appropriate transparency.

GI Partnering

Within the larger area of institutional GIS, a significant amount of attention has come to be focused on GI partnering activities. Needs for cooperative efforts emerge as local and regional governments collect or develop spatial data about their jurisdictions. As communities across an entire nation cooperate and as countries around the world work together, large spatial data infrastructures (SDIs) are being constructed to facilitate broad access to these many data sources. The tantalizing possibility of seamless access to the enormous volume of locally-produced spatial data from around the world is nearly irresistible. A parallel effort to develop a GSDI has also become a dominant element in discussions of SDI issues. Whether local or global, GI partnering can be both rewarding and challenging with technical and institutional problems. Particularly notable is the recent dramatic emergence in the USA of national and homeland security issues in preventing access to or in the dissemination of significant public data sources.

GI partnering often requires a dramatic degree of cooperation between organizations and individuals. The sharing of spatial data is a cornerstone of these relationships (Onsrud and Rushton 1995). Cooperative and collaborative relationships can be very difficult and unpredictable as they introduce a variety of human factors into a seemingly technical process, including cultural conflicts (Hofstede 2001) and interpersonal issues like trust (Harvey 2003). The technical difficulties and political barriers expose these high visibility long-term processes to a great deal of criticism (for example, Hissong 2002) and reflective review (Tosta 1999, Lachman, Wong, Knopman, and Gavin 2002) often asking whether such an SDI can ever be built for a nation as large and complex as the United States.

The 3 Cs of regional and state partnering: consortia, collaboration, and cooperation

Much of the early history of GIS is interwoven with stories of local government system development. Some of the early examples reflect the need to coordinate and cooperate internally in order for municipal or county agencies to satisfy their mandates. One broad set of examples of these efforts are those represented in the MPLIS literature. Research in MPLIS (and previously multi-purpose cadastres) looked at the sharing of land information between departments in a single organization and the resulting institutional complications ranging from data standards to database structures to power struggles to professional cultural conflicts (National Research Council 1980, 1983, Niemann and Moyer 1988). Many of the lessons learned from these internal organizational efforts are often applicable for larger regional, national, and global efforts.

Regional (sub-national) consortia and collaborative efforts are providing an important test bed for clearinghouses and other cooperative efforts. One interesting type of collaboration that has become somewhat common is voluntary consortia of communities who bind themselves together for collaboration in the acquisition, development, and dissemination of GI. While often initiated as a means for creating a sufficient mass to acquire a bulk of data or imagery as a reasonable price, the result can be a small region united under common data standards and lasting formal and informal relationships between communities and agencies. The emergence of regional planning agencies, although often resisted by local governments, has created unusual opportunities for data sharing. Examples like Portland's Metro (Bosworth, Donovan, and Couey 2002) and San Diego's SanGIS (Regional Urban Information System 1997) demonstrate how cooperation and collaboration can lead to powerful regional GIS institutions.

In the USA, states have also become important loci for GI partnering efforts. While some examples of this continue to be coordination programs meant to satisfy the needs of a single agency, more often these programs signify a major shift towards collective efforts satisfying many goals. In some cases, like Wisconsin, the state legislature has actively worked to create a state-wide data infrastructure supported by the state but acted out primarily at the local level (Kuhlman and Niemann 1993). Other states have instead created an agency tasked with encouraging and facilitating cooperative efforts at the local level, but lack the political power or resources to address the situation comprehensively. Technical reports and professional literature have captured much of this activity (see the earlier section on status monitoring) and the development of a theoretical base for describing collaboration, consortia, and cooperation at local and state levels has begun (for example, Harvey 2001), but much remains to be done.

National and global spatial data infrastructures

With enormous amounts of data being produced and an understanding that locally-produced data is usually better than nationally-produced data, many members of the GIS community have pursued national and global SDIs aggressively. With the promise of access to seamless, timely data from countless local sources, NSDI and GSDI hold immeasurable promise, but may ultimately require more action than is reasonable to expect from the agencies involved. Nancy Tosta, formerly responsible for the US NSDI efforts, writes about the many issues that prevented more "action" from contributing to the development of the NSDI with human nature being a key obstacle (Tosta 1999).

Nationwide strategies to coordinate data and standardize production while continuing to allow local control of the processes and data are no longer uncommon. Masser (1998) details and compares a variety of national approaches with varying mixtures of top-down and bottom-up control. While some of the early significant interest in SDIs came from the USA with an interest in its NSDI (National Research Council 1993, 2001), prominent projects outside the USA are now advancing under the umbrella of GSDI projects. Elements of the US model (FGDC 1995, 1997) still seem to dominate with a popular interest in voluntary participation and selecting a set of common framework layers both serving as recurring themes. There certainly has been some clear leadership and cooperation for GSDI efforts from both the USA (particularly through the FGDC) and European nations (through mechanisms like EUROGI and INSPIRE) (Craglia and Masser 2003).

The basic visions for the US NSDI and the GSDI have come from a fairly simple expectation that if all of the freely available data were made available in a loosely coordinated mechanism then the results would be much more useful for many users than any one national or global data set. The early visions (like that of the National Research Council 1993) were optimistic, but with the 1994 establishment of a US Federal Geographic Data Committee tasked to create a NSDI, analyses began to recognize the dramatic nature of the institutional barriers that exist (NAPA 1998). While sometimes these grand and complex visions have drifted into something more pragmatic like a single "national map" (US Geological Survey 2001, National Research Council 2003) or "digital earth" (NASA Digital Earth Office 2000), SDIs should truly serve as an infrastructure constructed to provide access to many different (although sometimes conflicting) dynamic data sources.

An important development has been the formal creation of an organization calling itself GSDI, with the stated goal of supporting "all societal needs for access to and use of spatial data" (GSDI 2004, p. 1). It has become quite difficult to track the many individual and multi-national efforts that contribute to this growing movement accurately, but there is documentation of over 50 countries with national spatial data clearinghouses (Crompvoets and Bregt 2003). Despite the common understanding of SDIs as being, in part, a digital network serving up data, it is conceivable that the GSDI could be considered quite successful if it simply continues to serve as an informal network of spatial data-producing nations and organizations who meet regularly, exchange ideas, make connections, and produce materials to gently guide along global spatial data development and access (for instance, Nebert 2001).

A few key themes emerge from among these many efforts to create GIS institutions to facilitate access to spatial data. Data standards are frequently a focus of SDI discussions because of their importance for the technical exchange and integration of the many data sets comprising or passing through the SDI. Clearly a greater awareness of infrastructure issues and mechanisms will be necessary for further continuing advances in this area. Questions linger about educating producers and owners of spatial data about the benefits they will receive and incentives that can be used to encourage participation. Finally, simple as it may seem, improved communication practices (including GIS-specific forms of communication like GI metadata) are frequently noted as a crucial element that is often overlooked in developing and maintaining the relationships that underlie SDI networks.

Discussion

Institutional issues and characteristics

While institutional GIS appears as a relatively new idea, both the idea and notable examples have existed for long enough to begin identifying important patterns. After all, there do exist some notable historic exceptions to the project-based bias of early systems. For example, much of the early history of the Canadian Geographic Information System (CGIS) in the 1960s seems clearly designed to emphasize a long-term system addressing long-term issues (Tomlinson 1997, Niemann and Niemann 1999). In fact, the CGIS was institutionalized with an explicit interest in providing a stable base for an ongoing series of public decisions.

Examples of institutional GIS are often shaped by some of the same structural factors that affect many types of institutions. Institutions can be viewed as being shaped by their power structure and the manner in which it empowers or limits the individuals within the institution. Similarly, institutions require an appropriate policy or rules structure – formal or informal – that fits the goals and culture within and around the system. These rules and power structures often shape many of the other important structural factors cited when describing these systems including issues of management, communication, access, and standards.

Of the previously described structural issues, access has emerged as a notable issue of great interest throughout the GIS community. Access directly impacts the potentially irresolvable debate over privacy. As an issue, privacy (see Chapter 29 by Cho in this volume) has the ability to illustrate the ways that a community's expectations can dominate over long-standing practices within an institution. Decisions about the access structures and policies will determine participation and ultimately shape the public support for a system and its applications.

Communication serves as another significant structural element within institutional GIS. Rogers' adoption and diffusion work stressed the importance of communication in the wide adoption of an innovation. But communication remains critical in a complex network of data users, producers, and maintainers. Since so many of these networks rely on relationships with a human element, communication is central to protecting and preserving these connections. Communications can work in both a personal and a technical way, as is often proved by the widely varying forms of metadata that different offices use.

Finally, institutional GIS may rely on an appreciation of the fact that it is, indeed, institutional. For spatial data infrastructures to succeed, they rely on an ongoing effort from local data producers across a wide (perhaps global) scale. Moreover, these local data producers need to recognize that while they may focus on their own data and its quality, they have to be aware of the larger infrastructure and their place within it. Their day-to-day work must continue with its internally focused attention informed by a larger awareness of the data infrastructure and its rules about data standards, access, and timeliness.

Looking ahead

A recognition of the importance that GI partnering and institutional GIS will hold in the future is demonstrated by the placement of these issues in the list of research priorities chosen by the US University Consortium for Geographic Information Science (UCGIS). In developing its research agenda, UCGIS identified Insitutional Aspects of SDIs (Tulloch 2002) and Geographic Information Partnering (Nedovic-Budic 2002) as two key areas in which it should direct attention. Unfortunately, while it does show that these are high priorities within the GIS research community, it does not assure that they will be successful in getting further attention from the funding agencies or the professional GIS community.

Institutional GIS and GI partnering play a critical role in almost any scenario for the future of GIS (National Research Council 1997). While the emergence of new technologies (for example, wireless applications and internet mapping) will have untold impacts on the future of GI and its applications, GI partnering and the defining GIS institutions will ultimately determine how these impacts play out. Preparing institutions for the different GI applications that will be needed is a difficult task, but efforts to promote an appropriate awareness in both users and decision-makers will certainly be rewarded.

Barriers to institutionalization are common and need to be better understood. Instead of simply raising awareness, more needs to done to understand the many motivations behind data sharing, coordination, and partnerships. The varying motivations – mandates, idealism, generosity, politics, expectations of access in return – behind data sharing become great barriers to advancing large SDI projects. These projects are often undertaken with little or no formal recognition of the complexity this adds to its networks of data providers and users.

Increasingly, institutionalized GIS requires "advanced decision making" – agency leaders decide far in advance of unforeseen problems to invest in a system, at some later time their community can be rewarded for their commitment. While it no longer takes decades to establish these systems, institutionalization often outlasts elected political administrations. This demonstrates the importance of educating leaders to the long-term and unanticipated outcomes that make their decisions so significant. It may also speak to a need for public and agency expectations that could be raised in ways that would make it difficult for elected officials to resist.

We also need to demand more of these institutionalized systems. It should not be enough that our public agencies are developing the data, because they also need to be institutionalizing the process of disseminating them. Included in this are public access policies that free the data to serve the public better, and engage in practices that improve the usability of these data (like maintaining metadata). As participants in these larger policy debates we should also demand that agencies seek improved coordination and cooperation in data development and use. Institutionalized GIS means that these agencies can develop interdependent policies, but it also requires trust.

Educators need to raise student awareness of the ultimate potential for a permanent spatial information infrastructure system developing, integrating, and supplying local, state, and federal data. Some students may go on to help raise expectations, while some may apply the lessons by developing institutionalized GIS. But educators can hope that some students will find unanticipated opportunities and leverage their communities' institutionalized GIS to shape their landscapes in ways that we could never imagine.

REFERENCES

Anderson, C. S. 1996. GIS development process: A framework for considering the initiation, acquisition, and incorporation of GIS technology. *URISA Journal* 8(1): 10–26.

Antenucci, J. C., Brown, K., Croswell, P. L., Kevany, M. J., and Archer, H. 1991. *Geographic Information Systems: A Guide To The Technology.* New York: Van Nostrand Reinhold.

Arizona Geographic Information Council. 1992. *Arizona Geographic Information Council (AGIC) Survey of Existing Digital Spatial Data and Future Needs.* Phoenix, AZ: Data Management Division, Arizona Department of Administration.

Azad, B. and Wiggins, L. L. 1995. Dynamics of inter-organizational geographic data sharing: A conceptual framework for research. In H. J. Onsrud and G. J. Rushton (eds) *Sharing Geographic Information.* New Brunswick, NJ: Center for Urban Policy and Research: 22–43.

Bosworth, M., Donovan, J., and Couey, P. 2002. Portland Metro's dream for public involvement. In W. Craig, T. Harris, and D. Weiner (eds) *Community Participation and GIS*. London: Taylor and Francis, pp. 125–36.

Budic, Z. 1994. Effectiveness of geographic information systems in local government planning. *Journal of the American Planning Association* 60: 244–63.

Budic, Z. 1993. GIS use among southeastern local governments. *URISA Journal* 5(1): 4–17.

Campbell, H. J. 1999. Institutional consequences of the use of GIS. In P. A. Longley, M. F. Goodchild, D. J. Maguire, and D. W. Rhind (eds) *Geographical Information Systems* Vol. 2. New York: John Wiley and Sons, pp. 621–31.

Campbell, H. and Masser, I. 1992. GIS in local government: Some findings from Great Britain. *International Journal of Geographical Information Systems* 6: 529–46.

Campbell, H. and Masser, I. 1995. *GIS and Organizations: How Effective are GIS in Practice?* London: Taylor and Francis.

Chan, T. O. and Williamson, I. P. 1999. The different identities of GIS and GIS diffusion. *International Journal of Geographical Information Science* 13: 267–81.

Chrisman, N. R. 1987. Design of geographic systems based on social and cultural goals. *Photogrammetric Engineering and Remote Systems* 53: 1367–70.

Coffeen, S. 1994. Using GIS to support the "reverse urbanization" process: A case study of the Presidio. *URISA Conference Proceedings* 9: 517–26.

Cook, R. N. and Kennedy, J. L. 1966. *Proceedings of the Tri-State Conference on a Comprehensive Unified Land Data System (CULDATA)*. Cincinnati, OH: College of Law, University of Cincinnati.

Cowen, D. J. 1994. The importance of GIS for the average person. In *Proceedings of the First Federal Geographic Technology Conference on GIS in Government: The Federal Perspective*, Washington DC, USA. Fort Collins, GIS World Books, pp. 7–11.

Craglia, M. and Masser, I. 2003. Access to geographic information: A European perspective. *URISA Journal* 15(1): 51–9.

Craig, W. and Johnson, D. 1997. Maximizing GIS benefits to society. *Geo Info Systems* 7(3): 14–8.

Craig, W., Harris, T., and Weiner, D. (eds). 2002. *Community Participation and GIS*. London: Taylor and Francis.

Crain, I. K. and MacDonald, C. L. 1984. From land inventory to land management. *Cartographica* 21: 40–6.

Crompvoets, J. and Bregt, A. 2003. World status of national spatial data clearinghouses. *URISA Journal* 15(1): 43–50.

Croswell, P. L. 1989. Facing reality in GIS implementation: Lessons learned and obstacles to be overcome. *URISA Conference Proceedings* 4: 15–35.

Croswell, P. L. 1991. Obstacles to GIS implementation and guidelines to increase the opportunities for success. *URISA Journal* 3(1): 43–56.

Cullis, B. J. 1995. *An Exploratory Analysis of Responses to Geographic Information System Adoption on Tri-Service Military Installations*. Vicksburg, MS: US Army Corps of Engineers Waterways Experiment Station.

Curry, M. R. 1998. *Digital Places: Living with Geographic Information Technologies*. New York: Routledge.

de Man, E. 2003. Cultural and institutional conditions for using geographic information: Access and participation. *URISA Journal* 15(1): 29–33.

Dobson, J. 1983. Automated geography. *Professional Geographer* 35: 135–43.

Dobson, J. 1993. The geographic revolution: A retrospective on the age of automated geography. *Professional Geographer* 45: 431–9.

Dueker, K. J. and Kjerne, D. 1989. Multipurpose cadastre terms and definitions. In *Proceedings of the American Society for Photogrammetry and Remote Sensing and American*

Congress on Surveying and Mapping. Bethesda, MD: American Society for Photogrammetry and Remote Sensing and American Congress on Surveying and Mapping, pp. 94–103.

Eason, K. 1988. *Information Technology and Organisational Change*. London: Taylor and Francis.

Epstein, E. and Marble, D. (uncredited eds). 2002. *Geographic Information Science & Technology in a Changing Society: A Research Definition Workshop*. Columbus, OH: Center for Mapping and School of Natural Resources, Ohio State University.

Ezigbalike, I., Coleman, D., Cooper, R., and McLaughlin, J. 1988. A land information system development methodology. *URISA Conference Proceedings*. Bedford Park, IL: Urban and Regional Information Systems Association 3: 282–96.

FDGC. 1995. *Development of a National Digital Geospatial Data Framework*. Washington, DC: Federal Geographic Data Committee.

FDGC. 1997. *Framework Introduction and Guide*. Washington, DC: Federal Geographic Data Committee.

Foresman, T. W. (ed.). 1997. *The History of Geographic Information Systems: Perspectives from the Pioneers*. Upper Saddle River, NJ: Prentice Hall.

Gillespie, S. R. 1994. Measuring the benefits of GIS use: Two transportation case studies. *URISA Journal* 6(2): 62–7.

GSDI. 2004. *Global Spatial Data Infrastructure Newsletter* (January 2004). Reston, VA: GSDI Secretariat.

Goodchild, M. F. 1997. What next? Reflections from the middle of the growth curve. In T. W. Foresman (ed.) *The History of Geographic Information Systems: Perspectives from the Pioneers*. Upper Saddle River, NJ: Prentice Hall: 369–82.

Handy, C. 1993. *Understanding Organizations*. New York: Oxford University Press.

Hart, D., Koch, T., Moyer, D. D., and Niemann, B. J. 2001. *Land Information Modernization Activity in Wisconsin: Impacts, Status and Future Tasks*. Madison, WI: Land Information and Computer Graphics Facility, University of Wisconsin-Madison (report to Wisconsin Legislature).

Harvey, F. 2001. Constructing GIS: Actor networks of collaboration. *URISA Journal* 13(1): 29–37.

Harvey, F. 2003. Developing geographic information infrastructures for local government: The role of trust. *Canadian Geographer* 47: 28–36.

Hissong, F. 2002. The Sisyphean death of the US NSDI: Oh, Prometheus . . . Prometheus? *Earth Observation Magazine* 11(11): 31.

Hofstede, G. 2001. *Culture's Consequences: Comparing Values, Behaviors, Institutions and Organizations across Nations*. Thousand Oaks, CA: Sage.

Huxhold, W. E. 1982. An evaluation of the City of Milwaukee automated geographic information and cartographic system in retrospect. In *Proceedings of Seminar on Land-Related Information Systems, Municipal/Provincial Integration*, Toronto, Canada.

Jankowski, P. and Nyerges, T. 2001. *Geographic Information Systems for Group Decision Making*. London: Taylor and Francis.

Jankowski, P. and Nyerges, T. 2003. Toward a framework for research on geographic information-supported participatory decision-making. *URISA Journal* 15(1): 9–17.

Keen, P. 1981. Information systems and organizational change. *Communications of the ACM* 24: 24–33.

Kishor, P., Niemann, B. J., Moyer, D. D., Ventura, S. J., Martin, R. W., and Thum, P. G. 1990. Lessons from CONSOIL: Evaluating GIS/LIS. *Wisconsin Land Information Newsletter* 6(1): 1–13.

Kuhlman, K. and Niemann, B. J. 1993. Modernizing land records: Tracking GIS/LIS technology in Wisconsin and beyond. *URISA Journal* 7(1): 60–3.

Lachman, B., Wong, A., Knopman, D., and Gavin, K. 2002. *Lessons for the Global Spatial Data Infrastructure: International Case Study Analysis.* Santa Monica, CA: RAND Science and Technology Policy Institute.

Larsen, B., Clapp, J., Miller, A. H., Niemann, B. J., and Ziegler, A. L. 1978. *Land Records: The Cost to the Citizen to Maintain the Present Land Information Base: A Case Study of Wisconsin.* Madison, WI: Department of Administration, Office of Program and Management Analysis.

Masser, I. 1998. *Governments and Geographic Information.* London: Taylor and Francis.

Masser, I., Campbell, H., and Craglia, M. (eds). 1996. *GIS Diffusion: The Adoption and Use of Geographical Information Systems in Local Government in Europe.* London: Taylor and Francis.

Masser, I. and Onsrud, H. (eds). 1993. *Diffusion and Use of Geographic Information Technologies.* Dordrecht: Kluwer.

Minnesota Governor's Council on Geographic Information. 1994. *Minnesota GIS Survey of GIS Organizations, Geographic Data Files, and Data Needs.* Minneapolis, MN: Minnesota Governor's Council on Geographic Information and Minnesota GIS/LIS Consortium.

Morton, M. (ed.). 1991. *The Corporation of the 1990s: Information Technology and Organizational Transformation.* New York: Oxford University Press.

Moyer, D. D. 1993. Economics of MPLIS: Concepts and tools. In P. M. Brown and D. D. Moyer (eds). *Multipurpose Land Information Systems: The Guidebook.* Washington, DC: Federal Geodetic Control Committee 15.1–15.25.

NAPA. 1998. *Geographic Information for the 21st Century: Building a Strategy for the Nation.* Washington, DC: National Academy of Public Administration.

NASA Digital Earth Office. 2000. *What is Digital Earth?* Washington, DC: National Aeronautics and Space Administration.

National Research Council. 1980. *Need for a Multipurpose Cadastre.* Washington, DC: National Academy Press.

National Research Council. 1983. *Procedures and Standards for a Multipurpose Cadastre.* Washington, DC: National Academy Press.

National Research Council. 1993. *Toward a Coordinated Spatial Data Infrastructure for the Nation.* Washington, DC: National Academy Press.

National Research Council. 1997. *The Future of Spatial Data and Society.* Washington, DC: National Academy Press.

National Research Council. 2001. *National Spatial Data Infrastructure for the Nation.* Washington, DC: National Academy Press.

National Research Council. 2003. *Weaving a National Map: Review of the U.S. Geological Survey Concept of The National Map.* Washington, DC: National Academy Press.

Nebert, D. (ed.). 2001. Developing Spatial Data Infrastructures: The SDI Cookbook Global Spatial Data Infrastructure (GSDI), USA, Version 1.1. WWW document, http://www.gsdi.org.

Nedovic-Budic, Z. 2002. Geographic Information Partnering. Washington, DC: University Consortium for Geographic Information Science Research Brief (available at http://www.ucgis.org/priorities/research/2002researchPDF/shortterm/m_gi_partnering.pdf).

Nedovic-Budic, Z. and Pinto, J. K. 1999. Understanding inter-organizational GIS activities: A conceptual framework. *URISA Journal* 11(1): 53–64.

New Jersey State Mapping Advisory Committee. 1997. *New Jersey GIS Resource Guide*, 1997. Trenton, NJ: New Jersey State Mapping Advisory Committee (HTML-formatted document distributed on New Jersey Geographic Information System).

Niemann, B. J. and Moyer, D. D. (eds). 1988. *A Primer on Multipurpose Land Information Systems.* Madison, WI: University of Wisconsin-Madison, Institute for Environmental Studies Report No. 133.

Niemann, B. J. and Niemann, S. 1999. Roger F. Tomlinson: The father of GIS. *Geo Info Systems* 9(3): 40–1, 43–4.

North, D. C. 1990. *Institutions, Institutional Change, and Economic Performance*. New York: Cambridge University Press.

Obermeyer, N. J. and Pinto, J. K. 1994. *Managing Geographic Information Systems*. New York: Guilford.

Onsurd, H. J., Calkins, H. W., and Obermeyer, N. J. (eds). 1989. *Use and Value of Geographic Information: Initiative Four Specialist Meeting Final Report*. Santa Barbara, CA: National Center for Geographic Information and Analysis Technical Report.

Onsrud, H. J. and Pinto, J. K. 1993. Evaluating correlates of GIS adoption success and the decision process of GIS acquisition. *URISA Journal* 5(1): 18–39.

Onsrud, H. J., Pinto, J. K., and Azad, B. 1992. Case study research methods for geographic information systems. *URISA Journal* 4(1): 32–44.

Onsrud, H. J., Pinto, J. K., and Azad, B. 1993. *Testing Technology Transfer Hypotheses in GIS Environments Using a Case Study Approach*. Santa Barbara, CA: National Center for Geographic Information and Analysis Technical Report No. 93–8.

Onsrud, H. J. and Rushton, G. J. (eds). 1995. *Sharing Geographic Information*. New Brunswick, NJ: Center for Urban Policy Research, Rutgers University.

Openshaw, S. 1991. A view on the crisis in geography, or using GIS to put Humpty-Dumpty back together again. *Environment and Planning* A 23: 621–8.

Pickles, J. (ed.). 1995. *Ground Truth: Social Implications of Geographic Information Systems*. New York: Guilford.

Regional Urban Information System. 1997. *RUIS Strategic Plan*. San Diego, CA: Regional Urban Information System.

Renner, P. J. Y. and Waldon, J. L. 1995. *Kentucky Fish and Wildlife Resource Information Management Survey*. Frankfort, KY: Kentucky Department Fish and Wildlife Resources.

Rogers, E. M. 1993. The diffusion of innovations model. In I. Masser and H. Onsrud (eds) *Diffusion and Use of Geographic Information Technologies*. Dordrecht: Kluwer, pp. 9–24.

Rogers, E. M. 1995. *Diffusion of Innovations*. New York: The Free Press.

Sandberg, B. 1992. *A Geographic Information Systems User Survey: The Emerging GIS Community in Michigan*. East Lansing, MI: Center for Remote Sensing, Michigan State University.

Sheppard, E. 1995. GIS and society: Towards a research agenda. *Cartography and Geographic Information Systems* 22: 5–16.

Sinton, D. 1991. Reflections on 25 years of GIS. *GIS World* 4: special insert.

Smith, N. 1992. History and philosophy of geography: Real wars, theory wars. *Progress in Human Geography* 16: 257–71.

Steinitz, C. 1993a. GIS: A personal historical perspective. *GIS Europe* 2(5): 19–22.

Steinitz, C. 1993b. GIS: A personal historical perspective – Part 2: A framework for theory and practice in landscape planning. *GIS Europe* 2(6): 42–5.

Steinitz, C. 1993c. GIS: A personal historical perspective – Part 3: The changing face of GIS from 1965–1993. *GIS Europe* 2(7): 38–40.

Taylor, P. and Overton, M. 1991. Commentary: Further thoughts on geography and GIS. *Environment and Planning* A 23: 1087–94.

Tomlinson, R. 1997. The Canada Geographic Information System. In T. Foresman (ed.) *The History of Geographic Information Systems: Perspectives of Pioneers*. Upper Saddle River, NJ: Prentice-Hall, pp. 21–32.

Tosta, N. 1999. NSDI was supposed to be a verb: A personal perspective on progress in the evolution of the U.S. National Spatial Data Infrastructure. In B. Gittings (ed.) *Integrating Information Infrastructures with GI Technology*. Philadelphia, PA: Taylor and Francis, pp. 13–24.

Tulloch, D. L. 1999. Theoretical model of multipurpose land information systems development. *Transactions in GIS* 3: 259–83.

Tulloch, D. L. 2002. Institutional aspects of SDIs. Washington, DC: University Consortium of Geographic Information Science Research Brief (available at http://www.ucgis.org/priorities/research/2002researchPDF/shortterm/l_institutional_gis.pdf).

Tulloch, D. L. and Epstein, E. F. 2002. Benefits of community MPLIS: Efficiency, effectiveness, and equity. *Transactions in GIS* 6: 195–212.

Tulloch, D. L. and Fuld, J. 2001. County-level production of framework data: Pieces of a National Spatial Data Infrastructure? *URISA Journal* 13(2): 11–21.

Tulloch, D. L. and Niemann, B. J. 1996. Evaluating innovation: The Wisconsin Land Information Program. *Geo Info Systems* 6(10): 40–4.

Tulloch, D. L., Niemann, B. J., Ventura, S. J., and Epstein, E. F. 1996. Measuring GIS/LIS progress in local governments: Land records modernization and its outcomes. In *Proceedings of the Seventeenth International ESRI User Conference*, Palm Springs, CA, USA: CD-ROM.

Tulloch, D. L. and Robinson, M. 2000. A progress report on a U.S. national survey of geospatial framework data. *Journal of Geographic Information Systems* 27: 285–98.

Tulloch, D. L. and Shapiro, T. 2003. The intersection of data access and public participation: Impacting GIS users' success? *URISA Journal* 15(2): 55–60.

US Geological Survey. 2001. *The National Map: Topographic Mapping for the 21st Century*. Reston, VA: Office of the Associate Director for Geography, US Geological Survey.

Ventura, S. J. 1995. The use of geographic information systems in local government. *Public Administrative Review* 55: 461–7.

Wisconsin Land Information Board. 1995. *1994 Annual Status Report Survey to the Wisconsin Land Information Board*. Madison, WI: University of Wisconsin-Madison, Land Information and Computer Graphics Facility.

Wunderlich, G. and Moyer, D. D. 1984. Economic features of land information systems. In B. J. Niemann (ed.) *Seminar on the Multipurpose Cadastre: Modernizing Land Information Systems in North America*. Madison, WI: University of Wisconsin-Madison, Institute for Environmental Studies Report No. 123: 183–202.

Zaltman, G., Duncan, R., and Holbek, J. 1973. *Innovations and Organization*. New York: John Wiley and Sons.

Chapter 26

Participatory Geographic Information Systems

Daniel Weiner and Trevor M. Harris

It is now twenty years since Chrisman (1987) provided one of the very early contributions to what subsequently became known as the Geographic Information System (GIS) and Society debate by proposing that GIS be developed in the context of social, economic, and ethical needs. In 1991, both Pickles (1991) and Edney (1991) presented papers concerning the transforming nature of GIS technology on society at the Applied Geography Conference in Toledo and the topic was also discussed at a 1991 Asso-ciation of American Geographers (AAG) Annual Meeting session on GIS and Society. In the same year a sharp exchange of words between Openshaw (1991) and Taylor and Overton (1991) occurred over the future role of GIS in the discipline of Geography and Openshaw's view of GIS as the superglue to put "Humpty Dumpty" (that is to say, Geography) back together again.

It was the Friday Harbor, Washington conference, however, sponsored by the National Center for Geographic Information and Analysis (NCGIA) that brought the academic GIS and social theory communities together, face to face, for the first time with the explicit purpose of discussing the potential societal impacts of GIS. The positive attitude of the Friday Harbor conference significantly moved the debate forward in identifying salient issues and putting in place institutional procedures and forums for continued discussion. Out of the Friday Harbor conference was born NCGIA Initiative 19 and the subsequent workshop in Minnesota (Harris and Weiner 1996) and the special edition of *Cartography and Geographic Information Systems* on GIS and Society published in January, 1995. Central to the broader dissemination of the GIS and Society discussion was, of course, the publication of the seminal book entitled *Ground Truth: The Social Implications of Geographic Information Systems* (Pickles 1995).

Public Participatory GIS (PPGIS) was one of the more substantive methodological and political themes to arise out of the GIS and Society debate. Specific focus on the intersection of participatory development and geospatial technologies was formalized in an NCGIA Varenius initiative *Empowerment, Marginlization and Public Participation GIS* (http://www.ncgia.ucsb.edu/varenius/varenius.html). This resulted in the publication in 2002 of a collection of essays on *Community Participation and*

Geographic Information Systems (Craig, Harris, and Weiner 2002). This book contains 28 chapters from 48 contributing authors, comprises 10 conceptual chapters and 18 case studies, and provides a valuable perspective on developments in the field of PPGIS. In this review we draw upon the findings of the book, complemented by our own personal research and field-based experiences. This chapter was written soon after the book was published and important additional research on Participatory GIS has been subsequently published (see, for example, Fox, Suryanata, and Hershock 2005, Hawthorne, Dougherty, Elmes, et al. 2006, IIED 2006, Jankowski and Nyerges, Chapter 27 in this volume, Schlossberg and Shuford 2004).

Initially we seek to define what is meant by PPGIS; especially as the term has evolved through several variant forms that reflect refinements important to the field. Despite these variant forms, however, the focus remains on systems and projects that link communities with GIS and other geospatial technologies. Subsequent to identifying the current forms of PPGIS, the review focuses on exploring PPGIS methods, the nature of participation in PPGIS, and the regional and domain areas in which these systems have been developed.

Defining Participatory Geographic Information Systems

The concept of PPGIS arose primarily out of the broader discussions on GIS and Society as discussed above. This origin is significant because of the influence of the debate on the nature of "early" PPGIS projects and their subsequent forms. Significantly, PPGIS arose not from an applied field of inquiry, or even out of participatory development, but from a distinctly academic GIS and Society debate. Considerable discussion at the NCGIA Initiative 19 workshop in Minnesota focused on how GIS might methodologically address the issues raised by the social theoretic critique (Harris and Weiner 1996). Scenarios based on existing GIS software and usage were considered (labeled version 1.0) along with possible modified usage of software currently available (labeled version 1.2). Much of the discussion, however, revolved around speculation about alternative forms of GIS that would bring about a substantive shift toward addressing core GIS and Society issues. These alternative systems were labeled GIS 2. Envisaging an alternative system in the context of differential public access to data, hardware, software, and expertise, to questions of distorted knowledge databases and the perceived exclusion of community knowledge and to broad issues of democratic GIS-based decision-making, led ensuing discussion to consider the development of PPGIS. PPGIS provided a suitable vehicle that would address the social theoretical critique of GIS and contribute to the methodological development of GIS and the move toward GIS 2. At a subsequent workshop held at the University of Maine the term PPGIS was more formally defined (Schroeder 1996).

In 1998, Project Varenius brought together a number of researchers, again primarily from the academic community, with experience in the implementation of PPGIS projects (Craig, Harris, and Weiner 1999). The intent of the workshop was to move beyond theory and to explore the methodological and participatory issues thrown up by the field-based experiences and project development. The workshop papers were subsequently updated and formed the core of a new book (Craig, Harris, and

Weiner 2002). Subsequent variations in terminology have appeared that reflect subtleties in the conceptualization of PPGIS. Thus Community-integrated GIS (CiGIS)
(Harris and Weiner 1998) specifically seeks to position PPGIS in the context of the
political economy of differential access to resources, data, and knowledge representation. Developing PPGIS without consideration of how such systems are to be
resourced or sustained misses a central concern of the GIS and Society critique. CiGIS
was designed to broaden public participation in GIS decision-making based on the
assumption that state and local government agencies, along with NGOs, will provide the resources and the infrastructure to support such efforts. In this respect the
core of a CiGIS is focused on how communities gain access to such systems and how
their knowledge might complement traditional top-down GIS data typical of more
traditional systems (see the early part of Chapter 28 by Dana in this volume for an
application illustrating these issues). Further to this, the term PPGIS has increasingly
been shortened to Participatory GIS (PGIS). Specifically, this term acknowledges
not just the involvement of the public in the development and use of such systems
but that of other governmental and non-governmental agencies and private sector
members of a community as well. In addition, the term Participatory GIS places
the greater emphasis on participatory methods in the production of a GIS. For the
remainder of this chapter we use the preferred term of PGIS.

It is notable that much of the PGIS work of the 1990s, perhaps epitomized by
the collection of case studies in Craig, Harris, and Weiner (2002), are primarily
located in the GIS and Society literature. The studies invariably originate from an
academic environment, are steeped in GIS and Society themes, are project-based,
and are driven by a concern to pursue social justice issues as well as research PGIS
themes. In more recent years, several other trends can be discerned. One such trend
mirrors developments in GIS itself and especially the move toward Geographic
Information Science (GISc). Use of the term Participatory Geographic Information
Science (PGISc) (MacMaster 2002; see also Pickles 1999 and Sheppard, Couclelis,
Graham, Harrington, and Onsrud 1999) seeks to position PGIS less as a focus on
the information system than on the science of spatial information. While the term
PGISc has been used extensively in panel discussions and has been used in general
discussion, there is as yet no clear definition of what such PGISc would comprise,
nor has a demonstration project been developed that distinguishes between PGIS
and PGISc. In essence, PGISc acknowledges that PGIS is situated within the evolving field of GISc. In one respect this fits well with the emergence of PGIS that are
as much dependent on related geospatial technologies, such as remote sensing and
the Internet, as they are on GIS *per se*. However, the positioning of PGIS in the
context of a science paradigm would appear to raise epistemological and philosophical
issues that are even further at odds with the GIS and Society issues and alternative
conceptualizations of space and place that the debate raised.

A further trend in PGIS may be identified as a result of the diffusion of PGIS
concepts among a broader user community. This should perhaps be expected as the
field becomes more widely known. In particular, as PGIS becomes more distant from
the incubating academic environment of its origins, and less situated within the
broader GIS and Society discussion, so the nature of PGIS will reflect such distancing.
This change is perhaps best reflected in the Annual **Urban and Regional Information
Systems Association** (URISA) PPGIS conferences that started in 2002, and a 2005

Mapping for Change International Conference on Participatory Spatial Information Management and Communication that was held in Nairobi (IIED 2006).

PGIS Methods and Forms of Participation

PGIS explicitly situate GIS within participatory research and, as a result, local knowledge is incorporated into GIS production and use. In blending local knowledge with "expert" information, PGIS projects have in common the application of geospatial technologies to address concerns articulated by community participants. As a result, data products and the scale of analysis must be appropriate for the needs of the participating community, and community data production and access is assumed.

Given the characteristics of PGIS outlined above, it is surprisingly difficult to explicitly identify its core methods. Indeed, therein lies the challenge. PGIS methods can range from conventional field-based participatory development methods that primarily utilize digital mapping, to Internet-dependent spatial multimedia. Digital cartography remains the core of many PGIS efforts as the "power of maps" is extended to participant communities. Such has been the extent of community mapping initiatives under the name of PGIS that there have been calls to change the term to Community Mapping Systems, for in many PGIS projects very little GIS functionality is used beyond the mapping component. However, PGIS methods also include geo-visualization, sketch mapping, satellite image and air photo interpretation, GPS transect walks, mental mapping exercises, spatial multimedia, and, in the case of some contemporary work, even virtual GIS (Harris, Alagan, and Rouse 2002). GPS transect walks are sometimes utilized to locate community resource access and for landscape interpretation. In cases where communities do not have access to such resources, mental mapping with tracing paper overlaid on GIS products has proven to be very useful. Mental maps are a useful method for communities to express their perceptions of landscape and to collect socially differentiated community local knowledge that can be digitized and analyzed as GIS data (Harris and Weiner 2002). They are a platform for individual and community "spatial story telling" (Aitken 2002). Sketch maps are also a proven development planning method for investigating community perceptions of local landscapes and for communities to articulate their spatial needs, aspirations, and frustrations without the constraints of cartographic accuracy or the constraints of boolean logic (Sheppard 1993). An important challenge in the production of PGIS is, therefore, the integration of valuable qualitative local knowledge such as sound, voice, text, photos, and video with GIS. Spatial multimedia systems are becoming increasingly common and the Internet provides a powerful technology for performing such integration (Harris, Weiner, Warner, and Levin 1995, Kingston 2002, Shiffer 2002).

It is of some interest to explore the range of approaches used in the recent compendium of case studies that link community participation with GIS (Craig, Harris, and Weiner 2002). An exploration of the case studies suggests that three dominant components are apparent: participatory methods, the Internet, and GIS (Figure 26.1). The respective contribution of each component in the case studies indicates that four of the 18 case studies make extensive use of the Internet. Interestingly, this does not

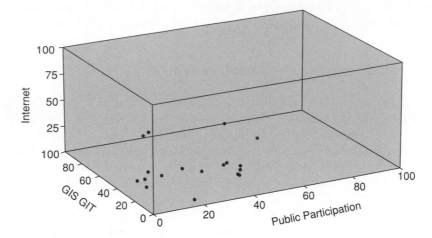

Fig. 26.1 The relations of GIS, participation, and the Internet in the case studies (Craig, Harris, and Weiner 2002)

imply the use of Internet GIS but of peer-to-peer capability for data, information, and e-mail exchange between communities often separated by considerable distances. Participatory methods, as one would expect, are extensively used and are familiar to a range of academic disciplines and development practitioners. Of all the three dimensions, it is GIS that exhibits the least complexity. While a few projects have sought to augment GIS through multimedia technology, very few go beyond digital mapping. It could be postulated that as PGIS evolves so there will be a migration toward the far top right of the box, though this assumes that greater technological and participatory complexity will produce more effective participatory systems. As earlier comments should indicate, this is by no means an acceptable assumption and it could be that the simplest mapping system could still produce very effective solutions.

Regardless of the specific PGIS methods used, community participation is essential for successful project implementation. Importantly, participation is not a uniform homogenous method and it is the *nature of participation* employed that is central to understanding PGIS. Community participation has long been a mantra in development planning and field-based academic research. Unfortunately, most participation associated with development planning can stand accused of being *participation as legitimization*, and such models could equally well apply in PGIS. Thus, community meetings are held, local input is gathered, the GIS is generated, reports are produced, and top-down planning under the legitimating guise of PGIS is maintained. In this context, participation could assist in legitimizing decisions that are not necessarily popular within impacted communities.

In the academic world, participation has come to designate a configuration of qualitative methods designed to understand complex social processes in ways that conventional quantitative methods cannot achieve. Efforts to hear the voices of "ordinary" people and to represent "local knowledge" are well intentioned, but in many instances these are forms of participation for publication, in which academics undertake research to produce books and journal articles while leaving the subject communities with little (if any) tangible benefits. This type of participatory practice tends to be exploitative. Popular participation, in the sense that we use it, is an attempt to locate community participation within the context of local and democratic configurations of power within civil society (Levin and Weiner 1997). Participatory processes then become part of the structures of everyday life, and ordinary people are able to express their opinions as openly as possible. There are, therefore, inherent contradictions in the marriage of community participation with GIS. Participation is a highly contested "development" theory and practice that is compounded still further with the introduction of geospatial information technologies.

PGIS projects are, at their core, political because they attempt to broaden access to digital spatial information and empower historically disempowered people and communities. PGIS projects are also political because they involve community participation, which is again essentially a political process. This suggests that understanding the politics and associated power relationships of PGIS projects is critical in order to unpack their impacts, wherever and however implemented. Participatory GIS is a reflection of the politics of the builders and users of such systems, although these politics extend beyond the local impacts on participating and non-participating communities. The issue of who has access to GIS and who benefits from such systems is, therefore, a fundamental yet complex question.

PGIS Implementation

The coming together of community participation with GIS, and geospatial technologies more generally, is global in its scope and occurring in a diversity of social and environmental contexts (Abbot, Chambers, Dunn, et al. 1998, Craig, Harris, and Weiner 2002). GIS are being integrated in communities in a wide variety of ways, to serve many purposes, and with various degrees of effectiveness (Leitner, McMaster, Elwood, McMaster, and Sheppard 2002). Some communities use GIS to administer and manage territory under their control (Elwood 2002, Walker, Leitch, De Lai 2002, Kwaku-Kyem 2002, Jordan 2002, Bond 2002) and to make informed input into local planning processes (Sieber 2002, Parker and Pascual 2002, Ventura, Niemann, Suttphin, and Chenoweth 2002, Kingston 2002, Bosworth, Donovan, and Couey 2002). There are also cases where GIS have helped communities to develop their own spatial strategies and policies (Sawicki and Burke 2002, Tulloch 2002, McNab 2002, Laituri 2002, Harris and Weiner 2002). Bosworth, Donovan, and Couey (2002) show the multiple ways government can make data available to communities, while Kingston (2002) and Ventura, Niemann, Suttphin, and Chenoweth (2002) demonstrate how PGIS is rapidly merging with the Internet. Some of the earliest applications involved indigenous natural resource mapping in arctic and tropical regions within the Americas (Marozas 1993; see also Chapter 28 by Dana in this volume). There is also a rapidly growing network of planning professionals interested in PGIS in the context of neighborhood revitalization and urban planning (Aitkin and Michel 1995; Craig and Elwood 1998; Talen 1999, 2000; Leitner, McMaster, Elwood, McMaster, and Sheppard 2002; Sawicki and Peterman 2002). Environmental groups are also experimenting with community GIS applications to promote environmental equity and address environmental racism

(Kellog 1999, Sieber 2000). Furthermore, NGOs, aid organizations, and governmental agencies are linking communities with GIS as they seek to promote more popular and sustainable development projects (Hutchinson and Toledano 1993, Obermeyer and Pinto 1994, Gonzalez 1995, Harris, Weiner, Warner, and Levin 1995, Weiner, Warner, Harris, and Levin 1995, Dunn, Atkins, and Townsend 1997, Mitchell 1997, Elwood and Leitner 1998, Jordan and Shrestha 1998, Kwaku-Kyem 1999, Rambaldi and Callosa 2000, Weiner and Harris 2003).

One of the greatest difficulties involved in implementing a PGIS is the incorporation of complex and socially differentiated information within a GIS. Harris and Weiner (2002) confronted this problem using socially differentiated mental maps with particular themes, and incorporating that information into a spatial multimedia database. Al-Kodmany (2002) employed an innovative graphic design method to extend GIS to incorporate block-specific community views. But community organizations do not necessarily represent the views of the majority of community members. Kwaku-Kyem's (2002) case study in Ghana identifies the common contradictions inherent in practices of community participation. For example, women are often excluded, some people are intimidated by the technology, clans have difficulty collaborating, and the existing power structure is often disinterested in empowering citizens. Bosworth, Donovan, and Couey (2002) use a communication pyramid to show that many people choose not to get involved in community activity, but clearly there are aspects of organizations and technology that tend to systematically exclude some individuals.

A further point is that participation in PGIS should not be viewed solely as a method but as a process. Walker, Leitch, De Lai, et al. (2002) demonstrate that communities working together to create a GIS center helped resolve many conflicts among the participating groups. Process was also a central theme of Jordan's (2002) case study in Nepal and the study by Meridith, Yetman, and Frias (2002) in Canada and Mexico. The latter identified "second order cybernetics" whereby people working together became more aware of their situation, and thus made personal adaptations to accommodate community needs and desires. Many PGIS projects do not include community participation as part of the GIS production process. For example, some community information needs can be met by conventional maps and reports delivered by a government service center on compact disk or over the Internet. Casey and Pederson (2002) refer to this as a "public records GIS," and many cities and counties now provide this type of public data inventory. For example, the Data and Policy Analysis Group of the Atlanta Project provides sophisticated maps to assist local committees in understanding the nature of prioritized community issues, and to help develop policy recommendations (Sawicki and Burke 2002). Such an approach does not, however, fulfill the needs of what Casey and Pederson call "community-based GIS" that should provide not only relevant local data but be capable of performing spatial analysis for participating communities.

But what have been the impacts of this rapid merging of community participation with geospatial technologies? How have PGIS initiatives contributed to individual and community empowerment? And, are there cases of PGIS disempowerment? While it is still very early in the adoption process, and the nature of PGIS is in flux, there is evidence of success stories in the areas of crime prevention, housing condemnation and renovation, smart growth and land-use planning; natural resource management, and the conservation and preservation of indigenous territories. PGIS give communities access to advanced IT map and/or spatial analysis applications and include their voices in the resulting GIS products. This contributes to community empowerment through digital countermaps that communicate community spatial stories that are integrated into local decision-making politics. Furthermore, the process of participation itself can be empowering, as communities are "experts" about the local landscape and teachers of the development practitioners. In New Zealand, for example, "the construction of community-based GIS has provided the iwi (tribes) with the opportunity to identify meaningful applications of their particular areas from their own perspective rather than applications based only on Euro-American models of resource management and land use. Tribal people were not only the GIS users, but the GIS designers as well" (Laituri 2002, p. 273).

Parker and Pascual document a similar empowerment process in a San Francisco gentrification case study:

> It is unlikely that the Planning Department would have engaged in such a detailed study and invited public participation had it not been of the actions of a very informed and sophisticated group of community activists. At the core of this effort was the GIS-generated living neighborhood map, which empowered the community, educated community members, and offered a means by which people could shed their individual opinions and judgments in order to see the situation for what it truly was. By both improving the quality of information available and providing a means for people to work together, the living neighborhood map allowed people to stop reacting based on emotion, hearsay, and opinions and develop a more credible and powerful voice with which to argue in the public arena for their rights as a community. (Parker and Pascual 2002, p. 64)

There is also evidence that PGIS mapping of indigenous lands is empowering participant communities in the United States. Bond concludes that:

> . . . many tribes are using geospatial technology to manage their natural and cultural resources. The Yakima Tribe in Washington State, the White Mountain Apache in Arizona and the Salish Kootenai in Montana utilize GIS technology to inventory, analyze, map and make decisions regarding tribal resources. Examples of such resources include timber production, grazing and farm land, water rights, wildlife, native plants, cultural sites, environmental data and hazardous site monitoring, historical preservation, health and human resources. (Bond 2002, p. 292)

In the case of inner city housing decay and renovation, a Minneapolis community organization (PPNA) produced a PGIS to:

> . . . address several critical housing issues in Powderhorn Park. They have been used most extensively in making plans for housing improvement in the neighborhood. For example, in 1998, PPNA's Housing and Land-Use committee relied extensively on analysis of these data to inform residents in making housing improvements . . . the organization was able to, for the first time, conduct comprehensive geographic analysis of housing issues . . . For instance, a staff member's analysis showing the concentration of dilapidated rental properties along the neighborhood's major transportation corridors inspired the committee to design assistance programs specifically targeted to the improvement of rental properties . . . (Elwood 2002, p. 81)

However, this Minneapolis case study also demonstrates how marginalization can occur simultaneously with empowerment:

> Neighborhood deliberations about housing improvements increasingly utilize language that requires detailed knowledge of City or County housing policies . . . In the past, residents relied on visual attributes in their descriptions of housing conditions . . . For residents who have experience in planning or housing issues, either through their community activities or their professional employment, these changes are not problematic. For residents without such expertise, these changes constitute a significant barrier to their participation in PPNA's planning efforts. In Powderhorn Park, the residents who are most affected by this shift in language and expertise are those who have traditionally been marginalized from neighborhood organizations – people of color, renters, senior citizens, and non-native English speakers. (Elwood 2002, p. 85)

In this case, disempowerment was observed through the reconfiguration of established community groups. Changes in the planning discourse associated with PGIS altered existing community power relationships. Disempowerment can also take place when government agencies limit data access to community groups that are deemed to be too radical. Sieber (2002), for example, came to the conclusion that in California:

> . . . research demonstrated that environmental groups use and value GIS . . . All [five] groups exhibit an indigenous demand for GIS, backed by a history of scientific and technical knowledge. A leader committed to championing GIS innovation, so critical in government implementation, emerged in each of the cases. Improvisation demonstrated that resources did not represent a barrier to GIS implementation. This was not the case for acquiring digital data, however, which favored groups engaged in proactive and non-confrontational agendas. (Sieber 2002, p. 169).

The politics of data access were also observed in a project in the Ashanti Region of Ghana. A PGIS was developed to help build collaborative forest management and was successful in involving local "forest user groups" (Kwaku-Kyem 2002). But resistance from local elites was also observed as the "rich and powerful people in the community objected to the open and participatory uses of GIS. Some were particularly resentful of the inclusion on the forest committee of representatives of local farmers. The resentment about participatory uses of GIS could hinder future adoptions of the technology for the empowerment of less privileged groups" (Kwaku-Kyam, 2002, p. 229).

Unequal access to the Internet – and computer skills in general – also empower and disempower simultaneously and are important considerations in Internet-based PGIS initiatives. For example, Ventura, Niemann, Suttphin, and Chenoweth (2002) are involved in a very successful land use planning project in Wisconsin, in one of the fastest growing counties in the region. They are sponsoring local land-use forums and innovative uses of the Internet with WebGIS to involve residents in county land-use planning. Community participation was impressive, but:

> . . . citizens come into the process with a relatively high level of computer acumen. Ninety-two percent of the participants in the land use forums had home computers and 70 percent had Internet access. Though these were citizens motivated to become

involved in county-wide land use issues, approximately the same percentages of access to computers and the Internet were recorded in the survey of all households in the Town of Verona (Ventura, Niemann, Suttphin, and Chenoweth 2002, p. 122).

A similar pattern was observed in Portland (Oregon) in a project concerned with growth management through land use and transportation planning (Bosworth, Donovan, and Couey 2002). Their innovative Metro-Map is an interactive web-based application for accessing GIS data layers of metropolitan Portland and to enable public input. As with the Wisconsin example, participation is high, but is biased towards community members with home computer and Internet access.

These case studies provide examples of how a diversity of PGIS applications impact community power relations and local decision-making. They also demonstrate the political nature of PGIS. To date, however, we have seen only glimpses of this empowerment/disempowerment process. As a result, the specific mechanisms by which PGIS empower and disempower people and communities remains a fundamental area for PGIS research.

SUMMARY AND CONCLUSIONS

The marriage of community participation with GIS remains in its infancy. Nevertheless, much has been learned and observed in the decade since the initial GIS 2 was conceptualized. In this chapter, we offer some reflections about PGIS that can be summarized with ten broad observations:

1 PGIS emerged out of an academic debate but have now become a spontaneous merging of community development initiatives with new information technologies. Differential access to digital geospatial technologies and the (re)discovery of the power of maps are driving their rapid diffusion.

2 There are many forms of PGIS and they are emerging in all world regions. PGIS projects simultaneously empower and marginalize individuals and communities, contribute to the development of placed-based methodologies, promote more inclusive community spatial decision-making, and incorporate multiple realities of landscape.

3 PGIS in core industrialized regions are increasingly Internet-based spatial multimedia systems, whereas in underdeveloped regions they tend to be participatory development projects with an IT component. Furthermore, PGIS in the periphery tend to be participatory development projects with a GIS interface whereas in the core they tend to be GIS projects with a participation interface.

4 There has been a tendency to overstate the technical complexities of PGIS projects while the importance of political context – which is scale dependent – has been underestimated.

5 There is a need to distinguish between PGIS research that is academically intentioned and projects with the primary objective of community empowerment. The intermediary technical expert remains a substantial part of the PGIS process.

6　PGIS are purposefully value-laden and redefine the meaning of "accuracy."

7　PGIS are characterized by the creation and use of countermaps and narratives that help represent complex socio-economic, cultural, and political landscapes. Whether such mapping can be translated into real power and political influence remains to be seen though there are now intriguing case studies in which PGIS augment place-specific political struggles by "jumping scale" (Aitken 2002, Stonich 2002).

8　An unintended but important and valuable outcome of early PGIS is that they provided a platform for integrating qualitative and quantitative information. This is significant for social scientists because of the historic dualism between researchers who employ qualitative methods and those who employ quantitative methods, and because of the methodological difficulties in merging the two. PGIS highlights place, and does so in ways that conventional GIS normally do not.

9　Most current PGIS projects do not utilize GIS functionality for advanced spatial analysis and are essentially participatory digital mapping projects. Internet mapping systems are likely to play a significant role in future PGIS initiatives. Internet-based PGIS will further complicate the definition of a community and the practices of participation. Virtual communities present significant opportunities and challenges as participation is broadened, but becomes placeless. Community participation from the home computer will ultimately transform PGIS in ways that we do not yet understand.

10　The primary challenges for the successful implementation of PGIS projects are a complex web of technological, social, and political factors that are locally constituted and promote more inclusive spatial decision-making. To date, there has been no systematic evaluation of the contribution of PGIS to local and regional spatial planning. This is understandable given that PGIS is only now penetrating the administrative and bureaucratic structures of planning agencies, development organizations, universities, NGOs, and the private sector. The monitoring and evaluation of projects over a longer time span will provide insight into the effectiveness of such implementations (Barndt 2002).

We have proposed in this chapter that GIS and Society debates concerning the diffusion of geospatial technologies shaped early PGIS applications. There is no universal PGIS model, but rather place-based methods that navigate local politics and production relations. There is some concern, however, that the rapid growth of PGIS projects might be submerging a critical theory of GIS. PGIS is not a panacea, and must not undermine the robust debate and research agenda concerned with the political economy, epistemology, philosophy, and contested practices of GIS.

REFERENCES

Abbot, J., Chambers, R., Dunn, C., Harris, T., de Merode, E., Porter, G., Townsend, J., and Weiner, D. 1998. Participatory GIS: Opportunity or oxymoron. *PLA Notes* 33: 27–34.

Aitken, S. C. 2002. Public Participation, technological discourses and the scale of GIS. In W. J. Craig, T. M. Harris, and D. Weiner (eds) *Community Participation and Geographic Information Systems*. London: Taylor and Francis, pp. 357–66.

Aitkin, S. and Michel, S. 1995. Who contrives the "Real" in GIS? Geographic information, planning, and critical theory. *Cartography and Geographic Information Systems* 22: 17–29.

Al-Kodmany, K. 2002. GIS and the artist: Shaping the image of a neighborhood through participatory environmental design. In W. J. Craig, T. M. Harris, and D. Weiner (eds) *Community Participation and Geographic Information Systems*. London: Taylor and Francis, pp. 320–29.

Barndt, M. 2002. A model for evaluating public participation GIS. In W. J. Craig, T. M. Harris, and D. Weiner (eds) *Community Participation and Geographic Information Systems*. London: Taylor and Francis, pp. 346–56.

Bond, C. 2002. The Cherokee Nation and tribal uses of GIS. In W. J. Craig, T. M. Harris, and D. Weiner (eds) *Community Participation and Geographic Information Systems*. London: Taylor and Francis, pp. 283–93.

Bosworth, M., Donovan, J., and Couey, P. 2002. Portland Metro's dream for public involvement. In W. J. Craig, T. M. Harris, and D. Weiner (eds) *Community Participation and Geographic Information Systems*. London: Taylor and Francis, pp. 125–36.

Casey, L. and Pederson, T. 2002. Mapping Philadelphia's neighborhoods. In W. J. Craig, T. M. Harris, and D. Weiner (eds) *Community Participation and Geographic Information Systems*. London: Taylor and Francis, pp. 65–76.

Chrisman, N. R. 1987. Design of geographic information systems based on social and cultural goals. *Photogrammetric Engineering and Remote Sensing* 53: 1367–70.

Craig, W. J. and Elwood, S. 1998. How and why community groups use maps and geographic information. *Cartography and Geographic Information Systems* 25: 95–104.

Craig, W. J., Harris, T. M., and Weiner, D. 1999. *Empowerment, Marginalization and Public Participation GIS*. Santa Barbara, CA: National Center for Geographic Information and Analysis, Varenius Specialist Meeting Report.

Craig, W. J., Harris, T. M., and Weiner, D. (eds). 2002. *Community Participation and Geographic Information Systems*. London: Taylor and Francis.

Dunn, C., Atkins, P., and Townsend, J. 1997. GIS for development: A contradiction in terms? *Area* 29: 151–9.

Edney, M. H. 1991. Strategies for maintaining the democratic nature of geographic information systems. *Papers and Proceedings of the Applied Geography Conferences* 14: 100–8.

Elwood, S. 2002. The impacts of GIS use for neighborhood revitalization in Minneapolis. In W. J. Craig, T. M. Harris, and D. Weiner (eds) *Community Participation and Geographic Information Systems*. London: Taylor and Francis, pp. 77–88.

Elwood, S. and Leitner, H. 1998. GIS and community-based planning: Exploring the diversity of neighborhood perspectives and needs. *Cartography and Geographic Information Systems* 25: 77–88.

Fox, J., Suryanata, K., and Hershock, P. (eds). 2005. *Mapping Communities: Ethics, Values, Practise*. Honolulu, HI: East West Center.

Gonzalez, R. M. 1995. KBS, GIS and documenting indigenous knowledge. *Indigenous Knowledge and Development Monitor* 3(1) (available at http://www.nuffic.nl/ciran/ikdm/3-1/articles/gonzalez.html).

Harris, T., Alagan, R., and Rouse, J. 2002. Geo-visualization approaches to PGIS decision-making in Environmental Impact Assessment. In *Proceedings of the First Annual URISA Public Participation GIS (PPGIS) Conference*, Rutgers University, NJ, USA. Bedford Park, IL: Urban and Regional Information Systems Association, pp. 180–3.

Harris, T. M. and Weiner, D. 1996. *GIS and Society: The Social Implications of How People, Space and Environment are Represented in GIS*. Santa Barbara, CA: National Center for Geographic Information and Analysis Scientific Report for Initiative No. 19.

Harris, T. M. and Weiner, D. 1998. Empowerment, marginalization and community-integrated GIS. *Cartography and Geographic Information Systems* 25: 67–76.

Harris, T. M. and Weiner, D. 2002. Implementing a community-integrated GIS: Perspectives from South African fieldwork. In W. J. Craig, T. M. Harris, and D. Weiner (eds) *Community Participation and Geographic Information Systems*. London: Taylor and Francis, pp. 246–58.

Harris, T. M., Weiner, D., Warner, T., and Levin, R. 1995. Pursuing social goals through participatory GIS: Redressing South Africa's historical political ecology. In J. Pickles (ed.) *Ground Truth: The Social Implications of Geographic Information Systems*. New York: Guilford, pp. 196–222.

Hawthorne, T. L., Dougherty, M., Elmes, G., Fletcher, C., McCusker, B., Pinto, M., and Weiner, D. 2006. Beyond the public meeting: Building a field-based Participatory Geographic Information System for land use planning in Monongalia County, West Virginia. In S. Balram and S. Dragicevic (eds) *Collaborative Geographic Information Systems*, Hershey, PA: Idea Group Inc., pp. 43–65.

Hutchinson, C. F. and Toledano, J. 1993. Guidelines for demonstrating geographical information systems based on participatory development. *International Journal of Geographical Information Systems* 7: 453–61.

IIED. 2006. Special issue on mapping for change: Practice, technologies and communication. *Participatory Learning and Action* 54: 1–150.

Jordan, G. 2002. GIS for community forestry user groups in Nepal: Putting people before the technology. In W. J. Craig, T. M. Harris, and D. Weiner (eds) *Community Participation and Geographic Information Systems*. London: Taylor and Francis, pp. 232–45.

Jordan, G. and Shrestha, B. 1998. *Integrating Geomatics and Participatory Techniques for Community Forest Management: Case Studies from Yarsha Khola Watershed*. Katmandu, Nepal, International Centre for Integrated Development.

Kellog, W. 1999. From the field: Observations on using GIS to develop a neighborhood environmental information system for community-based organizations. *URISA Journal* 11(1): 15–32.

Kingston, R. 2002. Web-based PPGIS in the United Kingdom. In W. J. Craig, T. M. Harris, and D. Weiner (eds) *Community Participation and Geographic Information Systems*. London: Taylor and Francis, pp. 101–12.

Kwaku-Kyem, P. 1999. Examining the discourse about the transfer of GIS technology to traditionally non-western societies. *Social Science Computer Review* 17: 69–73.

Kwaku-Kyem, P. A. 2002. Promoting local community participation in forest management through a PPGIS application in Southern Ghana. In W. J. Craig, T. M. Harris, and D. Weiner (eds) *Community Participation and Geographic Information Systems*. London: Taylor and Francis, pp. 218–31.

Laituri, M. 2002. Ensuring access to GIS for marginal societies. In W. J. Craig, T. M. Harris, and D. Weiner (eds) *Community Participation and Geographic Information Systems*. London: Taylor and Francis, pp. 270–82.

Leitner, H., McMaster, R. B., Elwood, S., McMaster, S., and Sheppard, E. 2002. Models for making GIS available to community organizations: Dimensions of difference and appropriateness. In W. J. Craig, T. M. Harris, and D. Weiner (eds) *Community Participation and Geographic Information Systems*. London: Taylor and Francis, pp. 37–52.

Levin, R. and Weiner, D. 1997. *No more tears: Struggles for land in Mpumalanga, South Africa*. Trenton, NJ: Africa World Press.

MacMaster, R. 2002. GIS and Society. Panel presentation at Annual Meeting of the Association of American Geographers, Los Angeles, CA, USA.

McNab, P. 2002. There must be a catch: Participatory GIS in a Newfoundland fishing community. In W. J. Craig, T. M. Harris, and D. Weiner (eds) *Community Participation and Geographic Information Systems*. London: Taylor and Francis, pp. 173–91.

Marozas, B. A. 1993. A culturally relevant solution for the implementation of geographic information systems in Indian Country. In *Proceedings of the Thirteenth Annual International ESRI User Conference*, San Diego, CA, USA. Redlands, CA: Environmental Systems Research Institute, Inc., pp. 365–81.

Meredith, T. C., Yetman, G. G., and Frias, G. 2002. Mexican and Canadian case studies of community-based spatial information management for biodiversity conservation. In W. J. Craig, T. M. Harris, and D. Weiner (eds) *Community Participation and Geographic Information Systems*. London: Taylor and Francis, pp. 205–17.

Mitchell, A. 1997. *Zeroing In: Geographic Information Systems at Work in the Community*. Redlands, CA: ESRI Press.

Obermeyer, N. and Pinto, J. 1994. *Managing Geographic Information Systems*. New York: Guilford.

Openshaw, S. 1991. A view on the GIS crisis in geography, or, using GIS to put Humpty-Dumpty back together again. *Environment and Planning A* 23: 621–8.

Parker, C. and Pascual, A. 2002. A voice that could not be ignored: Community GIS and gentrification battles in San Francisco. In W. J. Craig, T. M. Harris, and D. Weiner (eds) *Community Participation and Geographic Information Systems*. London: Taylor and Francis: 55–64.

Pickles, J. 1991. Geography, GIS, and the surveillant society. *Papers and Proceedings of Applied Geography Conferences* 14: 80–91.

Pickles, J. (ed.) 1995. *Ground Truth: The Social Implications of GIS*. New York: Guilford.

Pickles, J. 1999. Arguments, debates, and dialogues: The GIS-social theory debate and concerns for alternatives. In P. A. Longley, M. F. Goodchild, D. J. Maguire, and D. W. Rhind (eds) *Geographical Information Systems: Principles, Techniques, Management, and Applications*. New York: John Wiley and Sons, pp. 49–60.

Rambaldi, G. and Callosa, J. 2000. *Manual on Participatory 3-Dimensional Modeling for Natural Resource Management* Vol. 7. Manila: Philippines Department of Environment and Natural Resources.

Sawicki, D. S. and Burke, P. 2002. The Atlanta Project: Reflections on PPGIS practice. In W. J. Craig, T. M. Harris, and D. Weiner (eds) *Community Participation and Geographic Information Systems*. London: Taylor and Francis, pp. 89–100.

Sawicki, D. S. and Peterman, R. 2002. Surveying the extent of PPGIS practice in the United States. In W. J. Craig, T. M. Harris, and D. Weiner (eds) *Community Participation and Geographic Information Systems*. London: Taylor and Francis, pp. 17–36.

Schlossberg, M. and Shuford, E. 2004. Delineating public and participation in PPGIS. *URISA Journal* 16(2): 15–26.

Schroeder, P. 1996. Report on Public Participation GIS Workshop. In T. Harris and D. Weiner (eds) *GIS and Society: The Social Implications of How People, Space and Environment are Represented in GIS*. Santa Barbara, CA: National Center for Geographic Information and Analysis Technical Report No. 96–7.

Sheppard, E. 1993. GIS and society: Ideal and reality. In *Proceedings of the NCGIA Geographic Information and Society Workshop*, Friday Harbor, WA, USA. Santa Barbara, CA: National Center for Geographic Information and Analysis.

Sheppard, E., Couclelis, H., Graham, S., Harrington, J., and Onsrud, H. 1999. Geographies of the information society. *International Journal of Geographical Information Science* 13: 797–823.

Shiffer, M. J. 2002. Spatial multimedia representations to support community participation. In W. J. Craig, T. M. Harris, and D. Weiner (eds) *Community Participation and Geographic Information Systems*. London: Taylor and Francis, pp. 309–19.

Sieber, R. E. 2000. Conforming (to) the opposition: The social construction of geographical information systems in social movements. *International Journal of Geographical Information Science* 14: 775–93.

Sieber, R. E. 2002. Geographic information systems in the environmental movement. In W. J. Craig, T. M. Harris, and D. Weiner (eds) *Community Participation and Geographic Information Systems*. London: Taylor and Francis, pp. 153–72.

Stonich, S. C. 2002. Information technologies, PPGIS, and advocacy: Globalization of resistance to industrial shrimp farming. In W. J. Craig, T. M. Harris, and D. Weiner (eds) *Community Participation and Geographic Information Systems*. London: Taylor and Francis, pp. 259–69.

Talen, E. 1999. Constructing neighborhoods from the bottom up: The case for resident generated GIS. *Environment and Planning B* 26: 533–54.

Talen, E. 2000. Bottom-up GIS. A new tool for individual and group expression in participatory planning. *APA Journal* 66: 279–94.

Taylor, P. J. and Overton, M. 1991. Further thoughts on geography and GIS. *Environment and Planning A* 23: 1087–94.

Tulloch, D. L. 2002. Environmental NGOs and community access to technology as a force for change. In W. J. Craig, T. M. Harris, and D. Weiner (eds) *Community Participation and Geographic Information Systems*. London: Taylor and Francis, pp. 192–204.

Ventura, S. J., Niemann Jr. B. J., Suttphin, T. L., and Chenoweth, R. E. 2002. GIS-enhanced land-use planning. In W. J. Craig, T. M. Harris, and D. Weiner (eds) *Community Participation and Geographic Information Systems*. London: Taylor and Francis, pp. 113–24.

Walker, D. H., Leitch, A. M., De Lai, R., Cottrell, A., Johnson, A. K. L., and Pullar, D. 2002. A community-based and collaborative GIS joint venture in rural Australia. In W. J. Craig, T. M. Harris, and D. Weiner (eds) *Community Participation and Geographic Information Systems*. London: Taylor and Francis, pp. 137–52.

Weiner, D. and Harris, T. 2003. Community-integrated GIS for land reform in South Africa. *URISA Journal* 15: 61–73.

Weiner, D., Warner, T., Harris, T. M., and Levin, R. M. 1995. Apartheid representations in a digital landscape: GIS, remote sensing, and local knowledge in Kiepersol, South Africa. *Cartography and Geographic Information Systems* 22: 30–44.

Chapter 27

Geographic Information Systems and Participatory Decision Making

Piotr Jankowski and Timothy L. Nyerges

Decision making is a fundamental part of Geographic Information System (GIS) use, particularly in the context of decision-making groups that are increasingly becoming agents of change within a variety of organizational, community, and societal settings. Many of the intended changes, for example those involving implementation of public-private policies about the use of common-pool resources, such as water, are mired in controversy. Locational conflict often arises due to peoples' differences in values, motives, and/or locational perspectives about what is to be accomplished. In such situations, conflict, and therefore negotiation management with shared decision making, is a fundamental concern in coming to consensus about choices to be made. Dealing with location-based resource conflicts in an open manner is becoming more important as stakeholder participation increases in land use, natural resource, and environmental decision making. The idea of participatory decision making is as old as democracy. Participatory decision making has exemplified the democratic maxim that those affected by a decision should participate directly in the decision-making process (Smith 1982).

In the domain of local decisions affecting land use, transportation, access to public spaces, and use of natural resources public participation was limited for a long time to gathering input about decision alternatives that had already been prepared. More recently the idea of public participation in decision making has been enjoying a renaissance. Terms such as participatory spatial decision making, place-based decision making or community participation in decision making, have been used to reflect the idea of people taking an active role in decision-making processes concerning their locales. The renewed interest in public participation has been largely driven by both the growing demand of communities for a greater say in decisions impacting on their daily lives and the growing involvement of various agents helping communities increase their participation in local decision making. Many of these efforts have been only partially successful due to a gap between communities' knowledge, skills, and resources and those of problem domain experts and decision makers.

Using geographic information technology to support public participation in locational decisions dates back to the beginning of the 1990s. In the Geographic Information

Science (GISc) community much of it transpired through a series of research workshops and specialist meetings sponsored by the National Center for Geographic Information Analysis (NCGIA) and organized in the USA between 1993 and 1998. Meetings in the USA were paralleled by workshops in Europe. In the planning community a non-profit organization – PlaceMatters – that promotes the development of decision support tools for participatory community planning and design has been organizing conferences on "Tools for Community Decision Making and Design" since 1998.

In this chapter we present concepts and methods for participatory decision making supported by geographic information technologies (GIT). We begin the chapter with an overview of empirical research on collaborative/group decision-making since this serves as the precursor to research on the use GIT in support of participatory spatial decision making.

Conceptual and Empirical Foundations for Participatory Spatial Decision Making

There are many arguments in the literature of decision making suggesting that more participatory arrangements are likely to lead to greater effectiveness of decision processes – irrespective of which criterion of measuring the effectiveness is employed (Coenen, Huitema, and Toole 1998). Collaborative learning, shaping of preferences, and social construction of meaning through interpretative processes are seen as significant benefits of more open decision processes. Similarly, processes that involve parties in conflict are seen as increasing the credibility and perceived legitimacy of resulting decisions. Benefits attributed to more open decision processes include consideration of diverse perspectives and interests resulting in better understanding of others' views, and improvement in communication and cooperation through sharing values and interests (Schneider, Oppermann, and Renn 1998).

Group Support Systems

Development of group support systems (GSS) technology (Coleman and Khanna 1995), including group decision support systems (GDSS) (DeSanctis and Gallupe 1987, Hwang and Lin 1987), as well as theoretical and empirical studies of its use (Gray, Alter, DeSanctis, et al. 1992, Jessup and Valacich 1993, Chun and Park 1998), have been carried out in the management and decision sciences since the early 1980s. Most of the empirical studies have been laboratory experiments conducted in conference room settings with subject groups using GDSS software. A major purpose of the experiments was to understand the implications of using decision-support software for group decision processes and decision outcomes. Chun and Park (1998) suggest that the results of these experiments, although mixed and inconclusive, provide valuable insights into the effects of using group decision support software on group performance, group member attitudes, level of participation, and group conflict. They further suggest, predictably, that group performance, represented by decision time, depend on the familiarity of users with GDSS tools. They also suggest that there is no advantage of using GDSS for improving decision quality in simple problems. However, the use of GDSS becomes advantageous in complex decision problems. User attitudes measured by user satisfaction with the decision process depend strongly on the presence or absence of a group facilitator. The presence of a facilitator enhances users' satisfaction with computer-supported decision processes. Another interesting finding concerns heightened conflict level among groups supported by GDSS, to the contrary of an earlier supposition suggesting that the anonymity of electronic communications would increase the number of interpersonal exchanges and reduce the chance of one or a few "strong" individuals dominating a meeting. Yet, the anonymity feature, enabled by computer network-driven GDSS, can help embolden group members to communicate more forcefully, thus increasing the perception of conflict in the group (Chun and Park 1998). In contrast to laboratory experiments, the results of the few field studies reported in the literature are more consistent and positive. They demonstrate both the increased decision quality and shortened meeting time when using GDSS as compared to conventional meetings (Chun and Park 1998). They also demonstrate high user satisfaction and enhanced decision confidence independent of prior user experience with a GDSS.

Spatial Decision Support Systems

On the heels of the above research, during the 1990s GIS and their offspring – spatial decision support systems (SDSS) – were suggested as GIT aids to support decision making for groups, including groups embroiled in environmental conflict (Godschalk, McMahon, Kaplan, and Qin 1992, Armstrong 1993, Faber, Knutson, Watts, et al. 1994, Faber, Wallace, and Cuthbertson 1995, Couclelis and Monmonier 1995, Heywood, Oliver, and Tomlinson 1995). Clearly, research concerning collaborative decision making for geographically oriented public policy problems continues to gain momentum (Shiffer 1992, Densham, Armstrong, and Kemp 1995, Nyerges, Moore, Montejano, and Compton 1998a, Nyerges, Oshiro, and Dadswell, Reitsma 1996, Reitsma, Zigurs, Lewis, Wilson, and Sloane 1996, Kingston, Carver, Evans, and Turton 2000, Jankowski and Nyerges 2001b, Craig, Harris, and Weiner 2002). Reducing the complexity of a decision process by reducing the cognitive workload of decision makers is one goal of developing GIS capabilities for participatory decision-making. Reducing cognitive workload will hopefully lead to a more thorough treatment of information, exposing initial assumptions more clearly, and subsequently resulting in more participatory decisions (Obermeyer and Pinto 1994).

A fruitful approach to developing GIS capabilities for participatory decision making is to learn through experimental studies about human–computer–human interactions during collaborative decision processes. Reitsma, Zigurs, Lewis, Wilson, and Sloane (1996), Stasik (1999), Jankowski and Nyerges (2001a), report on laboratory experiments and Nyerges, Jankowski, Tuthill, and Ramsey (2006), and Jankowski, Nyerges, Robischon, Tuthill, and Ramsey (2006) report on field experiments with GIS tools that support participatory decision making. In terms of comparability across the experiments, their results underscore that groups would rather have facilitators and/or chauffeurs help them work through problems. Individuals having access to technology in the Reitsma experiment were frustrated with the complexity of the software. In land use planning experiments conducted by Stasik, the role of facilitator was partially fulfilled by a computer server, which managed the collaboration process. The participants expressed a strong preference for making a complex collaborative spatial decision support software that they used as easy as a standard WWW browser. Many of them essentially wanted the software to look and act as a standard WWW browser. In the laboratory experiment concerned with the selection of habitat restoration sites in the Duwamish Waterway of Seattle, Washington, Jankowski and Nyerges (2001a) found out that different phases of the decision process had two different levels of conflict: an exploratory–structuring phase characterized by a lower level of conflict and an analytic–integrating phase characterized by a high conflict level. The higher level of conflict during the analytic–integrating phase tells us that decision aids aimed at conflict management are likely to help participants work through conflict; especially now that conflict is recognized as a necessary part of making progress in public decision problems. The same two authors and their colleagues conducted more recently a field experiment study of groups using spatial decision support tools in developing decision alternatives for conjunctive administration of water resources in the Boise River Basin in Idaho. They found that decision support tools, such as tables, maps, graphs, and Structured Query Language (SQL) queries, if operated only by a facilitator, are likely to result in group members engaging more in deliberation than in data analysis. Additionally, they observed that full facilitation, including both the group process and handling of decision support software by a facilitator, promotes a higher level of consensus, most likely because there are more opportunities for deliberation among participants. However, fewer options are explored than in a situation, in which facilitation involves primarily a group decision process and group members operate decision support tools. Such participant-driven decision support settings promote generation and exploration of more options at the expense of group consensus.

The aforementioned studies about participatory decision making for geographically oriented public policy problems offer guidelines for the design of GITs for participatory decision making, such that:

1   The tools for participatory decision making should offer decisional guidance to users in the form of an agenda.
2   The tools should not be restrictive but rather allow the users to select tools and procedures in any order.
3   The tools should be comprehensive within the realm of spatial decision problems and thus offer a number of decision space exploration tools and evaluation techniques.
4   The user interface should be both process and data oriented allowing an equally easy access to task-solving techniques as well as maps and data visualization tools.
5   The tools should be capable of supporting facilitated meetings and hence allow for the information exchange to proceed among group members and between group members and the facilitator. It should also support space- and time-distributed collaborative work by facilitating information exchange, electronic submission of solution options, and voting.
6   Tool functionality should be capable of supporting participatory exploratory-structuring and analytic-integrating phases of a decision process.

Frameworks for using information technologies to support participatory decision making

There is a wide array of frameworks about decision making. The literature includes managerial (Simon 1977), public participation (Renn, Webler, Rakel, Dienel, and Johnson 1993), landscape planning (Steinitz 1990), environmental (Dobson, Urban, and Kelly 1998) and community-based (Electrical Power Research Institute 1998) decision-making frameworks.

Simon (1977) recognizes that the four steps: intelligence, design, choice, and review are essential tasks as part of individual decision making in an organizational context. Renn, Webler, Rakel, Dienel, and Johnson (1993) have used a three-step process: criteria development, options generation, and options evaluation, in public participatory decision making in both the USA and Germany to help recommend environmental policy. Steinitz (1990) sees six steps in modeling for landscape planning, which include representation models, process models, evaluation models, change models, impact models, and decision models. Dobson, Urban, and Kelly (1998) describe a sequence of tools for data gathering and analysis that could include more or fewer steps depending on the situation at hand. The Electrical Power Research Institute (1998) recommends using their SmartPlaces Series E GIS (oriented to community development decision making) as part of a ten-step process. The National Research Council effort concerning analytic–deliberative decision making about risk-oriented hazards suggested that multiple groups be involved in such processes, and that the processes should be determined by the groups being convened. The point is that each of the above strategies consists of a variable number of steps that are appropriate for a variety of decision situations.

There is something systematic about each of the steps in the above decision frameworks. Bhargarva, Krishnan, and Whinston (1994) articulated the idea that Simon's (1977) steps of intelligence, design, and choice each had within them an iterative process of intelligence, design, and choice. That is, when undertaking an intelligence step for decision making, it is natural that people pursue intelligence (that is, gathering information), design (that is, organizing information), and choice (that is, selecting information). The same would occur for the subsequent steps of design and choice, that is to say each has within it, intelligence, design, and choice. As such, complex decision processes are recursive at two levels: a conceptual level involving process steps, and a task level involving details of each step. Jankowski and Nyerges (2001b) used these two levels to formulate a *macro–micro framework* for organizing participatory decision processes supported by GIT. The *macro* component of the framework is flexible and depends on the decision problem at hand, information needs, participation dynamics, goals, and objectives of decision process conveners and participants, and organizational context. Any of the frameworks mentioned above could be used to provide steps for the macro component. Although different macro-steps are likely to be appropriate for different contexts, a micro-strategy is nonetheless at play for each macro-step associated with a work task. That micro-step strategy, encouraged by fundamentals of human information processing, includes gathering, organizing, selecting, and reviewing information.

The macro–micro perspective allows us to appreciate that every macro-phase in a macro strategy can have a different set of information needs, based on the collective needs of the micro-step activities. Consequently, a macro–micro decision strategy motivates the requirements for decision support tools. Such information needs and the associated decision support tool requirements can only be addressed by a good understanding of the decision situation at the time and place (context) within which it occurs.

Methods for GIS-supported Participatory Decision Making

The macro–micro framework can be used to formulate a specific macro–micro decision strategy guiding the selection of participatory decision support tools. One such strategy, fitting many participatory decision problems, can be formulated by synthesizing the Renn, Webler, Rakel, Dienel, and Johnson (1993) three-step public-participation decision process with the Simon (1977) three-step process for the macro-level of a macro–micro decision strategy. The decision strategy consists of three macro phases – Intelligence on Criteria, Design of Options Set, and Choice of Options (Simon 1977, Renn, Webler, Rakel, Dienel, and Johnson 1993); each phase composed of four micro activities: Gather, Organize, Select, and Review (Simon 1977, Bhargarva, Krishnan, and Whinston 1994). It is possible to use this strategy to relate types of decision support methods and tools to the phase activities as presented in Table 27.1 (Jankowski and Nyerges 2001b).

Each cell in the table contains a specific method/tool, which we identified as potentially useful for supporting a specific phase activity. Some methods/tools address Phase 1, as for example with *representation aids* that organize discussions and/or present information about goal(s) and objective(s). Other methods/tools might help with structuring a problem in Phase 2, in the sense of designing an approach for analysis and generating options. This is where GIS plays a significant role, starting out with the computations necessary to manipulate spatial characteristics of decision options. Then, in Phase 3, decision analysis methods help with evaluating options, as in the case of *choice models* and *judgment refinement techniques*. Consequently, a suite of methods/tools is likely to be needed to address complex decision problems, since no single method/tool addresses all phases.

In the intelligence phase of the decision process the Activity Phases 1A–1D are meant to encourage the articulation of values, objectives, criteria, constraints, and standards for characterizing solutions. The methods and tools potentially useful for these activity phases are:

1 **Information management and structured-group process techniques** that help gather participant input on values underlying the decision process from which goals and objectives can be determined.
2 **Representation aids** that help organize goal and objectives into structures that can be analyzed and symbolized using text and graphics.
3 **Group collaboration support methods** that help select criteria on the bases of objectives, and articulate those criteria in terms of measurements.
4 **Group collaboration support methods** that aid in the review of criteria, constraints, and standards for option generation.

Table 27.1 Methods and tools for GIT-supported participatory decision-making derived macro–micro decision strategy

Micro – Decision Strategy Activities	Macro – Decision Strategy Phases		
	1 Intelligence about values, objectives, and criteria	2 Design of a feasible option set	3 Choice about decision options
A Gather	Participant input on values, goals, and objectives using information management, and structured-group process techniques	Data and models (GIS and spatial analysis, process models, optimization, simulation) to generate options	Values, criteria and feasible decision options using group collaboration support methods
B Organize	Goals and objectives using representation aids	Synthesis of decision options using structured-group process techniques and models	Values, criteria, and feasible decision options using choice models
C Select	Criteria to be used in decision process using group collaboration support methods	Decision options from outcomes generated by group process techniques and models	Goal- and consensus-achieving decision options using choice models
D Review	Criteria, resources, constraints, and standards using group collaboration support methods	Decision options and identify feasible options using information management and choice models	Recommendation(s) of decision options using judgment refinement techniques

The intelligence phase can be supported by open group discussion, but more commonly stakeholder interviews have been performed to gather information about values. The goal is to draw out ideas and synthesize them for each stakeholder group. In organizing the material from stakeholder interviews *information management techniques* can be useful. An example of a useful technique is "value tree analysis" (von Winterfeldt 1987). Value trees are hierarchical (tree-like) representations of values, objectives, and criteria, where values are roots, objectives are nodes to branches, and criteria are leaves at the ends of branches. Objectives stem from values as concerns that are to be considered. Criteria, stemming from objectives, are measurable characteristics for evaluating performance of options. The "tree" is a way of organizing ideas/issues into a hierarchy of values, objectives, and criteria. Once the values, objectives, and criteria have been identified for each stakeholder group and represented as a "value tree," then it is recommended that the individual trees be consolidated into an overall tree so that all groups know where they "stand" in regards to overall concerns. Some roots, branches (nodes) might be shared and others might not.

If the process of eliciting values, objectives, and criteria were to take place in a collaborative setting, rather than through stakeholder interviews, then *structured-group process techniques* could likely be useful. One such structured-group process technique is the Technology of Participation (ToP) developed and used by the Institute for Cultural Affairs (Spencer 1989). ToP helps people share an understanding of what is at issue by attempting to generate consensus during strategic planning through a four-step process that aims to (1) generate and cluster ideas from everyone, (2) identify constraints and barriers that might impinge on those issues, (3) prioritize issues in line with the constraints, and (4) flesh out a plan developed from the prioritized issues. Value trees can be constructed off-line by processing stakeholder interviews, or they may be constructed on-line with support of a structured-group process technique.

In the review activity, a decision is made as to whether a single, consensus-based set of criteria will be used or multiple (stakeholder) sets of criteria will be used in the design phase involving options generation. If a single set, then which set, or a combined set? This review might involve negotiation among the groups to see which criteria, hence objectives, hence values, move forward to the next phase. In a sense this is actually negotiation of a problem definition, as certain criteria (called attributes in a GIS context) will describe options differently to how other criteria (attributes) would describe them. This negotiation leads to specifying types of options that would be considered as feasible solutions to the problem and can be aided by *group collaboration support methods* such as Delphi combined with mediation. Delphi is a method for systematically developing and expressing the views of participants. The method involves a series of questioning rounds. Structured questionnaires are submitted to the participants listing pros and cons of the criteria being considered. The results of the responses are described in terms of numbers of respondents who hold each position, and are then returned to the participations along with a new questionnaire especially developed for the next round. After several cycles of judgment and feedback, there is usually some convergence toward a common set of goals. If difficulties appear, a mediation process can be used to help arrive at an agreement. The mediator must be neutral and perceived as such by all participants. The participants work toward reaching a consensus. The third party makes suggestions for possible compromise positions and otherwise helps the participants to negotiate.

In the design phase (Phase 2) it is appropriate to allow each group to generate the options they feel can address the goal of the decision problem in line with the criteria that have been established. To do this, groups would:

1 Gather *data* about the outcomes of criteria to be used as a basis for option generation, that is, criteria which drive the generation of scenarios that become decision options; different types of *models* can also generate the data;
2 Organize and apply an approach for generating feasible options using one or more *structured-group process techniques* or *computer models* – structured-group process techniques may include brainstorming, Delphi, or ToP; models may range from suitability models implemented in GIS to optimization models generating decision option scenarios that satisfy the decision objectives;
3 Select the full array of options to be considered, despite immediate constraints, resources, and standards identified in activity-phase 1D;

4   Review the option set(s) in line with constraints, resources, and standards from Activity Phase 1D and select feasible decision options. Information management tools and choice models can support the selection of feasible options.

In the design phase, groups work in Activity Phase 2A to identify the basis for creating options. Primary attribute(s) are used to identify options, differentiating among fundamental choices first, and then identifying secondary attributes that are used in determining "feasibility." For example, a "land" parcel is a primary attribute that once set, allows a decision analyst to examine the entire range of land parcels for feasible options. In Activity Phase 2B various techniques can be used to process the secondary attributes (for example, parcel size, cost per hectare, tax rate, etc.) to establish a feasibility (minimum threshold) for each option to be included in the feasible set. Examples of these techniques include decision option enumeration through group conversation, standards articulation and minimum thresholds, GIS analysis, and/or modeling such as location-allocation optimization analysis. No single tool has surfaced that specifies the "best way" to identify options, but GIS is a tool that has grown in use because it allows large data sets to be processed to find feasible options. The use of a technique must properly match the character of a decision situation to be effective. For example, a location-allocation model can be used to identify "optimal" options under varying conditions of an objective function. Those options might be the "best cases" to be used in starting a search for decision options, relaxing the objective function to one of satisficing rather than optimizing. Thus, the challenge of Activity Phase 2B is to have participants recognize which kind of problem they are facing, in addition to what data are available, in order to provide the most effective decision support for option generation. The Activity Phase 2C has participants listing the full array of feasible options. This list might come from various stakeholder groups, or it might have been generated as the result of considering different parameters and constraints in models describing a decision situation. During Activity Phase 2D participants from different groups can come together to share an understanding of the different options that each sees as feasible options for the problem. The activity is one of reviewing the principal goal of the process, understanding that certain equitable concerns might need to be addressed. The equitable concerns might in fact be "geographic area" related, but are likely idiosyncratic to the decision situation at hand. Because of these differences it is possible to conceive of various scenarios that might be generated, each scenario having an option set associated with it. An example here can be a transportation infrastructure improvement scenario comprised of a number of specific improvement projects. The result of Phase 2 could be a plan describing one or more scenarios each with option set(s) for consideration in Phase 3.

The choice phase (Phase 3) involves evaluation of the option set(s), within one or more scenarios. In general the evaluation process involves the following phase activities:

1   Gathering information using group collaboration support methods about how to proceed with evaluation of decision options based on values and criteria articulated in Phase 1 and options identified in Phase 2;
2   Organizing an approach to evaluating decision options using choice models;

3 Selecting a prioritized/ordered list of options, using choice models, consensus
   building, negotiation, mediation, or arbitration;
4 Reviewing how the recommendation(s) address decision situation goal(s) in line
   with values consideration, using judgment refinement techniques.

In activity phase 3, decision participants gather information from Phase 1 in terms
of values and criteria to be considered and from Phase 2 in terms of option set(s)
for one or more scenarios. At this time it is possible that various stakeholder
groups might elect to enter into discussion about what values, objectives, criteria,
resources, constraints, and option lists have been identified to this point before
evaluation (in many cases a negotiated evaluation) takes place. A value and criteria
assessment is a good lead into the prioritization of options that takes place within
a group. Next, in Activity Phase 3 participants might use GIS and/or multiple
criteria choice models. Standard GIS operations only allow performing a non-
compensatory evaluation of decision options. That is, a threshold range(s) is set
for one or more criteria, and all options that fall within that range pass through
the "filter." With choice models, a compensatory (tradeoff) analysis can be performed. In compensatory analysis, when evaluating a decision option, a satisfactory
outcome for a high priority criterion can "compensate" for less than satisfactory outcomes on lower priority criteria. For example, less than desirable size of a given land
parcel can be compensated by a favorable tax rate. In Activity Phase 3 "selecting a
choice" can be accomplished through option evaluation using choice models, consensus voting, negotiation, mediation, and/or arbitration. Consensus voting combines
the prioritized lists of all separate stakeholder groups. Such a consensus vote might
be the start of a negotiation process. Negotiation might involve taking the overall
list of priorities as a start and then discussing how one option might move above
another if certain of the criteria were weighted in a different way. Option sets
might be devised to help with this negotiation. Such sets could be developed as
an "equity agreement" whereby certain options from a given area are discussed in
relation to each other, but the options in different sets are not discussed in relation
to each other. That is, "if you get one, then I get one, and so does she." Alternatively, the participants might turn to mediation or arbitration to affect a decision.
When negotiation does not work, mediation using a third party go-between might
prove beneficial. The third party helps to balance the negative feelings between
the (commonly) two perspectives (sides). Having more perspectives complicates the
mediation process even further. Discussing the differences using "decision aids" can
help externalize the disagreements before agreement is reached. Regardless of what
party establishes the recommended decision outcome, a review takes place. Such a
review can be supported by judgment refinement techniques. Judgment refinement
and amplification techniques facilitate a "what-if" analysis of impacts on the selected
decision option in response to changes in assumptions and baseline conditions.
A common approach to facilitate judgment refinement is sensitivity analysis. Many
implementations of multiple criteria evaluation techniques provide sensitivity analysis
capability by which a user can change his/her preferences and observe the consequences
of such change on the solution. During the review of recommended choice(s) in
Activity Phase 3, the options are discussed in terms of the original values devised
from Phase 1.

The above decision process is but one of many normative processes that can be established as an agenda agreed upon by participants – or at least the convener and those responsible for the process. A different set of macro phases might shorten or lengthen the process agenda in different situations. In any case, no matter how many macro phase steps might exist, the micro activities for each phase set the stage for requirements analysis of information needs.

CONCLUSIONS

Both large and small groups are increasingly acting as agents of change in society. Since externalities resulting from the impacts of peoples' activities on each other and the environment are increasing, and approval of corrective action is increasingly participatory, GIS support for participatory decision making is likely to increase substantially. All techniques that are invented for single user GIS, can be used in a participatory setting. When we add the need for communications technology to structured participation technology to decision modeling software technology, and all that to the standard GIS technology, we then arrive at the functionality needed to support groups. Systematic evaluation of that technology is needed to ensure that the software integration problem is tackled in a manner that efficiently, effectively and equitably supports groups both large and small. Systematic evaluation through empirical studies provides evidence to improve the guidelines for participatory system designs.

REFERENCES

Armstrong, M. P. 1993. Perspectives on the development of group decision support systems for locational problem solving. *Geographical Systems* 1: 69–81.

Bhargarva, H. K., Krishnan, R., and Whinston, A. D. 1994. On integrating collaboration and decision analysis techniques. *Journal of Organizational Computing* 4: 297–316.

Chun, K. J. and Park, H. K. 1998. Examining the conflicting results of GDSS research. *Information and Management* 33: 313–25.

Coenen, F., Huitema, D., and Toole, L. J. O. 1998 *Participation and Environment*. Dordrecht: Kluwer.

Coleman, D. and Khanna, R. 1995. *Groupware: Technologies and Applications*. Upper Saddle River, NJ: Prentice Hall.

Couclelis, H. and Monmonier, M. 1995. Using SUSS to resolve NIMBY: How spatial understanding support systems can help with the "not in my back yard" syndrome. *Geographical Systems* 2: 83–101.

Craig, W. J., Harris, T., and Weiner, D. 2002. *Community Participation and Geographic Information Systems*. London: Taylor and Francis.

Densham, P. J., Armstrong, M. P., and Kemp, K. 1995. *Report from the Specialist Meeting on Collaborative Spatial Decision Making*. Santa Barbara, CA: National Center for Geographic Information and Analysis.

DeSanctis, G. and Gallupe, R. B. 1987. A foundation for the study of group decision support systems. *Management Science* 33: 589–609.

Dobson, J. E., Urban, R. D., and Kelly, J. L. 1998. *EnvironAid: A Proposed Conceptual Design for a National Environmental Decision-making Information Infrastructure*. Knoxville, TN: National Center for Environmental Decision-Making Report No. 98-05.

Electrical Power Research Institute. 1998. Smart Places: Intelligent Solutions for Commerce. WWW document, http://www.smartplaces.com.

Faber, B., Knutson, J., Watts, R., Wallace, W., Hautalouma, J., and Wallace, L. 1994. A Groupware-enabled GIS. In *Proceedings of the Eighth Annual Symposium on Geographic Information Systems (GIS'94)*, Vancouver, British Columbia. Fort Collins, CO: GIS World Inc.

Faber, B., Wallace, W., and Cuthbertson, J. 1995. Advances in collaborative GIS for land resource negotiation. In *Proceedings of the Ninth Annual Symposium on Geographic Information Systems (GIS'95)*, Vancouver, British Columbia. Fort Collins, CO: GIS World Inc.

Godschalk, D. R., McMahon, G., Kaplan, A., and Qin, W. 1992. Using GIS for computer-assisted dispute resolution. *Photogrammetric Engineering and Remote Sensing* 58: 1209–12.

Gray, P., Alter, S. L., DeSanctis, G., Dickson, G. W., Johansen, R., Kraemer, K. L., Olfman, L., and Vogel, D. R. 1992. Group Decision Support Systems. In E. A. Stohr and B. R. Konsynski (eds) *Information Systems and Decision Processes*. Los Alimitos, CA: IEEE Press: 75–135.

Heywood, D. I., Oliver, J., and Tomlinson, S. 1995. Building an exploratory multi-criteria modeling environment for spatial decision support. In P. Fisher (ed.) *Innovations in GIS*. London: Taylor and Francis, pp. 127–36.

Hwang, C. and Lin, M. 1987. *Group Decision Making Under Multiple Criteria*. New York: Springer-Verlag Lecture Notes in Economics and Mathematical Systems No. 281.

Jankowski, P. and Nyerges, T. 2001a. GIS-Supported collaborative decision making: Results of an experiment. *Annals of the Association of American Geographers* 91: 48–70.

Jankowski, P. and Nyerges, T. 2001b. *Geographic Information Systems for Group Decision Making*. London: Taylor and Francis.

Jankowski, P., Robischon, S., Tuthill, D., Nyerges, T., and Ramsey, K. 2006. Design considerations for collaborative, spatio-temporal decision support systems. *Transactions in GIS* 10: 335–54.

Jessup, L. and Valacich, J. 1993. *Group Support Systems: New Perspectives*. New York: Macmillan.

Kingston, R., Carver, S., Evans, A., and Turton, I. 2000. Web-based public participation geographical information systems: An aid to local environmental decision-making. *Computers, Environment and Urban Systems* 24: 109–25.

Nyerges, T., Moore, T. J., Montejano, R., and Compton, M. 1998. Interaction coding systems for studying the use of groupware. *Journal of Human-Computer Interaction* 13: 127–65.

Nyerges, T., R. M., Oshiro, C., and Dadswell, M. 1998. Group-based geographic information systems for transportation site selection. *Transportation Research C: Emerging Technologies* 5: 349–69.

Nyerges, T., Jankowski, Tuthill, D., and Ramsey, K. 2006. Collaborative water resource decision support: Results of a field experiment. *Annals of the Association of American Geographers* 96: 699–725.

Obermeyer, N. and Pinto, J. 1994. *Managing Geographic Information Systems*. New York: Guilford.

Reitsma, R. 1996. Structure and support of water-resources management and decision making. *Journal of Hydrology* 177: 253–68.

Reitsma, R., Zigurs, I., Lewis, C., Wilson, V., and Sloane, A. 1996. Experiment with simulation models in water-resources negotiation. *Journal of Water Resources Planning and Management* 122: 64–70.

Renn, O., Webler, T., Rakel, H., Dienel, P., and Johnson, B. 1993. Public participation in decision making: a three-step procedure. *Policy Sciences* 26: 189–214.

Schneider, E., Oppermann, B., and Renn, O. 1998. Implementing structured participation for regional level waste management planning. *Risk: Health, Safety and Environment* 9: 379–95.

Shiffer, M. J. 1992. Towards a collaborative planning system. *Environment and Planning B* 19: 709–22.

Simon, H. 1977. *The New Science of Management Decision* (3rd edn). Englewood Cliffs, NJ: Prentice-Hall.

Smith, L. G. 1982. Alternative mechanisms for public participation in environmental policy-making. *Environments* 14: 21–34.

Spencer, L. 1989. *Winning through Participation*. Dubuque, IA: Kendall/Hunt.

Stasik, M. I. 1999. Collaborative Planning and Decision Making under Distributed Space and Time Conditions. Unpublished PhD Dissertation, Department of Geography, University of Idaho.

Steinitz, C. 1990. A framework for theory applicable to the education of landscape architects (and other design professionals). *Landscape Journal* 9: 136–43.

Von Winterfeldt, D. 1987. Value tree analysis: An introduction and an application to offshore oil drilling. In P. R. Kleindorfer and H. C. Kunreuther (eds) *Insuring and Managing Hazardous Risks: From Seveso to Bhopal and Beyond*. Berlin: Springer-Verlag: 349–85.

Chapter 28

Surveys of People and Place

Peter H. Dana

This chapter uses examples from recent participatory mapping projects in Central America to illustrate the dynamic interplay between conceptions of people and place and the methods used to survey them. Surveyors, census takers, and map makers have long been characterized as agents of power. J. B. Harley suggested in a workshop twenty years ago that "as much as guns and warships, maps have been the weapons of imperialism" and that "surveyors marched alongside soldiers, initially mapping for reconnaissance, then for general information, and eventually as a tool of pacification, civilization, and exploitation of the defined colonies" (Harley 1988, p. 282). More than a dozen years ago Benedict Anderson reminded us, in the context of colonial nation-building, that the map and the census "shaped the grammar that would in due course make possible 'Burma' and 'Burmese,' 'Indonesia' and 'Indonesian'" (Anderson 1991, p. 185). Mark Monmonier coupled the idea that "the map is the perfect symbol of the state" with his caustically worded assertion that "maps made it easy for European states to carve up Africa and other heathen lands, to lay claim to land and resources, and to ignore existing social and political structures" (Monmonier 1991, p. 90). In the past few years individuals, communities, and special interest groups of all kinds have accepted Denis Wood's proposition that "the interest the map serves can be yours" (Wood 1992, p. 182). In 1995 Peter Poole reported on 60 projects around the world applying geomatics to bolster land claims, manage resources, gather traditional knowledge, and mobilize indigenous groups (Poole 1995). Since then many more projects have been undertaken with varying levels of success. The past decade has seen an increasing awareness of the influence that use and misuse of spatial data can have on the welfare of people and place (Chapin, Lamb, and Threlkeld 2005). Phrases such as *counter-mapping*, *participatory mapping*, *community participation in Geographic Information Systems (GIS)*, and *public participation in GIS* are now commonplace in the field and in the literature of the social sciences, representing a wide range of activities that have begun to restore the balance of power in the realm of mapping, surveying, and census taking (see Chapters 26 and 27 of this volume by Weiner and Harris and Jankowski and Nyerges for additional information about participatory GIS and GIS and participatory decision making, respectively).

Increasingly, geographers have shown an interest in "mapping the landscape of identity" (Knapp and Herlihy 2002). Ethnicity and landscape are tightly coupled and in some ways exhibit similar characteristics. Just as ethnicity is a concept that shifts with time, space, and point of view, so too landscape is a complex and fluid notion. It is not surprising that ethnicity and identity often reflect context and circumstance. It is no less common for ideas about the use, value, ownership, and possession of land to be unstable, especially when people and places are threatened. The problem, then, for geographers and other social scientists, is how to survey the landscapes of people and place in an appropriate manner when the very subjects of study are so difficult to characterize.

The premise of this chapter is that surveys of people and place are not simply the end result of GIS projects; they are a part of a process through which ideas of ethnicity and landscape are formulated, explored, altered, and articulated. Mapping people and place is a complex process involving choices in methods and techniques that can impact the ways in which territory is perceived as well as the ways in which territory is eventually portrayed or used as the basis for spatial analysis.

Central American Case Studies

To illustrate the interactive relationship between methods and ideas of territory, examples are used from two participatory mapping projects in Nicaragua and Honduras completed over a period of years in the early 2000s (Figure 28.1).

From the spring of 1997 to the fall of 1998, a World Bank funded project to research and analyze the land claims of indigenous communities on the Caribbean coast of Nicaragua was carried out under the direction of the Caribbean Central American Research Council (CCARC – known prior to 2004 as CACRC). This project resulted in the mapping of the claims of 127 indigenous, Garifuna, and Creole communities, many of which had not been depicted on maps before (Dana 1998, Gordon, Gurdián, and Hale 2003). CCARC directed another study in 2002, along the northern Caribbean coast and Mosquitia of Honduras, mapping the claims of the Garifuna and Miskitu communities. Both projects were participatory in nature, linked to land rights and indigenous groups, and were accomplished with the help of GIS and Global Positioning System (GPS) technologies.

Representations of Territory in the GIS Process

The idea of defining territories with specified boundaries is not a modern concept. Clay maps and petroglyphs from as long ago as 2,500 BC appear to show property lines and field boundaries (Harvey 1980). Cadastral maps may have been among the first preserved maps and their history suggests that control of bounded space is a very old notion. Modern notions of national, provincial and other administrative boundaries, and the private and communal property defined within them come out of a long tradition of bounding space. Rivers, coastlines, and mountain ridges have often provided natural linear features with which to bound territory. The less well-defined midpoints of impenetrable regions of wetlands, deserts, or jungles have been used as convenient divisions between adjacent groups. Where territory is not well

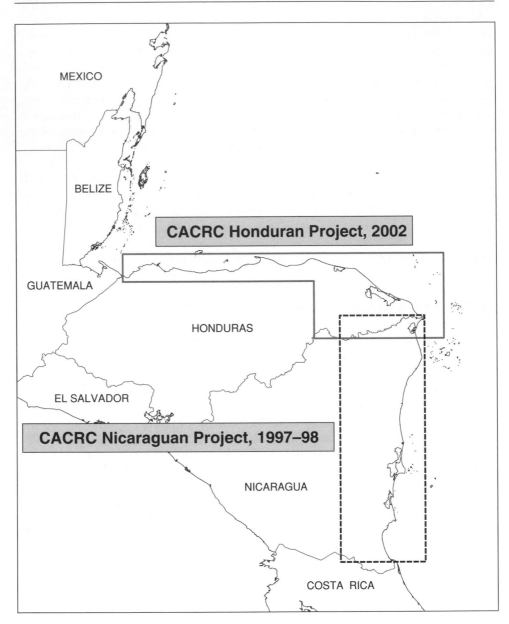

Fig. 28.1   CCARC Central American mapping projects

defined by natural or artificial linear boundaries, divisions between places reflect the patterns on the land made by settlements and land use practices.

When maps became tools of empire, natural features, populated places, and locations in empty spaces on maps were connected with intangible lines that eventually became realized boundaries. In the last few centuries, with the advent of celestial positioning technologies, boundary lines have been constructed with meridians, parallels, or paths between points defined by longitudes and latitudes. Then and now, these conceptual boundaries very often have no relationship to people and places on the ground. In some cases people living on or near national borders, park boundaries, or concession limits may have little or no knowledge of the existence of these lines. More often they have little control over the forces that put these lines on the map and that control events from afar.

Territory is sometimes defined or claimed within an indeterminate spatial extent. Land use patterns, historical claims, built environments, and the significance of certain locations can implicitly define space without requiring boundary lines, either natural or artificial. For communities, ethnic groups, indigenous organizations, or settlers in regions where neither private property nor administrative district boundaries have significant impact, territorial claims may be based on land use within only vaguely agreed-upon limits.

GIS practitioners are familiar with the field and object views of space. Points, lines, areas, and volumes defined by points and directions are the fundamental objects of vector-based GIS. Spaces partitioned into two- or three-dimensional compartments filled with attribute values are the fields of raster-based representations of geographic entities. While modern GIS platforms can handle both approaches, the raster and vector views of the world still impact the way we view territory.

Territory without specified boundaries or limits is not easily defined within a GIS platform. While we are used to grouped and nested polygons, with enclave and exclave features sometimes conceptualized as *islands* or *lakes*, space is usually mapped in vector-based GIS with defined polygons. In raster-based GIS, fields are explicitly filled with contiguous cells that represent unique identifiers signifying connection with a particular territory.

The fluid and seasonally-dependent territories of nomadic groups are also difficult to represent in GIS platforms, whether raster- or vector-based. While both methods can easily handle multiple layers of territorial identification, neither lends itself to views of territory that are not based in concepts of defined, delimited, and demarcated extent. People living within territories neither encompassed by natural boundaries nor constrained by neighboring community extents may not be able to fit their ideas of the land into the bounded spaces of GIS polygons or defined grid cell contents.

We only have to look at the complicated multiple polygons of hunting regions in British Colombia in the "Maps and Dreams" of Hugh Brody (1982) or the "Fifty versions of the Great Plains boundary " in Rossum and Lavin (2000, p. 546) to see the difficulties in trying to represent convoluted space, real or imagined, in GIS. When surveying more complex environments with multiple ethnicities and contested landscapes the difficulty of representing a multiplicity of ideas of territory within the same space can become almost unmanageable.

The Interaction between Method and Result

When territory is not well defined or when its definition is contested, the measurement techniques selected for the survey and the representational models chosen for analysis may influence the ways in which territory is ultimately perceived and defined. In some cases the effects of mapping and conceptions of territory may be beneficial to all concerned. In other situations there may be unintended and unforeseen consequences that can result in an increase in the level of contestation. The hope is that with attention to appropriate measurement and analysis methods and their effect on notions of territory we might be able to do a better job than with methods less in touch with the territories under study.

The vector view of territory is of space bounded by lines and populated by objects at locations of interest. This point, line, and area view of the world gives special significance to coordinates, points in Cartesian space, and to linear boundaries even where none may exist on the ground being mapped. Populated places with considerable spatial extent and complex shapes are often represented on maps by single points. Wetlands, streams, and river networks are sometimes signified by a single line. Intricate distributions of variously defined ethnic groups are located within databases and on maps with discrete linear boundaries. The vector approach often ignores those elements of the territory that do not easily lend themselves to vector descriptions.

Raster-based approaches to mapping usually classify space through rectangular grids of regularly spaced cells, each of which has an attribute described with thematic or numeric attributes. Raster views of territory, especially those that model with multiple layers of grids, allow a conception of territory that is content or use-based rather than one defined by spatial limits. Land use, land cover, ethnicity, elevations, rainfall amounts, and other complex continuous or semi-continuous spatial distributions lend themselves to raster display and analysis. The raster view does not easily handle those cases where territory is defined by the points and lines of political boundaries or where rivers and coastlines form natural linear boundaries. Since each cell in a conventional raster layer can only signify a single attribute value, the raster approach often results in the implication that in a single thematic layer, each point on the ground can only be classified in a single, discrete way.

In most mapping projects, some combination of these vector and raster views is helpful. Conventional land surveying techniques have long been based on the measurement of distances and angles. These are points defined by angles and radii in a local polar coordinate system, resolved using simple planar geometry into points, lines, and areas. GPS receivers measure points in space defined within some local rectangular coordinate system or by latitude and longitude. The result of these measurements made with varying degrees of precision and accuracy should more reasonably be thought of as centers of probability clusters, rather than the dimensionless mathematical point they emulate. Lines are constructed on the map as connections between points. Areas are formed from boundaries made from lines. The noise, biases, and blunders found in surveying and mapping technologies propagate from their original measurement into the line and area features found on final maps. Lines on maps should be considered as having some fixed or variable width rather than as infinitely thin mathematical constructions. Areas defined as mapped regions also share the uncertainty of the points and lines that define them.

When mapping contested lands, approximations may be appropriate even if higher levels of accuracy are possible and affordable. During the 1997–98 mapping period in Nicaragua the US Government was still applying the intentional degradation of the civil service called Selective Availability (SA) to GPS signals. SA added biases of many tens of meters to each satellite signal in a pseudo-random manner that

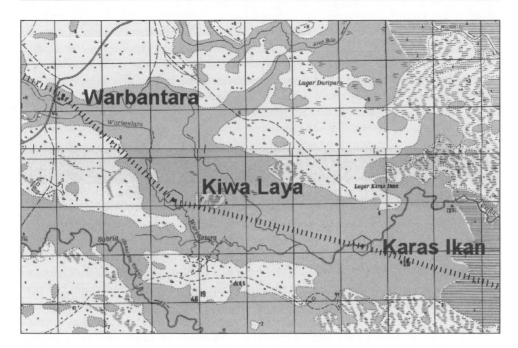

Fig. 28.2 CCARC Nicaraguan map detail with large point symbols and line hatching indicating survey imprecision

caused time-varying biases resulting in horizontal position errors of about 100 m. For this project, mapping using 100 m precision was a desirable feature. Rather than working with a precision that implied some finality and legal weight we preferred a precision that implied approximation. To display this appropriate and useful imprecision, we used large point symbols and hatched boundary lines on the final maps (for example, Figure 28.2).

In a place where territory is not well-defined, methods used in mapping can change views of territory. Local conceptions of territory can be unconstrained by conventional ideas about extents and boundaries. When territory is unmarked and unbounded, a mapping project may result in the territorializing of people and place, changing, for good or bad, the perception of landscape in fundamental ways (Sack 1986). During the Nicaraguan project mapping process many of the 127 communities selected for inclusion in the study chose not to make claims or be mapped as individual communities. Communities, faced with fractures and contested individual boundaries, formed regional groupings that melded into *bloques*, larger regions within which internal boundaries were not measured or mapped. Figure 28.3 illustrates the complex tiling and overlaps in the final maps.

The result of this mapping project was a tiling of territory within the study area that left few of the un-mapped regions sometimes considered "empty" and designated as "state land." Since the purpose for the study was to set out initial claims, not to produce land titles or final demarcation we expected overlaps between claims. Overlaps were common, and are now the subject of considerable negotiation between communities.

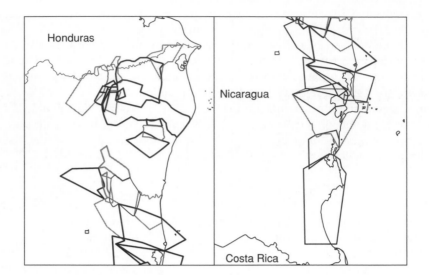

Fig. 28.3   Tiling and overlaps in CCARC Nicaraguan *Bloques*

Point markers signify extents of territories. This does not imply that these monuments are necessarily points along a polygonal boundary. In early Asian empires these point features were often regarded as the end points of conceptual lines emanating from a power center (Anderson 1991, p. 172). In Miskitu settlements along the Caribbean coast of Nicaragua monuments sometimes mark community extent without implying any fixed line between them. If a mapping project is designed to delimit territory with regional boundaries, points previously conceptualized as isolated markers may become the vertices of polygons. Neighboring communities may end up with boundary lines between them that they neither needed nor wanted.

This happened at the beginning of the CCARC Nicaraguan study. We conducted a pilot project with the permission and assistance of the community of Krukira. After community meetings in which ideas of territory and community boundaries were discussed, we set off to find and measure important boundary turning points with GPS. Included were a coco palm at Barra Sahnawala on the coast and the remains of a concrete/rebar monument at Yulu Tingni. That night on a small notebook computer, I prepared the first project draft map (Figure 28.4). The next day community members agreed on the placement of all the monuments but were surprised by my choice of boundary line placement. They all agreed that the straight line I had placed between Barra Sanawala and Yulu Tingni was quite wrong. There was no agreement on how the line might be better placed and no suggestion on the placement of other turning points. In short, the assumption that the boundary should follow a vector between two points was inappropriate for the local conceptions of territory.

Where communities do want boundary lines, the methods used to mark them can change their placement. If paper maps are used, the features selected for inclusion on those maps by the agency that produced them can influence the placement of boundary turning points and connecting lines. If remote sensing is used, features

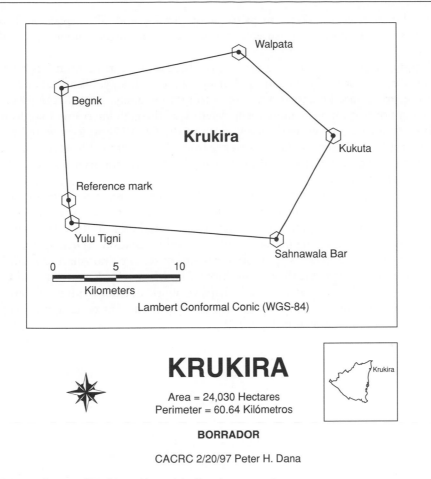

Fig. 28.4 Draft map of Krukira with straight lines between points

with recognizable visual patterns on aerial photographs or particular electromagnetic signatures in satellite imagery may take precedence over sacred sites, traditional hunting regions, or the sources for medicinal plants, that are important to communities but utterly lost to those technologies.

If a project selects traditional surveying equipment to help define boundaries, the resulting boundary lines may well reflect the line-of-sight path that the demarcation team follows as they proceed, often an easier and simpler path than the more difficult transect through swamps, jungles, and difficult terrain that might have formed the boundary of the community were a less arduous method chosen.

Appropriate Technologies

Measurement technologies should be selected that respect and faithfully reflect local perspectives on territory. Approaches to mapping people and place range from informal interviews and sketch maps to remote sensing and GIS displays. In between these is a broad range of measurement tools and information gathering techniques (Poole 1998).

Mapping from interviews, narratives, or discussions can be useful as part of a mapping project or as the fundamental mapping methodology. Historical and current land use, from specific agricultural zones to more vaguely defined regions of cere-monial importance, may be more easily determined through interviews than from any analysis of imagery or ground survey. Both of the CCARC projects were based on ethnographies gathered by trained investigators. These detailed histories of occupa-tion and descriptions of past and present land use were important for establishing the validity of land claims.

Sketch maps, from those drawn in the earth during an informal discussion to those carefully drafted on waterproof paper during a visit to a new area, can be useful at every stage of a mapping process. Often suggested as a tool in the map design process, informal sketching has been called "graphic ideation" (Dent 1999, p. 239). "Compilation worksheet" is the more formal term for the draft map produced dur-ing the map design process (Robinson, Morreson, Muehrcke, Kimerling, and Guptill 1995, p. 426). We often used sketch maps to work out conceptions of territorial boundaries and land use such as the sketch map made during a community visit to Auka, Honduras shown in Figure 28.5.

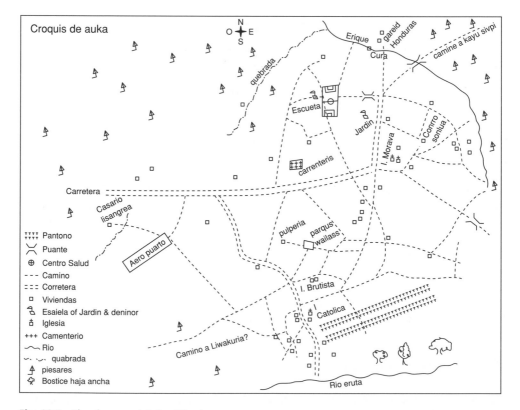

Fig. 28.5 Sketch map of Auka, Honduras

Base maps are general-purpose reference maps that define the basic geographic features of a region (Muehrcke, Muehrcke, and Kimerling 2001, p. 188). Thematic maps produced in a participatory mapping project can be made by overlaying new information on a base map. Base maps can be produced especially for a mapping project, but in most places some form of base map already exists. Existing base maps need to be used with care. They have often been produced by agencies with military or commercial agendas. The places and the place names on maps may reflect these agendas more than they do the ground truth. Roads connecting mines and ports may be present while paths more important to community members may not be shown at all.

Base maps that exist in digital form may be expensive or tightly controlled by mapping agencies. Scanning paper maps (especially in color and in detail) is expensive and sometimes difficult without special equipment. All maps contain omissions, generalizations, and exaggerations. Base maps produced by and for groups not concerned with community conceptions of territory may not contain any useful features to aid in community mapping. Unfamiliar grids and coordinate systems, combined with toponyms in the language of the mapping agency may have little meaning to community members and may make mapping more difficult rather than aid in participatory mapping. Appropriate base maps that reflect the conceptions of territory held by participants may be difficult or impossible to find.

The base maps we used in the CCARC projects suffered from some of these limitations. Maps of both Nicaragua and Honduras are available as 1:50,000-scale topographic sheets. In general they use Spanish names and often do not contain information important to indigenous communities. They almost never contain monument symbols or boundaries that represent communities. Produced usually by military or resource management agencies, existing maps sometimes reduce communities to just a few symbols on a map. Paths, health centers, churches, and other features are often not portrayed on these reference maps. Compare the sketch map of Auka in Figure 28.5 to the Honduran government topographic map of the same region in Figure 28.6. With the contour lines removed, the latter would be a very empty map.

In the CCARC projects, we tried to keep pre-existing maps out of the process of community discussions of land tenure. Maps sometimes influence decisions by deflecting interest from actual territory in favor of the simplified versions they depict. Administrative boundaries, particular roads and streams selected for mapping, settlement patterns that may no longer reflect the status of neighboring communities, and even the grid pattern superimposed on the mapped ground can shape views of the land. Once conceptions of territory had been established there were many times when maps were useful in the field particularly in navigating from place to place once sketch maps and discussions had determined community extents. Scanned maps became the base maps on which community boundaries were placed for final display. CCARC had to scan, crop, and georegister 103 Nicaraguan map sheets. In Honduras we had access to a complete set of 1:50,000-scale maps already scanned and stored in MrSID format.

There are many time-honored methods that are sometimes overlooked when a GIS-based project is planned. There are still plenty of places where 120–240 volt AC power for recharging a battery is not available. Solar-based recharging systems

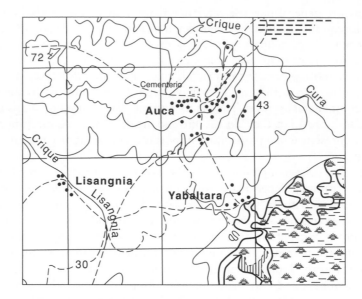

Fig. 28.6   Topographic map of Auka, Honduras

are heavy, fragile, and may not be able to handle the power requirements of multiple devices. There are places where even finding a few "AA" batteries is a challenge. The Caribbean coast of Nicaragua and the Mosquitia of Honduras are challenging environments. Outside of the towns of Bilwi, Bluefields and Puerto Lempira, electrical power is intermittent or unavailable. Finding a set of "AA" batteries is difficult or impossible in many places. These Caribbean coastal regions experience some of the highest annual rainfalls in the tropics. This is not a place for keyboards, electrical cables, rechargeable batteries, or dependence on computers. For both projects we designed very "low-tech" measurement approaches. Participants carried waterproof boxes into the field containing low-cost water-resistant GPS receivers, extra batteries, waterproof notebooks, pencils, and inexpensive magnetic compasses (Figure 28.7).

The magnetic compass is a valuable tool in mapping projects. At the least it can keep investigators oriented in the field. Relative angular measurements can be made to a few degrees of arc irrespective of magnetic variation, allowing the compass to act as an alidade. When combined with the practiced pacing of distances, the compass can be an effective tool for a reconnaissance survey. Magnetic variation, the difference between true north and the direction of the magnetic poles is marked on most topographic map sheets. Variation changes over time and so absolute direction requires up-to-date variation information. Local anomalies can cause significant local changes in variation so projects should be wary of absolute measurements made with magnetic compasses. Use of a magnetic compass within a range of 1,000 km of the magnetic pole is not advised. An important consideration for participatory mapping projects is the problem of magnetic dip. The north pole of a magnet points toward the magnetic north pole which is not in a direction horizontal to the local level surface, but rather down, in a direct line to the pole through the Earth. The

Fig. 28.7 Waterproof kit contents: GPS receiver, field notebook, pencils, compass, and extra batteries

needle of a compass must be adjusted for this "dip" to keep it from scraping the base of the compass as it turns. Magnetic compasses are adjusted by the manufacturer to work within specific latitude ranges (five degree latitude ranges are common). There are newer "international" or "global" compasses with a separate needle and magnet that work at all latitudes. During both the Nicaraguan and Honduran projects where obstructions made GPS measurements at boundary points, or, as in the case of small islands that had to be measured offshore from small boats where it was impossible to reach the boundary point, participants included compass bearings and estimated distance from the GPS receiver to the point. Final point positions were computed in the GIS from the measured point and the distance and magnetic bearing corrected for local variation.

The alidade and plane table can still be effective mapping tools, especially for mapping in the field under conditions that might make one-time mapping the only practical method. Based on the intersections of lines of sight from two or more points, the alidade is simply a ruler with a sight on it, usually an optical telescope attached to a metal rule. The alidade is placed on top of a piece of paper attached to a "plane table," a drafting table on a tripod. Leveled and aligned along a baseline, the plane table supports the alidade which is pointed at objects to be mapped. Moved from place to place, the intersection of lines to common points marks their position on the paper map as the process continues. By combining intersecting lines for more than three points the process allows the mapper to estimate error and correct for mistakes while still in the field. Without batteries, computers, or communication systems, these are still viable tools for participatory mapping projects. As a point-demarcation process, the technique lends itself to conceptions of territory that are based on discrete point features.

Positioning with sextant and celestial almanacs is a means of finding latitude and longitude at sea. On land, with the stable platform of a leveled tripod, latitude and longitude can be determined to within a few meters. Requiring precision optical vertical angle measurements, access to precise time, and accurate almanacs describing positions of objects in the sky, these methods are still used by surveyors to establish points of beginning and base lines. They are perhaps best used to establish a few geodetic points where GPS or other more accurate and more easily used methods are not available.

The surveyor's theodolite is an optical and mechanical angle measuring device that allows horizontal and vertical angle measurements with respect to a baseline and a local level plane defined by gravity. Combined with a tape (or chain) for measuring distance, the theodolite can measure the metes and bounds of a survey traverse. Theodolites are expensive, require calibration, and take considerable training and practice to use effectively. Damage that occurs in the field will not likely be repaired without expert help. The traverse approach to area boundaries with such instrumentation requires a clear and level line of sight from one traverse point to another. The tape must be physically connected from point to point, making boundary surveys across swamps, through jungles, and along rivers difficult.

Where electrical power can be used and where access to the supporting supplies of batteries, cables, computers and disks can be found, one can manage projects with all of the tools available to the modern mapper or surveyor. Where temperature, humidity, and security allow their continued use, they can be effective tools. One sharp jolt from a fall or a bump in the road can ruin or damage an expensive device required for project completion. Perhaps worse, difficult conditions can cause an instrument to become inaccurate, often without any indication of a problem until after the survey. The digital theodolite can measure angles and automatically record them on a small handheld calculator or data collector. They require battery recharging and considerable care. They come with or without laser pointers to facilitate alignment with targets. Visible laser alignment aids can only be used within 50 m of so of the instrument. Infrared laser measurements of distance can be made with Electronic Distance Measurement (EDM) devices. Prism reflectors placed at a target return a coded laser signal to the EDM device, and range is calculated from travel time. These devices can measure ranges as long as 4 km under good atmospheric conditions. Normal operating range limits are in the 2 km range. Range accuracies are typically 2 mm plus two parts per million (4 mm total over a 2 km distance). Newer reflectorless devices can measure the return signal from a target without a prism, but are only useful for distances of less than 100 m. Use of non-prism EDMs can introduce blunder errors when the reflected signal is not from the intended target. In the field, small EDM devices can replace or augment the tape or chain to provide distances without the requirement for a physical path between one point and another. Measuring across rivers or gorges can be simplified through the use of EDM.

The Total Station, an electronic theodolite with digitally encoded angle measurements combined in a single instrument with EDM, can measure and record both angles and distances. Usually coupled with handheld data collectors these instruments, which can measure angles to between 2 and 5 arc-seconds, have largely replaced the conventional theodolite and chain. The Total Station is expensive (in the range of US$10,000), heavy (10 kg), and requires regular recharging of batteries.

Binoculars with built-in magnetic compasses can provide relative angle measurements over long distances. Costing in the range of US$500 these devices are sold to mariners and so can often be found in waterproof configurations requiring no batteries (some require battery power to illuminate the compass); they can be valuable aids to mapping projects. Incorporating EDM laser techniques within handheld binocular packages, rangefinding binoculars can be used as lightweight total stations. They can be used to measure distances over ranges as long as 2 km to accuracies of 2–3 m. These do require batteries, but for field work and preliminary surveys they can be effective tools in mapping projects. Laser rangefinders with approximate angle measuring capabilities (about a half a degree of arc) cost about US$3,000 and can be used as handheld replacements for the Total Station when high accuracy is not required. They make useful accessories for GPS projects that require offset range and bearing measurements.

GPS Approaches

The fundamental concept of GPS is the estimation of three-dimensional point position in a geodetic reference system from measurements of the relative arrival times of satellite signals. As a point positioning system, GPS lends itself to projects in which territory is conceived of as discrete points which form either turning points in polygonal boundaries or represent the centers of places. GPS can be used to measure points, paths between points, and the spatial extent of areas. There are three major categories of GPS methods. Unaided code-phase GPS is the use of the civil GPS service with a single receiver that tracks the GPS codes, resulting in horizontal accuracies of between 2–20 m. Differentially corrected code-phase GPS can provide horizontal accuracies of between 1–2 m within 300 km of a suitable reference station supplying corrections to the GPS signals. Carrier-phase GPS is always differential in nature and can provide position accuracies of a few centimeters within 10–30 km of a suitable reference station. These three techniques differ significantly in their accuracy, cost, and complexity.

Unaided code-phase GPS

The US Department of Defense maintains GPS. A civil service is provided that allows unrestricted access to part of the GPS signal structure. The C/A code (coarse acquisition code) contains timing edges that are used within a GPS receiver to measure relative arrival times from each tracked satellite. The timing edges are coded with a repeating sequence of bits such that each millisecond can be resolved into 1,023 distinct intervals. Fractions of an interval can be measured with precisions of a few nanoseconds (three nanoseconds is about one meter of range). System information, satellite clock corrections, and satellite orbital data are contained within the Navigation Message broadcast by each satellite. The GPS receiver uses the C/A code arrival times and the Navigation Message from at least four satellites to resolve the common clock offset from the relative arrival times in order to compute the three-dimensional position of the receiver. GPS receivers typically track all the satellites their antennas can acquire and produce position reports in latitude, longitude, and height for display or recording.

Ionospheric and tropospheric signal delays, clock and orbital data errors, and local reflections result in position accuracies of between 2–20 m depending on the number of satellites tracked and the geometry of the receiver-antenna with respect to the satellite positions in space. Because the satellites are always moving, this geometric relationship and the combination of satellites in view are always changing. All GPS receivers must have a clear and unobstructed view of the sky in order to acquire and track the necessary GPS satellite signals.

Differential code-phase GPS

A special purpose GPS receiver at a precisely known location can compute the differences between the relative ranges it measures and those predicted for that location. These differences are the basis for Differential GPS (DGPS) corrections that can mitigate the bias errors from atmospheric delays and system errors that are common to a reference receiver and a remote receiver up to a few hundred kilometers away. The DGPS corrections are computed for each satellite and are range-domain corrections. They are not position-domain corrections. One cannot simply shift the position of a remote receiver based on the position error measured at the reference receiver unless both receivers are tracking the same set of satellites at the same time, have the same geometric relationship to the satellites, and are using identical clock and orbital data sets. This commonality in the position domain is almost impossible to achieve. DGPS works by providing the remote receiver with range and rate of range change corrections for each satellite. These are then used to correct the measured ranges used in the position solution in the remote receiver. When the reference receiver is close to the remote receiver much of the bias error is common and can be differentially removed. Local reflections (multipath) and receiver-induced errors cannot be removed by DGPS. In situations where the remote receiver geometry or signal strength is insufficient DGPS cannot help.

DGPS corrections can be applied in real-time if a radio link is available to send reference station corrections to the remote receiver. This requires licensing and maintaining a radio link that can reach the remote receiver. There are an increasing number of public and private sources for real-time DGPS corrections. The US government operates the Wide Area Augmentation System (WAAS). This system is based on a network of reference stations that are used to produce correction signals that are transmitted to receivers over communications satellite links in a format that is relatively simple for GPS designers to incorporate into GPS receivers. Where a WAAS-enabled GPS receiver can track either of the two WAAS satellite signals, it can usually estimate position to within 2–3 m. Care must be taken not to use WAAS corrections outside of the regions for which corrections have been calculated. In parts of Central America, WAAS signals can be received but their use can actually degrade rather than enhance position accuracy. Low frequency beacons are used by many nations to provide coastal coverage for maritime use and can be used inland as well where signal strengths allow. Most of the US coast and the east and west coasts of Canada are covered by transmitters operated by the US and Canadian Coast Guard. Several commercial communications-satellite based DGPS services offer coverage over much of the world.

For post-processed DGPS, a receiver capable of saving DGPS data files is needed along with a local DGPS reference station equipped to provide files for post-processing. Dedicated DGPS reference receivers can be established at a cost of about US$20,000 including the computer required to log data for post-processing. While there are hundreds of Community Base Stations (CBS) located throughout the world, locating one and obtaining permission to access to the data files can be difficult. Recent versions of Trimble's Pathfinder Office software include Internet addresses for hundreds of cooperating CBSs, but there are hundreds more that are not part of this list.

Carrier-phase surveying

Special purpose GPS receivers can track both the code edges of the civil GPS signals and the microwave signals that carry them. Tracking the "carrier" signal makes it possible to measure relative arrival times to accuracies of about a centimeter. The technique is limited because these carrier-phase measurements are difficult to resolve into relative ranges. Unlike the C/A code edges there is no marking that differentiates one carrier wavelength cycle from another. In addition, the signal-to-noise ratios for carrier-phase measurements are lower than for C/A code-phase measurements. Special hardware is required to track these low level signals and complex software methods are required to resolve wavelength ambiguities.

Static techniques have been developed that can provide centimeter relative ranges with respect to the position of a reference receiver within 30 km of the remote receiver. Software must have access to the continuous measurement of carrier phase at both receivers. These systems cost from US$10,000 to US$25,000 and require considerable training to operate. Real-time-kinematic (RTK) techniques make it possible to move the remote receiver while making relative position measurement with respect to a reference receiver. RTK techniques require a special RTK reference receiver within 10 km of the remote, a continuous radio link between reference and remote, and the simultaneous tracking of five or more satellites at both receivers. These restrictions, the additional need for training, and the expense (these RTK systems cost more than US$50,000), make RTK difficult to use in many projects.

Data attribute collection

Whatever GPS technique is used, data attribute collection must be considered before, during, and after a mapping project. Names, distinguishing features, topological connections, entity attributes, times, dates, and quality control information are all required in the data collection phase of a project. While inexpensive (some around US$100), recreational GPS receivers have very limited capabilities for recording more than a time, date, position, and a name. Complex mapping projects will require some form of data collection method, devices, and procedures. Professional GPS equipment suites usually include some form of data attribute collection software and hardware.

These systems consist of a dedicated computer or a program resident in a notebook or handheld computer. Attribute collection software platforms have functions for defining data dictionaries or forms for entering attributes. A data dictionary pre-defines fields and menus that enable the operator to quickly and efficiently select from attribute choices or fill in alpha-numeric records. Data collection software will usually attach time, date, geometry, and tracked satellite lists to each record. Newer palm-sized devices that attach to GPS receivers can be programmed to accept data attributes in both text and graphic format. Figure 28.8 shows a "Pocket PC" running ESRI's ArcPad software connected to an inexpensive WAAS enabled DGPS receiver with external antenna.

For both CCARC projects, where power and maintenance of electronic equipment posed a serious problem, we developed a "low-tech" data collection approach. We held workshops in Nicaragua and Honduras teaching the fundamentals of geodesy, GPS, and GIS. Investigators and community participants then went into the field to existing or new boundary turning points and recorded position information and

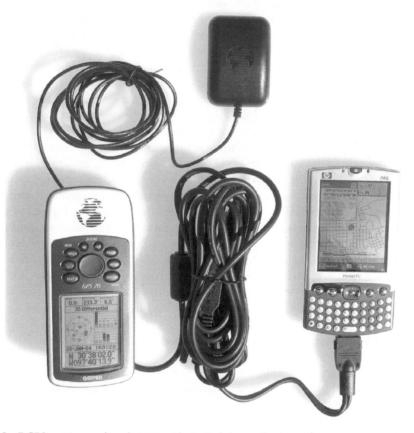

Fig. 28.8   DGPS receiver and pocket PC with ArcPad data collection software

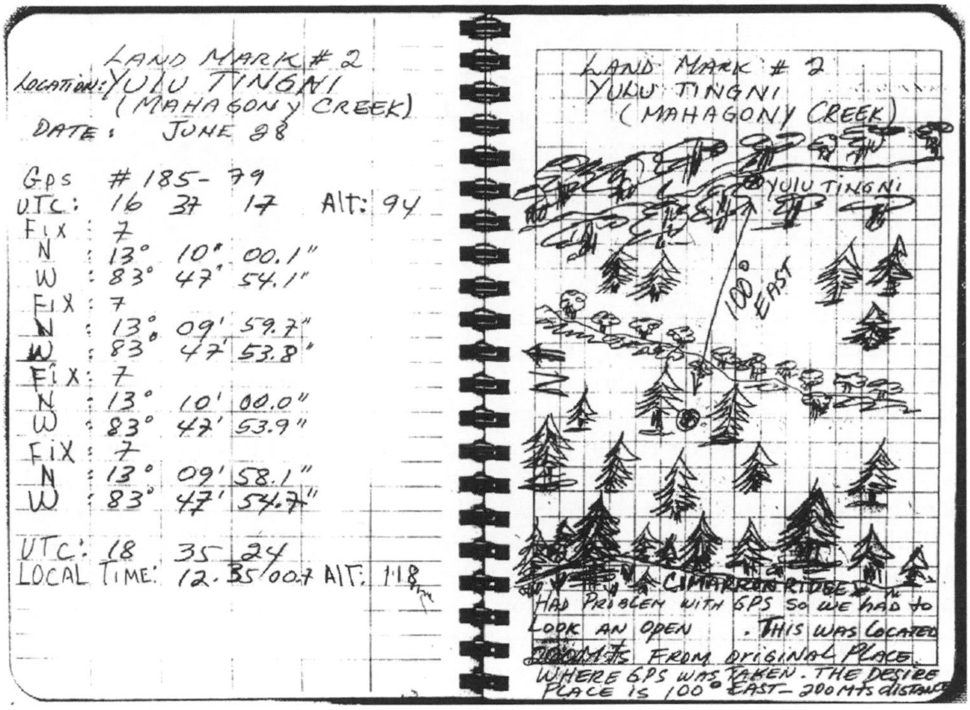

Fig. 28.9 GPS waterproof notebook pages

sketches of locations in waterproof notebooks (Figure 28.9). Crucial in this process was the construction of preliminary community maps with the order of points carefully marked. To define a specific polygon the order of the points is required. In both projects we experienced difficulties when investigators and participants had trouble defining point order unambiguously.

Because most of the communities we helped map seemed to perceive boundary turning points as an appropriate way to define territory, determining the locations of these points was central to the projects. Because the land claims resulting from these projects were contested between communities and between communities and other agencies it was important that communities should show that they really did occupy and use these territories right up to the boundaries. Participants understood the symbolic importance of occupying and physically measuring turning points. They often made extraordinary efforts to reach and measure points at the extent of their lands. For the most part investigators were able to reach points by boat, vehicle, or on foot. Most of these accessible points could be directly measured while others required GPS measurements from a few hundred meters away. Combined with compass bearings and distance estimates, these indirect GPS readings were converted later in the GIS to boundary point position. In many communities, natural line features such as streams and coastlines formed community limits. In order to incorporate these boundaries into community maps, we had to establish the geodetic

Puntos

⊙ GPS Directo

⊗ GPS Indirecto

✕ GPS Registrado

✳ Estimado

Fig. 28.10   Position types: Direct, indirect, registered, and estimated

position of these features, often misplaced on maps or missing in digital databases. In these cases we asked participants to measure enough points along streams and coastlines to establish their correct position in space. Then we were able to place the natural features on georegistered maps. Where turning points were inaccessible to participants because the terrain was too difficult or because land owners prevented access, we encouraged participants to use maps or other descriptions to "mark" points they could not reach to occupy safely.

With lessons learned in Nicaragua we developed a methodology for point measurement and description in the Honduran project that seemed to handle these different cases quite well. We defined four point types: (1) *Direct GPS* points, where the receiver was co-located with the point; (2) *Indirect GPS* points where GPS was not possible at the point but could be measured at some distance and direction away; (3) *GPS Registered* points that were established from maps and databases after GPS measurements established the position and orientation of natural features; and (4) *Estimated* points, with a geodetic position estimated from maps or imagery. This last set of points provides the least authority in establishing land claims. Figure 28.10 shows each of these point types and their symbolization on final maps.

Point information, including place names, position, and the all-important point order in completed field notebooks was entered into Excel spreadsheets for importation into GIS. Offset range and bearing computations and geodetic datum shifts were all accomplished within the GIS process. In workshops conducted in Nicaragua and Honduras, investigators produced draft maps of community boundaries on 1:250,000-scale scanned base maps.

Draft maps with measured community boundaries were returned to communities for land use annotation. In this validation phase, communities verified or asked for revisions of boundaries or additional turning points. Land use, crucial for the development of land claims, is a difficult concept in regions of multiple and contested uses for territory. Land use is also perhaps the most difficult attribute to collect in the field. There are so many conflicting, overlapping, and contested uses for land that any attempt to map land use is bound to be a compromise. For the CCARC projects, focused on land claims, land use was determined and categorized differently than it often is within the resource-management tradition of land cover/land use schemes. We were as much interested in land significance as in vegetation cover or the planned use of lumber product by forestry agencies, so we developed, with the help of the communities, a set of land use categories that had significance to communities, representing their traditional and customary land uses, irrespective of what remote sensing might decide through the analysis of electromagnetic

Leyenda Para El Contenido Ethnográfico Del Mapa

Mineria
Main nani pliska

Ganadería
Daiwan sahwaia pliska

Madera
Dus nani pliska

Valor histórico-cultural
Blasi kulkan bri pliska

Pesca
Inska pliska

Lugares segrados religiosos
Huli tasbaya kulkan bri pliska

Agricultura permanente
Dus aiura pliska

Reserva Ecológica
Tasba ritska apahkaia pliska

Agricultura anual
Insla pliska

Interacción-social recreativa
Impaki ris takaia pliska nani

Caseria
Antin pliska

Infraestructura de
comunicación y transporte
Yabal tara an briks nani

Fig. 28.11   Land use categories and symbols for Nicaragua

signatures. Many so-called *land use* schemes include categories that are more appropriately termed *land cover*. For instance, forested land is a category that does little to suggest the hunting and fishing activities or fallow status that might be the actual land use. To help establish and defend land claims CCARC categories included places of historical significance, social gathering places, and ecological reserves. Figure 28.11 shows the land use categories in Spanish and Miskitu developed by community members for Nicaragua. Figure 28.12 shows the land use categories for Honduras in their Spanish/Garifuna and Spanish/Miskitu versions.

Map Production and GIS Analysis

GIS were used as the basis for map-making in both projects. MapInfo was selected for Nicaragua because at that time it was the only entry-level platform that could handle coordinate system and geodetic datum shifts without special scripting. In Honduras, ESRI's ArcView 3.x was used in many government, resource agency, and non-governmental organization settings. To maintain a compatibility with a variety of organizations and to make distribution of project results easier, we selected ArcView 3.x, which by 2002 had added rudimentary coordinate system and datum conversion capabilities to earlier versions.

Uso de la Tierra

Gul sikbaia plikisa
Mineria/Guiriseros

Dus nani pliska
Madera

Inska miskaia pliska
Pesca

Insla pliska
Agricultura

Antin pliska
Caseria

Bip sahwaia pliska
Ganaderia

Blasi kulkan bri pliska
Valor Historical-Cultural

Huli tasbaya kulkan bri pliska
Lugares Sagrados Religiosos

Tasba ritska apahkaia pliska
Reserva Ecologico

Impaki ris takaia pliska nani
Inter-Accion Social Recreativo

Yabal tara an briks nani
Infraestructura de Comunicacion y Transporte

Sika wahaia an wakia
Plantas Medecinales

Lata tani wina kirb anka
Turismo

Dus nani yus muni diara paskaia
Material de Construccion

Daiwan alki pliska
Recoleccion Iguana Verde, otros Animales

Uso de la Tierra

Catei gebeti linda
Minería

Fulansu
Madera

Ouchahani
Pesca

Ichari
Agricultura

Agaliuhani
Casería

Pouteu
Ganadería

Walangante
Valor Histórico-Cultural

Gabusandu
Lugares Sagrados Religiosos

Davirugu
Reserva Ecologico

Anaguni ungua
Inter-Acción Scoial Recreativo

Bunague
Infraestructura de Comunicación y Transportación

Hiduru lun arani
Plantas Medecinaies

Luchudigutiña
Turismo

Tubuñe Muna
Material de Construcción

Agaliujani
Recolección Iguana Verde, otros animales

Fig. 28.12 Land use categories and symbols for Honduras

Scanned 1:50,000-scale maps were used as base maps. Geodetic datum and co-ordinate system conversions were accomplished to keep all layers in a common coordinate system. Community boundary points with name, longitude, latitude, and point order were processed with a MapBasic or Avenue script that resulted in

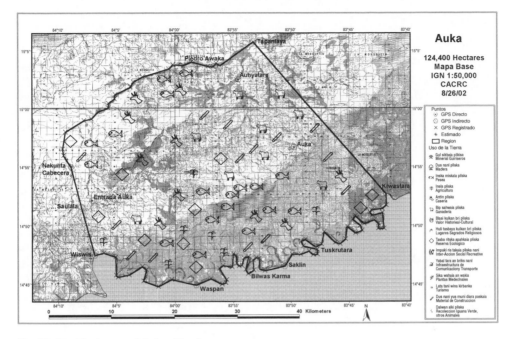

Fig. 28.13 Final map of Auka, Honduras

point and polygon files for use in the GIS process. Land use symbols were then transferred from the draft maps to point databases. Named boundary turning points, community boundary polygons, and land use symbol point files were mapped over the scanned base maps. GIS were used to produce area and perimeter values and to compute the area of overlaps between communities. Figure 28.13 illustrates a completed community map for the Honduran Mosquitia.

GIS were the basis for mapping and were used for measurements of distance and area, but a GIS-based project offers much more. Once information is captured in georeferenced layers, all sorts of analysis become possible. As an example, we found that land use, so critical for land claims, was often very differently designated by adjoining communities along boundaries and within overlaps. Figure 28.14 shows the areas of matching and non-matching land use in the overlap between two Honduran communities.

Because all our project information was in GIS form, we were able to use *allocation (proximity)* and cost-weighted allocation analysis to quantify the spatial extent implied by land use point symbols placed on maps. We were able to use spatial statistics such as the kappa coefficient to evaluate percentages of land use correlation in overlaps or along boundaries between communities. Using GIS processes we measured the relationship between regions of matching land use and proximity to streams that forms much of the visual pattern in Figure 28.14. We have since found similar land use differences in Nicaragua and in adjoining communities in Belize as depicted in another participatory mapping project (Toledo Maya Cultural Council and Toledo Alcaldes Association 1997).

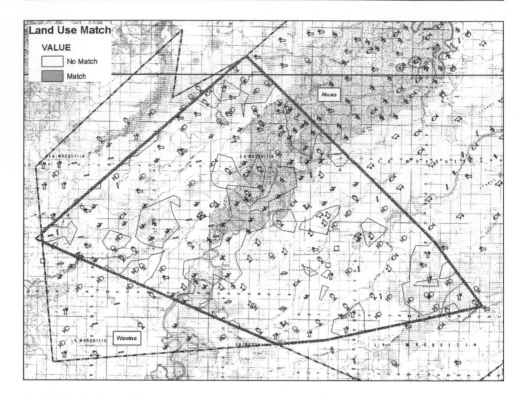

Fig. 28.14 Land use differences in overlap between Ahuas and Wawina, Honduras

CONCLUSIONS

Surveys of people and place should attempt to discover and use methods and technologies that reflect local conceptions of territory. Boundary and point-based conceptions of territory lend themselves to point measurement methods, vector carto-graphic representations, and vector-based GIS processing. Where land use defines territory, remote sensing, raster-based mapping, and raster-based GIS may be more appropriate. Solid lines defining territory may be inappropriate where territory is contested or poorly defined. In many cases mapping processes and notions of territory will interact with each other.

The selection of mapping technologies such as interviews, sketch maps, and remote sensing that favor land use mapping may be more appropriate for some projects than boundary-based technologies such as conventional traverses or GPS measurements that favor bounded conceptions of territory. Relative position measurements may require complete boundary traverses where such point-to-point access to land is dangerous or costly. Injudicious selection of base maps, such as selecting a base map produced by an agency of an unpopular government, can change a locally controlled participatory mapping project from one with an appearance of self-determination to one that requires that its community boundaries be based on the product(s) of a faction competing for land rights. Selection of high-precision survey techniques may require hundreds of point measurements to accomplish a boundary survey that would more appropriately be accomplished with just a few points in a reconnaissance survey. Contested boundaries might be better measured with approximate methods to avoid confrontations and resistance to modification with respect to the claims of neighboring communities.

Map projections, geodetic datums, and coordinate systems should be selected with respect to scale and accuracy requirements while respecting local independence from the reference systems often imposed by colonial powers or military mapping agencies. Surveying methods that use independent measurements of absolute position may be difficult to repeat or to link to physical features on the ground, making legal land titling impossible in a context in which local laws require reference to physical ground features or existing boundary markers. In the CCARC projects, for example, there is often no direct link between monuments on the ground and the GPS-derived points. This makes legal land tenure claims using the CCARC data sets difficult, lessening the chance that these approximate boundary claims will result in disputes over exact boundary placement.

Problems of supply, maintenance, and replacement of sophisticated electronic equipment, while not unique to mapping in remote areas, are often insurmountable where budgets typical of participatory mapping projects are limited. Climatic conditions, often extreme in tropical or polar environments, can cause failure of instrumentation on the ground. Cloud cover can render some remote sensing methods useless. Language-based processes embedded in modern electronic devices may not be suitable for use in many parts of the world. GIS processes bring to a project assumptions about space embedded in the software. Current GIS are limited in their ability to handle temporal change, uncertain boundaries, and multiple or contested land use. If local notions of territory are fluid and shifting, an entirely new or modified GIS platform might be required to handle complex or different notions of territory.

Ideas of territory and the mapping of territory are intertwined in an inseparable relationship. The ways in which territory is perceived influences its portrayal on maps, and methods used to make maps can change notions of territory. Any GIS-based mapping project is at the juncture of this process. Successful surveys of people and place require a synthesis of appropriate technologies and local participation.

REFERENCES

Anderson, B. 1991. *Imagined Communities*. London: Verso.

Brody, H. 1982. *Maps and Dreams*. New York: Pantheon Books.

Chapin, M., Lamb, Z., and Threlkeld, B. 2005. Mapping Indigenous lands. *Annual Review of Anthropology* 34: 619–38.

Dana, P. H. 1998. Nicaragua's "GPSistas": Mapping their lands on the Caribbean coast. *GPS World* (September): 32–42.

Dent, B. D. 1999. *Cartography: Thematic Map Design*. Boston, MA: McGraw Hill.

Gordon, E. T., Gurdián, G. C., and Hale, C. R. 2003. Rights, Resources, and the Social Memory of Struggle: Reflections on a study of indigenous and black community land rights on Nicaragua's Atlantic coast. *Human Organization* 62: 369–81.

Harley, J. B. 1988. Maps, knowledge and power. In D. Cosgrove and S. Danials (eds) *The Iconography of Landscape: Essays on the Symbolic Representation, Design, and Use of Past Environments*. Cambridge: Cambridge University Press, pp. 277–312.

Harvey, P. D. A. 1980. *The History of Topographical Maps: Symbols, Pictures, and Surveys*. New York: Thames and Hudson.

Knapp, G. and Herlihy, P. 2002. Mapping the Landscape of Identity. In G. Knapp (ed.) *Latin America in the 21st Century: Challenges and Solutions*. Austin, TX: University of Texas Press, pp. 251–68.

Monmonier, M. 1991. *How to Lie with Maps*. Chicago: University of Chicago Press.

Muehrcke, P. C., Muehrcke, J. O., and Kimerling, A. J. 2001. *Map Use: Reading, Analysis, Interpretation* (4th edn). Madison, WI: JP Publications.

Poole, P. 1995. Land Based Communities, Geomatics, and Biodiversity Conservation: A Survey of Current Activities. *Cultural Survival Quarterly* 18(4): 74–6.

Poole, P. 1998. Appropriate geomatic technology for local earth observation. *Yale F & ES Bulletin* 98: 156–66.

Robinson, A. H., Morreson, J. L., Muehrcke, P. C., Kimerling, A. J., and Guptill, S. C. 1995. *Elements of Cartography* (6th edn). New York: John Wiley and Sons.

Rossum, S. and Lavin, S. 2000. Where are the Great Plains? A Cartographic Analysis. *Professional Geographer* 52: 543–52.

Sack, R. D. 1986. *Human Territoriality: Its Theory and History*. Cambridge: Cambridge University Press.

Toledo Maya Cultural Council and Toledo Alcaldes Association. 1997. *Maya Atlas: The Struggle to Preserve Maya Land in Southern Belize*. Berkeley, CA: North Atlantic Books.

Wood, D. 1991. *The Power of Maps*. New York: Guilford Press.

Chapter 29

Geographic Information Science, Personal Privacy, and the Law

George C. H. Cho

The relationship between Geographic Information Science (GISc), personal privacy, and the role of regulation and self-regulation is at best nebulous, and uncertain. While tidy minds might wish a strict dividing line between what is private and what is public in reality there seems to be a "zone" of privacy. This zone delineates those parts of our lives that provide strategic information about ourselves, our financial status, health, and education. At different times what might be private information may meld into what might be considered public. The tools of GISc deal with geospatial information in which spatial relationships are *the* fundamental data. But, GISc may also handle a diverse range of personal information from the truly "personal" ones through to those of a more general nature – age, gender, height, home address, social security number, marital status, religion, and so on. By its very nature the tools of GISc allow these kinds of information to be collected, manipulated, displayed, and transmitted cheaply, easily, and speedily. But the very capabilities of geospatial tools in information analysis and transfers have raised a multitude of novel and interesting information and personal privacy issues.

Professor Arthur Miller of the Beckman Center for Internet and Society, Harvard Law School has described privacy as an intensely, perhaps uniquely, personal value. The word stems from a Latin root *privare* which means "to separate." To want privacy is to want to be separate, to be individual. Another meaning of the Latin is "to deprive," and privacy also means leaving something behind.[1]

In the common law world the claim to privacy as a right rather than merely an "interest" to be protected is a relatively recent development because the protection of privacy in the past has been *ad hoc*. The Australian Constitution has no vested power over privacy protection while the common law protects privacy rights indirectly. For example, the law of defamation, negligence, and passing off give a semblance of an overarching umbrella, as does the shield provided by contract including the duty of confidence. Likewise, in the USA the origins of the privacy right may be traced to a law review article by Warren and Brandeis (1890). In a famous dissenting opinion, Judge Louis Brandeis in 1928 reiterated the right to be left alone as "the most comprehensive of rights and the right most cherished by civilized men."[2] Based on the principle of the right to be left alone, US law has developed along the lines of a common law right and those rights found under Amendments to the US Constitution.

In this chapter, the first section examines the doctrinal issues that provide the basis of discussion of the nature and structure of the problem of personal privacy and geospatial technologies. Then in the second section the legal, regulatory, and policy framework that underlies the source of this interest is evaluated in the context of considering privacy as a "right" in some circumstances. In the third section a review of the different geospatial technologies that promote intrusiveness, enhance privacy protection, or are sympathetic to privacy protection is undertaken.

Nature of the Problem

There are three interrelated matters that need to be resolved initially. The first is whether Geographic Information Systems (GIS) are a threat to personal privacy. The second matter is that there seems to be a lack of understanding of privacy issues, just as there is, for instance, some fuzzy thinking about whether we are attempting data protection on the one hand or protecting the privacy of that information. Finally, there are, inevitably, some ethical questions in the use of geospatial technologies especially where privacy issues are involved and what the "right" thing to do might be.

GIS is not personal data intensive

Flaherty (1994), in relating his experiences with privacy protection and GIS in a national and provincial context, has said that the technology is "not personal data intensive" and that information privacy issues are *always* solvable by applying fair information practices.

GIS has the power to integrate diverse information from multiple sources. Some of the data are of a personal nature where individuals may be identified or identifiable, while others are of a spatial nature that may be used to locate individuals through geocoded data such as a home address. The privacy threat is from the new inferences that may be made by correlating geographic information with personal information.

Data on geographical location when combined with other data transforms GISc into a powerful tool for tracking, storing, and analyzing personal information. Tracking, data integration, and analysis capabilities give Geographic Information (GI) technologies the potential to be more invasive of personal privacy than many other technologies.

Dobson (1998) promotes the view that geo-information in combination with personal information clearly poses a privacy threat. But there is a tension between the two fundamental values of privacy on the one hand and the public's right to know on the other. For example, the differences between satellites monitoring a farmer's use of water and the public's acceptance of CCTV and video cameras in shopping malls, buses, and taxis, in city streets and in dorms and apartments.

Lack of understanding of privacy issues

There seems to be a lack of understanding of privacy issues among many who use geospatial tools. For instance, there may be confusion in terms of the differences between data protection, privacy of information, and personal privacy.

Data protection, whether by legislation or by a code of practice, relates to the protection of data rather than about the people themselves. The protection is as much of a concern for both the data subject (a natural person and, in certain instances, corporations and legal persons) as the organization or agency that collects and processes the data in order to ensure legislative compliance.

The privacy of information relates to undertakings to keep the information private and the various interests individuals may have in controlling and significantly influencing the handling of these data. The privacy of information is often taken for granted and we willingly provide personal information to others who request the information because we believe that the receiver has undertaken to abide by some privacy guidelines. Here we may be equating private with confidential.

Informational privacy relates to knowledge about some person that is kept private and confidential. This is quite different from personal privacy where an individual may have chosen to be left alone and facts about that person shielded from view. The former is about the protection of personal information kept confidential and the latter is about a person acting and taking steps to keep personal facts private. Informational privacy becomes a problem where the datum may be traced to a person. Another dilemma is whether, while one may have divulged personal information for one reason or another, one's consent should be given again. The dilemma is particularly pointed when that information is either disguised in aggregate form or re-purposed in some other form so that new, synthesized data are vastly different from what was given by an individual initially.

Ethical use of geospatial technologies

A final issue is the ethical use of geospatial technologies, in which users and proponents of this potentially invasive technology need to be reminded of their heavy social responsibilities and to use the technology ethically and in an ethical manner in all ways and at all times (Ball 2003).[3] Geospatial technologies include GIS as a mapping tool for decision-making through to those technologies that amass data by spatial attributes including global positioning systems (GPS), and transponders and other intelligent computer chips embedded in some devices that can report location as well as an identity. The latter are in a class of intelligent spatial technologies that can declare both personal information as well as locational and device-specific information in response to a poll by another device either in a pre-established relationship or to a new, soon to be established, relationship.

Geospatial technologies are in daily use and have heightened locational information personal privacy concerns. For instance, in using geospatial technologies, an apparent legal fallacy may have arisen, if only by accident. The idea is that if there is a legal right to do something then it follows that it must be the right thing to do. So, if it is permissible to undertake data aggregation activities using a number of databases, then it is lawful to do so. But really, the issue is that the legal right must only be the starting point rather than the end point for justifying one's actions. The fact that something is legal does not mean it is either right or a wise thing to do. Thus, data taken out of context – *acontextual data* – and used in that sense may produce results that may be highly unjust and totally incorrect in particular cases. This is where ethical questions are raised that should be foremost in the thinking and practice of GI scientists. Some would like to consider ethics as a continuum in which there is both a duality of a right and a wrong way of undertaking activities as well as ethics being a way of dealing with a right way and a better way of doing things. In equitable jurisdictions and civil cases there may be a claim to a right, but it is the one who has the better claim that often wins out.

Today the right to be left alone is being vigorously defended, and there is resistance to increased surveillance in parts of our private lives. But it seems that the real threat is the creeping acquiescence to all sorts of intrusions without the accompanying public debate, information, and education. It used to be that when a video camera was installed, say in the computer lab, the spectre of George Orwell's *1984* and "Big Brother is watching you" was raised (Orwell 1990).

Today, it appears that we have grown accustomed to all sorts of cameras watching us in all kinds of circumstances. The reality TV genre of the "Big Brother" type where the public is invited to watch the antics of four or five couples displaying their private, sexual habits raises no public outrage. On the other hand, employers keeping watch on the Internet use of their employees and monitoring their e-mail traffic has raised civil liberty concerns. This can be particularly serious when these monitoring records are used as grounds for dismissal because of workplace abuse, which would no doubt raise both moral and ethical outrage. The really big question is: which is the greater sin – protecting personal privacy or surveillance?

It has been observed that the surrender to surveillance is now happening and it is taking place "one step at a time, and each step is attractive and relatively benign" (Dobson 2000, p. 24). Clarke (2001) has asked how such advanced, well-educated, well-informed free societies have been so myopic as to permit what he calls "dataveillance" to have taken place. The answer to him seems to be that these societies have been conditioned by Orwell's *1984*. In the past fifty years or so, technology has developed and delivered far superior surveillance tools than Orwell had imagined and to cap it all "we didn't even notice" (Clarke 2001).

So the message is that the public, users of GIS, and policy-makers have to be ever-vigilant requiring particular sensitivity in the application, design, and use of inherently privacy invasive technologies such as those embedded in a GIS.

Privacy: The Legal and Regulatory Framework

In the developed world, the legal and regulatory framework concerning personal privacy is characterized by its recency and *ad hoc* regulation and control. No less, its complexity and multiple dimensions have led to conflicting views as to how to structure such a framework. Legislative instruments are usually passed after much debate and discussion at various fora and legislative bodies. Such laws may be contrasted with the common law that is the result of litigation in courts that produces precedents that are subsequently followed by the courts. Some countries, however, have promulgated a civil code and these follow rules and regulations that have been codified. Examples of these countries are to be found in the European tradition – France, Germany, Austria.

To begin with the Australian situation, the Australian Constitution does not empower the Commonwealth Parliament to enact a general law for the protection of privacy throughout Australia.[4] However, the Australian government has been obligated to protect personal privacy stemming from various international covenants, agreements, and treaties to which Australia is a signatory, for example the Universal Declaration of Human Rights (UDHR) adopted in 1948.[5]

A "right to privacy" is absent in the US Constitution or the Bill of Rights. However, the US Supreme Court has interpreted a right to individual privacy under the First, Fourth, Fifth, Ninth, and Fourteenth Amendments.[6] Many privacy decisions in the US federal courts are based on the Fourth Amendment, which generally provides for the right of people to be secure in their persons, houses, papers, and effects against unreasonable searches and seizures.[7] In 1965, the US Supreme Court suggested that there may be "zones of privacy" implicit in the Bill of Rights in the leading case of *Griswold v. Connecticut*.[8]

Common law right to privacy

Since the majority decision in *Victoria Park*[9] it has generally been accepted that a cause of action for the breach of privacy does not exist in the common law of Australia any more than it existed in the common law of England.[10] The issue of a right to privacy was revisited in Australia recently in a High Court case.[11] In the UK the right of privacy of a corporation has been held to exist.[12] Also, more recently, privacy rights have also been extended to individuals drawn from the fundamental value of personal autonomy.[13] Courts in several other jurisdictions have also addressed the availability under common law of an actionable wrong of invasion of privacy – Canada, India, and New Zealand.[14] One Canadian court has recognized a general right to privacy and to protect privacy interests under the rubric of nuisance law.[15] In New Zealand the tort of invasion of privacy has been recognized and s 14 of the *New Zealand Bill of Rights Act* 1990 (NZ), while it does not confer a right to privacy, ensures the freedom of expression (Tobin 2000).

In the USA the tort based upon the right to privacy has been developed and is still evolving in response to the encroachments upon privacy by the media and others. The common law of tort has identified four activities that give rise to liability for the invasion of privacy. These are: (1) intrusion upon seclusion; (2) appropriation of name or likeness; (3) publicity given to private life; and, (4) publicity placing a person in a false light.[16] Some states do not, however, recognize such claims; for example, New York does not have a false light claim provision.[17] Other states protect a larger class of persons and private persons, as well as celebrities, such as the Californian and New York laws on misappropriation of name or likeness.

Privacy legislation – Australia

In Australia there are four federal acts that are of importance that deal directly with privacy: the Privacy Act 1988, Privacy Amendment Act 1990, Data Matching

Program (Assistance and Tax) Act 1991, and Privacy Amendment (Private Sector) Act 2000. Each of these pieces of legislation has been passed in response to specific obligations under the UDHR, International Covenant on Civil and Political Rights (ICCPR) 1966, Organization of Economic Cooperation and Development (OECD) Guidelines of 1980[18] and the United Nations (UN) 1990 Guidelines for the Regulation of Computerized Personal Data Files.[19]

Privacy legislation in the US

The Privacy Act of 1974, 5 USC §552 provides limited privacy protection for government-maintained databases. In general the Act prohibits any government agency from concealing the existence of a personal data record-keeping system. The Privacy Act applies to all collections of spatial data collected by federal agencies. The Federal Geographic Data Committee (FGDC) established under the Office of Management and Budget (OMB) has endorsed a policy on access to public information and the protection of personal privacy in federal geospatial databases.[20] In particular, the policy applies to all federal geospatial databases from which personal information may be retrieved.

In terms of direct legislation, Congress has passed a number of laws that affect the protection of privacy in general and those regulations applying to the online world in particular. Privacy protection legislation has been passed to address specific activities including financial information, education, health, electronic communications, video hiring habits, drivers' licenses, and children.

Many States in the USA have a general privacy act that mirrors the federal government's Privacy Act. In general the state acts are designed to control the information that a state agency or local government may gather on individuals and the manner of their use. In addition, most states have separate acts that address the protection of privacy in specific situations.

Other jurisdictions

The Privacy Act 1993 (NZ) favors the development of a tort for breach of privacy in respect of public disclosures of true private facts where the disclosure would be highly offensive and objectionable to a reasonable person of ordinary sensibilities.[21] Unsurprisingly, Privacy Acts that provide for a tort of privacy have been enacted in the provinces of British Columbia, Manitoba, and Saskatchewan in Canada.[22] A selection of various privacy acts and provisions in 20 countries can be found in Campbell and Fisher (1994).

The issue of privacy exemplifies the contrasting approaches of Australia, North America and Europe. In Europe, for example, private industry codes of conduct are tantamount to no regulation or certainly insufficient regulation. In Australia and North America, by contrast, self-regulation is seen as an effective way to achieve the balance between consumer privacy concerns and business needs.

Evolving fair information privacy principles

Privacy *per se*, serves several valuable functions and more generally it is the ability to control what other people can know about you. The right to keep identity information secret serves to help protect the individual from stalkers, abusive ex-spouses, and others whose company that individual may wish to avoid. In the online world, privacy makes identity theft – the wrongful use of a person's identifying information to obtain goods and services fraudulently – less likely while anonymity enables an individual to blow the whistle on wrongdoing without fear of retribution as well as affording protection for the "whistleblower." The handling of personal information would therefore need to give respect to the emerging "right" to privacy and its protection, and other competing interests that may require the free flow of information.

Over the past quarter of a century, governments in Australia, the USA, Canada, and Europe have examined and analyzed their "information practices," the safeguards for the collection and use of personal information, and the adequacy of privacy protection. Fair information practice principles were comprehensively articulated in the US Department of Health, Education and Welfare's seminal report entitled *Records, Computers and the Rights of Citizens* in 1973. Five core principles of fair information privacy protection gleaned from the various codes noted above relate to Notice–Awareness, Choice–Consent, Access–Participation, Integrity–Security, and Enforcement–Redress.[23] A particular example of an evolving and developing use of fair information practice principles is the EU Data Directive which marks a major shift in the shaping of global privacy protection but also demonstrates the impact of a foreign protective schema on the remainder of the world.

The EU Data Directive

The European Union promulgated comprehensive privacy legislation entitled *The Directive on the Protection of Individuals with regard to the Processing of Personal Data and on the Free Movement of such Data*[24] that became effective on October 25, 1998. Article 25 of the regulations requires that transfers of personal data take place only to non-EU countries that have an "adequate" level of privacy protection. This is designed to prevent the circumvention of the directive and the creation of "data havens" outside the EU.

Before entering into an agreement with a foreign country to allow free circulation of personal data outside the EU, an evaluation of the adequacy of data and privacy protection in that country has to be undertaken. Several countries have done this including Australia,[25] Switzerland, Hungary, the USA,[26] and Canada.[27] In the case of the USA, it has taken a long time to reach a decision since it related to a specific system applied in that country known as the "safe harbor" principle.[28]

A safe harbor system permits US companies to satisfy the European "adequacy" standard while maintaining their traditional self-regulatory approach to data protection. In July 2000 the European Commission approved the safe harbor framework as meeting the "adequacy" standard (Harvey and Verska 2001, Yu 2001).

The implications of the EU Data Directive for other countries, however, are uncertain. Australia's Privacy Amendment (Private Sector) Act 2000 (Cwlth) puts in place regulations for the use and handling of personal information by individuals and private companies. One of the NPPs obligates companies to secure all personal and sensitive electronic data that is stored, processed, or communicated in their software, systems, and networks (Handelsmann 2001).

The Data Protection Act 1998 (UK) replaced the Data Protection Act 1984 (UK) and took effect from 1999. The new Act governs the collection, processing, and use of data in the UK by any organization that requires registration with the Data Protection Commissioner. Transfer of data or permitting its access by anyone outside the EU, for example via the Internet, is heavily regulated and data controllers have to ensure compliance with the regulations governing such transfers.

The Canadian Personal Information Protection and Electronic Documents Act 2001 (PIPEDA) (Canada) came into effect on January 1, 2001 and regulates the use and collection of personal information (Krause 2001). The Act applies not only to Canadian companies but also potentially to any entity that collects personal information in Canada and/or personal information from Canadian residents. This enactment ensures that Canada complies with the EU Directive's standard for "adequacy" for the protection of privacy.

Geospatial Technologies and Privacy Implications

Geospatial technologies in common use that rely on both spatial attributes as well as other personal and feature information have the "ability" to track people, their shopping, and travel habits, the places that they go to for recreation and for what duration, and in some instances to make an inference of the purposes of that event. Here we consider location-based services (LBS) that rely on the key ingredients of time and space (see Chapter 32 by Brimicombe in this volume, for further discussion of these services). LBS may be considered to be no different from geodemographics, an information technology that enables marketers to predict behavioral responses of consumers based on statistical models of identity and residential location (Goss 1994, 1995).

LBS have become commonplace because of the use of geocodes and GPS and other mobile communication and tracking technologies. LBS inferentially involves the tracking of people through the use of credit card data that may result in profiling exercises, statistical modeling, and pattern analysis. More generally, GISc, using such technologies, may give the game away as to who we are, where we are, and what we have been doing either by way of speech, purchases, or simply being at a location. There is one view that without legislation to curb the (mis)use of such data there would be chaos. However, an equally compelling but opposing view is that there should be no legislation but rather just self-regulation by industry itself (Westin 1967, 1971).

Data aggregation and databases

Private sector commercial applications of both GIS and LBS are perhaps the fastest growing areas of business and this has fed the need for more data. For example, in order to maintain a competitive edge, marketers need good databases to make their decisions while simultaneously handling geographical data efficiently. Patterns, relationships, and trends become clearer when depicted visually in graphs, charts, and maps rather than just columns of numbers or text.

A database called EQUIS, developed by National Decision Systems, "maintains a database of financial information for over 100 million Americans on more than

340 characteristics including age, marital status, residential relocation history, credit card activity, buying activity, credit relationships (by number and type), bankruptcies, and liens. This information is updated continuously at a rate of over 15 million changes per day" (Curry 1992, p. 264).

In 1993 Equifax and National Decision Systems announced Infomark-GIS – a fully integrated GIS specifically designed for marketing applications and decision making (Equifax National Decision Systems 1993). There are other US companies too that are also engaged in the collection, processing, and storage of data pertaining to individuals.[29] These firms obtain consumer information from credit bureaus, public records, telephone records, professional directories, surveys, customer lists and other data aggregators.

In view of the commercial market for data, databases, and data aggregation services, Curry (1994) has suggested that the use of geospatial technologies will produce multi-faceted problems that would similarly require multi-dimensional solutions. The concerns raised include the fact that the technologies consist of and promote the widespread availability of unregulated data. This leads to the difficulty of regulating data matching that must take place if the geospatial tools are to produce meaningful results. Further, geospatial technology is inherently visual but this strength also exposes a major weakness in that it may produce map inferences that may be both statistically and ecologically fallacious. Finally, there is the altered expectation of privacy rights because the case law may promote an erosion of those aspects of life where a person can feel safe, secure from search and surveillance, and most importantly with information kept private.

Geospatial technologies are said to do best at the intersection of location, time, and content. But each of these elements tends to produce tensions of their own. For example, in regard to letting people know where they may be and keeping this fact hidden because they may not wish to be found. Here we have a technology that is employing the power of "place." GIS-based geostatistical models using locations and space have been used to study the home range of animals, for instance. In an analogous way, geographical profiling may be applied to the "home range" of predatory humans who may have inadvertently patterned habitual routes when stalking potential victims.

Leipnik, Bottelli, von Essen, et al. (2001) reported on the important role of geospatial technologies in investigating and gathering evidence of a locational nature in order to convict a serial killer. When the serial killer Robert Lee Yates was arrested in April 2000 in Spokane County, Washington, he is reported to have told his wife to "destroy the GPS receiver." This was because there were incriminating data on it showing the 72 waypoints associated with several journeys that had been used in disposing of the bodies of the victims. Indeed, this example demonstrates the extent to which GIS and GPS technologies are permeating society and their use – both by law enforcement agencies and criminals. However, a court appeal could challenge the validity and use of the GPS data. The argument could be that using the GPS to track suspects without their knowledge might involve an invasion of privacy rights as well as failing to meet the legal test of finding the "least obtrusive means" for police to gather information about a suspect. But, in counter argument, it may be submitted that sometimes consent needs to be conspicuously absent in cases in which the suspicions of the "quarry" are not to be aroused prematurely.

The question may be asked "Is there any special data protection or privacy issue associated with locational data?" The answer is in the affirmative but its explanation must be given indirectly. "Sensitive personal data" is regarded as data that identifies, among other things, a person's ethnic background, religion, political affiliations, or sexual habits. However, the location of a person is not considered sensitive personal data. Yet, when processing data on persons, especially when locations are involved, say in terms of visiting synagogues on a regular basis or particular areas of ethnic concentrations, it is arguable that sensitive data are being processed and unintended inferences are being made about a person's religion or ethnicity (Rowe and McGilligan 2001).

Location, tracking and dataveillance

Most geographers will understand location to mean the relative positions of entities in space and time and of events taking place. Locational information describes the whereabouts of a person or entity in relation to other known objects or reference points.

In this context, tracking refers to the plotting of the trail or sequence of locations within a space that is taken by an entity over a period of time. The "space" within which the entity's location is tracked can be a physical or a geographical space. However, such a space can also be "virtual" in cases in which that person may have had successive interactions in time with different people either simultaneously or at different times.

Data surveillance, abbreviated to *dataveillance*, is the systematic use of personal data in the investigation or monitoring of actions or communications of one or more persons. Conceptually there may be two separate classes – *personal surveillance* being the surveillance of identified persons for various purposes. This may include investigation, monitoring, or gathering information to deter particular actions by that person or particular behaviors of that person. *Mass surveillance*, on the other hand, is the surveillance of large groups of people, again for the purpose of investigation or monitoring, which may aid in the identification of persons of interest about whom a surveillance organization has cause for concern (Clarke 1999a).

Geospatial technology applications: home location

While geospatial applications based on remote sensing of the Earth on regional scales, and the use of GIS in very small-scale city planning are relatively well known, the application of such technologies to home location are less prominent. In general, utility companies may use home location data to track usage of power, gas, and water, whereas social security benefit providers and license administrators may make use of street addresses to tag locations for administrative purposes.

Telecommunication services to the home via the double-twisted copper wire provide home location information to the Public Switched Telephone Network (PSTN) of a phone utility. All phone communications to and from the home address can be recorded, stored, analyzed and made available to others. Telephone traffic data may be analyzed as call records and have been used for billing and invoicing purposes.

The data also give paired locational information of the origin and destination of calls (dyads) that may be analyzed for particular information.

More recently, home telephone services have facilities such as caller id, calling-line id (CLI), and calling-number display (CND) that give information of a caller, the caller's phone number, and the time of the call. Caller, line, and number identifications are now generally available and have been used by telecommunication companies, law enforcement agencies, and consumers as a means of screening calls.

Australia on Disk (AOD) is a directory of every residential and business phone number in the country. This comprises all 55 national phone directories giving a total of 1.3 million business listings and 6.9 million residential phone numbers in 2005. While there are obvious environmental benefits in not having to produce paper copies of telephone directories, there are equally potent privacy implications not to do so given that a disk containing residential information may be put to other than legitimate uses.[30] AOD does not permit reverse searching – that is, if you have a phone number and you want to search to find out the street address to which it is connected, the product will be unable to answer the query. However, AOD permits searches and printing of whole lists by whatever criteria are specified such as post codes, surnames, or "target" markets. The information obtained can then be used to individually address letters to be sent out or used in a phone market-ing campaign. An additional product, the *AOD Mapper* presents data and maps with areas of interest color coded to show their relative importance. The maps could identify hotspots where the residents are most likely to match particular profiles, as with, for example, the suit-buying yuppie, the private school targeting high-income earners with young families, or health insurance companies looking for low risk, single, professionals.[31]

Tracking movements of individuals over space

The tracking of individuals over borders has normally been controlled and monitored at immigration counters by checking and stamping national passports and by the use of identity cards. More recently, such travel documents have had electronic chips embedded in them that permit electronic scanning and automatically provide entry and egress at checkpoints. The data captured may yield patterns of entry and departure of citizens and visitors alike. At a micro-scale, movements within buildings may be monitored using video surveillance equipment. In combination with video evidence, movement over time and space within buildings, analyses using pattern recognition and/or matching algorithms may yield greater insights to movement patterns of individuals or groups of people and the most trafficked areas. However, while the system may provide greater effectiveness to the security of a particular building, the technology is highly intrusive of the privacy of its users.

The US Department of Transport (DOT) Intelligent Transportation Systems (ITS)[32] program has made extensive use of GIS technologies, along with surveillance and other computer technology, to provide in-vehicle mapping as well as both lagged and real-time transportation management services.

The ITS concept is one where there is an interactive link from a vehicle's elec-tronic system with roadside sensors, satellites, and a centralized traffic management system to constantly monitor each vehicle's location and the road traffic conditions.

More advanced systems include the reception of alternative road information in real-time via two-way communications, on-board TV screens, and mapping systems. While few will argue with the efficacy of such a system in regard to public safety and convenience there are other unintended outcomes that must be considered. Policy makers and developers of these systems must contend with its impact on individual privacy.

Highway surveillance and pattern recognition can potentially be both valuable supervisory tools for regulators as well as investigative and forensic weapons for law enforcement agencies.

Tracking transactions

There has been a variety of methods used to capture transaction data. These include cheques that carry data in MICR – magnetic ink character recognition – and turn-around documents with OCR – optical character recognition – where the form has already been filled-in automatically for the client to authenticate and return. Other means of data capture are magnetic strips, embossed data codes, bar coding, or a device with a location identity such as a phone socket in the home environment. The main applications are in financial transactions from deposits, to loan repayments, salaries, cash withdrawals at automatic teller machines (ATMs), use of credit/debit cards and electronic funds transfer-point of sale (EFTPOS), and others. These seemingly helpful and efficacious systems, however, turn previously unrecorded and/or anonymous activities into recorded-identified transactions. Even more important, the data are being aggregated in a far more intense manner than before and these data have an associated location tag in the data trail. There are also real-time location mechanisms built into some of these electronic transaction tools so that passive monitoring and surveillance can now be re-purposed for law enforcement activities.

Tracing communications

While locational information and address tags may be readily available when using the PSTN, mobile telephony services including the use of pagers, personal digital assistants (PDAs), analog/digital and satellite phones make the tracing of persons and locations difficult. With mobile telephony the tracing of a device and its usage either in real-time or logged-in message banks is more difficult because its location is constantly changing.

The global plan to give each home phone its own e-mail address has also raised privacy concerns. An alliance between the telecommunications industry and the Internet Engineering Task Force (IETF) is seeking to create a protocol called *ENUM*.[33] This protocol is used to map telephones, facsimile machines, and other devices from a phone number to the Internet. The plan is to use the domain name system (DNS) for the storage of E.164 numbers. E.164 is the International Telecommunications Union (ITU) standard that defines the format of telephone numbers, specifically for international subscriber dialler numbers (ISDN).[34] Thus, with ENUM you have a single electronic access point or address (International Telecommunications Union 2001). However, Clarke (2002) is not sanguine about this prospect because the effect of ENUM would establish a single unique contact number for each individual. If it were successful, it would represent a unique personal identifier, with all of the dangers to privacy and freedoms that this entails.

A combination of national security, law enforcement, and corporate marketing interests is pushing for the provision of handsets that are locatable to within a few meters; just as in the case of the location of callers to the emergency number 911 in USA or 112 in Europe.[35] The US Federal Communications Commission (FCC) has now imposed requirements on wireless carriers to provide emergency call services with far more precise locational identifiers. Handset-based solutions must be identified to within 50 m of a call while for network solutions the identification must be within 100 m.[36]

Convergence of locational and tracking technologies

By far the technology of most relevance for present discussion is GPS. This technology depends on a constellation of satellites to give positional information in four dimensions: latitude, longitude, altitude, and time. With the presidential edict of turning off selective availability (SA) – the purposeful degrading of positional information – users of GPS are now able to poll satellites for positional information and be given references to within a meter of their location.[37] Differential GPS (DGPS) uses the same technology except that locations are determined as a differential to the data received in addition to and relative to a surveyed point on the ground. Hence, the accuracy obtained by DGPS methods can be quite precise. Assisted GPS (AGPS) technology, on the other hand, has been developed in conjunction with information communication technology (ICT), which uses a server at a known geographical location in the network. This information reduces the time, complexity, and power required in determining location (see Chapter 28 by Dana in this volume for additional information on Global Positioning Systems).

With the advent of wireless communications, location information has come under scrutiny. In the USA the Telecommunications Act of 1996 included location information as Customer Proprietary Network Information (CPNI) that gave time, date, and duration of a call, and the number dialed. In the EU the Directive on Privacy and Electronic Communications (2002/58/EC) established a technology-neutral legal standard for privacy protection in the processing of personal data for all electronic communications. By the end of 2003 only four countries – Denmark, Sweden, Finland and Spain – had implemented the Directive.[38]

RFID is an abbreviation for radio frequency identification, a technology similar to bar code identification. With RFID the electromagnetic or electrostatic coupling in the radio frequency portion of the electromagnetic spectrum is used to transmit signals. RFID systems can be used just about anywhere, from clothing tags to missiles to pet tags to food – anywhere a unique identification system is needed.[39]

Privacy risks with locational and tracking technologies

There is a saying that "a man's home is his castle." However, this bastion of privacy is slowly being eroded not only in regards to a loss in physical terms but also in regards to a loss of privacy of the airwaves, and, perversely, thermal emissions from a property and visual monitoring.

With geospatial technologies there has already been litigation on the basis of such "trespass" to land properties. The grounds on which these cases have been argued include the idea of an "objective" expectation to privacy as may be found in the Fourth Amendment to the US Constitution or a "subjective" expectation as provided under case law interpretations as the following cases demonstrate.

In *Dow Chemical v. United States* (1986)[40] although the District Court held that the aerial photography was a "violation of Dow's reasonable expectation of privacy and an unreasonable search in violation of the Fourth Amendment," the US Supreme Court held that the "open field" doctrine applied to the case, and therefore there was no invasion of privacy. Similarly in *California v. Ciraolo*[41] the Supreme Court found that it was acceptable for the police to fly over a fenced-in backyard at an altitude of 1,000 feet to undertake monitoring. Also in *Florida v. Riley*[42] a court approved the use of a helicopter, hovering at 400 feet, to observe marijuana plants through a hole in the roof of a defendant's greenhouse.

The remit to observe from the air has also extended to monitoring emissions. In *United States v. Penny-Feeny*[43] a Hawaii District Court endorsed the police use in a helicopter of a forward looking infrared (FLIR) device to discern heat emissions from a garage to gather information on illegal activities. Beepers and identity tags have also been endorsed to track the location of individuals in motor vehicles and then use GIS to map and trace routes.[44]

In *United States v. Smith*[45] a US Court of Appeals for the Fifth Circuit has ruled that technological advances may be capable of expanding the legally protected range of privacy that individuals enjoy.

The discussion so far suggests that there may be three different types of technologies that either invade or enhance the privacy of individuals. Some are expressly privacy-invasive technologies (PIT) including "data-trail generation through denial of anonymity, data-trial intensification as in identified phones, stored valued cards (SVC), and intelligent transport systems (ITS), data warehousing and data mining, stored biometrics, and imposed biometrics" (Clarke 1999b). Then, since around the mid-1990s, there has emerged those that are expressly designed as privacy-enhancing technologies (PET) seemingly as a reaction and bid to reverse trends in technology that have hitherto been privacy invasive. Tools that have been developed to assist the protection of privacy interests include those that make individuals genuinely untraceable and anonymous. The third type of technology, labeled privacy sympathetic technologies (PST), are capable of delivering genuine anonymity. Such PST are further subdivided into anonymity services, pseudonymity services, and personal data protection.[46] These services use various devices that provide outright anonymity to those that encrypt identity tags which may only be shared with selected parties. In between there are pseudo-anonymous devices that give a semblance of anonymity but from which identity may be traced. Such technologies provide cogent reminders that we should always be conscious about maintaining a balance between privacy and other interests such as accountability.

To assuage consumer privacy concerns both legislators and those in the geospatial technologies industries are battling to find the right balance: a solution that provides PET and PST while at the same time preserving the functionality of the technology even if it may have the effect of PIT.

It may be appropriate here to make two final observations. First, location information should only be used to provide services to users. It will be inappropriate for location information to be used for secondary purposes, such as marketing based on an individual's location or by government for law enforcement and national security purposes as it may create the *dataveillance* of society without the necessary investigatory mandate. Second, while voluntary standards and codes of conduct may work well in Australia, UK, Canada, the USA, and elsewhere, in the EU the preferred route to stem the erosion of personal privacy is through the use of legislation and mandatory standards. Whatever the solution, the strong message is that laws and regulations specifying how consumers authorize access to their privacy need to be clear, consistent, and technology-neutral.

The use of legislation as a means of protecting personal privacy and the soft-touch, self-regulatory approach to such protection may reflect differences in cultures, histories, and philosophies. It may be that for economies like Australia, the UK, Canada and the USA there is reluctance for governments to interfere with market forces and they thus choose to have minimal legislation. Most of these countries use sectoral approaches that rely on a mix of legislation, regulation, and self-regulation, while the European approach is one of strict regulation. Moreover, a sceptical citizenry may be suspicious of, and loath to permit governments to have greater control of private information than is necessary. The soft-touch approach will permit a middle ground with industry self regulating itself as well as giving governments some control over privacy protection. It is still an open question whether the marketplace or legislation can decide what is best for the protection of personal informational privacy. As a special case, with globalization as well as the commodification of personal data, the EU Data Directive portends a different future.

Table 29.1 synthesizes the material presented in this section.

CONCLUSIONS

This chapter has canvassed the issues of personal and informational privacy and the use of geospatial technologies. The question of how invasive GI technology has been or can potentially be is answered in the negative, that is, the technology is not personal data invasive. Nevertheless, there is a need for vigilance as well as the ethical use of such technologies even if there are social benefits. The legal and regulatory framework governing the issues of privacy was then discussed using the Australian jurisdiction as the backdrop and contrasting it with the US regime. While there are no constitutional impediments in preserving a right to privacy in Australia, there are four so-called privacy laws supported by the common law in protecting the confidentiality and disclosure of personal information. Supplementing the law are industry codes of conduct and self-regulation that complete the privacy package.

Analyses of geospatial applications with regard to home location, the tracking of individuals over space, tracing financial transactions and communications has identified privacy risks inherent in the use of such technologies. The identified privacy risks may be categorized as invasive, enhancing, or sympathetic. The implications for user organizations is that geospatial technology applications are but one of the array of different kinds of surveillance and in particular that of dataveillance. Equally, technology providers should be reminded of these sorts of privacy issues and that they should genuinely strive for anonymity in the use of personal information when marketing their products.

Table 29.1 Privacy risk status and geospatial technologies

Geospatial technologies related to	Applications	Privacy status
LOCATION:		
Home address:	Utilities, benefit providers, licence administration	N/I
Fixed	Inhabitant registration systems	I
	White Pages, ex-Directory, silent numbers	E
	Reverse White Pages	E
	GPS/DGPS/AGPS	I
Mobile	Auto-reporting mobile devices	I
MOVEMENT		
Individual	Trans-border (passports)	N/I
	Credentialed building access	I
	Biometrics	I
	Transponders	N/I
Mass	CCTV	I
	Pattern recognition; pattern matching	N
	Intelligent Transport Systems and imposed identifiers (travel cards)	N/I
TRANSACTIONS		
Financial	Check data with magnetic ink character recognition	N
	Turnaround documents with optical character recognition	N
	Encoding, magnetic strips	N
Retail	Bar coding	N
	ATMs, EFTPOS, Credit/debit cards	I
	Loyalty schemes	I
	Smart cards, stored valued cards	E
	Real-time locater mechanisms	I
COMMUNICATIONS		
Fixed	PSTN traffic data, call records	I
	Real-time tracing, interception	I
Mobile	Mobile telephony, pagers, analogue/digital/satellite phones/PDA	N/I
	Personal phone numbers, ENUM	N
	Caller id, CLI, CND	I
CONVERGENT TECHNOLOGIES		
Information Technology	Voice data TCPIP/VoIP	N/I
	RFID computer wear	I
Mobile, on person	EPIB	I
	Identity tags (prisoners, children)	I

Privacy status: E – enhancing; I – intrusive, N – neutral.

The implications for policy makers including privacy and data protection commissioners is one where the tensions between economic rationalism and the social good is stretched and seemingly irreconcilable. But this need not be the case if governments are focused on both law and order, as well as striving for stability, consistency, and sensitivity that are supportive of privacy protection.

Geospatial technologies such as LBS may "push" content but at the same time "pull" in locational information. Their use should not have a chilling effect on personal behaviors or actions. That effect may only be apparent where there is the danger of the *acontextual* use of personal information and data. Hence, it is imperative that the idea of a "zone of privacy" around one's personal and private affairs should be fostered and encouraged so that the onus is on those who intrude into the zone to justify their conduct. This zone will then demarcate a boundary to a private and a "public" area with a nebula in between where everyone can interact and relate with each other and in which technology can be freely used. Privacy need no longer be "too indefinite a concept to sire a justiciable issue" (Tapper 1989, p. 325).

While technology will continue to be both a problem and a solution, technological advances such as LBS, geoinformatics, and GISc will continue to push the privacy envelope. But, technological means alone cannot help manage and enhance privacy protection, legislation, corporate policy, and social norms may, in the final analysis, eventually dictate the use of location information generated from tracking devices and geospatial technologies.

Fair information practices are the cornerstone of many privacy laws today. However, these practices may be found wanting, especially when they have to deal with data manipulation using disparate databases joined together in geospatial technologies such as a GIS. The solutions may lie in a mix of international standards, self-regulation, legislation, and government policy. While the harmonization of laws and regulations and achieving consistency of privacy protection, especially across all jurisdictions, is very difficult, yet, international standards must of necessity emerge. One way forward would be to keep canvassing for a global convergence of privacy regulation. It is may be counter-productive for each country to impose a separate privacy regime. We are all responsible for keeping an eye on the world in order to prevent abuse of surveillance technologies not by government regulation but by a mutual, shared responsibility for the world in which we live and disdain for those who abuse and misuse the privilege (Waters 2000).

ENDNOTES

1   See Beckman Center for Internet and Society and in particular the course syllabus for "Privacy in Cyberspace" at http://eon.law.harvard.edu/privacy99/syllabus.html.

2   Olmstead v. US, 277 US 438 (1928).

3   See Legal Information Institute at http://lii.law.cornell.edu/ and the *Concise Oxford Dictionary* (Oxford, Clarendon Press) for definitions of *ethics* relating to morals, treating of moral questions; morally correct, honourable.

4   Australian Law Reform Commission (ALRC) 1983 *Report No. 23, Privacy*. Canberra: AGPS, p. xliv.

5   UDHR 1948 *Universal Declaration of Human Rights*, December 10, 1948, Article 12 at http://www.un.org/Overview/rights.html and Article 17 of the ICCPR 1976 *Inter-*

national Covenant on Civil and Political Rights (New York, United Nations) at http://www.privacy.org/pi/ intl_orgs/un/international_covenant_civil_political_rights.txt.

6 The First Amendment guarantees freedom of communications and the expression of ideas; the Fourth Amendment guarantees freedom from unreasonable search and seizure, including (in some cases) electronic, aural, visual, and other types of surveillance; the Fifth Amendment guarantees freedom from self-incrimination, and guarantees due process of the law with regard to the Federal government; the Ninth Amendment recognizes that rights not specified in the Constitution are vested with the people; and the Fourteenth Amendment guarantees due process and equal protection of the law with regard to the states.

7 The Fourth Amendment of the US Constitution provides that "[t]he right of the people to be secure in their persons, houses, papers, and effects, against unreasonable searches and seizures, shall not be violated, and no Warrant shall issue, but upon probable cause, supported by Oath or affirmation, and particularly describing the place to be searched, and the persons or things to be seized."

8 Griswold v. Connecticut, 381 US 479, 14 L.Ed.2d 510, 85 S. Ct. 1678 (1965).

9 Victoria Park Racing and Recreation Grounds Co Ltd v. Taylor (1937) 58 CLR 479; 43 ALR 597 (HCA).

10 R v. Khan [1997] AC 558 at pp. 582–3.

11 Australian Broadcasting Corporation v. Lenah Game Meats Pty Ltd [2001] HCA 63, November 15, 2001.

12 R v. Broadcasting Standards Commission; *ex parte* British Broadcasting Corporation; [2000] 3 WLR 1327; [2000] 3 All ER 989.

13 Douglas v. Hello! Ltd [2001] 2 WLR 992; [2001] 2 All ER 289 per Sedley LJ at p. 120.

14 Aubrey v. . . . ditions Vice-Versa Inc. [1998] 1 SCR 591; Govind v. State of Madhya Pradesh (1975) 62 AIR (SC) 1378; *P v D* [2000] 2 NZLR 591.

15 *Canadian Tort Law*, 6th edn (1997) at 56; Aubry v. Duclos (1996) 141 DLR (4th) 683.

16 *Restatement of the Law (Second) Torts* § 652A.

17 See Howell v. New York Post Co., 596 N.Y.S.2d 350; 612 N.E.2d 699 (Ct. App.) (1993).

18 OECD 1980 *Guidelines on the Protection of Privacy and Transborder Flows of Personal Data, Recommendation by the OECD Council of 23 September 1980.* See http://www.oecd.org/document/18/0,2340,en_2649_34255_1815186_1_1_1_1,00.html, http://www.oecd.org/document/18/0,2340,en_2649_34255_1558954_1_1_1_1,00.html, and http://www.oecd.org/documentprint/0,2744,en_2649_201185_15589524_1_1_1_1, 00.html. See also EEC 1980 *Council of Europe Convention for the Protection of Individuals with regard to Automatic Processing of Personal Data*, Brussels, EEC and Council of Europe 1981 *Council of Europe Convention for the Protection of Individuals with regard to Automatic Processing of Personal Data*, Brussels, CE (at http://www.privacy.org/pi/intl_orgs/coe/dp_convention_108.txt).

19 United Nations 1990 *Guidelines for the Regulation of Computerized Personal Data Files* at http://www.datenschutz-berlin.de/gesetze/internat/aen.htm.

20 See FGDC Policy on Access at http://www.fgdc.gov/Communications/policies/policies.html.

21 Bradley v. Wingnut Films Ltd [1993] 1 NZLR 415; *P v D* [2000] 2 NZLR 591.

22 See Lord v. McGregor (2000) 50 CCLT (2d) 206; [2000] BCSC 750.

23 See Information Infrastructure Task Force Privacy Working Group at http://www.ntia.doc.gov/ntiahome/privwhitepaper.html. See also Federal Trade Commission (FTC), *Privacy Online: A Report to Congress* (1998) at http://www.ftc.gov/reports/privacy3/index.htm and FTC *Privacy Online: Fair Information Practices in the Electronic Marketplace* (May 2000) at http://www.ftc.gov/os/2000/05/index.htm#22.

24   European Union Data Directive 1995 *Directive on the Protection of Individuals with regard to the Processing of Personal Data and on the Free Movement of such Data*, Brussels, European Commission, Directive 95/46/CE, *Official Journal of the European Commission* (L 281) and at http://ec.europa.eu/justice_home/fsj/privacy/docs/95-46-ce/dir1995-46_part1_en.pdf and http://ec.europa.eu/justice_home/fsj/privacy/docs/95-46-ce/dir1995-46_part2_en.pdf.

25   See Australia, *Privacy Amendment Act* 2000 (Cwlth) approved December 22, 2000; http://www.privacy.gov.au.

26   *Official Journal of the European Commission* L 215 of August 25, 2000, pp. 1, 4 and 7, respectively.

27   See http://europa.eu.int/comm/external_relations/canada/summit_12_99/e_commerce.htm and *Official Journal of the European Commission* L 002, 04/01/2002, pp. 0013–6. See also the EU "adequacy" standard agreement at http://europa.eu.int/comm/internal_market/en/dataprot/wpdocs/wp39en.pdf.

28   Safe Harbor Principles are available at http://europa.eu.int/comm/internal_market/en/dataprot/news/shprintiples.pdf.

29   A search of those firms classified under the Standard Industrial Classification (SIC) code 7374 for companies engaged in marketing and business research services yielded approximately 50 companies.

30   Each year in Australia Sensis/Telstra produces about 55,000 tonnes of directories or 18 million sets of Yellow Pages and White Pages of which 80 percent is recycled. But that still means about 11,000 tonnes go into landfills each year. See Lowe (1994).

31   Because of an ongoing litigation between Sensis (Telstra) and Australia On Disc, both parties reached an out of court settlement that forced the publisher of AOD to immediately stop distributing the product as of December 29, 2003. See http://www.bsgi.biz/aodspecial.htm and Hull (1994).

32   Also known as Intelligent Vehicle Highway Systems (IVHS), and used interchangeably here.

33   See Internet Engineering Task Force website at http://www.ietf.org.

34   IETF 2000 "E.164 Number and DNS" RFC 2916 September 2000 at http://www.ietf.org/rfc/rfc2916.txt.

35   The 112 emergency number is incorporated in the international global system for mobile communications (GSM) and can be dialled from anywhere in the world where there is GSM coverage with the call transferred to that country's primary emergency call service number. The 106 emergency calls service number is for the exclusive use of text-based telecommunication users, especially the hearing or speech impaired. See Australian Communications Authority (ACA) 2004 *Location, Location, Location*, January, Melbourne, ACA.

36   See FCC requirements at http://www.fcc.gov/911/enhanced/.

37   The White House, Office of the Press Secretary 2000 "Statement by the President regarding the United States' decision to stop degrading global positioning system accuracy," May 1 at http://www.ostp.gov/html/0053_2.html.

38   In Japan the guidelines on the protection of personal data in telecommunications business established a clear standard for consent to use of location information. In May 2003 the Diet passed a package of bills known as the Personal Data Protection Law that codifies the requirement for informed opt-in consent (Ackerman, Kempf, and Miki 2003).

39   See Webopedia definition of RFID at http://www.webopedia.com.

40   106 S. Ct. 1819, 90 Led 2d 226 (1986).

41   106 S. Ct. 1809 (1986).

42   488 US 445 (1988).

43   773 F. Supp. 220 (D. Haw. 1991).

44   United States v. Knotts 460 US 276 (1983); United States v. Karo 468 US 705 (1984).

45   No. 91-5077 5th Cir. November 12, 1992.
46   Anonymity services make use of identifiers for the purpose of records or transactions
     whereas pseudonymity services are those where one cannot, in the normal course of events,
     be associated with a particular individual. When an person uses a pseudo-identifier then
     a digital persona, *e-pers* or a *nym* is born; this is the model of an individual's published
     personality, based on data, maintained by transactions and intended for use as a proxy
     for that individual. See Clarke (1999b).

REFERENCES

Ackerman, L., Kempf, J., and Miki, T. 2003. Wireless Location Privacy: Law and Policy in the
    U.S., E.U., and Japan. WWW document, http://www.isoc.org/briefings/015/index.shtml.
Ball, M. 2003. Concerning ourselves with privacy. *GIS World* 16(2) (available at http://
    www.geoplace.com/gw/2003/0302/0302ed.asp).
Campbell, D. and Fisher, J. (eds). 1994. *Data Transmission and Privacy*. Dordrecht: Martinus
    Nijhoff.
Clarke, R. 1999a. Introduction to Dataveillance and Information Privacy, and Definition of
    Terms. WWW document, http://www.anu.edu.au/people/Roger.Clarke/DV/intro.html.
Clarke, R. 1999b. The Legal Context of Privacy-enhancing and Privacy-sympathetic Techno-
    logies. WWW document, http://www.anu.edu.au/people/Roger.Clarke/DV/Florham.html.
Clarke, R. 2001. The end of privacy: While you were sleeping . . . surveillance technologies
    arrived. *AQ: Journal of Contemporary Analysis* 73: 9–14.
Clarke, R. 2002. ENUM. WWW document, http://www.anu.edu.au/people/Roger.Clarke/
    DV/Enum.html.
Curry, D. J. 1992. *The New Marketing Research Systems: How to Use Strategic Database
    Information for Better Marketing Decisions*. New York: John Wiley and Sons.
Curry, M. R. 1994. In Plain and Open View: Geographic Information systems and the Problem
    of Privacy. WWW document, http://www.spatial.maine.edu/tempe/curry.html.
Dobson, J. 1998. Is GIS a privacy threat? *GIS World* 11(7): 34–5.
Dobson, J. 2000. What are the ethical limits of GIS? *GeoWorld* 13(5) (available at http://
    www.geoplace.com/gw/2000/0500/0500g.asp).
Equifax National Decision Systems. 1993. *InfoMark-GIS: Tomorrow's Technology for Today's
    Business Success*. Atlanta, GA: Equifax, Inc.
Flaherty, D. H. 1994. Privacy Protection in Geographic Information Systems: Alternative
    Protection Scenarios. WWW document, http://www.spatial.maine.edu/tempe/flaherty.html.
Goss, J. D. 1994. Marketing the new marketing: The strategic discourse of Geodemographic
    Information Systems. In J. Pickles (ed.) *Ground Truth: the Social Implications of
    Geographic Information Systems*. New York: Guilford Press: 130–70.
Goss, J. 1995. We know who you are and we know where you live: The instrumental ration-
    ality of Geodemographic Systems. *Economic Geography* 71: 171–98.
Handelsmann, A. 2001. Strategies for Complying with Australia's Privacy Principles. WWW
    document, http://www.gigalaw.com/articles/2001/handelsmann-2001-11-p1.html.
Harvey, J. A. and Verska, K. A. 2001. What the European Data Privacy Obligations Mean for
    U.S. Businesses. WWW document, http://www.gigalaw.com/articles/harvey-2001-02-p1.html.
Hull, C. 1994. "Privacy question in phone-book CDs," *The Canberra Times*, August 1, p. 13.
International Telecommunications Union (ITU). 2001. ENUM. WWW document, http://
    www.itu.int/osg/spu/enum/index.html.
Krause, B. 2001. An Overview of the Canadian Personal Information Protection and Elec-
    tronic Documents Act. WWW document, http://www.gigalaw.com/articles/2001/krauser
    2001-02.html.

Leipnik, M., Bottelli, J., von Essen, I., Schmidt, A., Anderson, L., and Cooper, T. 2001. Coordinates of a Killer. WWW document, http://www.geoinfosystems.com/1101/101spokane.html.

Lowe, S. 1994. "Indecent disclosures," *The Sydney Morning Herald*, March 21, p. 47.

Orwell, G. [1949] 1990. *Nineteen-Eighty-Four*. New York: New American Library, Inc.

Rowe, H. and McGilligan, R. 2001. Data protection: Location technology and data protection. *Computer Law and Security Report* 17: 333–5.

Tapper, C. 1989. *Computer Law* (4th edn). Harlow: Longman.

Tobin, R. 2000. Invasion of privacy. *New Zealand Law Journal* 216.

Warren, S. and Brandeis, L. 1890. The right to privacy. 4 *Harvard Law Review* 193.

Waters, N. 2000. GIS and the bitter fruit: Privacy issues in the Age of the Internet. *Geoworld* 6(5) (available at http://www.geoplace.com/gw/2000/0500/0500edg.asp).

Westin, A. F. 1967. *Privacy and Freedom*. New York: Atheneum.

Westin, A. F. 1971. *Information Technology in a Democracy*. Cambridge, MA: Harvard University Press.

Yu, P. 2001. An introduction to the EU Directive on the Protection of Personal Data. WWW document, http://www.gigalaw.com/articles/2001/yu-2001-07a-p1.html.

Chapter 30

Geographic Information Systems in Education

Joseph J. Kerski

As Geographic Information Systems (GIS) quietly transformed decision making in universities, government agencies, industry, and nonprofit organizations, demand for GIS education has mushroomed. During the 1970s, GIS education, along with the development of GIS software, was proceeding at the Computer Graphics Laboratory at Harvard University in the USA and the Experimental Cartography Unit of the Royal College of Art in the UK. Advanced students and professors learned how to integrate traditional theories about spatial information, computational geometry, and computer science into a set of basic concepts useful for the computer processing of spatial information (Coppock and Rhind 1991).

From these beginnings, GIS technology developed more rapidly than the corresponding educational opportunities. Once GIS became a rewarding commercial venture, software vendors established extensive training programs in their own software. During these early years, education about GIS was largely synonymous with professional development, focusing on those who had already completed their formal university education. People learned about GIS to become more familiar with software tools so that they could *apply* GIS methodology on the job. GIS professional development mirrored the development of GIS itself, beginning with natural sciences in the 1970s, expanding to urban planning and business during the 1980s, and by the 1990s into virtually every major career path.

Between 1985 and 1992, the advent of Idrisi, MapInfo, PC Arc/Info and ArcView desktop software diffused GIS within organizations. Increased computing capabilities and more powerful software attracted additional users. Spatial data sets became more accessible and available, beginning with LANDSAT and SPOT satellite imagery, USGS Digital Line Graphs, Digital Elevation Models, commercial satellite imagery, UK Ordnance Survey data, Digital Orthophotoquads, and Census TIGER files (see Chapter 1 by Cowen in this volume for additional details about present-day spatial data resources). GIS users finally had a rich source of data to use as base layers with data sets they collected themselves.

By 1992, a research base for GIS had been established with strong ties to the disciplines of geography, cartography, geodesy, computer science, and remote sensing.

Fig. 30.1 PC monitor in Texas high school showing series of themes for Africa

With the maturing of Geographic Information Science (GISc) (Goodchild 1992), education became more complex. People were still interested in learning about GIS *applications* to address real-world societal issues and problems. However, others developed an educational framework to learn about GISc as a discipline. Others examined GIS education in the framework of research about GISc. Still others sought to use GIS as a tool and method in education, to teach geography, environmental studies, history, and other disciplines.

Today, GIS education is in demand more than ever as spatial tools have become widely available as desktop clients and over the web. The integration of GIS, Global Positioning Systems (GPS), and remote sensing tools into standard office productivity software and in everyday devices such as mobile telephones and in-vehicle navigation systems fuels the demand. This chapter examines the history and spectrum of GIS education, including the major developments and organizations involved, and opportunities for educating oneself in GIS.

GIS in education can be conceptualized in four distinct ways: (1) professional development, (2) research *about* GIS, (3) *teaching about* GIS, and (4) teaching and learning *with* GIS. Teaching and learning about GIS (after Sui 1995) takes place on three well-developed levels involving thousands of people, organizations, and programs: (1) professional development, or skill-building, (2) research about GIS, and (3) teaching about GIS. The goal of teaching and learning about GIS is to become familiar with the theories concerning GISc and the acquisition of skills to manage GIS and operate GIS software. Courses emphasize topics such as topology, data structures, database management, map scale and projections, data quality, and generalization.

By contrast, teaching and learning *with* GIS is smaller in scope. It focuses not on GIS, but on the disciplines that are home to the issues being addressed, using GIS as tools and spatial analysis to understand the Earth. This group of practitioners has a research base and has experienced modest growth since 1990.

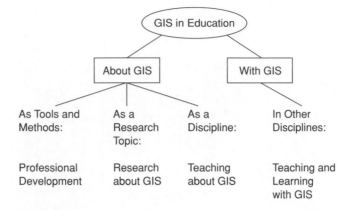

Fig. 30.2 Dimensions of GIS education

Much GIS education occurs outside of educational institutions in government agencies and private GIS software companies. Inside educational institutions, teaching *about* GIS dominates at the university level, where courses in methods and theory of GIS are taught. However, it has made considerable inroads in various disciplines across university campuses during the past ten years, and many courses and programs in both IT and the environmental sciences now incorporate GISc concepts and tools. Teaching *with* GIS began and still dominates at the primary and secondary level, where it is used as an instructional method in established subject content areas. Through such initiatives as the National Institute for Technology in Liberal Education, teaching with GIS is expanding at the university level in history, language, business, and even art.

GIS vocations continue to expand, including system administrators and managers, application developers, database designers, spatial analysts, and researchers. The types of jobs a GIS professional may do suggests an equally long list of knowledge areas that might be addressed in GIS education: general awareness of the technology and applications, systems operation, computer science, management skills, spatial information, and geographic science, to name a few. The appropriate combination of these different knowledge areas varies with the professional activity. For example, a system operator needs to know much about system operation, but may not require extensive exposure to management skills.

The combinations of skills and knowledge may be best understood by identifying pedagogic dimensions that need to be considered when planning and implementing a GIS curriculum (Figure 30.2). One dimension contrasts teaching technical *skills* in system operation against teaching basic *concepts*. This is often referred to as *training* (skills) versus *education* (skills *and* concepts). A second dimension contrasts emphasis on GIS *theory* versus an emphasis on the *applications* of GIS. A third dimension places education about the *management* of GIS at the opposite pole from education about the *use* of GIS. While these dimensions overlap, recognition of the fundamental pedagogic differences between each is useful during course planning.

We now turn to the first component of teaching about GIS, the one that many people associate with GIS education – professional development.

GIS Education as Professional Development

Need

Most people who want to learn about GIS for professional development reasons have two motivations. One is to get a job. Despite the ups and downs of most information technology (IT), GIS has been expanding at over 15 percent per year since the 1980s. Many students believe that having a record of successful completion of one or more GIS courses will help them obtain their first job. Other people seek out GIS as a possible new career field when their former professions become obsolete.

However, a much larger number of people now find that they need to learn about GIS because the technology is becoming part of their current job. In addition to those who want to learn about GIS, there are also many who *need* to learn about GIS but do not realize it. Politicians, administrators, and the general public increasingly interact with GIS or its products. They need to know what GIS is and how it can be used so that they can make informed decisions about acquiring and maintaining it in their organizations. How can education, awareness, and training be designed to meet their needs? How can we create a spatially literate society? Such a society would ultimately provide the base necessary to support the types of research and teaching that the GISc community advocates.

Therefore, in terms of professional development, three groups of people exist who want or need to study GIS: (1) people who want to be GIS experts; (2) people who want to know about GIS as a supplement to their job duties; and (3) people who *should* know about GIS because they will make decisions about its implementation or because they are using information derived from a GIS.

Historical development

Up through 1990, GIS was not considered as a discipline, but rather as a powerful new set of computer tools. The overriding need of those seeking GIS education was professional development. To use GIS on the job they needed some theoretical background, but the bulk of the training required was commands and procedures specific to a particular type of software.

Owing to the complexity of GIS software, most GIS trainers emphasized one type of software and one type of data model because early GIS software tended to emphasize either the vector or raster model. GIS software companies taught much GIS, and therefore, training tended to develop along software lines, dividing the user community to this day.

Because of the lack of base data, GIS professional development also emphasized collecting data. Digitizing tablets and large format scanners were ubiquitous in early labs, and many GIS professionals today have grim memories of beginning their careers crouched over these large pieces of equipment!

Professional development training was conducted in hands-on mode in expensive computer laboratories. Even graphics monitors such as the Tektronix were specially built for GIS users and cost thousands of dollars. Consequently, the subject was largely taught by employees of large government organizations and universities that could afford the training facilities. GIS was run on mainframe and minicomputers in raised-floor, closed-door computer laboratories; the software was difficult to learn, dominated by punch card and command-line input. GIS naturally became the specialty of just a few staff in each organization.

The GIS community quickly realized that personal networking was essential to each person's success as a GIS professional. AutoCarto and GIS/LIS provided annual professional development opportunities for GIS professionals. These conferences provided networking through workshops, presentations, and exhibits. Professional societies such as AM/FM International (Automated Mapping and Facilities Management) from the utilities industry, later the Geospatial Information Technology Association (GITA), and Urban and Regional Information Systems Association (URISA) from city planners interested in GIS began holding their own conferences.

Current status

Since 1990, professional development has been characterized by growth, specialization, and increased reliance on web-based tools. Hundreds of consultants working alone or in large companies provide training for tens of thousands of users annually. No longer confined to a limited number of expensive laboratories, GIS training takes place in many venues – on laptop computers, mobile hand-held devices in the field, and in field vehicles. Blurring boundaries between GIS and Remote Sensing and Computer Aided Drafting (CAD) has further widened GIS professional development. Many trainers focus on specific GIS applications such as law enforcement or business applications.

Online training is commonplace. ESRI's Virtual Campus now offers over 60 courses and has seen 250,000 users since its launch in the late 1990s. The UNIGIS International Network is a consortium of more than 20 universities spread across 14 countries on four continents, offering graduate degrees and certificates with over 1,400 students enrolled at any one time. Pennsylvania State University, the University of Redlands, University of Denver, and others offer distance courses and degrees in GIS.

Today's GIS professional development still tends to divide along a single brand of GIS software. The field is dominated by ESRI (ArcGIS and others), Intergraph (GeoMedia), MapInfo, Caliper (Maptitude), and Clark Labs (Idrisi). Accordingly, GIS software companies still provide much of the training. Finally, professional development still emphasizes hands-on acquisition of skills needed to accomplish specific tasks using a chosen brand of software.

GIS education as professional development is an enormous industry, involving private companies; regional, national, and international conferences; professional societies; government agencies; and nonprofit organizations. In a 1990s survey, 49 percent responded that they had received their GIS training on the job, 22 percent from a workshop, 19 percent from university courses or their university degree, and 4 percent through a certificate program (Huxhold 1999). Companies involved in GIS education range from sole proprietorships to large companies such as ESRI and Intergraph.

Despite web-based meetings and training, face-to-face GIS conferences continue to serve as one of the primary means of GIS professional development. GIS is not just about hardware, software, and methods, it is about *people*. Human networking is the primary reason for the rapid growth of GIS throughout society, made more rapid by the enthusiastic nature of the GIS user community that is eager to share information and expand their field.

As evidence, the annual ESRI International User Conference has grown in 25 years to 13,000 attendees from 150 countries who give over 1,200 papers annually. The Association for Geographic Information's (AGI's) conference in the UK, and India's annual MapIndia event are just a few of the global events that occur each year. GeoPlace, Intergraph, MapInfo, URISA, and GITA's annual GIS conferences each attract thousands, and, in addition, most regions and states hold their own GIS conferences and workshops. The annual conferences of the Association of American Geographers (AAG), American Planning Association, and others now include significant GIS strands. Theme-based GIS conferences such as in health and national security are increasing.

Content

In 1993, the US Department of Labor added "GIS Specialist" to its Dictionary of Occupational Titles (Huxhold 1999). Its skill sets form the basis of much GIS professional development: (1) information technology, including database design, data management, user interface design, requirements definition, graphics, and image processing; (2) geography and cartography; (3) GIS technology; (4) other skills, including coordinate geometry, topology, and presentation and training skills.

In all areas of education a debate exists about how much theory should be taught versus skills. For GIS professionals, while the emphasis is on skills, theory provides context. For example, changing map projections in GIS menus does not make sense unless the operator understands why map projections are necessary and their inherent advantages and disadvantages. When spatial data are in digital form, the potential for misuse increases with the numbers of users and applications. Because of the power of the computer to manipulate, combine, and derive data, users can perform these even if the combinations and derivations make no sense, and hence, theory is included in education.

International GIS education

As most of the development of GIS began in North America and the UK, GIS education diffused to the world using the model developed by these countries. As free trade agreements mature in the Americas and in Europe, professionals in all fields begin to cross national borders. As a result, people are asking how we can ensure that GIS professionals are sufficiently qualified for international work, and how qualifications are recognized internationally.

Regional diversity in GIS education is suggested with some countries placing GIS within the "cadastral" sphere and others placing it more in the "geographic" context. Disciplinary differences are often reflected in the choice of data model. For example, GIS specialists working with transportation systems require topologically structured data models while ground water specialists will more likely depend upon raster models. Clearly, each GIS professional must determine his or her own educational needs given this diversity. This has led some professional organizations to campaign for a professional development framework that indicates what a GIS professional should know. Once formalized, individuals can use the framework to organize their own educational requirements. GIS certification is one attempt to meet this need.

Certification

Over the past decade, the GIS community has been asking itself whether it should establish a professional certification in the field. Many argued that because GIS represented a specialized body of knowledge, had a mission, had formal organizations, required specialized training, and had its own culture, it qualified as a profession (Huxhold 1999) as identified by Pugh (1989) and Obermeyer (1993). Increased recognition, salary, and knowledge are cited as benefits to certification. Efforts by URISA, AGI, the International Standards Organization, University Consortium of Geographic Information Science (UCGIS), ASPRS, and others culminated in the establishment of a certification process. The nonprofit GIS Certification Institute (www.gisci.org) operates the program; 2003 saw the first graduates of the program and the North Carolina Information Coordinating Council became the first state to endorse GISCI's certification program in December 2004.

GIS Education as Research about GISc

Research about GISc provides guidance *to* professional development and becomes the theoretical basis for GISc. Because this provides the content of what GISc *is*, research about GISc therefore dictates what GIS is *taught*, and what GIS is *learned*. Research about GISc represents the breadth of GIS – software, teaching, learning, tools, and methods. There is a difference between how GIS is used *for* research and how it is used *in* research. In some cases, GIS itself is the focus of research, and in others, GIS is a tool and way of thinking that assists research in other domains. Research in teaching and learning about GIS forms an important part of the agenda of the investigators, many of whom have cognitive science backgrounds.

In 1988, the National Science Foundation in the USA established the National Center for Geographic Information and Analysis (NCGIA) at the University of California-Santa Barbara, State University of New York at Buffalo, and University of Maine. Through the mid 1990s, NCGIA researchers coordinated much of the initial research in GISc.

The UCGIS was launched in 1995 to serve as an effective, unified voice for the GISc research community, to foster multi-disciplinary research and education, and to promote the informed use of GISc for the benefit of society.

Some of the rapidly expanding GISc research focuses on the analysis, design, visualization, and generation of various forms of maps, and how people think about their geographic surroundings. Software developers use the research generated by these scientists to make GIS more powerful and easier to use. The education priorities of these researchers include emerging technologies for delivering GIS education, including distance education, supporting infrastructure to ensure that education happens, access and equity (Pisapia 1994), alternative designs for curriculum content and evaluation, professional education, learning with GIS, certification, and accreditation.

Beginning in 2000, the NCGIA, Association of American Geographers (AAG), Association of Geographic Information Laboratories in Europe (AGILE), and UCGIS have organized a biennial conference entitled, simply, "GIScience." The GIScience 2000, 2002, 2004, and 2006 conferences became a key means of the education of the GISc research community.

GIS Education as Teaching about GISc

By the mid 1980s, it was clear that there was a growing scarcity of people educated in this rapidly spreading technology. Only one textbook in the field existed, that of Burrough (1986). This led to the initiation of a US National Science Foundation funded curriculum development project that culminated in the development of the *NCGIA Core Curriculum in GIS*. The project was founded on the premise that growth of GIS education opportunities could be encouraged by the preparation and distribution of materials designed to help university instructors develop introductory courses. The Core Curriculum project sought to bring together and formalize a wide-ranging set of topics, and included volumes focused on introductory, technical, and application issues in GIS. By 1995, over 1,300 copies of the 1,000-page curriculum had been distributed to faculty and practitioners in over 70 countries (Kemp 1998). The *NCGIA Core Curriculum in GIS* became the recognized guideline that educators should use in teaching about GISc and for several years, it was really the *only* comprehensive curriculum in the field.

Curriculum development is a difficult, labor-intensive process, particularly so in a new, complex field such as GISc. Jenkins (2000) describes curriculum development as "an interaction between aims and objectives, methods of assessment, teaching methods and content." Educationalists often insist that curriculum development must be driven by a set of objectives – "what one expects students to know or do as a result of a particular course." Nevertheless, other efforts made to determine generic, "core," or "model" GIS curricula began. The Royal Institution of Chartered Surveyors (RICS) in the UK funded a workshop that produced a remarkably similar syllabus for the teaching of GIS. The Canadian geomatics industry survey was one of many conducted to determine the basic skill set for geomatics practitioners. The first extensive study to determine the necessary components for an international European postgraduate GIS course was conducted by the Technical University of Vienna in 1993. Intensive formal courses were organized by international agencies such as the United Nations Institute for Training and Research, by national training agencies such as FORMEZ in Italy and by universities such as the Technical University in Vienna (Kemp 1995).

Teaching about GISc is almost exclusively the domain of higher education. Those who teach about GISc in educational institutions face difficulties that some in industry and government do not experience. GIS involves a great investment of time for university faculty, as they not only have to teach the concepts and skills, but also must prepare and test lessons, download and format spatial data, and a myriad of other technical and pedagogical tasks. This confines it to small numbers of faculty.

Furthermore, as GIS education becomes more popular, diffuse, and directed at an increasing diversity of audiences, there is less cohesiveness and agreement on what needs to be taught and when.

Despite these challenges, GIS education has become well established in geography departments and/or programs in nearly every university and community college in North America, Australia, and Europe, is expanding in other departments on these continents and added to universities in Asia, South America, and Africa. In some universities, GIS is introduced in a geography fundamentals course, where students are exposed to a range of geographic techniques. In most universities, GIS is taught as a skills course, in a manner similar to the way statistics or computer sciences are taught to non-majors. Some graduate programs in GIS focus on learning how to apply the software and completing a detailed project, whereas others concentrate on theory.

A three-volume reference book on *GIS Principles and Applications* was published in 1991 (Maguire, Goodchild, and Rhind 1991) and widely used. Today, with at least 60 major GIS textbooks now in existence, GIS is starting to be considered a foundation subject across a broad range of disciplines. Indeed, there is more evidence that spatial literacy and geographic problem solving are core educational needs throughout society, as documented in the report on GIS across the curriculum published in 2006 (NRC 2006).

What should GIS courses teach?

Any education program must begin with a clear recognition of the audience it is intended to reach and their needs. While learning about the technology drives many to learn GIS, all GIS educators must impress upon their students important basic principles that transcend the technology. First, because they are making maps, GIS specialists need to know some principles of cartography – map projections, scale, and symbology. Producing a choropleth map has become simple, but when is a quantile classification preferable over a Jenks natural breaks classification?

Second, GIS specialists should learn something about geography, the science of space and place on the Earth's surface. Geographers describe the changing patterns of places in words, maps, and images, explain how these patterns come to be, and unravel their meaning. GIS is powerful because it depends upon and emphasizes the very principles geographers have been studying for centuries.

Third, geographers have long been developing sophisticated quantitative techniques for analyzing how location affects components of the human and physical environment. Since spatial analysis predated GIS, the GIS specialist needs to understand these techniques so they can think "outside the box" when grappling with complex problems. Geographically referenced data is unique because of spatial autocorrelation that recognizes that places close together are more similar than places far apart. Many traditional statistical techniques that assume independence between observations or samples are invalid when used on geographic data. This problem has led statisticians to develop a wide range of spatial statistical techniques, which require knowledge of the underlying theory before they can be properly interpreted for other people's use (see Chapter 22 by Jacquez in this volume for examples of spatial statistical techniques that can be implemented inside GIS).

A GIS curriculum should include spatial information concepts, skills, and representation, determining and representing location, modeling reality, data sources, development and applications of GIS, and needs analysis. It should emphasize database issues, spatial data, spatial analysis, systems design, organizational implementation, project management, and GIS in society. GIS curricula should include a practical project of the student's own choosing.

GIS education reinforces skills such as information retrieval, independent work, working comfortably with a computer, organizing team projects, working with professionals in the field, solving problems, managing a project, educating others, speaking and writing, using graphics, and, perhaps most important of all, being critical of and knowing the limitations of information, particularly on a computer.

We do not yet know with complete certainty what core skills and knowledge are needed by those who use GIS. Furthermore, the approach changes depending on the discipline. One needs to consider the spatial concepts relevant within those other disciplines and how they are implemented, modeled, and used there. The UCGIS has adopted this view in developing the model curricula in Geographic Information Science and Technology (GIS&T). The current draft emphasizes curricular paths that incorporate a number of common features but lead to significantly different outcomes for the undergraduate. The goal is curricula that will result in more highly and relevantly educated graduates, greater consistency in GIS&T degree-granting programs, and increased communication across academic disciplines with an interest in GIS&T (UCGIS Model Curriculum Task Force 2003).

GIS Education as Teaching with GIS

Why teach with GIS?

The goal in teaching *with* GIS is to use GIS to help students understand a problem or issue in geography, chemistry, biology, environmental science, mathematics, business, history, and other disciplines (Barstow 1994, Fazio and Keranen 1995, McGarigle 1997). Factors encouraging the use of GIS in education include the educational standards movement, pedagogy, workforce development (Braus 1999), and holding public education fiscally accountable.

Educational systems in many countries have been moving toward educational content standards and national curricula, specifying what students should know and be able to do at certain benchmark educational levels. For example, the UK's National Curriculum, and the USA's national content standards in geography (Geography Education Standards Project 1994), social studies (National Task Force for Social Studies Standards 1994), science (NRC 1996), and technology (International Society for Technology in Education 2000) state that students must use real-world tools in the same "hands-on" manner as a scientist would to solve real-world problems. Because GIS was created as a problem-solving tool, it finds a natural home in the curriculum.

Furthermore, educators have progressed toward a model of "inquiry-based" instruction that emphasizes hands-on, research-based learning experience. Inquiry draws upon learning theory known as constructivism, which holds that rather than being transferred from teacher to student, knowledge is constructed by the learner based on his or her own experiences and making connections (Driver, Asoko, Leach, Mortimer, and Scott 1994). The USA's national geography standards state that "the power of a GIS is that it allows us to ask questions of data" (Geography Education Standards Project 1994, p. 256). Students using this inquiry approach form research questions, develop a methodology, gather, and analyze data, and draw conclusions.

The use of GIS in education is increasingly viewed as active learning that engages students in critical thinking skills. Students are attracted to the visual learning that GIS offers. Its interdisciplinary character is appealing: Many argue that inter-disciplinary education, rather than teaching each subject in isolation, may be a more effective means to help students solve problems (Jacobs 1989). Some educators feel that implementing GIS into the curriculum may encourage students to examine data from a variety of viewpoints (Furner and Ramirez 1999, Sarnoff 2000). GIS allows "authentic assessment," in which instructors evaluate student performance based on assessing student portfolios or projects, rather than examinations.

Educators worldwide bemoan the fact that young people feel disengaged from local and national decision-making. Some view GIS as helping students to engage in community-based issues such as the siting of landfills, urban sprawl, water quality, crime, energy, and transportation, and thus to achieve excellence in citizenship education (Kerski 2004).

Rather than using GIS to add *more* information to a project, educators advocate that it can help students to use information more effectively to arrive at a decision by using generative, rather than inert knowledge (Dede 1995). GIS is used to make decisions given incomplete information, inconsistent objectives, and uncertain consequences, reflecting the world outside of the classroom. Sustainable lessons that can be easily used, modified, and transferred are those most in demand.

Concurrent with these developments in pedagogy has been a renewed emphasis on preparing students to become productive members of the workforce. The US Labor Secretary's Commission on Achieving Necessary Skills (SCANS) stated that the most effective way to teach skills is in the context of an established subject (US Department of Labor 1991). SCANS competencies include identifying and using resources, working with others, and understanding complex interrelationships (Hill 1995a, b). These activities mirror the types of tasks that students engage in when they use GIS, termed "authentic practice." School-to-career and workforce development grants are more commonplace and likely to include GIS in the "tool belt" that educators believe students should have by the time they graduate. Information literacy, computer literacy, and the ability to integrate information from different sources are increasingly considered important to teach. GIS is one of the few tools to fully take advantage of the full power of the computer.

Challenges in teaching with GIS

> The trouble with education . . . is that the best teaching methods are in fact the most difficult. (Piaget 1929)

Challenges in teaching with GIS have slowed its implementation in the primary, secondary, and university curricula. Challenges lie more with the structure of educational systems themselves than on software and hardware. Indeed, with the price of a school site license of ArcView GIS for one year at only US$75 on a CD distributed with the *Mapping Our World* book (Malone, Palmer, and Voigt 2002), cost is the least of the concerns. Technology is a "process; a systematic blend of people, materials, methods, and machines" (Ely, Foley, Freeman, and Scheel 1992). In addition, GIS is by definition a *system* – it is both a technology and a set of methods. Integrating GIS into classroom practice is a complex process. Guided inquiry entails changes in the social organization and management of the classroom (Powell 1999) and involves unpredictable results and student-directed learning that many educators are not equipped to deal with (Audet and Paris 1997). The lack of training at the preservice stage slows implementation, because most instructors learn about GIS at the *inservice* stage – when they are already on the job (Boehm, Brierley, and Sharma 1994, Bednarz and Audet 1999).

A survey of 1,520 secondary school teachers who own GIS software in the USA showed that student projects were community-based, fieldwork-based, interdisciplinary, and open-ended, involving ill-structured problems with real-world data (Kerski 2003). However, it also revealed that the challenges facing educators include the lack of training, lack of time in the curriculum, limited access to the computer lab, lack of software for Macintosh computers, inadequate understanding about spatial concepts, and the curricular "fit" of GIS. This study and others (Alibrandi 2003, Baker and Case 2003) made it clear that geotechnologies in education must be used within the context of reform for long-term impact (Means 1994).

Although educational content standards point to the use of tools such as GIS, most assessment instruments emphasize memorizing facts. Using GIS goes beyond facts, focusing on understanding spatial patterns, linkages, and trends, and if students are not scoring higher on standardized tests because of GIS, its use will be restricted (Kerski 2003).

A lack of research on the effectiveness of geotechnologies identified in a 1967 study persists, slowing advancements in teaching with GIS: "the research-oriented geographer and educator have paid scant attention to assessing the effectiveness of one tool over another" (Gross 1967). The organizers of the first conference on educational GIS asked "What is the learning that GIS allows that other ways do not?" (Salinger 1994). Administrators and educators want to see evidence of the benefits of GIS to student learning before they will invest time and effort necessary to implement it (Downs 1994).

Status of teaching with GIS in the curriculum

Despite these challenges, teaching with GIS continues to expand, especially at the primary and secondary level (Figure 30.3). More science teachers in the USA use GIS than geography teachers, reflecting the relative importance of science in the educational curriculum (Kerski 2003); the situation is exactly the opposite in the UK and New Zealand. Using Binko's (1989) four stages of learning – awareness, understanding, guided practice, and implementation – GIS is barely in the awareness phase for most teachers. In a diffusion of innovation model (Rogers 1995), GIS is primarily used by the "early adopters." The publication of the first two textbooks containing curriculum (Malone, Palmer, and Voigt 2002, 2003) hastened curricular adoption.

Fig. 30.3 Student displaying the results of his GIS analysis in 3D

Today, two different philosophies of software use dominate. The first is to use existing off-the-shelf GIS software. Proponents of this method believe that students receive the most educational and career benefits by using the same tools as those used in the workplace. The second group believes that using software created for educational use provides students the benefits of spatial thinking without the technological demands and complexity that accompanies industry-grade GIS software. Examples include Digital Worlds from Canterbury Christ Church University College in England, MyWorld from Northwestern University, and GEODESY (Radke 1999).

Organizations promoting GIS in education are grant-funded organizations, private companies, individuals, professional societies, universities, and government agencies. Active grant-funded organizations include the Center for Image Processing in Education, the Technological Education Research Center (TERC), the University Corporation for Atmospheric Science (UCAR), the Digital Library for Earth Systems Education (DLESE), the Global Learning and Observations to Benefit the Environment (GLOBE) project, the Missouri Botanical Garden, Kansas GIS (www.kangis.org), and the Upper Midwest Aerospace Consortium. Professional societies such as the AGI and Royal Geographical Society in the UK and GITA in the USA include outreach staffs who conduct training and loan GPS equipment to schools. ESRI and Intergraph have staffed education teams since the 1990s. University researchers and government organizations such as the Ordnance Survey in the UK and USGS in the USA are also active in promoting GIS with grant funding, training, curriculum development, and technical support. Individuals and private companies such as GISetc. concentrate on providing GIS training for educators. Informal education – a term for that which occurs outside walls of formal educational institutions – is increasingly important as evidenced by ESRI's support of geotechnologies for the estimated 400,000 National 4-H groups across the USA.

These organizations view the advancement of GIS in education as critical for creating the spatially literate populace needed to solve critical twenty-first-century problems. The approach of all of these organizations is not "How can we get GIS into the curriculum?" but "How can GIS help meet curricular goals?"

The effect of teaching with GIS on education

GIS alters communication patterns and traditional roles of students and teachers, with a greater time spent on small group instruction, coaching, working more closely with weaker students, cooperation, peer-to-peer mentoring, student-to-teacher mentoring, and authentic assessment based on products and progress (Robison 1996, Walker, Casper, Hissong, and Rieben 2000). There is a shift from *covering* material to *sampling* material, and from unilaterally declaring *what is worth knowing* to *discovering what is important*. There is value in requiring students to "dig out" information rather than handing it to them. It is one of the few tools to take advantage of computer, relational, and content skills.

Students are becoming more adept at GIS software, but not necessarily at asking inquiry-based geographic questions. The teacher's role is therefore critical to learning with GIS. Teachers are more likely to adopt GIS if they have previous computer experience, a problem-solving approach, a geographic perspective, a positive attitude towards work change, and active networking and communication skills (Kerski 2003). Instructors need lessons and guidelines, but software proficiency is only a means to the end goal of teaching how to think geographically, scientifically, and environmentally. GIS is an *enabling technology*.

CONCLUSIONS

The demand for GIS education continues to grow. The expansion of GIS technology and applications into many different career paths ensures a continuous supply of new workers in the field and sustained demand for professional development. They need general training and support followed by specialized training in their own application area. These are people who use or would like to use GIS on the job, and seek additional training in the field to become more proficient at GIS tools and methods. New versions of the software and new ways to use the technology means that existing users also require ongoing professional development. Educators are increasingly using GIS for teaching and learning, opening up new demand for professional development for these educators to learn how to teach *with* GIS.

The diversity of the field of GISc is reflected in the diversity of its educational applications and issues. GISc is a growing enterprise and the need for educated practitioners will continue to drive the development of new education opportunities worldwide.

ACKNOWLEDGEMENTS

Comments from Dr Karen Kemp were used to improve the initial draft of this chapter.

REFERENCES

Alibrandi, M. 2003. *GIS in the Classroom: Using Geographic Information Systems in Social Studies and Environmental Science.* Portsmouth, NH: Heinemann.

Audet, R. H. and Paris, J. 1997. GIS implementation model for schools: Assessing the critical concerns. *Journal of Geography* 96: 293–300.

Baker, T. and Case, S. 2003. The effects of GIS on students' attitudes, self-efficacy, and achievement in middle school science classrooms. *Journal of Geography* 102: 243–54.

Barstow, D. 1994. An introduction to GIS in education. In *Proceedings of the First National Conference on the Educational Applications of Geographic Information Systems (EdGIS)*, Washington, DC, Cambridge, MA, TERC, pp. 14–9.

Bednarz, S. W. and Audet, R. H. 1999. The status of GIS technology in teacher preparation programs. *Journal of Geography* 98: 60–7.

Binko, J. B. 1989. *Spreading the Word.* Washington, DC, National Geographic Society.

Boehm, R. G., Brierley, J., and Sharma, M. 1994. The Bete Noire of geographic education: Teacher training programs. In R. S. Bednarz and J. F. Petersen (eds) *A Decade of Reform in Geographic Education: Inventory and Prospect.* Indiana, PA: National Council for Geographic Education: 89–98.

Braus, P. 1999. ABCs of GIS: Education Professionals are Discovering that GIS Software Provides a Powerful Classroom Resource. WWW document, http://directionsmag.com/features.asp?FeatureID=6.

Burrough, P. A. 1986. *Principles of Geographical Information Systems for Land Resources Assessment.* Oxford, Oxford University Press.

Coppock, J. T. and Rhind, D. W. 1991. The history of GIS. In D. J. Maguire, M. F. Goodchild, and D. W. Rhind (eds) *Geographical Information Systems: Principles and Applications.* Harlow: Longman, pp. 21–43.

Dede, C. 1995. The evolution of constructivist learning environments: Immersion in distributed, virtual worlds. *Educational Technology* 35(5): 46–52.

Downs, R. M. 1994. The need for research in geography education: It would be nice to have some data. *Journal of Geography* 93: 57–60.

Driver, R., Asoko, H., Leach, J., Mortimer, E., and Scott, P. 1994. Constructing scientific knowledge in the classroom. *Educational Researcher* 23: 5–12.

Ely, D. P., Foley, A., Freeman, W., and Scheel, N. 1992. Trends in educational technology 1991. In D. P. Ely (ed.) *Educational Media and Technology Yearbook 1992.* Englewood, CO: Libraries Unlimited: 1–29.

Fazio, R. P. and Keranen, K. 1995. Mapping a course with GIS. *Science Teacher* 62: 16–9.

Furner, J. M. and Ramirez, M. 1999. Making connections: Using GIS to integrate mathematics and science. *TechTrends* 43: 34–9.

Geography Education Standards Project. 1994. *Geography for Life: National Geography Standards.* Washington, DC: National Geographic Society.

Goodchild, M. F. 1992. Geographic information science. *International Journal of Geographical Information Systems* 6: 31–46.

Gross, H. H. 1967. *Research Needs in Geographic Education.* Normal, IL: National Council for Geographic Education.

Hill, A. D. 1995a. Geography standards, instruction, and competencies for the new world of work. *Geographical Education* 8: 47–9.

Hill, A. D. 1995b. Projections and perceptions. Editorial comments: Learning for the new world of work through geographic inquiry. *Geographical Bulletin* 37: 65–7.

Huxhold, W. E. 1999. Certifying GIS Professionals. WWW document, http://www.urisa.org/GIS_CERT_PRES/sld001.htm.

International Society for Technology in Education. 2000. *National Education Technology Standards for Students: Connecting Curriculum and Technology.* Eugene, OR: International Society for Technology in Education.

Jacobs, H. 1989. *Interdisciplinary Curriculum: Design and Implementation.* Alexandria, VA: Association for Supervision and Curriculum Development.

Jenkins, A. 2000. The relationship between teaching and research: Where does geography stand and deliver? *Journal of Geography in Higher Education* 24: 325–51.

Kemp, K. K. 1995. Teaching and Learning about GIS. WWW document, http://www.ptr.poli.usp.br/labgeo/congressos/simp3.html.

Kemp, K. K. 1998. The NCGIA core curricula in GIS and remote sensing. *Transactions in GIS* 2: 181–90.

Kerski, J. J. 2003. The implementation and effectiveness of geographic information system technology and methods in secondary education. *Journal of Geography* 102: 128–37.

Kerski, J. J. 2004. GIS for citizenship education. In A. Kent and A. Powell (eds) *Geography and Citizenship Education: Research Perspectives.* London: University of London Institute of Education (Papers from the IGUCGE London Symposium in April 2003).

Maguire, D. J., Goodchild, M. F., and Rhind, D. W. (eds) 1991. *Geographical Information Systems: Principles and Applications.* Harlow: Longman.

Malone, L., Palmer, A. M., and Voigt, C. L. 2002. *Mapping Our World: GIS Lessons for Educators.* Redlands, CA: ESRI Press.

Malone, L., Palmer, A. M., and Voigt, C. L. 2003. *Community Geography: GIS in Action.* Redlands, CA: ESRI Press.

McGarigle, B. 1997. High school students win national awards with GIS. *Government Technology* 9: 1.

Means, B. 1994. *Technology and Education Reform: The Reality behind the Promise.* San Francisco, CA: Jossey-Bass.

NRC. 1996. *National Science Education Standards.* Washington, DC: National Academy Press.

NRC. 2006. *Learning to Think Spatially: GIS as a Support System in the K-12 Curriculum.* Washington, DC: National Academy Press.

National Task Force for Social Studies Standards. 1994. *Expectations of Excellence: Curriculum Standards for Social Studies.* Washington, DC: National Council for the Social Studies.

Obermeyer, N. J. 1993. Certifying GIS professionals: Challenges and priorities. *URISA Journal* 5: 67–75.

Piaget, J. 1929. *The Child's Conception of the World.* London: Routledge.

Pisapia, J. 1994. *Technology: The Equity Issue.* Richmond, VA: Metropolitan Educational Research Consortium Research Brief No. 14.

Powell, J. C. 1999. The Relationship between Teachers' Beliefs and the Use of Reform-oriented Science Curriculum Materials. Unpublished PhD Dissertation, University of Colorado, Boulder.

Pugh, D. L. 1989. Professionals in public administration. *Public Administration Review* 49: 1–8.

Radke, S. L. 1996. GEODESY: An Educational Series for Youth. WWW document, http://gis.esri.com/library/userconf/proc96/TO350/PAP316/P316.HTM.

Robison, L. 1996. Plotting an island's future. *Geo Info Systems* 6: 22–7.

Rogers, E. M. 1995. *The Diffusion of Innovations* (4th edn). New York: Free Press.

Salinger, G. L. 1994. Remarks at the First National Conference on the Educational Applications of Geographic Information Systems. In D. Barstow, M. D. Gerrard, P. M. Kapisovsky, R. F. Tinker, and V. Wojtkiewicz (eds) *First National Conference on the Educational Applications of Geographic Information Systems (EdGIS) Conference Report.* Cambridge, MA: TERC Communications: 123.

Sarnoff, H. 2000. Census 1790: A GIS Project. WWW document, http:// www.cssjournal.com/ sarnoff.html.

Sui, D. Z. 1995. A pedagogic framework to link GIS to the intellectual core of geography. *Journal of Geography* 94: 578–91.

UCGIS Model Curriculum Task Force. 2003. Straw Report: Model Curricula Draft. WWW document, http:/www.ucgis.org/priorities/education/strawmanreport.htm.

US Department of Labor. 1991. *Secretary's Commission on Achieving Necessary Skills (SCANS), Blueprint for Action: Building Community Coalitions.* Washington, DC: Government Printing Office.

Walker, M., Casper, J., Hissong, F., and Rieben, E. 2000. GIS: A new way to see. *Science and Children* 37: 33–40.

Part VII   Future Trends and Challenges

The final four chapters take the book full circle in that they examine future trends and challenges. The first chapter in this group, Chapter 31 by Christopher B. Jones and Ross S. Purves, examines the role of the World Wide Web in moving Geographic Information Systems (GIS) out from their organization- and project-based roles to meet people's personal needs for geographic information. The various elements of web-based GIS are discussed, with special emphasis on the opportunities they provide for querying and visualizing geographic information, the provision of maps and geographic data, and the role of the Web as networked infrastructure for GIS. This discussion is followed by a review of web-GIS technology, the role of open standards for GI Services, and some examples of the functionality of specific web-GIS. The chapter concludes by considering some of the ways in which the Web itself is being extended to become geographically intelligent when searching the content of the Web itself.

The second chapter in Part VII, Chapter 32 by Allan J. Brimicombe examines the emergence of location-based services (LBS) as an important new application of GIS. This chapter explores some of the data implications of LBS, how LBS users are positioned so that the system knows where they are, and how queries are handled. The chapter concludes by reviewing some applications of LBS and some predictions as to what the future might hold for these types of services.

In the third chapter in this final set (Chapter 33), Michael F. Goodchild seeks to identify the challenges and issues that are likely to guide Geographic Information Science (GISc) research for the next decade or more. This chapter starts out by reviewing the research agendas published by various groups of scholars over a span of nearly 25 years and moves from these initiatives to discussions of the need to improve our understanding of the nature of the geographic world and the broader themes within science and the extent to which they might serve as the basic research challenges for GISc. Goodchild concludes this chapter with a discussion of the Digital Earth concept and why this might serve as a grand challenge – a theme that is capable of directing research to a common, distant, but imaginable end – in the foreseeable future.

Chapter 34 is the last chapter in this group and in the book, and in it Andreas Reuter and Alexander Zipf examine the developments that are likely to occur in the next 10 to 15 years along three dimensions: concepts and methods, applications, and platforms. Reuter and Zipf assume that the push towards integration will continue and that GIS will be instrumental in integrating data and services from heterogeneous sources into a uniform architecture, using new concepts and methods, delivering these services via the Grid, and thereby promoting and facilitating a whole range of new applications. Such a future points to the need for continued growth and innovation in GISc in the years ahead.

Chapter 31

Web-based Geographic Information Systems

Christopher B. Jones and Ross S. Purves

The World Wide Web has brought about major changes in the way Geographic Information Systems (GIS) are used and in the way in which they are implemented. GIS were dominated until the early 2000s by the creation of software technology and geographic data resources dedicated to the needs of professional users of geo-information. The typical GIS has been an isolated collection of technology and data, purchased for and installed within the confines of an individual organization. The Internet and the World Wide Web were rapidly recognized to have the potential to transform this closed world view of GIS by, for example, "dramatically increasing the applications of GIS . . . through integration of mapping, GIS and non-spatial information technologies . . . to create new forms of representation and new ways to address problems important to society" (MacEachren 1998, p. 575).

As a communication network the Web caters equally to the needs of commerce and industry, and to individual members of the public, irrespective of their personal or work-based affiliations. In the context of GIS these communication facilities are being exploited in several ways. They serve to link together different organizations and parts of the same organization, and also open access to geographic information services and functionality to a wide community of users. GIS are growing therefore from their original organization and project-based roles to meet people's personal needs for geographically-specific information. In doing so, they serve to increase awareness and participation in developments and activities at local and regional levels. From its very beginnings the Web has incorporated spatial information, with an early paper describing the concept of the World Wide Web (Berners-Lee, Cailliau, Groff, and Pollermann 1992) including the "authors coordinates" as examples of the information which might be served by the, then hypothetical, World Wide Web.

The Web provides access both to text, and other "unstructured" media, and to interactive services for retrieving specialized information or data from online databases. Many types of information are geographically referenced and most services have a geographical dimension, based either on the location of the service itself or on the user of the service. The geographical dimensionality of information has therefore introduced a requirement for aspects of the Web to become spatially-intelligent, in the sense of being able to understand and respond to requests for geographically-specific information and in being aware of the location of individuals. Furthermore, the distributed infrastructure of the Internet, on which the Web is based, can enhance the effectiveness of the traditional in-house GIS. This is reflected in the possibility of world-wide access to geographic data and to remote geographic data processing facilities and in allowing members of an organization to retrieve and maintain geographical information from multiple locations, whether office or field-based.

In the following section of this chapter we examine the elements of web-based GIS with regard to the introduction of maps to the Web, the provision of facilities for querying and visualizing geographical information, the provision of geographic data, and the role of the Web as networked infrastructure for GIS. This is followed by a review of web GIS technology, with reference to open standards for GI Services and some examples of the functionality of specific web GIS. The chapter concludes by considering the way in which the Web itself is being extended to become geographically intelligent for purposes of searching the content of websites.

Elements of Web-based GIS

Maps on the Web

From the earliest days of the World Wide Web, the facility to display images was exploited to present maps which provide geographical context to information. A very early interactive map server was developed by Xerox Parc and used by many services to display simple web maps (Putz 1994, Towers and Gittings 1995). Businesses often use maps on the Web to show their location, while news agencies such as the BBC (http://news.bbc.co.uk/) use maps to help people understand where events are taking place. Websites such as the BBC are constantly being updated, but new items referred to on the home page will often include maps. The simplest types of web map are static, non-interactive images, and this is the norm for these contextual maps. The standard facilities of HTML, the original mark-up language for web documents, can, however, be used to provide some degree of interactivity, whereby the map includes clickable icons, or hot spots, which provide hyperlinks to information about the highlighted location or map symbol (van Elzakker 2001).

Tourist maps such as those of Paris (http://www.paris.org/Maps/MM/) and Washington DC (http://maps/mapnetwork.com/wctc/dispmap.asp?map=1) contain such hyperlinks which lead to further web pages containing text and images, and sometimes more maps, relating, for example, to museums and monuments. A variation on the use of clickable hot spots is to provide pull-down menus that allow the user to select some particular type of associated information. This is found on the web map of the London underground transport system (http://www.visitlondon.com/tubeguru/), in which users can point to stations on a map and select menu items about timetables and associate transport networks.

The requirement for navigational and other geographical contextual information about commercial and public services can be met by web mapping services such as MultiMap (http://www.multimap.com/), Map 24 (http://www.mapsolute.com/) and

MapQuest (http://www.mapquest.com/) which can provide a link from a company web site to one of their online maps, which might then contain a symbol marking the location of the advertised facility. Yellow pages services can link all their entries directly to these mapping services, or a business directory may be integrated with the web mapping service, as in Google Maps (http://maps.google.com/), which operates as a stand-alone facility or can be called via an application programming interface (API). These sorts of web mapping services provide extensive geographical coverage, such as for Europe, the USA, or, indeed, the world. Users of these sites can obtain maps of an area of interest by several methods, such as pointing to and zooming into a displayed map, or by entering an address, place name, or postcode. The latter method will then present a map centered on the specific location and provide facilities for zooming and panning around the map. The maps themselves are typically retrieved from a database of web maps at different levels of detail but with fixed geographic content. Thus the features displayed on the map are predetermined and do not adapt significantly to the interests of the user except with regard to location and level of detail.

In addition to these navigation-oriented websites there are other services on the Web that deliver essentially static maps and remotely-sensed images relating to many different domains. The National Libraries of Scotland maintains a collection of maps drawn by Timothy Pont in the 1580s and 1590s that are among the earliest existing maps of Scotland. These maps have been scanned as high resolution images, and are made publicly available on the Web as jpegs at http://www.nls.uk/pont/ and constitute an amazing set of examples of early cartography (Figure 31.1). Such mechanisms

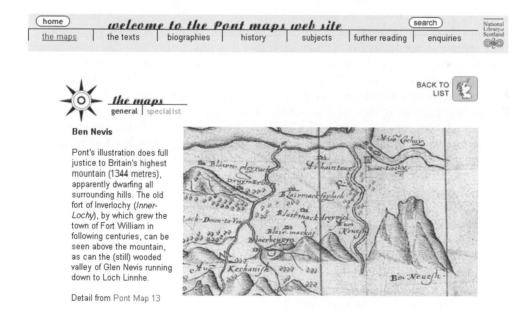

Fig. 31.1 Example from the Pont map collection
Copyright National Libraries of Scotland

transform access, with materials previously only accessible at a single location now available to anyone, anywhere who has a connection to the Internet.

The US Geological Survey (USGS; http://www.usgs.gov) is a rich source of maps which can be accessed either directly or via sites such as TerraServer-USA (http://terraserver-usa.com/), which specializes in aerial image data. Maps illustrating emergency situations such as due to floods and earthquakes, can be found at (http://www.reliefweb.int/w/map.nsf/home/).

Querying and visualizing geographic information on the Web

GIS are used in many types of organizations, including local government, utility companies, environmental monitoring, mineral exploitation, marketing, healthcare, and the police. See, for example, Part 4 (Applications) of Longley, Goodchild, Maguire, and Rhind (1999) for an extensive set of examples of the application of GIS in the private and public sectors.

There is a role for web GIS in all these areas. For some organizations the Web is a means of public access to information, while for others it provides a network infrastructure that enables employees to access central databases remotely, either from offices in multiple geographic locations or, for fieldwork activities such as land surveying or utility maintenance, from "the field" using wireless communications.

Local government organizations usually have a responsibility to make information about their services available to the public and, as a result, web interfaces to GIS have provided a tremendous opportunity for raising awareness of local services and, in some cases, of consulting with the local population. In the website for Greenwood County, South Carolina in the USA (http://165.166.39.5/giswebsite/default.htm) the user is given options for layer-based selection of map features representing, for example, boundaries for census and school districts, utility services, topographic features, soil types, and recreation areas. The various features can be then identified interactively. In the UK the use of web GIS promotes what is referred to as e-government and is reflected in many municipal websites that seek to improve communication with the local residents. For example, the city of Rotherham's website allows users to employ a postcode as a search key for information such as refuse disposal, schools, local taxation, and planning (development) applications. The website for the UK town of Huntingdon (http://www.huntsdc.gov.uk/) allows users to submit planning applications online, in coordination with a national planning portal.

A particularly effective example of the use of web GIS in communicating local information is seen in the context of monitoring transport systems. The Houston TranStar website (http://traffic.houstontranstar.org/incmap/) provides up-to-date information on road traffic in Houston with an interactive map that displays current traffic speeds and the presence of road accidents. Clicking on the map provides local details including graphs of past and present road speeds and live images of the traffic on the relevant section of road. Speeds of vehicles are determined using an automatic vehicle identification (AVI) system in which transponder tags within individual vehicles are detected by roadside tag readers. The average speeds of tagged vehicles can then be calculated by monitoring the progress of individual vehicles. For more examples of transport GIS see Peng and Tsou (2003).

Although many municipal websites provide what is essentially a one-way channel of communication, it has been recognized for some time that web GIS have the potential to facilitate public participation in decision making (Obermeyer 1998, MacEachren 2000, Carver 2001). This potential role in facilitating democratization is illustrated by the Virtual Slaithwaite website (Kingston 2002; http://www.ccg.leeds.ac.uk/slaithwaite).

Another perspective on the role of GIS in decision making is provided by the municipal websites that promote industrial development. For example, following massive local job losses associated with the closure of a shipyard, the city of Vallejo in California developed a web-GIS (http://www.ci.vallejo.ca.us/) which supports those searching for vacant properties within the city boundaries (Figure 31.2). A user may specify the characteristics of suitable properties (for example, size and cost) and can view information about existing businesses in the area. There is also an option of querying the demographic profile of the inhabitants of the sub-areas of the region, which might be used to predict take up of some proposed new services.

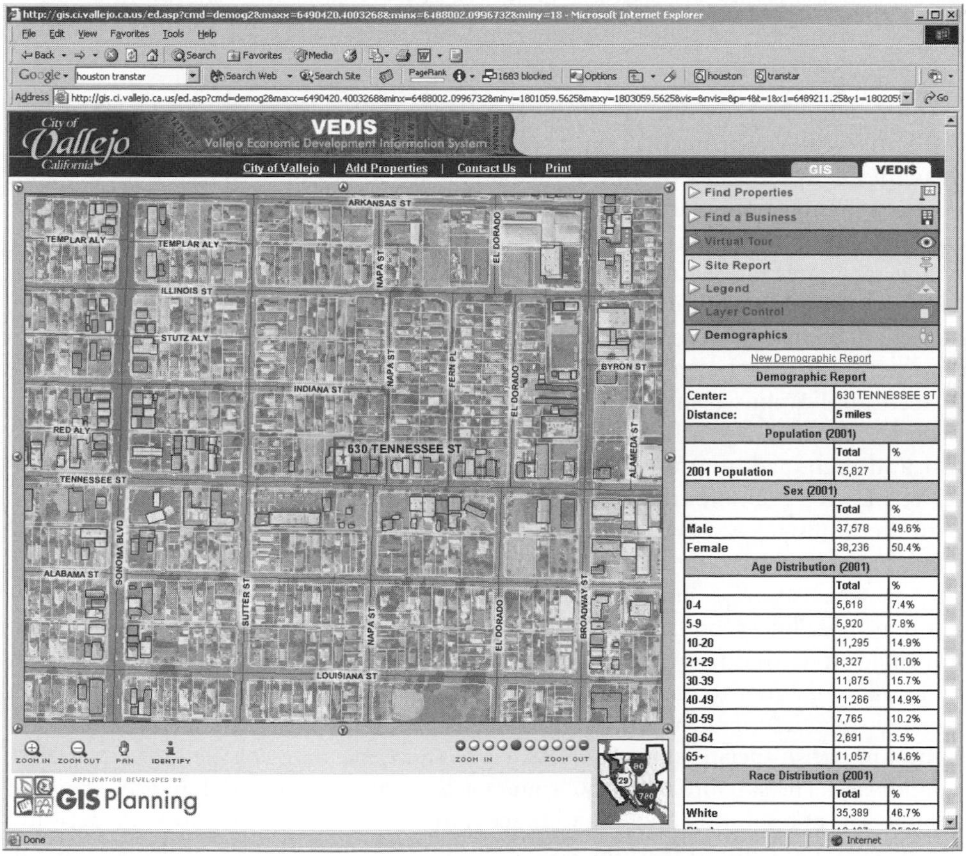

Fig. 31.2   City of Vallejo economic development information system
From http://gis.ci.vallejo.ca.us/

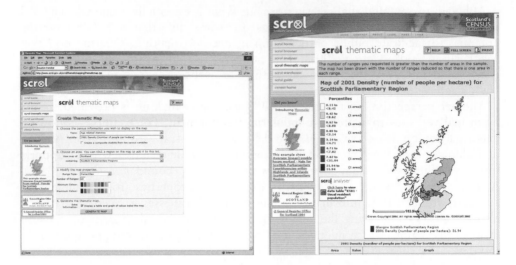

Fig. 31.3 Options for thematic map and resulting presentation of Scottish census data
2001 Census data supplied by the General Register Office for Scotland; © Crown Copyright

In the UK, the provision of census data via the Web is another example of the promotion of e-government. Census results from Scotland, England, and Wales are available as dynamically produced thematic maps in response to user queries. Some of the mapping is available as Scalable Vector Graphics (SVG) (see "Presentation and interaction with geographic information on the Web," below, for more on SVG) which facilitates interactive map query of individual areas for their local statistics. The Scottish Census Results Online website (http://www.scrol.gov.uk/scrol/common/home.jsp) enables the user to specify combinations of variables to plot and to exert considerable control over the presentation of these data (Figure 31.3).

Such applications come very close to delivering the functionality expected by a professional working with GIS, and providing that response times are sufficiently fast and high-quality printouts are available (through downloadable PDF or post-script files for example) they can replace a desktop GIS for many purposes.

In the Louisiana Statewide GIS (http://atlas.lsu.edu/) demographic data are made available in combination with topographic information. This quite extensive web-site presents scanned maps, vector boundary data, satellite images, and aerial photographs. Interactive maps are used to access recent and historic demographic information derived from user-selected census data on a parish by parish level. The maps allow the user to control the use of color symbols to indicate the values of the selected statistics for each parish, while the user can also click on individual parishes to obtain parish-specific statistics. There are interactive maps of the historical parish boundaries. The satellite imagery "tour" of the state is based on Landsat Thematic Mapper 30 m resolution scenes and is accompanied by informative textual accounts of physical and cultural features visible in the individual images. It is also possible to download geographic data on several themes, including biological, socio-economic, infrastructure, and geophysical topics.

There are many more examples of web GIS applications in the above application areas and in others such as emergency response, agriculture, leisure activities, and utility maintenance (see Peng and Tsou 2003 for additional examples).

The Web as a source of spatial data

We have described above some uses of web GIS to communicate geographical information, usually visualized as some form of map. An increasingly important use of web GIS is, however, in locating and downloading spatial data for use in desktop GIS and other programs. From the approach of the USA, where many data sets are made freely available, to that of many European countries, where much data is only available after payment of a licensing fee, the accessibility of data sets online varies greatly. The vast array of spatial data (and data with a spatial component) has also led to the development of digital libraries and portals, which act as clearinghouses for large collections of related spatial data (see Chapter 1 by Cowen in this volume for additional discussion of these types of innovations).

For example, the USGS has developed a spatially seamless data server (http://seamless.usgs.gov/) to deliver a wide range of its data holdings. Users can specify particular data sets and extents for retrieval through an intuitive map interface and, depending on the resulting data volumes, a file is made available for transfer either over the Internet or as a CD-ROM (Figure 31.4). Such approaches are gradually becoming a commonplace technique, and as they proliferate several challenges are becoming apparent.

Chief among these challenges is the distribution and use of accurate and up-to-date metadata (Green and Bossomaier 2002). As users combine data sets, the potential increases for generating inappropriate or erroneous results as a consequence of differing source scales, collection periods and data qualities. The data sets delivered by the USGS data server described above come bundled with Federal Geographic Data Committee (FGDC) metadata (FGDC 1997). Production and delivery of FGDC metadata for all US government bodies was made a legal requirement by an Executive Order from the President in 1994 (http://www.fgdc.gov/).

Digital libraries such as the Alexandria Digital Library (ADL) project set out to provide access via the Web to georeferenced materials such as maps and images (Andresen, Carver, Dolin, et al. 1995; http://www.alexandria.ucsb.edu/). An essential aspect of this provision is the maintenance of geographic metadata, such as an enclosing minimum bounding rectangle and the names of settlements and states, in association with the stored resources. The tasks of generating the metadata and of searching for the resources are facilitated by the presence of gazetteers and ontologies of spatial information (Hill, Frew, and Zheng 1999, Jones, Abdelmoty, Finch, Fu, and Vaid 2004) that provide the link between place names and geographic coordinates. Digital libraries aim to store not only explicit geospatial data (such as topographic or elevation data), but also any data which can be georeferenced – for example, documents describing events at a particular time (Buttenfield 1997). Another geolibrary, which illustrates this diversity of materials is the Go-Geo project that has developed a service allowing users in UK Higher Education to "discover" geographic and geographically-referenced data via the Web (Reid, Higgins, Medyckyj-Scott, and Robson 2004). Figure 31.5 shows an

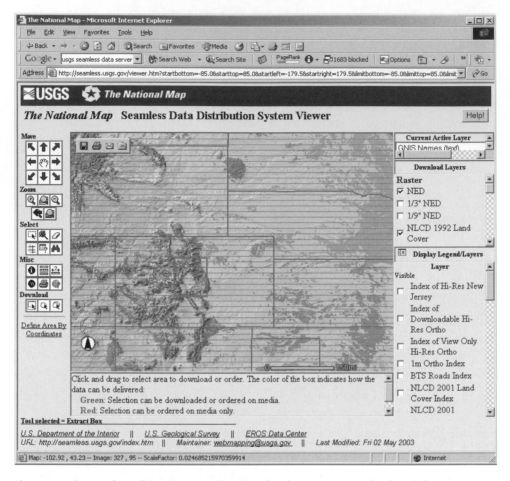

Fig. 31.4   The interface of the USGS seamless data distribution system with selected data highlighted in green (i.e. darkest shade in map window)
Data available from US Geological Survey, at the National Center for EROS, Sioux Falls, SD, USA

example of a search for research project reports related to Edinburgh. In "A Spatially Aware Web," below, we pursue the subject of web search of spatially referenced information in more detail.

The Web as an integral part of GIS

Earlier in this section we showed how the Web is playing an important role in using GIS to communicate geographically-referenced information to a very wide audience. In addition to this role as a presentation medium, the Web and associated Internet technology is also starting to become an integral element of the maintenance, and indeed of the structure, of some GIS. At the time of writing the most obvious aspect of this is in organizations that require their employees to be located at multiple locations for purposes of data acquisition and maintenance. Other aspects which can

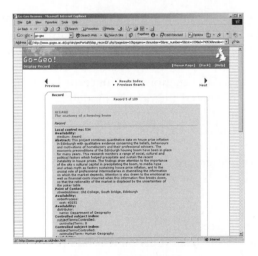

Fig. 31.5 An example search for data with GeoXWalk
© EDINA National Data Centre

be expected to develop further are the use of the Web to access data for immediate use at remote data servers and the use of the Web to access GIS functionality in a distributed manner.

Intra-organization maintenance of GIS

Some organizations operate in an essentially distributed manner in the sense that employees may work at multiple locations in order to acquire and update data for input to a GIS. Primary examples of this area are in topographic and other environmental survey organizations and in the utility and facilities maintenance industries.

In the case of geographic survey organizations, it has been commonplace for many years to employ mobile computers to record data in the field. In the past, the surveyor may have carried a copy of the relevant part of the spatial database and recorded new data on a disk on the mobile device. The updated components would then have been transported in some way to a central office. With the arrival of the Internet it became possible to transfer data quickly to a central data server from a local Internet connected office. With the advent of wireless Internet communications technologies it has become possible to transmit to and receive data directly from a main spatial data server irrespective of location, provided that it is within range of a wireless communications infrastructure. In the latter scenario the surveyor can download data directly to their mobile device and then send back new data in the course of an online session. The mobile device is simply a remotely located component of the GIS.

Similar efficiencies are gained by the use of the Internet in organizations that may have multiple geographically distributed offices all contributing to a central spatial database. In these situations, which might not depend so much on mobile communication technology, the Internet effectively closes the geographical gap between different parts of the same organization.

Inter-organization distributed GIS

The idea of distributed functionality of a GIS within an organization as outlined above is well established. A further development of distributed functionality, so far much less well established, is that in which different functional components of a GIS may be located within different organizations. In such a case one organization may provide specialized GIS functionality concerned, for example, with network analysis, geostatistics, or spatial interaction modeling which another organization accesses on demand. Thus the user organization may have some data on which they wish to perform the particular analysis.

The converse of this situation is one in which an organization may have relevant GIS processing power but require access to some particular data resource on which to perform an analysis in its own right, or which may need to be integrated with an existing "in-house" set of data.

A third possibility in this context is that of an organization that requires access both to remote data resources and to remote functionality. In this situation, it could be that the user organization has no permanent GIS functionality or data store, but accesses both types of resource on demand.

This view of distributed GIS leads to the concept of GI Services and it is one that depends very much upon effective interoperability at the level of both data and functionality. Peng and Tsou (2003) give a good introduction to the ramifications of this approach. It is a very interesting prospect in that it abandons the traditional rather monolithic view of GIS in favor of a much more flexible service-based approach. Its viability is yet to be proven and, as can be envisaged, it raises important questions regarding the willingness of some organizations to hand over responsibility for both functionality and data access to third parties, as well as issues such as ensuring consistency of provision of functionality and data for a particular organization over protracted periods of time.

Implementing Web-based GIS

In this section, we introduce the technical foundations for web GIS. We identify the main components of typical systems, before describing briefly some of the contributions of the Open GIS Consortium to the development of geographical web services and giving some examples of how these concepts are employed in a few web GIS applications.

Architectures for Delivering Web Services

The Internet is essentially a series of protocols, or agreements, that allow computers acting as servers to be accessed from remote locations by other computers acting as clients (for example, Spainhours and Eckstein 2002). The protocol most commonly used on the World Wide Web is known as HTTP (hypertext transfer protocol). Through this protocol a client (commonly a web browser) uses a URL (uniform resource locator) to ask a server to send data back to the client. Usually these data are sent in the form of HTML (hypertext mark up language), which is then rendered in the web browser according to the browser's interpretation of the mark-up commands. HTML is concerned with describing how contained text and images, and the associated hypertext link URLs, should be formatted and presented with regard to factors such as size, color, and type of text font, and the form of paragraphs and tables. Thus HTML contains both content and specification of the styling of that content (Raggett, Le Hors, and Jacobs 1999).

There are many possible ways to implement web servers to respond to requests from clients. The most simple web servers store HTML documents as files that are transmitted in response to a request from the client. However, for websites with a large amount of content this approach lacks flexibility with regard to access and update of particular types of information and with regard to the way in which information is presented to the user. These shortcomings are overcome by using a database management system to store information *content* which can then be retrieved selectively in response to a user request and formatted with a common *styling* before being transmitted to the client as HTML. This solution separates content from styling, which as we will see is also an important concept in delivering maps over the Internet. It is normally combined with facilities for the client to solicit the user's requirements via menus and text boxes. This user input is then sent to the server in order to formulate a query to the database.

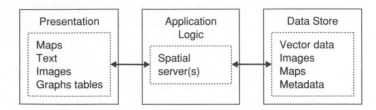

Fig. 31.6 Three-tier web GIS architecture

Three-tier architecture for Web GIS

Delivering both maps and data to clients over the Internet can be accomplished in many different ways. At the most basic level a website could store a collection of HTML pages that contain maps in the form of GIF or JPEG images, examples of which were presented above in "Maps on the Web." This suffers from the inflexibility referred to above. In practice a web GIS will usually generate a map in HTML or another markup language on the fly in response to a user query. The typical web GIS architecture consists of separate components, or tiers, responsible for presentation, application logic (also called business logic), and data management (see, for example, ESRI 2004).

The presentation tier is responsible for displaying data to the user and interacting with user requests, for example to pan and zoom a map, or a query to a particular feature. The application logic translates a request from the presentation tier for a map or data into a query for data from one or more data sources. This query formulation is normally performed by what are referred to as spatial servers or map servers, of which there may be several types differentiated according to the type of data that they retrieve. The application tier may also deal with load balancing between these servers (ensuring that performance is not compromised by making excessive demands upon an individual server). Furthermore, it can handle authorization and authentication requirements and some aspects of data processing and analysis associated with the required data. The data management tier is where the data are stored, generally in databases, although sometimes in simple file structures. The data management tier can consist of several different types of database and data representation and they may be located either on the same computer as the application logic or on different, possibly remote computers. Figure 31.6 illustrates the architecture of a typical web GIS solution.

Interoperability and the OpenGIS Consortium

The implementation of these web GIS components requires specialized software. However, it should be apparent to a user familiar with GIS that some of the functionality of these components can be found as elements of many desktop GIS. A key task in web GIS is to separate the components to allow *interoperability* between application logic, data, and presentation tiers. Thus the functionality and location of each component is relatively independent of the other, but they can all communicate effectively. Interoperability has long been a holy grail for GIS users, as anyone who has spent long afternoons converting between data stored in different formats will be all too aware of. The basic concepts of interoperability involve the development of a set of standard *interfaces* specifying which operations can be performed by particular services (as opposed to how they are implemented), in combination with standard languages and formats for representing and transferring information.

A host of different proprietary and open standards exist to deliver spatial data to web browsers. *Proprietary solutions* are those which are specific to an individual software application or data format and generally require users to buy a specific piece of software. *Open standards* are free to use and are used by both commercial and non-commercial developers. The OpenGIS Consortium (OGC) was founded in 1994 to develop open interfaces and standards to allow geographic software and data to interoperate. One of the early contributions of OGC was to define an XML (extensible Markup Language) "vocabulary" for representing and communicating spatial and non-spatial properties of geographic information on the Web. This is the Geographical Markup Language (GML), which is described in more detail by Knoblock and Shahabi in Chapter 10 of this volume.

Open Web Services Framework (OSF)

A key concept in the OGC's vision for interoperability is that of services which provide various types of functionality and data (Cuthbert 1999). The OSF is intended to facilitate interoperability between services through the use of standard interfaces and to facilitate publishing and discovering the presence of services. The OSF categorizes services as registry, data, portrayal, processing, and application services (Figure 31.7).

Registry services are concerned with maintaining information about data and services, that is to say metadata, that may be employed to discover the existence and the nature of these resources. They are linked with the use of catalogs that store metadata. Once a client has established the existence of relevant data, it may employ a *data service* to access the data from its respective database (or "repository"). The OGC Reference Model (OpenGIS 2003) refers specifically to four types of data service, namely Feature Access Services (for example, the OGC Web Feature Service WFS), Coverage Access Services, Sensor Collection Services, and Image Archive Services.

OSF *processing services* are concerned with operations upon geographic data and include services such as those for coordinate transformation, image registration, transformations between spatial data models, geocoding, and route finding. *Portrayal services* provide functionality for displaying geographic data and images and include the generation of maps and terrain model visualizations. Specific types of portrayal services may be dedicated to maps, coverages, and to the requirements of mobile devices.

OSF *application services* typically make use of the previous services of registry, data, and portrayal for specific applications such as resource discovery, map viewing, image exploitation, web-based sensor access, and mobile location-based services. An application service may consist of a chain of other services such that, for example, a registry service identified a data access service which was employed to retrieve some contents of an application-specific data repository, prior to user display in a portrayal service.

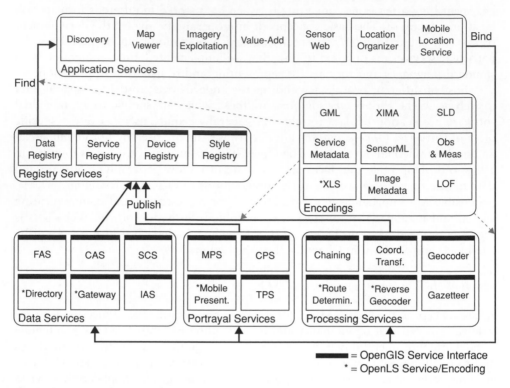

Fig. 31.7 Open web services framework
From OpenGIS, 2003; Copyright 2006 OGC

We now describe two of the most familiar OGC interfaces, the Web Map Service, which is a portrayal service, and the Web Map Feature Service, which is a data service.

Web Map Services (WMS)

A WMS delivers geo-referenced mapping to a client browser in respond to a request (OpenGIS 2002, 2003). Maps are in the form of an image, in either raster or vector form, which portrays an underlying set of spatial data. A WMS is required to implement two operations:

- GetCapabilities: A request to a WMS for its capabilities delivers XML metadata specifying the possible options that can be used when making a request for a map;
- GetMap: Requests that a WMS deliver a map to the client according to some given parameters.

A GetMap request would typically specify at least the geographical bounding box of the map (for example, $56° \rightarrow 57°N$; $3°W \rightarrow 2°W$) to be produced and the corresponding image size and format (for example, 200×200 pixels, jpeg), and the names

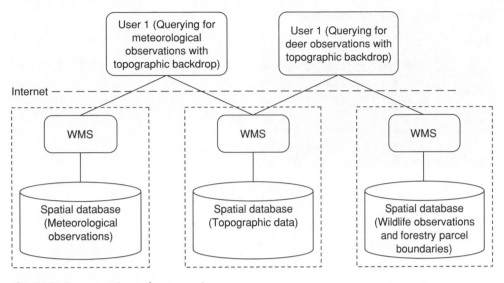

Fig. 31.8 Interoperating web map services
After ESRI

of the layers to be mapped (for example, roads, railways, lakes). Since a WMS may be capable of producing transparent maps it is possible to overlay the results of a query to one WMS on top of another – for example with the topographic data from a National Mapping Agency WMS providing a backdrop to wildlife sightings produced at a local WMS. Figure 31.8 illustrates such a process.

Web Feature Servers (WFS)

WFS deliver not maps but data in response to a request (OpenGIS 2002). These data are delivered in the form of GML and consist of "simple" geometric primitives in the form of points, lines, and polygons and collections of these primitives. A WFS must also implement a GetCapabilities command describing what features it may deliver. A WFS must also implement two further operations:

- DescribeFeatureType: Describes the nature of a feature;
- Get Feature: Delivers a feature according to some request, which may be spatially constrained.

Since a WFS delivers data, and a WMS delivers portrayals of data it is possible to chain together a WMS with a WFS to provide a service to deliver mapping on the Internet. Such an approach makes clear the separation of content from styling.

A WMS may provide the user with options, through a set of named styles, to deliver a number of different portrayals of the same data. Such predefined styles may be useful in, for example, a corporate environment, where a number of different map styles are used in different applications. Much greater flexibility is provided through the solution described above where a WFS and WMS interact. Here, the WMS can have extended capabilities which allow the client to pass it a Styled Layer

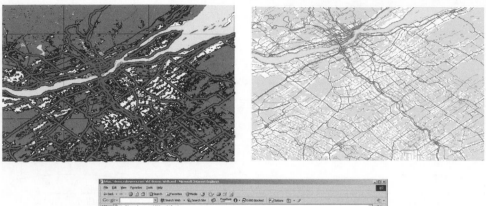

Fig. 31.9 Two representations of the same data using different style layer descriptors and an extract from the SLD for the representation given on the right
All from http://www.cubewerx.com

Descriptor (SLD). This SLD is essentially an XML template which specifies how layers retrieved from the WFS should be drawn, before being returned to the client through the WMS. This last solution allows a consistent user-defined styling to be applied for portrayal of data from a variety of sources (Figure 31.9).

Open standards as described above have many attractions and have been widely adopted by vendors (see http://www.opengis.org/resources/?page=products for additional details). Indeed, open standards do not in any way preclude proprietary solutions, but rather specify a set of operations which will provide compliance with the standard. However, proprietary solutions are equally, if not more, common on the Internet for a variety of reasons, often related to the provision of specifialized functionality and performance. For example, ESRI's *ArcIMS* (Internet Map Server) provides presentation, business logic and data management tiers. If these individual tiers implement open standards, for example the WMS for purposes of presentation, it becomes possible for them to interoperate with other compliant services to provide seamless mapping or data to the user.

Presentation and interaction with geographic information on the Web

Generally the first realization of a map (or data set) is not exactly what the user required. They may wish to refine their choice by zooming in or out, panning to include another location, turning on or off specific layers of data, or choosing to add symbols to the rendered map. For an example of the techniques required to deliver "intelligent zooming," see Cecconi and Galanda (2002).

The user may also wish to query attributes of data presented or even produce alternative visualizations of the same data, perhaps in the form of graphs or charts. In order to perform any of these operations the user must interact with the data through the user interface.

The implementations of user interfaces for web GIS take two basic forms: so called thin clients and thick clients. A thin client is usually confined to gathering information from the user, submitting it as a query to the server, and rendering the document that is retrieved from the server. Thus, when the user pans or zooms, a request is sent to the server and a new map is delivered. Thick clients move some of the work from the server to the client. Here, for example, vector data may be delivered to the client which is then responsible for rendering the data as an image (by means for example of a Java applet). With such an approach it is possible to deliver data to the client as a continuous stream in anticipation of a request. Bertolotto and Engenhofer (2001) describe an approach to progressive delivery of a stream of such data to maximize response times. For instance, after an initial request for data, information about the surrounding area may be also delivered to the thick client, on the not unreasonable assumption that the user is likely to pan around within the initial selection. Whether a thin or think client is used, it is vital that response times are short and controls are intuitive. Examples of client- and server-based web mapping tools are given in Medyckyj-Scott and Morris (1998) and Jankowski, Stasik, and Jankowska (2001).

Both of the examples shown in Figure 31.10 deliver raster data to the client. Thus, if the user desires information about an individual feature, a query for a location must be made, and attribute information to be displayed for all features at that location returned from the server. A much more flexible solution, where attribute information related to features is required, is provided by the use of vector data, for example through Scalable Vector Graphics (SVG) (see http://www.adobe.com/svg/main.html and Ferraiolo, Fujisawa, and Jackson 2003 for additional details). SVG delivers an XML-based data format to a client, which can be visualized through a plug-in. The SVG data contains the specification of how the data are to be symbolized on the display. It is possible to produce SVG by applying a styling to GML data. Figure 31.11 shows SVG representing Ordnance Survey *MasterMap* data. As the user moves the mouse around the screen, different objects are highlighted and a unique identifier is displayed below.

Much more complex interaction with SVG is possible, and Figure 31.12 shows an example developed at the ETH Cartography Department with a wide range of options to vary the rendering of the map. See the SVG series of conferences for a wide range of examples of the use of the SVG (http://svgopen.org/). All of the examples described here focus on 2D visualization of spatial data. However, as described by

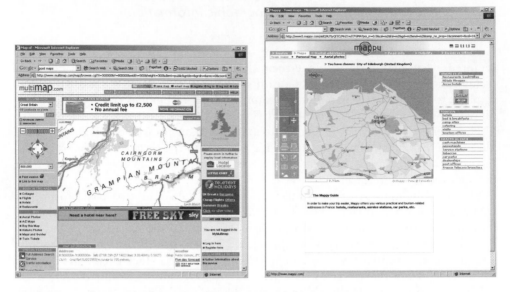

Fig. 31.10 A thin client (http://www.multimap.com) and a thick client (http://www.mappy.com). Note controls for panning and zooming in both cases. The thick client solution provides for display of a subset of the imagery that it stores
Image reproduced by permission of Multimap.com and Collins Bartholomew

Fig. 31.11 Ordnance Survey MasterMap data displayed as SVG
From http://www.ordnancesurvey.co.uk/oswebsite/xml/resource/index.html; © Crown Copyright, Ordnance Survey mapping, all rights reserved

Cartwright (1999), there are many ways of visualizing spatial data. For example, Purves, Dowers, and Mackaness (2002) described techniques to integrate georeferenced 3D visualizations of terrain with 2D mapping and navigation tools using a combination of Java, HTML, and the Virtual Reality Modeling Language (VRML),

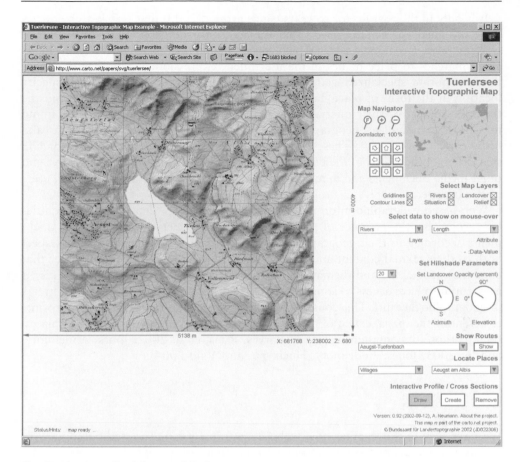

Fig. 31.12   Example of the use of SVG
From http://www.carto.net/papers/svg/tuerlersee/; with permission of Andreas Neumann, ETH Zurich;
background data ©SwissTopo

a mark-up language developed specifically for producing 3D visualizations (Ames,
Nadeau, and Moreland 1997). However, 3D GIS has been slow to develop on the
Web with limitations imposed by plug-in development and lack of a universally
agreed standard, together with difficulties in interaction and navigation for users.

A Spatially Aware Web

GIS are normally associated with processing highly structured geographic data,
spatially referenced by means of geometric coordinates. At present the Web is particularly effective in retrieving unstructured, usually textual, information. Much of
this is spatially referenced implicitly, in that web documents often refer to places
that provide context for their subject matter. The place information may occur, for
example, within addresses of businesses or within free text descriptions in which
people, events, and social and environmental processes may be associated with
geographic location designated by place names and landmarks.

It is only in the past five or so years (since around 2000) that web search engines have provided no facilities specialized for access to the geographic dimension of web resources. Geographic search would depend upon finding documents that included exact matches to place names in the user query expression. A few search engines do incorporate specialized geographic search facilities that let the user specify location in the form of place names, addresses, or post codes separately from the thematic subject of their query. An example of such functionality may be found in Google's specialized geographic search facilities (http://local.google.com/ and http://earth.google.com/). These types of web search provide a new form of GIS that differs radically from the conventional form in focusing on access to text and other media documents and in making their functionality freely available to all users of the Internet.

Web-based geographic information retrieval technology depends upon facilities for recognizing the presence of place names in a user query; classifying web documents according to their geographic context; indexing the documents with respect both to their textual content and to space; and ranking relevance in a manner that combines geographic and thematic context. In order to achieve this functionality it is essential to have access to some form of geographic ontology based on gazetteers or geographic thesauri. This could contain lists of place names, geometric footprints representing the spatial extent of places, and some spatial relationships between places, such as those of containment (Jones, Purves, Ruas, et al. 2002, Jones, Abdelmoty, and Fu 2003, Jones, Abdelmoty, Finch, Fu, and Vaid 2004).

CONCLUSIONS

We have seen in this chapter how web GIS is transforming the traditional view of GIS as a specialist professional resource to one that can be shared by a much wider community of people who have interests in geographic information. The Web is also helping major users of "in house" GIS technology, such as local government and environmental agencies, to promote their activities effectively, and in some cases to allow the public to participate in decision making. For the professional users of GIS the Web provides the potential for convenient access to remote data sources and remote geographic processing power. It facilitates mobile work modes for topographic and environmental survey organizations in which data may be acquired and updated in field operations that involve direct communications with centralized or distributed data servers. The opportunities for distributed access to geographic information can be exploited by the general public using location-aware in-vehicle and hand-held devices, to obtain information about local services and tourist attractions.

At the time of writing, web GIS technology is very much in a state of flux, due to rapid developments in Internet technology. While providing many opportunities the Web is also raising interesting research challenges that must be addressed if these opportunities are to be fully realized. These include solving problems of compatibility and hence interoperability between disparate data sources and between geographic information software. Progress will also need to be made in improving the quality of user interfaces of mobile devices and geographic web search facilities, for example to accommodate the map display limitations of small devices and the need for people to be able to communicate with geographic web services using natural language terminology. There is also a need to find better ways to categorize web documents automatically and accurately with regard to their geographic relevance.

REFERENCES

Ames, A. L., Nadeau, D. R., and Moreland, J. L. 1997. *VRML 2.0 Sourcebook*. London: John Wiley and Sons.

Andresen, D., Carver, L., Dolin, R., Fischer, C., Frew, J., Goodchild, M. F., Ibarra, O., Kothuri, R., Larsgaard, M., Manjunath, B., Nebert, D., Simpson, J., Smith, T., Yang, T., and Zheng, Q. 1995. The WWW Prototype of the Alexandria Digital Library. In *Proceedings of the International Symposium on Digital Libraries (ISDL '95)*, Tokyo, Japan, pp. 17–27. (Available at http://www.dl.slis.tsukuba.ac.jp/DLW_E/.)

Berners-Lee, T., Cailliau, R., Groff, J., and Pollermann, B. 1992. World-wide web: The information universe. *Electronic Networking: Research, Applications and Policy* 1: 74–82.

Bertolotto, M. and Egenhofer, M. J. 2001. Progressive transmission of vector map data over the World Wide Web. *GeoInformatica* 5: 345–73.

Buttenfield, B. P. 1997. Delivering maps to the information society: A digital library for cartographic data. *In Proceedings of the Seventeenth Conference of the International Cartographic Association*, Stockholm, Sweden. Stockholm: International Cartographic Association, pp. 1409–16.

Cartwright, W. E. 1999. Extending the map metaphor using web delivered multimedia. *International Journal of Geographical Information Science* 13: 335–53.

Carver, S. 2001. Public participation using web-based GIS. *Environment and Planning B* 28: 803–4.

Cecconi, A. and Galanda, M. 2002. Adaptive zooming in web cartography. *Computer Graphics Forum* 21: 787–99.

Cuthbert, A. 1999. OpenGIS: Tales from a small market town. In A. Vokovski and K. E. Brassel (eds) *Interoperating Geographic Information Systems*. Berlin: Springer-Verlag Lecture Notes in Computer Science No. 1580: 17–28.

ESRI. 2004. *ArcIMS 9 Architecture and Functionality White Paper*. WWW document, http://downloads.esri.com/support/whitepapers/ims_/arcims9-architecture.pdf.

FGDC (Federal Geographic Data Committee). 1997. *FGDC Standards Reference Model*. WWW document, http://www.fgdc.gov/standards/refmod97.pdf.

Ferraiolo, J., Fujisawa, J., and Jackson, D. 2003. *Scalable Vector Graphics (SVG): 1.1, Specification*. WWW document, http://www.w3.org/TR/2003/REC-SVG-20030114/.

Green, D. and Bossomaier, T. 2002. *Online GIS and Spatial Metadata*. London: Taylor and Francis.

Hill, L. L., Frew, J., and Zheng, D. 1999. Geographic names: The implementations of a gazetteer in a georeferenced digital library. *D-Lib Magazine* 5(1) (available at http://www.dlib.org/dlib/january99/hill/01hill.html).

Jankowski, P., Stasik, M. I., and Jankowska, M. A. 2001. A map browser for an Internet-based GIS data repository. *Transactions in GIS* 5: 5–18.

Jones, C. B., Purves, R., Ruas, A., Sanderson, M., Sester, M., van Kreveld, M., and Weibel, R. 2002. Spatial information retrieval and geographical ontologies: An overview of the SPIRIT project. In *Proceedings of the Twenty-fifth Annual ACM International SIGIR Conference on Research and Development in Information Retrieval (SIGIR '02)*, Tampere, Finland, pp. 387–8.

Jones, C. B., Abdelmoty, A. I., and Fu, G. 2003. Maintaining ontologies for geographical information retrieval on the Web. In R. Meersman, Z. Tari, and D. C. Schmidt (eds) *Proceedings of the OTM Confederated International Conferences, CoopIS, DOA, and OOBASE*, Catania, Italy: Springer Lecture Notes in Computer Science No 2888: 934–51.

Jones, C. B., Abdelmoty, A., Finch, D., Fu, G., and Vaid, S. 2004. The SPIRIT spatial search engine: Architecture, onologies and spatial indexing. In M. Egenhofer, C. Freksa, and H. Miller (eds) *Geographic Information Science: Third International Conference, GIScience 2004*. Berlin: Springer Lecture Notes in Computer Science No 3234: 125–39.

Kingston, R. 2002. Web-based PPGIS in the United Kingdom. In W. J. Craig, T. M. Harris, and D. Weiner (eds) *Community Participation and Geographic Information Systems*. London: Taylor and Francis, pp. 101–12.

Longley, P. A., Goodchild, M. F., Maguire, D. J., and Rhind, D. W. 1999. *Geographical Information Systems*. New York: John Wiley and Sons.

MacEachren, A. M. 1998. Cartography, GIS and the World Wide Web. *Progress in Human Geography* 22: 575–85.

MacEachren, A. M. 2000. Cartography and GIS: Facilitating collaboration. *Progress in Human Geography* 24: 445–6.

Medyckyj-Scott, D. J. and Morris, B. 1998. The virtual map library: Providing access to Ordnance Survey digital map data via the WWW for the UK higher education community. *Computers, Environment and Urban Systems* 21: 31–45.

Obermeyer, N. 1998. The evolution of public participation GIS. *Cartography and Geographic Information Systems* 25: 65–6.

OpenGIS Consortium. 2002. OpenGIS Web Feature Service Implementation Specification 1.0.0.2002. WWW document, http://www.opengeospatial.org/docs/02-058.pdf.

OpenGIS Consortium. 2003. OpenGIS Reference Model 0.1.2.2003. WWW document, http://www.opengis.org/docs/03-040.pdf.

Peng, Z.-R. and Tsou, M.-H. 2003. *Internet GIS: Distributed Geographic Information Services for the Internet and Wireless Networks*. Hoboken, NJ: John Wiley and Sons.

Purves, R. S., Dowers, S., and Mackaness, W. A. 2002. Providing context in Virtual Reality: The example of a CAL for mountain navigation. In P. F. Fisher and D. Unwin (eds) *Virtual Reality in Geography*. London: Taylor and Francis: 175–89.

Putz, S. 1994. Interactive Information Services Using the World-Wide Web, Hypertext in the First International World-Wide Web Conference. WWW document, http://www.parc.xerox.com/istl/projects/www94/iisuwwwwh.html.

Raggett, D., Le Hors, A., and Jacobs, I. 1999. HTML 4.01 Specification. WWW document, http://www.w3,org/TR/html401/.

Reid, J., Higgins, C., Medyckyj-Scott, D., and Robson, A. 2004. Spatial data infrastructures and digital libraries: Paths to convergence. *D-Lib Magazine* 10 (available at http://www.dlib.org/dlib/may04/reid/05reid.html.

Spainhours, S. and Eckstein, R. 2002. *Webmaster in a Nutshell* (3rd edn). Cambridge: O'Reilly.

Towers, A. L. and Gittings, B. M. 1995. Earthquake monitoring and prediction. A case study of GIS data integration using the Internet. In P. F. Fisher (ed.) *Innovations in GIS 2*. London: Taylor and Francis: 233–43.

Van Elzakker, C. P. J. M. 2001. Use of maps on the Web. In M. J. Kraak and A. Brown (eds) *Webcartography: Developments and Prospects*. London: Taylor and Francis: 37–52.

Chapter 32

Location-based Services and Geographic Information Systems

Allan J. Brimicombe

A century ago, motoring was about as young as Geographic Information Systems (GIS) are today. When veteran cars were new, roads were neither numbered nor signposted. Most towns, villages, and roads did not announce their names on signs as one entered them. In fact, finding one's way around became so troublesome that motoring associations such as the AA and RAC sprang up not just to assist with mechanical breakdowns but also to assist in wayfinding with maps and signage, and this continues to be their dual function. Today we take road signage including all the highway code mandatory and advisory notices for granted – and what a lot there are! Next time you go out in a car, just count how many signs you pass. Given the volume of traffic, could we be trusted today to drive safely without them? Given the huge analog information infrastructure we have surrounding just one form of transport – motoring – could GIS today be on the cusp of ubiquity (Li and Maguire 2003) that motoring was a century ago? Suppose all our wayfinding infrastructure for all forms of transport (including pedestrian) and for all forms of information was digital and accessible through an electronic mobile device, and anytime we wanted information tailored to where we were and what we were doing, all we would have to do would be to consult it. Suppose, further, that the in-built intelligence could tell us things we would like (or ought) to know even without being asked, just based on who we are, where we are, where we've been, and what time of day it is. Could we then get through the day without GIS? Welcome to the brave new future of location-based services!

In this chapter, I will set the context for the emergence of location-based services (LBS) as an application of GIS. LBS can then be defined and placed alongside other GIS-based technologies. I will be exploring some of the data implications of LBS, how LBS users are positioned so that the system knows where they are, and how queries can be expedited. I then explore some applications of LBS and conclude by reading some of the signposts as to what lies on the road ahead. LBS is a newly emerging technology, and as with most other technologies we cannot be sure where it will lead us. All I can do here is reveal what is known, cut through the inevitable hype, and scan the horizon of our possible futures. But one thing is for sure, LBS is an application of exciting potential which integrates nearly all aspects of Geographic Information Science (GISc) as presented in the other chapters of this book.

Context: Technological Convergence

We live in a networked society founded upon modern information and communication technologies (ICT) linking computers with computers, computers with individuals and between individuals on an unprecedented scale. For example, in 2003 an average of 55 million text messages were sent every day from mobile phones in Britain. We appear to have reached the stage where our understanding of and interaction with the real world is mediated through ICT and computers and any further evolution of our species may well be symbiotic with technology. According to Castells (1989, 1996), the ICT revolution of the late 1980s and 1990s has re-structured the material basis of society such that wealth creation is information-centered: information is both a raw material and an output of production as a tradable commodity. Technology time lines, such as those of the National Research Council (1997) and Brimicombe (2003), chart the pace of development. It is sobering to consider that the first IBM PC was launched in 1982, mobile telephones came on the market in the mid 1980s and the World Wide Web (WWW) was launched in 1990 – all of these now ubiquitous technologies and yet emergent within the lifetime of most readers of this chapter. In 2003, mobile phone subscribers will have topped one billion world wide (IDC 2001), pushing ahead of the number of people with online Internet access (ComputerScope 2001). As a commodified product, information tends to have a time-bound currency, a sort of "use by date" beyond which its value or its ability to generate wealth or other utility dramatically declines. This can be both in absolute terms (knowing when to buy or sell some shares on the basis of new information) or in relative/cyclic terms (knowing today's menu at nearby restaurants). Thus the ability to access information at anytime, anywhere, and in any circumstance is not only a desirable end for which organizations and individuals are willing to pay but is now achievable. Fixed-line and wireless broadband offers to make us switched-on and online permanently.

That the technological innovations of the late twentieth century have changed forever the way people work, communicate, relax or otherwise live their lives is beyond doubt. But this trajectory is far from finished. A wide and ever growing range of consumer products contain microprocessors. Whether it be a mobile phone, a DVD player or a car engine, there is likely to be a flow of digital data and the running of software algorithms to make them work properly. In recent years, the convergence and miniaturization of technologies has lead to significant blurring of the more traditional means of accessing electronically-stored information. Thus static devices (for example, the desktop PC) can be serviced by both fixed-line and wireless networks; wireless connectivity has released our ties to static devices for going online thus allowing greater mobility (for example, there are, at the time of writing, in excess of 100 publicly accessible wireless hotspots in central London where laptop and handheld computers can go online); the Internet is accessible from both static and mobile devices (for instance, mobile phones); some new models of mobile phone contain most of the functionality of a personal digital assistant (PDA); PDAs can dial-up for a data connection, and so on. A mobile phone is now a device for the two-way communication of voice, music, data, text, graphics, photos, and video. The increasing mobility of information-access technologies and their ability to be location-aware, that is to be able to have their geographic position fixed continuously and automatically, is giving rise to the possibility of customizing information services and their content based on the location of the recipient.

Defining Location-Based Services

The technological niche occupied by LBS is best discussed in relation to Figure 32.1 where it falls at the intersection of GIS, the Internet, and new information and communication technologies (NICTs).

NICTs are defined here as small portable electronic devices that can be location-aware and incorporate wireless ICT. Examples of such devices would be mobile phones, in-vehicle navigation devices, PDAs, pocket and tablet PCs, certain types of data loggers, and even laptop PCs, though these latter few devices would need the appropriate peripherals to be location-aware and have wireless communication in order to qualify. What is meant by location-awareness is further discussed below under the heading "Queries for LBS," but put simply, it is the ability to know the geographic location of a device (and hence its user) either through means internal to the device (such as integral GPS) or external to the device (such as interfaced GPS or wireless network triangulation).

While the acronym GIS is commonly associated with the software tools for storing, analyzing, and displaying spatial data (for example, ArcView, MapInfo), a broader systems view would extend to the organizational and institutional arrangements that support, for example, the collection of spatial data and dissemination of information products that are equally important as aspects of GIS. On the other hand, spatially-enabled databases such as Oracle Spatial that can be used for the storage and query of spatial data would not on their own be regarded as GIS. This raises the debate as to whether GIS as we currently know them really are at the heart of

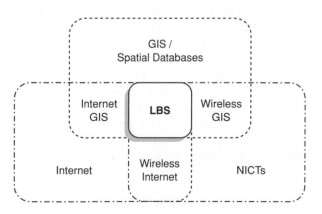

Fig. 32.1 The technological niche of LBS
After Brimicombe 2002

LBS, whether LBS can be adequately driven in the main by spatial databases or whether new hybridized forms of GIS will evolve for LBS. There is insufficient space in this chapter to fully explore this debate (but see Brimicombe and Li 2007) though a flavor can be gained from the sections that follow.

Wireless Internet refers to the ability of 2.5G and 3G mobile phones (and other NICTs) to access web pages, email, and other Internet services through a wireless connection. The distinction between Internet GIS and wireless GIS is a fine one, but one which is nevertheless made (see, for example, Braun 2003). The term Internet GIS as used here covers web-based and online GIS and refers to client-side access to GIS hosted on a server where the access is facilitated via the Internet. Usually the client-side device (PC) is online with an interface to the server-side GIS through a web browser. In the case of mobile GIS, GIS software and data are resident client-side on a mobile device that can then be taken into the field; examples are ESRI's ArcPad and MapInfo's MapX Mobile which run on PDAs. However, data are normally up- or down-loaded periodically on visiting an office (or home) where the mobile device can be attached to the organization's main network. Thus mobile GIS allow field mobility but not real-time access to data and are increasingly used for data collection. Wireless GIS are an extension of mobile GIS for which access to an organization's databases can be real-time via wireless telecommunications and/or local wireless networks. This means that current versions of a database can be accessed at all times as well as having updates initiated from the field. Thus wireless GIS have tended to focus on business operations such as those concerning work orders and inspections. With future increases in bandwidth and improved roaming ability, GIS and other software may become resident server-side with a browser interface client-side thus blurring, though not completely dissolving, the current distinction between Internet GIS and wireless GIS.

With LBS at the intersection of what has been defined above, they will include aspects of all the technologies just discussed. LBS, however, are in the early stages of development with current research establishing both the feasibility and the demand (Leonhardt, Magee, and Dias 1996, Laurini, Servigne, and Tanzi 2001, Sage 2001), and therefore any definition given to LBS at this time should not be taken as definitive but as reflecting our current knowledge of their intended purpose. At their heart, then, location-based services are the delivery of data and information services where the content of those services is tailored to the current or some projected location of the user. It is usually implied that the user is on the move and the delivery of services is to a NICT. The data or information delivered are not restricted to being spatial but could be any information whether conveyed as text, voice, images, or indeed as maps. The important spatial component refers to the known or projected location of the individual requesting the service. Thus, for example, a request by an individual for information might be "what are today's opening hours for Marks & Spencer?" with the response giving today's opening and closing times for the nearest branch of Marks & Spencer to where the individual is (as text or voice) accompanied perhaps by a map showing the location of the branch relative to the individual. Requests for information can of course be spatial – "where can I find . . . ?, "how do I get to . . . ?" – and it is therefore envisaged that major applications of LBS will be in wayfinding and navigation, though by no means being restricted

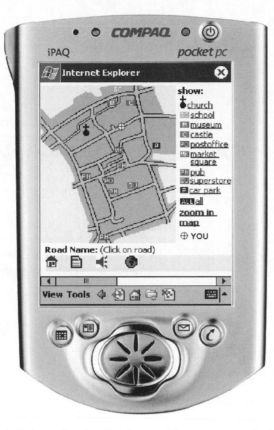

Fig. 32.2 Experimental PDA interface for LBS that provides map, text and voice wayfinding instructions
Copyright C. Li, used with permission

to these. An LBS experimental interface for wayfinding applications incorporating voice, text, and maps is given in Figure 32.2.

With the definition for LBS comes its basic system architecture, which is discussed in relation to Figure 32.3. The system architecture has five broad components:

- The real world in which the user is situated and about which information is required tailored to the user's current or projected location;
- The choice and nature of NICT employed by the user for receiving LBS;
- The positioning systems, such as GPS, which geographically locate the NICT;
- The wireless network through which the query and LBS response are communicated between NICT and server;
- The GIS, database, and distributed services and data components, including Internet connections to the World Wide Web, used to resolve the query and format the response. The data used to resolve the query are, of course, derived from the real world, thus implicitly closing the loop within the system.

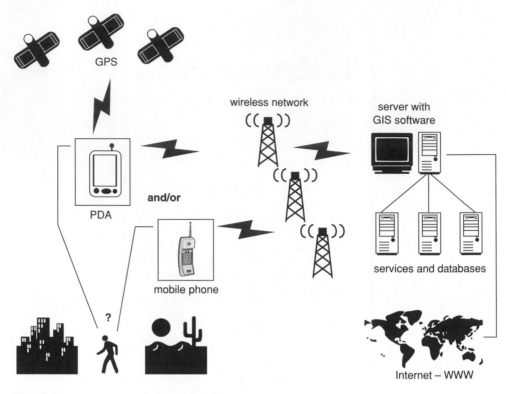

Fig. 32.3 Basic system architecture for LBS

Data Implications

Data are key to any information service, but LBS have implications for data over and above that which is currently required for GIS (Tsou and Buttenfield 2002, Grejner-Brzezinska, Li, Haala, and Toth 2004, Smith, Kealy, Mackaness, and Williamson 2004). In this section three aspects in particular will be discussed: the data themselves; their collection, and their organization. Data and information access via queries will be discussed in the section entitled "Applications" below.

Demand for services inevitably depends on the utility that can be derived from their purchase. In the case of LBS, this will largely depend on the currency, granularity, and fitness-for-purpose of the information provided for decision-making as well as response times and the comprehensibility of the products delivered. Bearing in mind the definition of LBS, requests for services are likely to be highly local in nature and therefore data will need to be fine-grained (high resolution) and up-to-date. This will inevitably mean that data sets will grow dramatically as the granularity is increased and as additional attributes are added. Consider, for example, the sizes of some typical data sets for the UK given in Table 32.1. These are indicative numbers of records, not file sizes. For comparison, taking Greater London as an example, while the Ordnance Survey's Code-Point for the 196,512 postcodes is 12 MB as a text

Table 32.1 Indicative size of some typical LBS data sets for the UK (2003)

Data category	Indicative size (approximate no. of records)
Addresses	24 million
Postcodes	1.7 million
Street gazetteer	764,000 entries (Navtech)
Financial & legal services	190,000 banks, building societies, estate agents, etc.
Hotel, restaurants, pubs	162,000 locations
Local services	230,000 local service and retail companies
Recreation and culture	87,000 locations

file, Address-Point for the 3.6 million addresses is 745 MB as a text file. LBS would require address-level data to be effective, particularly for urban applications. The classes of data that would be required for LBS would typically include: (1) base mapping (particularly street and building level data); (2) points-of-interest (landmarks, destinations), services-of-interest (timetables, availability); (3) navigation information (intersections, signage); (4) real-time information (traffic conditions, weather advisories/warnings); and (5) imagery (aerial, terrestrial, virtual).

As can be seen from the example given in the previous paragraph, it is inevitable that the move towards LBS and its need for high resolution data will result in substantial bulking of spatial databases. The classes of data (themes) accessible through LBS would need to be broadened beyond those of traditionally focused GIS applications. The spatial coverage of data would need to be national, even international. Many of the attributes of interest would not be resident in databases but would come from web pages (similar to a Google or Yahoo search). This makes the integration of data for LBS into a single database repository very unlikely. Indeed, LBS as a viable set of technologies is dependent upon the current trajectory towards interoperable, open systems software (for example, Buehler and McKee 1998), the use of distributed services and software components as facilitated by CORBA, Java, and .NET (Peng and Tsou 2003), and the use of distributed data objects (Tsou and Buttenfield 2002) as an evolution of distributed databases in a networked environment. While some core data sets, such as road networks, may be kept centrally at the main server coordinating an LBS provider's service, all other data will be distributed over intranets and the Internet for which there will be protocols and agreements covering syndication, access, use, and charging. In such a distributed environment, with large and diverse data sets to consider in quickly resolving an information request, metadata play a key role (Flewelling and Egenhoffer 1999). These descriptions of the content and quality of data sets are necessary because of the impracticality of browsing through large and distributed data sets in order to comprehend them. While metadata content standards exist (for example, FGDC 1997) and form the basis for online data clearinghouses, such content advises on data reliability at the point of purchase while providing only limited indication of information potential when combined with other data sets in resolving a query to provide sufficient utility or fitness-for-use (Brimicombe 2003). LBS providers will need to compile (and continually update) metadata databases that reflect user perspectives of information utility for a range of user contexts (for example, daytime versus night-time) so as to automate rapid data selection and combination.

Another implication of the LBS data market will be the need for faster update cycles and the means of achieving this. The currency of information is inevitably a key determinant of its utility. Fortunately, the bundle of technologies that are making LBS a reality are also making real-time data collection and update a reality. Beyond mobile and wireless GIS as technologies for efficient field data collection (in conjunction with GPS and remote sensing), there is the telemetry (by fixed-line or wireless links) of data collected at fixed and mobile installations. Examples of these would include the automated collection and transmission to an operations center of meteorological data, traffic data, river flow, and water level, transport of hazardous materials or the progress of a bus through a congested urban area. Laurini, Servigne, and Tanzi (2001) have classed this as TeleGeoInformation and noted that it has both spatial and temporal components. Grejner-Brzezinska, Li, Haala, and Toth (2004) consider that we are witnessing a paradigm shift in the acquisition, processing, and management of spatial data through a fusion of geoinformatics (geomatics), telecommunications, and mobile computing to produce an emerging discipline of telegeoinformatics (see also Karimi and Hammad 2004). The step change from traditional mapping will result in automatically extracted and integrated vector geometry and imagery that result in high resolution 3D data sets with short update cycles. This will allow, for example, LBS users to access virtual urban models that permit realistic visualization.

Finally, the delivery of LBS to users can in itself be a means of data collection. The technology's location awareness means that the user's location can be tracked. This in itself may provide important contextual information that allow responses to queries to be better personalized particularly if there is sufficient user history to extract a personal profile or if a profile has been pre-customized by the user. Real-time feedback from users via their NICT on, say, the accuracy of information provided can be used as triggers for data update.

Locating the User

Key to LBS is knowing the location of the user so as to appropriately tailor service delivery. Many queries will be based on the current location of the user which then needs to be determined. Within the definition of LBS, however, location may not necessarily refer to the current geographical position of the user but may be some projected (future) location of the user. This latter scenario may take two forms: a medium- to longer-term projected time and location supplied by the user ("Are there any concerts in Edinburgh tonight?") or a short-term projection based on current position and trajectory (vector of movement, such as moving at 100 kph along a motorway). Again, in the latter example, current position of the user needs to be tracked. In this section we will consider global positioning system (GPS) and network-based techniques of locating the user. For a fuller treatment of this topic, see Grejner-Brzezinska (2004) or Brimicombe and Li (2007).

GPS consists of a constellation of 21 operational satellites (plus some spares) orbiting at an altitude of approximately 20,000 km. The first was launched in 1978 and there is a program of continual replenishment of the constellation as satellites go out of service. The important feature of the orbital configuration is that five to eight satellites should always be visible from any point on the Earth. Each satellite emits a time-coded signal which allows the receiver to work out the distance to the satellite. Because each satellite is of known position in its orbit, a minimum of three satellite signals is sufficient for a receiver to calculate its co-ordinate position and elevation on the Earth's surface. Up until May 2, 2000, these signals were intentionally degraded by the US military using Selective Availability. Since Selective Availability has now been switched off, a handheld GPS receiver (no bigger than a mobile phone) can achieve 2D accuracy of 3 to 5 m (see Chapter 28 by Dana in this volume for additional information about GPS technologies). Although PDAs can be readily linked with GPS receivers, the actual chip that receives the signal and carries out the calculations is now about the size of a postage stamp and can thus be readily incorporated into a PDA or mobile phone at the point of manufacture. This would seem ideal at not much extra cost in the hardware, but GPS has an important drawback: it only works when there is a clear view of the sky. In other words it will not work indoors, in covered car parks, or in tunnels, and will work poorly if the receiver is inside a car, train, woodland, or if it is in an "urban canyon" surrounded by high buildings. For this reason, service providers have been looking at a range of alternative techniques based on the telecommunications network itself.

In order to deliver services to mobile phones, providers (e.g. Orange, Vodaphone, Verizon) install a network of cellular base stations (transmitters and receivers) usually country-wide in order to achieve a commercially viable coverage. When a mobile phone is switched on it is registered with the nearest base station (so the provider knows how to route a service to it). The area covered by a base station is called a cell and each cell has an ID or cell global identity (CGI); as the user moves, so the registration is handed over to whichever cell becomes the nearest. Thus a provider always knows where an in-range user is by the cell-ID holding the registration. In urban areas the size of a cell might have a radius of 100 m but in rural areas this may increase to upwards of 30 km. While positioning by cell-ID is quick and inexpensive, it remains too coarse for most LBS. Cell-ID can be augmented by reference to a mobile phone's timing advance (TA) and the directional sector within a cell from which the signal is received. Since it is common for a number of mobile phones to be using a cell simultaneously at the same wavelength, each base station has a system for scheduling the timing of transmission to each mobile phone so they do not interfere with each other. Because distance will cause a delay in the time taken to receive a transmission, each phone is instructed by the base station to advance its transmission time so it arrives correctly for its scheduled time slot. This gives a rough measure of the distance a user is from a base station and can be further augmented by the directional sector. While this provides useful sub-cell location within large rural cells, its usefulness within small urban cells becomes marginal. Although registration is with one cell at a time, a mobile phone will have some communication with all cells within range so that a smooth handover of registration to any adjacent cell can be achieved. Using the time difference of arrival (TDOA) of the signal at three or more base stations allows triangulation of the 2D position. For 3G networks in urban areas, this would give a theoretical accuracy of about 20 m provided the mobile phone is within range of three base stations, which might not always be the case. While the discussion here on network-based positioning methods has been in the context of mobile phones, the same principles apply for any device on a wireless network.

GPS and network-based methods both have their advantages and disadvantages and cannot be expected to work well in all situations. One longer-term hybrid solution is to couple the two by augmenting GPS with cell-ID and transmitted GPS signals from network base stations to give a theoretical accuracy of 10 to 15 m (Bedford 2004).

Queries for LBS

Having requested information from some automated system, we are usually impatient for a response, more so if we are on the move and need to make decisions. As a rule, our patience tends to run out after about 10 seconds and this is taken as the upper limit, for example, in designing web usability (Nielsen 2000). For a high speed processor 10 seconds is ample time to process millions of instructions, but for an LBS query it is not very long at all. Principally, geographically-based queries take much longer to transact than queries on well-structured attributes. Other components that introduce time delay are: the initial transfer of the query from the user to the server, augmentation of the query with the specificity of the user's location, parsing of the query to identify/interpret which data sets are required to fulfill the query, the access time required for the server to obtain the distributed data components and any necessary software services, the formatting of the response (perhaps as a map visualization or as a table), and the final transmission of the response to the user.

The transmission time for substantial amounts of data as would be necessary, for example, in responding to a query with colored vector maps capable of interrogation or 3D virtual reality models, is being taken care of by advances in cellular network technologies. Current Global System for Mobile Communications (GSM) technology (2-G) would be "painfully slow" at 10 kbps (Peng and Tsou 2003, p. 458) while General Packet Radio Services (GPRS) (2.5-G) has a maximum data speed of 115 kbps. The projection for 3-G technologies when fully implemented will be 2 Mbps which is equivalent to ISDN lines. More than ample for any currently conceived LBS application will be 4-G, scheduled for around 2010 with a transmission rate of up to 100 Mbps. For a complete description of these technologies, see Peng and Tsou (2003).

Query augmentation is going to be an important element of LBS and is defined as the increase in specificity of a natural language query on the basis of additional known parameters and contextual information pertaining to the user who generated the query. If, for example, we return to the query "what are today's opening hours for Marks & Spencer?" this would need to be augmented with today's date (today could be a Sunday, a public holiday, or a day on which there are extended opening hours in the run-up to Christmas) and for the response to be tailored geographically, the query would need to be augmented automatically with the user's location (whether GPS-derived co-ordinates, cell ID, and so on). Furthermore, if by tracking the user's location it is evident that the user is moving along a railway line, then context would suggest the query did not pertain to the Marks & Spencer nearest to the user's current location and the server would need to clarify this with the user and appropriately augment the query on the basis of the response. Furthermore, suppose, for example, there was an archived request concerned with rail timetables and the purchase of a ticket and seat reservation for today's date, then the query might be automatically augmented with the user's destination. The augmented natural language query will then need to be translated into a formal query language that instructs a Database Management System (DBMS) on how to resolve the query and parsed to identify which data sets are required. Traditionally, GIS queries are constructed *ab initio* using formal language (such as SQL) or through the intermediary of a user interface that structures the query. Such formalisms directly (and exactly) specify the data sets (or GIS coverages) to be used (see Chapter 7 by Shekhar and Vatsavai in this volume for more information about database queries). Progress towards natural language queries in GIS is summarized by Wang (2003). Most natural language queries are fuzzy in nature – terms such as "near" or "on the way to" are vague – and will either need to be translated into the existing precise formalisms (for example, Wang 2003) or both the ambiguity of the query and the data sets need to be captured and resolved within the formalism of the query language and the spatial data structures on which they operate (see, for example, Brimicombe 1997, 1998). In parsing the query, use would have to be made of the metadata databases discussed above under the heading "Data Implications" in order to select relevant data sets, but recourse to Internet search engines and on-the-fly data mining may also be needed.

GIS-based queries of spatial relations are traditionally based on the topology of static objects. A body of GIS research has focused on defining and providing a mathematical framework of 2D topological relationships (for example, Egenhofer and Herring 1990, Mark and Egenhofer 1994). For a pair of region objects in 2D space there are eight basic spatial relations (separate, meets, partial overlap, covers, is covered by, is inside, contains, equal overlap) while for a linear object and a region object there are 19 basic spatial relations. When looking at temporal relationships between objects, such as "α happens before β" or "α happens at the same time as β," there are 13 basic topological relations (Peuquet 1999). But again, these are for temporally static objects. In LBS the problem is made more complex in that the user may not be static but dynamic. The nearest commercial GIS come to performing queries concerning dynamic objects is in resolving shortest path (or some other criteria) along a network. GIS queries will need to resolve new dynamic relations such as "α [dynamic] approaches β [static]" or "α [dynamic] and β [dynamic] converge." Theodoridis (2003) has recently devised a set of ten benchmark database queries for LBS with relations based on co-ordinates, distance, direction, nearest-neighbor, topology (including network), and buffer. Queries for LBS will continue to be an important research area.

Applications

True applications of LBS are at an early stage and it is inevitable that the emergence of any new technology is accompanied by considerable hype as providers attempt to attract customers and carve out a market share. With rapidly growing numbers of individuals using NICTs – already in excess of one billion – the market is potentially big business. Growth in LBS is, however, very much dependent on the roll out of 3G technologies which, to date, have been slow to materialize. Applications are likely to be as varied as any other technology that is flexible, scalable, and has wide appeal. There are a number of ways we could classify these applications. One is to differentiate "pull," that is user-initiated, from "push" which would be provider-initiated. An example of a pull application would be a request for traffic information. A push application might be the automatic broadcast of a relevant warning (for example, burst water main causing road closure and/or cut-off of water supply) or online advertising of "specials" and discounts as one comes into range of the relevant outlet. Whether we are ready to cope with what might be a flood of pushed online advertising segmented and differentiated by the geo-demographics of our home post-code is another matter! Another way to classify LBS applications is by broad areas of activity regardless of "push" or "pull." Thus, the presently envisaged areas of activity likely to take advantage of LBS for some, if not all, of the time are:

- Navigation, such as in-car navigation systems, which identify appropriate routes for a vehicle to be driven from some start point to a destination (or series of destinations);
- Wayfinding, which is the discovery of routes, modes of transport, and other spatially located objects by an individual on the move. The distinction between navigation and wayfinding varies between disciplines;
- Mobile commerce (m-commerce), which would incorporate both transactions made by individuals on the move via their NICT or the receiving of intelligence (advertising, marketing materials) of opportunities for transactions that are location specific;
- Real-time tracking of vehicle fleets, business associates, social contacts (to know when a friend is nearby) or one's family members (such as tracking children home from school by parents who are at work);
- User-solicited information for all kinds of business and social purposes such as weather forecast, traffic conditions, delays to trains and flights, film showings and ticket availability, menus, local maps, and so on;
- Location-based tariffs such as differentials in road pricing, pay-as-you-go car insurance, and similar schemes;
- Fulfillment of field-based work orders including deliveries, inspections, and data collection;
- Coordinating emergency and maintenance responses to accidents, interruptions of essential services and disasters;
- Artistic expression in the community.

Given the pace of development, this list should not be taken as exhaustive. Some of these may sound far-fetched, but they are based on services that are either under development or are offered now. For example, Japanese parents are already able to track their kids traveling home after school through the cell ID of children's mobile phones. In the UK, Vodaphone has started allowing third parties to offer LBS over its network. At present these are driven by text keywords sent using a short-code service with replies by a Short message service (SMS). Thus "club" sent to 80400 will return the address of the nearest club frequented by celebrities. In Germany, the city of Darmstadt offers a more sophisticated service: Da-Mobil (Janosch 2003). Through a wireless Local area network (LAN), enabled PDAs can download route maps providing navigation information, video, 3D visualizations and web pages on surrounding points of interest. Interactive in-car navigation systems are already a standard feature in luxury cars and a not so expensive optional extra available for all vehicles. In Times Square, passers-by can engage in a Yahoo cellphone game displayed on a building facade. Norwich Union, one of the UK's largest insurance companies, is collaborating with network provider Orange to develop a pay-as-you-go car insurance scheme so that premiums would be based on when, where, and how often a vehicle is used. The Urban Tapestries project (Lane 2003, http://research.urbantapestries.net) has conducted experiments in Bloomsbury, London, whereby individuals using wireless-enabled PDAs can create or follow virtual threads (annotated with text, sound, and images) which link individual experience and contexts to local features as a means of social and artistic expression. Perhaps the most bizarre application of LBS to date is a GPS-enabled phone strapped to hunting dogs in Finland so that the dogs can be tracked. The owner can tell what type of animal the dog is hunting from the dog's bark and the owner can relay orders to the dog. Not so much wayfinding as woof-finding!

Numerous applications for mobile devices are in various stages of development, offering different degrees of sophistication depending on the nature of the expected user and the class of NICT available to them. The next five to ten years will see an exponential expansion in the numbers of services offered, some free others to be paid for. Their common characteristic is that they are location-specific, tailored to where one is. These are location-based services.

Only the Beginning

This chapter has focused on a number of key technological aspects of LBS. Many of these depart from the usual discussions of GISc, and indeed GISc *per se* has not featured prominently. GIS as technology or toolbox was around for nearly 30 years before debates around the scientific foundation for such a substantive technology arose (see Goodchild 1992 for details). It was not until 1997, for example, that one of the leading research journals in the field changed its name from the *International Journal of Geographical Information Systems* to the *International Journal of Geographical Information Science*. Now we are reaching an important new phase: if science is about discovering new knowledge and making it available to society, engineering is the systematic and reliable application of that knowledge in society. At the heart of LBS is geographic location; at the heart of LBS technology is the application of most aspects of GISc for managing, processing, and delivering spatial information. LBS is therefore a prime example of an emerging field of endeavor which we can label: geo-information engineering. Like most engineered products, as lay users we are mostly unaware of the detailed scientific foundations. Thus we use bridges, lifts, and home appliances in the belief that they are safe and without a detailed knowledge of their engineering. The challenge for GIS is for ubiquitous applications, such as LBS, to emerge without users being particularly aware that they are engaging with GISc, that GISc can be a foundation science and technology in everyday engineered products. That the public can engage with GISc without really knowing it will mean the difference between GIS remaining a niche technology and emerging as a quotidian technology. I'm confident that the GISc community can rise to the challenge.

We should, however, end on a cautionary note. Tenner's (1997) revenge theory of technological innovation means that new technologies never ultimately solve the problem for which they were designed without creating new ones along the way. What is more, the new problems tend to be shifted in space and time becoming more hidden and therefore dangerous. But these are also opportunities for researchers and providers of LBS alike. As with other forms of engineering, new applications can challenge or question the science and any outstanding problems and failures must be carefully analyzed for new knowledge that is then fed back into research. We are, however, only at the beginning of LBS; a rich journey lies ahead.

REFERENCES

Bedford, M. 2004. Are you ready for the ride? *GEOconnexionUK* 3: 50–1.

Braun, P. 2003. *Primer on Wireless GIS*. Park Ridge, IL: Urban and Regional Information Systems Association.

Brimicombe, A. J. 1997. A universal translator of linguistic hedges for the handling of uncertainty and fitness-for-use in Geographical Information Systems. In Z. Kemp (ed.) *Innovations in GIS 4*. London: Taylor and Francis, pp. 115–26.

Brimicombe, A. J. 1998. A fuzzy co-ordinate system for locational uncertainty in space and time. In S. Carver (ed.) *Innovations in GIS 5*. London: Taylor and Francis, pp. 143–52.

Brimicombe, A. J. 2002. GIS: Where are the frontiers now? In *Proceedings of GIS 2002*, Manama, Bahrain, pp. 33–45.

Brimicombe, A. J. 2003. *GIS, Environmental Modelling and Engineering*. London: Taylor and Francis.

Brimicombe, A. J. and Li, C. 2007. *Location-Based Services and Geo-Information Engineering*. Chichester: John Wiley and Sons.

Buehler, K. and McKee, L. 1998. *The Open GIS Guide: Introduction to Interoperable Geoprocessing and Open GIS Specification*. Wayland, MA: Open GIS Consortium Inc.

Castells, M. 1989. *The Informational City: Information Technology, Economic Restructuring and the Urban-regional Process*. Oxford: Blackwell.

Castells, M. 1996. *The Information Age: Economy, Society and Culture: Volume 1, The Rise of the Network Society*. Oxford: Blackwell.

ComputerScope. 2001. *How Many Online?* Dublin: ComputerScope Ltd.

Egenhofer, M. J. and Herring, J. R. 1990. A mathematical framework for the definition of topological relationships. In *Proceedings of the Fourth International Symposium on Spatial Data Handling*, Zurich, Switzerland. Columbus, OH: International Geographical Union, pp. 803–13.

FGDC. 1997. *Content Standard for Digital Geospatial Metadata*. Washington, DC: Federal Geographic Data Committee.

Flewelling, D. M. and Egenhofer, M. J. 1999. Using digital spatial archives effectively. *International Journal of Geographical Information Science* 13: 1–8.

Goodchild. M. F. 1992. Geographical information science. *International Journal of Geographical Information Systems* 6: 31–45.

Grejner-Brzezinska, D. A. 2004. Positioning and tracking approaches and technologies. In H. A. Karimi and H. Hammad (eds) *Telegeoinformatics: Location-Based Computing and Services*. Boca Raton, FL: CRC Press, pp. 69–110.

Grejner-Brzezinska, D. A., Li, R., Haala, N., and Toth, C. 2004. From mobile mapping to telegeoinformatics: Paradigm shift in geospatial data acquisition, processing and management. *Photogrammetric Engineering and Remote Sensing* 70: 197–210.

IDC. 2001. *Erratic Signals: Worldwide Handset Market Forecast and Analysis, 2000–2005*. Framington, MA: International Data Group.

Jasnoch, U. 2003. GIS-based location services: A new service for the City of Darmasdt. *GeoInformatica* 6: 24–5.

Karimi, H. A. and Hammad, H. 2004. *Telegeoinformatics: Location-Based Computing and Services*. Boca Raton, FL: CRC Press.

Lane, G. 2003. Urban Tapestries: Wireless Networking, Public Authoring and Social Knowledge. WWW document, http://www.probiscus.org.uk/urbantapestries/Unis_WW_paper.html.

Laurini, R., Servigne, S., and Tanzi, T. 2001. A primer on TeleGeoProcessing and TeleGeo-Monitoring. *Computers, Environment and Urban Systems* 25: 248–65.

Leonhardt, U., Magee, J., and Dias, P. 1996. Location service in mobile computing environments. *Computers and Graphics* 20: 627–32.

Li, C. and Maguire, D. 2003. The handheld revolution: Towards ubiquitous GIS. In P. A. Longley and M. Batty (eds) *Advanced Spatial Analysis*. Redlands, CA: ESRI Press, pp. 93–210.

Mark, D. M. and Egenhofer, M. J. 1994. Modeling spatial relations between lines and regions: Combining formal mathematical models and human subjects testing. *Cartography and Geographic Information Systems* 21: 95–212.

National Research Council. 1997. *The Future of Spatial Data and Society*. Washington, DC: National Academy Press.

Nielsen, J. 2000. *Designing Web Usability*. Indianapolis, IN: New Riders.

Peng, Z. R. and Tsou, M. H. 2003. *Internet GIS: Distributed Geographic Information Services for the Internet and Wireless Networks*. Hoboken, NJ: John Wiley and Sons.

Peuquet, D. J. 1999. Time in GIS and geographical databases. In P. A. Longley, M. F. Goodchild, D. Maguire, and D. W. Rhind (eds) *Geographical Information Systems: Principles and Technical Issues*. New York: John Wiley and Sons, pp. 91–103.

Sage, A. 2001. Future positioning technologies and their application to the automotive sector. *Journal of Navigation* 54: 321–8.

Smith, J., Kealy, A., Mackaness, W., and Williamson, I. 2004. Spatial data infrastructure requirements for location based journey planning. *Transactions in GIS* 8: 23–44.

Tenner, E. 1997. *Why Things Bite Back: Technology and the Revenge of Unintended Consequences*. New York: Vintage Books.

Theodoridis, Y. 2003. Ten benchmark database queries for location-based services. *Computer Journal* 46: 713–25.

Tsou, M. H. and Buttenfield, B. P. 2002. A dynamic architecture for distributed geographic information services. *Transactions in GIS* 6: 355–81.

Wang, F. 2003. Handling grammatical errors, ambiguity and impreciseness in natural language queries. *Transactions in GIS* 7: 103–21.

Chapter 33

Geographic Information Science: The Grand Challenges

Michael F. Goodchild

Many chapters in this book have touched on aspects of the research agenda – the research questions that arise in the mind of a user of Geographic Information Systems (GIS); the research problems that need to be solved to enable the next generation of GIS technology; in essence the science behind the systems. Since 1992 there have been various broader efforts to define a more-or-less-complete agenda for Geographic Information Science (GISc), and many of them have been framed in terms of challenges to the research community. More broadly still, many fields of science have attempted to provide long-term motivation – and not incidentally to open sources of funding – by identifying grand challenges, themes that are capable of directing research to a common, distant, but imaginable end. President John F. Kennedy's famous challenge of 1960 – to "put a man on the Moon by the end of this decade" – resulted in an unprecedented peacetime integration of science, engineering, political support, and, of course, a successful ending. In a somewhat similar vein biologists have been calling for the completion of "the web of life," the identification of all of the species (only a fraction of which are currently known to science; May 1988). The mapping of the human genome is a similarly integrated effort across a multitude of laboratories and institutions, tied together by computer networks. Could there be a grand challenge in GISc that could drive a decade-long effort by the research community?

In the first part of the chapter I review the various published research agendas of GISc. I then discuss a further type of challenge: the need to understand the nature of the geographic world. The third major section examines broader themes within science, and the degree to which they might translate into research challenges for GISc and institutional challenges for the GISc community. The final major section discusses the concept of Digital Earth as a grand challenge for GISc.

The Research Agendas of GISc

The late 1980s

It seems appropriate to begin this review in the late 1980s, because several events at that time helped to dramatically alter the landscape of the mapping sciences. In the UK, the Department of the Environment's Committee of Enquiry into the Handling of Geographic Information (the Chorley Committee; Department of the Environment 1987) saw three specific stimuli: the rapidly falling costs of hardware, which had reduced the cost of entry into GIS and related activities from US$500,000 at the beginning of the decade to US$10,000 at the end; the advent of COTS (commercial, off-the-shelf) software to perform the basic operations of GIS; and rapid growth in the availability of spatially referenced digital data. In the USA, the National Science Foundation announced a competition for a National Center for Geographic Information and Analysis (NCGIA), to advance the theory and methods of GIS, to promote the use of GIS across the sciences, and to increase the nation's supply of experts in GIS.

Rhind (1988) presented a research agenda for GIS, identifying problems in what he termed the handling of geographic data: the volumes of data involved, the numerous types of queries that might be addressed, the prevalence of uncertainty in geographic data; the need for integration of data among organizations; and the lack of awareness of such issues as scale. He recognized that the solution of the more generic of these issues would come with time from mainstream information technology; but that issues that were more specific to the geographic case would have to be solved by an active research community focused on GIS. He saw a substantial role for knowledge-based or expert systems in the automated extraction of features from images, the integration of disparate data sets, the development of intelligent search procedures, the automation of cartographic generalization, the development of machine-based tutors, and the elicitation of knowledge from data. He also recognized the importance of research into better methods of visualization for geospatial data, the role of organizations, the legal issues of liability and intellectual property, and the costs and benefits of GIS.

The NCGIA research agenda (NCGIA 1989) has much in common with Rhind's, but already shows signs of a search for the more fundamental issues of GISc, in contrast to the practical issues of GIS. Five major research areas are identified:

- **Spatial analysis and spatial statistics,** the techniques used to model uncertainty in geospatial data, to mine data for patterns and anomalies, and to test theories by comparison with reality;
- **Spatial relationships and database structures,** addressing the representation of real geographic phenomena in digital form, and the interface between digital structures and human reasoning;
- **Artificial intelligence and expert systems,** reflecting Rhind's concern for the role of advanced machine intelligence in GIS operations;
- **Visualization,** and the need to advance traditional cartography to reflect the vastly greater potential of digital systems for display of geographic data;

- **Social, economic, and institutional issues,** the host of social issues surrounding GIS.

The NCGIA went on to propose 12 specific research initiatives within this general framework:

- **Accuracy of spatial databases,** focusing on error models for geographic data with strong links to the discipline of statistics;
- **Languages of spatial relations,** including principles of spatial cognition and linguistics;
- **Multiple representations,** the need to integrate representations of the Earth's surface at different scales and levels of generalization;
- **Use and value of geographic information in decision making;**
- **Architecture of very large GIS databases;**
- **Spatial decision support systems,** the design of systems to support decision-making by groups of stakeholders;
- **Visualization of the quality of geographic information** through methods that explicitly display information about the uncertainty associated with data;
- **Expert systems for cartographic design,** using intelligent systems to augment the skill of cartographers;
- **Institutions sharing geographic information,** including research on the impediments to sharing between agencies;
- **Temporal relations in GIS,** the extension of GIS data models to include time;
- **Space-time statistical models in GIS,** the extension of spatial analysis to include time;
- **Remote sensing and GIS,** researching the issues involved in the integration of data acquired by remote sensing with data from other sources.

Eventually, NCGIA sponsored a total of 21 research initiatives between 1988 and 1996; reports, papers, and other products are available at http://www.ncgia.org.

The 1990s

Very substantial progress was made on most of these topics in the years following their publication. In addition, four factors contributed to the evolution of these research agendas in the 1990s: (1) the continued arrival of new technologies, including most notably the WWW, the Global Positioning System, object-orientation, and mobile computing; (2) the broadening of the research community to include active participation by new disciplines, notably cognitive science, computer science, and statistics; (3) the trend away from technical issues of systems to fundamental issues of science; and (4) the recognition that certain topics were, in effect, dead ends. This last factor perhaps accounts for the virtual disappearance of expert systems, despite their prominence in Rhind's 1988 agenda.

In 1996 the recently formed University Consortium for Geographic Information Science (http://www.ucgis.org) published the first edition of its research agenda (UCGIS 1996), the result of a successful consensus-building exercise amongst the

30 or so research institutions that were then members (the number has since risen to more than 60). The agenda had 10 topics:

- **Spatial data acquisition and integration,** including new sources of remote sensing, ground-based sensor networks, and fusion and conflation of data from different sources;
- **Distributed computing** and the issues of integrating data and software over large heterogeneous networks;
- **Extensions to geographic representations,** addressing particularly the third spatial dimension and time;
- **Cognition of geographic information,** including studies of the processes by which people learn and reason with geographic data, and interact with GIS;
- **Interoperability of geographic information,** including research to overcome the difficulties of different formats and lack of shared understanding of meaning;
- **Scale** and the complex issues surrounding representations at different levels of detail;
- **Spatial analysis in a GIS environment,** advancing the analytic capabilities of GIS;
- **The future of the spatial information infrastructure** and the institutional arrangements that provide the context for GIS;
- **Uncertainty in geographic data and GIS-based analysis** including the modeling and visualization of data quality;
- **GIS and society,** the study of the impacts of GIS on society, and the societal context in which the technology is used.

UCGIS later added four emerging themes to the list:

- **Geospatial data mining and knowledge discovery,** the development of methods for extracting patterns and knowledge from very large data sources;
- **Ontological foundations of geographic information science,** addressing the fundamental components on which our knowledge of the Earth's surface is based;
- **Geographic visualization;**
- **Remotely acquired data and information in GI Science.**

The UCGIS completely revised this list in 2002, replacing it with lists of long-term research challenges and short-term research priorities.

The theme of visualization has been taken up and developed by the International Cartographic Association's (ICA's) Commission on Visualization and Virtual Environments. The four-part research agenda (MacEachren and Kraak 2001) includes:

- Cognitive and usability issues in geovisualization;
- Representation and its relationship with cartographic visualization;
- The integration of geographic visualization with knowledge discovery in databases and geocomputation;
- User interface issues for spatial information visualization.

More generally, the list of commissions of the ICA reflects the broad interests of the association, including many research topics.

The International Society for Photogrammetry and Remote Sensing also provides a useful insight into contemporary research needs, with particular emphasis on imaging systems. Its seven permanent commissions address:

- Sensors, platforms, and imagery;
- Systems for data processing, analysis, and representation;
- Theory and algorithms;
- Spatial information systems and digital mapping;
- Close-range vision techniques;
- Education and communications;
- Resource and environmental monitoring.

In 1998, NSF sponsored a workshop under its Digital Government Initiative to explore ways of improving geographic information services (National Computational Science Alliance 1999). The workshop made 10 recommendations, all aimed at advancing GISc research:

- Advance research efforts directed toward the study of optimizing geographic query mechanisms and incorporating geometry and spatial relational operations;
- Develop improved mechanisms for storing and representing time-varying geospatial data;
- Support research on integrating spatial data fusion from multiple agencies, distributed data, and multiple collection devices;
- Support research on multiple representations/interfaces focused on task-specific (procedural) workflow classes;
- Support research in developing algorithms for knowledge discovery applied to very large, frequently updated spatial data sets such as those derived from spaceborne Earth-monitoring sensors;
- Support research in the theory and methods of representing data with varying degrees of exactness and reliability;
- Support research in the context of decision-making to improve the representation of diverse data and the dynamics of geographic phenomena;
- Extend the promise of cognitive research to make geographic information technologies more accessible to inexperienced and disadvantaged users and also examine how government information policies affect access to and use of geospatial data for a broad spectrum of public and private sector stakeholders;
- Support research to examine commerce's issues in geospatial information such as preserving privacy despite geographic locators and breaking potential bottlenecks in distributing geographic information services due to GIS's unique workflow processes;
- Develop a Geospatial Digital Government Prototyping Center to create a network for testing and developing processes consistent with US priorities for geographic information technologies and services in the government workplace.

More recently, the Center for Mapping of The Ohio State University organized a 2002 workshop on Geographic Information Science and Technology (GIS&T) in a Changing Society. The perspective of the workshop was notably toward societal issues, and it identified six research areas:

- Geospatial data availability: its sources and influences;
- GIS&T workforce studies;
- Conditions associated with the adoption of GIS&T-based approaches;
- Spatial understandings, or the cognitive ability of people to work with spatial data;
- Cross and longitudinal studies of the use of GIS&T;
- Improved tools for the societal evaluation of GIS&T activities.

Finally, the National Research Council's Computer Science and Telecommunications Board reported in 2003 on a study of research needs at the intersection of GISc and computer science (NRC 2003). It identified two over-arching themes: the need for an integrative, multidisciplinary approach to research, and the need to address issues of policy. Within this context, it proposed eight research topics:

- Accessible location-sensing infrastructure, based on systems that know their location;
- Mobile environments, freeing users from the desktop;
- Geospatial data models and algorithms;
- Geospatial ontologies;
- Geospatial data mining;
- Geospatial interaction technologies;
- Geospatial for everyone, everywhere;
- Collaborative interactions with geoinformation.

Towards Synthesis

Although each of these studies and reports contains an extensive list of topics, several common themes are apparent, as well as several consistent trends through time. Mark (2003) has analyzed several of the lists, including the topics included in my 1992 paper in which I proposed the term "geographic information science" (Goodchild 1992).

A somewhat different approach to framing the research agenda was taken by NCGIA's Project Varenius, a research effort begun in 1996 to advance the fundamentals of GISc, with funding from the National Science Foundation (Goodchild, Egenhofer, Kemp, Mark, and Sheppard 1999, Mark, Freksa, Hirtle, Lloyd, and Tversky 1999, Egenhofer, Glasgow, Günther, Herring, and Peuquet 1999, Sheppard, Couclelis, Graham, Harrington, and Onsrud 1999). In this strikingly simple model, GISc was anchored by three concepts – the individual, the computer, and society – represented by a triangle, with GISc at the core. Research about the individual would be dominated by cognitive science, and its concern for understanding of spatial concepts, learning and reasoning about geographic data, and interaction with the computer. Research about the computer would be dominated by·issues of representation, the adaptation of new technologies, computation, and visualization. Finally, research about society would address issues of impacts and societal context. Many research issues would involve the interaction between the three corners of the triangle.

A Natural Science?

It will be clear from the previous section that the research agenda of GISc is diverse, covering issues of technology, society, and human cognition. This section introduces a fourth, which as been largely neglected to date – the dependence of GISc on an understanding of the nature of the Earth's surface.

Many decisions are made in the design of a GIS, and more generally the design of any technology that must process geographic information. They include decisions about data models, data structures, indexing schemes, and algorithms – and about the set of analytic routines and processing functions. Many of these decisions are, in turn, dependent on expectations about the nature of geographic information. For example, a decision to represent rivers as polylines (sequences of points connected by straight segments) is a compromise, balancing the disadvantages of representing a smoothly bending river with a series of straight lines and sharp corners against the advantages of using such a simple geometry (intersections between straight lines are much easier to compute than intersections between curves). Yet, while there clearly are expectations about the nature of geographic information, very few attempts have been made to research the topic systematically, or to assemble what is known in coherent fashion.

The best-known statement is probably Tobler's, generally known as the First Law of Geography: "all things are related, but nearby things are more related than distant things." This first appeared in a paper on urban growth in Detroit (Tobler 1970), and formed the subject of a recent forum in the *Annals of the Association of American Geographers* (see Sui 2004, Barnes 2004, Miller 2004, Phillips 2004, Smith 2004, Goodchild 2004, and Tobler 2004 for additional details). More formally it is a statement about the endemic presence of positive spatial autocorrelation in geographic information, and thus of the principle underlying the entire field of geostatistics.

The consequences of Tobler's First Law (TFL) for GIS design are profound. If it were not true, and nearby things were as different as distant things, then all forms of spatial interpolation would be impossible, along with the derivative processes of contour mapping and resampling. All advanced GIS data structures would be impossible, since there would be no basis for assuming that terrain could be represented as a mesh of triangles, or that points with similar characteristics could be grouped into polygons. One can go further and argue that a geographic world without TFL would be impossible to learn about or describe, since every point would be independent of its most immediate surroundings. There are of course exceptions, and TFL is not a deterministic law. It is possible, for example, for spatial independence to exist over distances in excess of what geostatisticians would define as the phenomenon's range, and it is possible for negative spatial autocorrelation to exist at certain scales (at the scale of the cells in a checkerboard, for example).

Anselin (1989) has argued that TFL, or the principle of positive spatial auto-correlation, is one of two endemic properties of geographic information. The other is spatial heterogeneity, or the tendency for properties to vary from one area to another over the Earth's surface. In the terms of spatial statistics, this is a first-order effect, or a property of places taken one at a time, while TFL describes a second-order effect, a property of places taken two at a time. For that reason it might be preferable if TFL were the Second Law, and spatial heterogeneity the basis of a First Law.

The consequences of spatial heterogeneity are also profound. If the Earth's surface is heterogeneous, it follows that standards and design decisions adopted in one region, and designed for the conditions of that region, will be different from those adopted in other regions. Spatial heterogeneity thus explains the lack of inter-operability between the various classification schemes used in geographic information, and the tension between local geodetic datums and global ones. It dictates that the results of any analysis will depend explicitly on the bounds of the analysis, and will change when the bounds change. It also makes a compelling case for the new place-based or locally centered methods of spatial analysis, such as Geographically Weighted Regression (Fotheringham, Brunsdon, and Charlton 2002) and LISA (Anselin 1995).

Additional generalizations can be made not so much about the nature of geographic information, as about the nature of its representation. Many geographic phenomena reveal more detail as they are examined more closely, and the rate at which additional detail is revealed is to some degree predictable. Mandelbrot (1982) termed such phenomena fractals, and showed that fractal properties were broadly characteristic of geographic information. They imply a degree of predictability in the effects of scale change, allowing better-informed design decisions to be made regarding hierarchical structures in GIS. I have argued that the endemic presence of uncertainty in GIS representations has profound impacts on GIS design, and leads to an entirely different approach to the representation of position (Goodchild 2002).

These cases all point to the need for GISc to address the nature of geographic information, and in effect to become in part a natural science, comparable to physics, chemistry, or biology, with its own unique domain of study in the natural world. Of course this vision of the domain of GISc must include phenomena that are of human origin, or are influenced by humans.

Broader Themes

Since its inception in the 1960s, GIS has become useful and almost indispensable in a vast range of human activities, from Earth science to human health, and from transportation to resource management. It is the enabling technology that has permitted utility companies to move to a higher level of efficiency in their management of distributed networks, and package delivery companies to save millions in delivery costs. Advances in GISc are essential to the further development of GIS, and the key to the success of the technology's next generation.

Despite its importance, however, GIS remains a comparatively small application of information technology, and as such it must rely on the larger mainstream for many developments. The relational and object-oriented databases now widely used in GIS are mainstream products, that would probably not have been developed if GIS was their only market; and at the same time GIS has relatively little influence on such developments.

In this context, one might argue that the future of GIS lies not in the specialized research agenda of GISc, but in broader research agendas that will determine the future of information technology. Many of the research agenda topics identified above are indeed more general than GIS, and one can expect a broader set of minds to be interested in them. To what extent, for example, are issues of ontology and interoperability unique to GIS, and to what extent are they common to a much larger domain? Semantic interoperability is a problem common to all applications that rely on the meaning of terms, and one might therefore expect solutions to come from a number of disciplines, not only from GISc.

This argument has two important implications. First, it suggests that GISc must continue to look more broadly to developments that may have potential for GIS. A case in point is the Grid, a term that encompasses a range of technologies and research efforts aimed at integrating the distributed computing resources of widely dispersed communities into transparent wholes. The bandwidth of the Internet now makes it possible for the components of computing – the processor, data, and software – to be distributed virtually anywhere, and for a user at a desktop to access what amounts to a previously unimaginable resource. Services need no longer be provided at the desktop, but can be invoked from servers located any-where on the network, and might include any of the functions currently performed locally by a GIS. Several GI Services are already available (the Geography Network, http://www.geographynetwork.com, includes a directory), but much research remains to be done to explore GIS applications of the Grid.

Second, it suggests that it is in the strategic interest of the GISc community to generalize its efforts wherever possible. One such generalization would be from geographic to spatial, to explore domains defined by spaces other than that of the Earth's surface. The space of the human brain, for example, has similarities and also important differences, that can lead both to new applications for GIS, and also to the cross-fertilization of research. While there is only one Earth, each human brain is different, requiring brain researchers to develop techniques for defining a generic brain, and for mapping each individual brain to it. One can ask whether other spaces display the same properties as are observed for geographic space, such as TFL, and whether information system design considerations are therefore the same or different for these other spaces.

A Grand Challenge: Digital Earth

The previously cited NSF workshop (National Computational Science Alliance 1999) asked whether grand challenges existed for GIS, and in its report listed four research themes that it felt might merit this distinction:

- To find ways to express the infinite complexity of the geographical world in the binary alphabet and limited capacity of a digital computer (the representation challenge);
- To find ways of summarizing, modeling, and visualizing the differences between a digital representation and real phenomena (the uncertainty challenge);
- To achieve better transitions between cognitive and computational representations and manipulations of geographic information (the user interface challenge); and
- To create simulations of geographic phenomena in a digital computer that are indistinguishable from their real counterparts (the modeling challenge; in effect a Turing test of GIS-based modeling).

All of these have intellectual depth, but somehow lack the compelling appeal of a mega-project. But the concept of Digital Earth (DE) perhaps has the ability to capture popular imagination. The term was coined by Gore in 1992 and elaborated in a much-quoted 1998 speech (http://www.digitalearth.gov/VP19980131.html):

> Imagine, for example, a young child going to a Digital Earth exhibit at a local museum. After donning a head-mounted display, she sees Earth as it appears from space. Using a data glove, she zooms in, using higher and higher levels of resolution, to see continents, then regions, countries, cities, and finally individual houses, trees, and other natural and man-made objects. Having found an area of the planet she is interested in exploring, she takes the equivalent of a "magic carpet ride" through a 3-D visualization of the terrain. Of course, terrain is only one of the numerous kinds of data with which she can interact. Using the system's voice recognition capabilities, she is able to request information on land cover, distribution of plant and animal species, real-time weather, roads, political boundaries, and population. She can also visualize the environmental information that she and other students all over the world have collected as part of the GLOBE project. This information can be seamlessly fused with the digital map or terrain data. She can get more information on many of the objects she sees by using her data glove to click on a hyperlink. To prepare for her family's vacation to Yellowstone National Park, for example, she plans the perfect hike to the geysers, bison, and bighorn sheep that she has just read about. In fact, she can follow the trail visually from start to finish before she ever leaves the museum in her hometown. She is not limited to moving through space, but can also travel through time. After taking a virtual field-trip to Paris to visit the Louvre, she moves backward in time to learn about French history, perusing digitized maps overlaid on the surface of the Digital Earth, newsreel footage, oral history, newspapers and other primary sources. She sends some of this information to her personal e-mail address to study later. The time-line, which stretches off in the distance, can be set for days, years, centuries, or even geological epochs, for those occasions when she wants to learn more about dinosaurs. (Gore 1992)

This vision of DE raises numerous problems (Goodchild 1999, 2000). First, it implies that data structures can be found that support a smooth zooming from resolutions as coarse as 10 km (whole-Earth view) to near 1 m (individual houses and trees). Research on this topic has been under way for many years, and implementations are now widely available (in ESRI's ArcGlobe and Google Earth, for example). Second, it presents enormous problems of data volume, since there are

5×10^{14} m^2 on the Earth's surface. It raises problems of visual rendering, since although some data (for example, terrain elevation) are easily rendered in three-dimensional views, others (for example, average income) would have to be communicated symbolically. Perhaps more problematic are the institutional issues, since DE would require smooth interoperability between data sets, and collaboration by numerous data suppliers and custodians. But all of these are, of course, exactly what is required for a grand challenge – the collaboration of many disciplines, agencies, and communities making progress towards a commonly held vision.

DE is presented by Gore as an educational tool, a way for a younger generation to acquire knowledge of the planet, and particularly of its environmental problems and ways in which they might be solved. Another major benefit of DE would lie in its ability to serve as an experimental environment, allowing planners to evaluate the consequences of management and development alternatives. One can imagine evaluating the consequences of a steady rise in atmospheric CO_2 using DE, as an alternative to the infinitely costly and dangerous experiment that humanity is currently conducting on the real thing (for a Japanese effort along these lines see http://www.nec.com/global/features/index9/index.html). GIS is currently used for this purpose, of course, but only over much smaller domains. Again, these are the kinds of massive benefits one would expect from the solution of a grand challenge.

CONCLUSIONS

In the previous sections I have implied numerous criteria for grand challenges, some explicitly and some implicitly. In this final section I will review these and examine the extent to which they are satisfied by the various proposals.

First, a grand challenge should focus many apparently disparate forms of research on a common goal. DE clearly satisfies this test, since it involves researchable issues that are both technical and institutional, and which touch on the disciplines of geography, computer science, cognitive science, and any of the many disciplines that study the processes responsible for the evolution of the Earth's surface.

Second, there are expectations regarding the magnitude of a grand challenge, whether measured in numbers of investigators, levels of funding, or numbers of papers produced. DE has already spawned several conferences, and research on many of its sub-problems continues throughout the GISc community, though often without explicit recognition of its relevance to DE. DE has many pseudonyms: Virtual Earth, Earth System, and Digital Globe all produce numerous WWW hits on DE-like projects. The Japanese Earth Simulator project alone represents an investment of several hundred million dollars.

Third, a grand challenge should capture the popular imagination, and thus political support. The results to date on this test are much less clear for DE, and also for GIS and GISc. While there is now very extensive name recognition of GIS, and its value is without question, its development clearly has not attracted the kind of widespread public attention accorded to the lunar landings, or even the mapping of the human genome. It is, quite simply, a very useful tool that we can no longer do without – and DE a very useful vision of where it might be headed in the years to come.

REFERENCES

Anselin, L. 1989. *What Is Special about Spatial Data? Alternative Perspectives on Spatial Data Analysis.* Santa Barbara, CA: National Center for Geographic Information and Analysis Technical Paper No. 89–4.

Anselin, L. 1995. Local indicators of spatial association: LISA. *Geographical Analysis* 27: 93–115.

Barnes, T. J. 2004. A paper related to everything but more related to local things. *Annals of the Association of American Geographers* 94: 278–83.

Department of the Environment. 1987. *Handling Geographic Information: The Report of the Committee of Enquiry Chaired by Lord Chorley.* London: Her Majesty's Stationery Office.

Egenhofer, M. J., Glasgow, J., Günther, O., Herring, J. R., and Peuquet, D. J. 1999. Progress in computational models for representing geographic concepts. *International Journal of Geographical Information Science* 13: 775–98.

Fotheringham, A. S., Brunsdon, C., and Charlton, M. 2002. *Geographically Weighted Regression: The Analysis of Spatially Varying Relationships.* New York: John Wiley and Sons.

Goodchild, M. F. 1992. Geographical information science. *International Journal of Geographical Information Systems* 6: 31–45.

Goodchild, M. F. 1999. Implementing Digital Earth: A research agenda. In G. Xu and Y. Chen (eds) *Towards Digital Earth: Proceedings of the International Symposium on Digital Earth, Beijing, China.* Beijing: Science Press, pp. 21–6.

Goodchild, M. F. 2000. Cartographic futures on a digital Earth. *Cartographic Perspectives* 36: 3–11.

Goodchild, M. F. 2002. Measurement-based GIS. In W. Shi, P. F. Fisher, and M. F. Goodchild (eds) *Spatial Data Quality.* New York: Taylor and Francis, pp. 5–17.

Goodchild, M. F. 2004. The validity and usefulness of laws in geographic information science and geography. *Annals of the Association of American Geographers* 94: 300–3.

Goodchild, M. F., Egenhofer, M. J., Kemp, K. K., Mark, D. M., and Sheppard, E. S. 1999. Introduction to the Varenius project. *International Journal of Geographical Information Science* 13: 731–46.

Gore, A. 1992. *Earth in the Balance: Ecology and the Human Spirit.* Boston, MA: Houghton Mifflin.

MacEachren, A. M. and Kraak, M. J. 2001. Research challenges in geo-visualization. *Cartography and Geographic Information Science* 18: 3–12.

Mandelbrot, B. B. 1982. *The Fractal Geometry of Nature.* San Francisco, CA: Freeman.

Mark, D. M. 2003. Geographic information science: Defining the field. In M. Duckham, M. F. Goodchild, and M. F. Worboys (eds) *Foundations of Geographic Information Science.* New York: Taylor and Francis, pp. 3–18.

Mark, D. M., Freksa, C., Hirtle, S. C., Lloyd, R., and Tversky, B. 1999. Cognitive models of geographical space. *International Journal of Geographical Information Science* 13: 747–74.

May, R. M. 1988. How many species are there on earth? *Science* 247: 1441–9.

Miller, H. J. 2004. Tobler's First Law and spatial analysis. *Annals of the Association of American Geographers* 94: 284–9.

NCGIA. 1989. The research plan of the National Center for Geographic Information and Analysis. *International Journal of Geographical Information Systems* 3: 117–36.

National Computational Science Alliance. 1999. *Toward Improved Geographic Information Services within a Digital Government: Report of the NSF Digital Government Initiative Geographic Information Systems Workshop.* Champaign, IL: University of Illinois at Urbana-Champaign.

NRC. 2003. *IT Roadmap to a Geospatial Future.* Washington, DC: National Academy Press.

Phillips, J. D. 2004. Doing justice to the law. *Annals of the Association of American Geographers* 94: 290–3.

Rhind, D. W. 1988. A GIS research agenda. *International Journal of Geographical Information Systems* 2: 23–8.

Sheppard, E. S., Couclelis, H., Graham, S., Harrington, J. W., and Onsrud, H. 1999. Geographies of the information society. *International Journal of Geographical Information Science* 13: 797–824.

Smith, J. M. 2004. Unlawful relations and verbal inflation. *Annals of the Association of American Geographers* 94: 294–9.

Sui, D. Z. 2004. Tobler's First Law of Geography: A big idea for a small world? *Annals of the Association of American Geographers* 94: 269–77.

Tobler, W. R. 1970. A computer movie: Simulation of population change in the Detroit region. *Economic Geography* 46: 234–40.

Tobler, W. R. 2004. On the First Law of Geography: A reply. *Annals of the Association of American Geographers* 94: 304–10.

UCGIS. 1996. Research priorities for geographic information science. *Cartography and Geographic Information Science* 23: 115–27.

Chapter 34

Geographic Information Science: Where Next?

Andreas Reuter and Alexander Zipf

The title question "Where next?" is based on a number of assumptions in (at least) three dimensions, the analysis of which will actually help in answering the question itself.

In order for the "where" to be meaningful there needs to be a frame of reference within which to determine the current position as well the target location and the route taking us from here to there. The "next" part of the question implies a temporal dimension, together with a unit speed of the clock ticks. And finally, we have to make assumptions about the momentum of the object under consideration, that is Geographic Information Science (GISc): Does the "where next?" mean we picture Geographic Information Systems (GIS) resting at some position, leisurely pondering which path to choose now, or do we assume a system in motion for which we try to extrapolate the trajectory?

Clearly, without making all those assumptions explicit, the chapter cannot possibly make much sense. On the other hand, all the choices that need to be made are subjective to a certain degree, biased by the background of the authors, and so it should not be too surprising if some readers disagree with the frame of reference chosen, others with the temporal units, and yet others with the calculation of the trajectory. But this is not a problem at all because, in order to disagree, one needs to consciously make one's own choices, define a frame of reference, etc., thereby coming up with a different prognosis – which is just as good as the authors'. And think about it: the real value of any of those technological predictions, extrapolations, forecasts, or whatever does not lie in what they say, but in the reactions they provoke in the readers' minds.

A simple approach towards answering the question would be to focus on the inner consistency of the present volume and maintain that of course GISc should strive to meet the grand challenges outlined in the previous chapter. That would keep this chapter very short indeed, but it would answer the wrong question. A grand challenge is typically defined as "a ca.15-year project with clearly defined criteria of success or failure, resulting in fundamental and radical advances in basic science or engineering." The title question, however, asks where next, and we certainly don't want to suggest that the clock of progress in GISc ticks in units of 15 years.

So our question boils down to the problem of which path to take in order to get there. Clearly, GISc is not an isolated field: on the contrary, it is closely interacting with other fast-moving disciplines such as Computer Science, and one can expect most progress in those areas where it is possible to leverage new results in other fields, provided they are in line with one's own agenda. Incidentally, a group of Computer Scientists around Tony Hoare has been discussing grand challenges for CS for some time, and it is interesting to note that at least two of the topics named are immediately relevant for our discussion. Quoting from a recent version of the list (http://www.nesc.ac.uk/esi/events/Grand_Challenges/index.html), we find (together with brief references to projects addressing those challenges):

- **Science for Global Ubiquitous Computing:** Within 20 years computers will be ubiquitous and globally connected, and academics and scientists believe they will be regarded collectively as a single Global Universal Computer (GUC). Professor Robin Milner of the University of Cambridge says the challenge is to work out who will program the GUC, who will benefit, how will they benefit, and how do we trust it.

- **Memories for Life:** In 10 to 20 years digital data and images that are unique to us will have grown substantially. These data will include digital pictures, emails, phone numbers and audio recordings. The project will seek to establish a way in which all these data can be securely stored and searched. Professor Aaron Sloman of the University of Birmingham says research will attempt to establish how someone can search all records and data stored on their system – whatever form that may take – in a secure environment, from wherever is convenient.

The first topic should be immediately obvious: Global ubiquitous computing will require deeply integrated facilities for locating devices, for navigation, and for many other types of spatial referencing.

The second topic might be less obvious, but is potentially even more interesting. Most of the data/information pertaining to a person's life has explicit (and more often implicit) spatial dependencies, which may not be important at the time of recording, but will be relevant for retrieval in various contextual settings.

So let us now define the "here" and "there" in terms of the themes appearing in the list of Grand Challenges, keeping in mind that "next" should refer to a foreseeable future rather than some 15+ years down the road. We are convinced that the "next" important developments in GISc will happen along the following three dimensions: concepts and methods, applications, and platforms. Of course, one can name additional dimensions, which might be interesting to consider, but for the current purposes, we will flatten them out and just take into account the ones mentioned. The remainder of the chapter will illustrate the anticipated developments in some detail, but let us briefly summarize what we are going to discuss:

- **Concepts and Methods:** GISc will incorporate formal models for describing deep semantics of a large variety of spatial phenomena. This will enable new levels of quality in spatial reasoning, thereby enabling common-sense modeling, which necessarily requires profound understanding of space.

- **Applications:** GISc will enable and support new applications. As an example consider the Virtual Telescope, which already has proved to be a research vehicle proper (www.cnn.com/2002/TECH/space/10/02/radio.telescope/).
- **Platforms:** The results of GISc will be delivered on and will influence the development of new platforms, such as the Grid. The NGG2-report (NGG2 2004) contains a number of scenarios illustrating the need for sophisticated spatial referencing. For instance, the idea of a mobile assistant with a large repertoire of user support functions not only requires locating the device in some coordinate system but also requires the automatic analysis of the device's (i.e. the user's) trajectory to support trip planning and scheduling – just to mention one example.

In the following discussion, we assume that all the developments under consideration will follow a trend that can be observed in many areas of IT and its supporting disciplines: Integration.

Many of the current efforts are directed at integrating information and services from different domains and contexts, such that they can be used and exploited in a uniform way. The success of XML and all its derivates is due to the strong momentum behind the idea of integration. Web services carry the idea from data to applications, and, again, integration is the name of the game. Semantic modeling in its many different guises aims at integration of information from heterogeneous sources, and even a well-established field such as databases is undergoing a transition, which some observers call a "revolution" (Gray 2004), because of the integration of data of arbitrary types into a single consolidated overall architecture.

So our main hypothesis is that GISc will, in the foreseeable future, participate in and contribute to the ongoing effort of large-scale information integration. GIS is well-positioned for that task, because it has a tradition of working towards the integration of heterogeneous data sets like raster and vector data with alphanumeric information.

Wrapping everything together in one sentence, we can say that GIS will be instrumental in integrating data and services from heterogeneous sources into a uniform architecture, using new concepts and methods, delivering these new services via the Grid, thus enabling a whole range of new applications.

Let us now consider the trends in each of the three dimensions in turn.

Concepts and Methods (Dimension 1)

Semantic geoinformation

In terms of concepts and methods, we expect the most interesting developments in approaches to modeling semantic aspects of space and spatial objects. So far, modeling has been largely restricted to mathematical formalisms and – to a certain degree – image processing techniques. However, for integrating geo objects with information from other domains and for reasoning about spatial phenomena, a deeper semantic representation of spatial relationships is needed. This is in line with observations from other fields, where the traditional specialized models are also found to be too narrow for integrating data with information from different domains. The most notable expression of this diagnosis is the development of what is summarily called the Semantic Web (Berners-Lee, Hendler, and Lassila 2001). Whereas in the current WWW one can search for data by specifying reference strings, the goal of the Semantic Web is to provide means for searching at a semantic level, that is to say for concepts rather than words, for relationships between those concepts, be they causal, temporal, or spatial (see Chapter 31 by Jones and Purves in this volume for some additional discussion of Web-based GIS capabilities).

Currently, a number of languages suited for that type of semantic modeling and querying are being developed. The latest achievement is the Web Ontology Language OWL which builds on the Resource Description Framework (RDF) (Smith, Welty, and McGuinness 2003). OWL allows the definition of *ontologies*, which are explicit formal descriptions of concepts or classes in a domain of discourse, which express a shared specification of a conceptualization. OWL thus provides the possibility of expressing and sharing information associated with people, events, devices, places, time, space, etc.

A number of current GIS issues are related to a lack of formal expression of semantics of the domain of discourse. The challenge, therefore, is to define reasonably general ontologies dealing with spatial phenomena expressed in OWL. These then need to be integrated into the current major standardization effort for GI services and data – the OGC open web services (OWS), as the technical fundamentals for Spatial Data Infrastructures (SDI). In particular, ontologies seem to be relevant in the area of Global Ubiquitous Computing (see the Applications section). This is because they are needed to model the diverse aspects of the context of a situation. Apart from that they are, of course, relevant for semantic interoperability due to the semantic heterogeneity of geographic information. This has already triggered intense research efforts (for example, Frank 1997, Winter 2000, Fonseca, Egenhofer, Agouris, and Câmara 2002). A spatial ontology primarily deals with physical space and spatial relations, but also with abstract spaces that can be mapped onto the physical primitives. Along with the research within GISc, related proposals for ontologies were developed outside the GISc field, like DAML-Space, OpenCyc, SUMO, or the Region Connection Calculus (RCC). One can also find examples of draft ontologies identifying relevant domains for integrated ubiquitous applications (for example, Chen 2004). Among others these usually include the following domains:

- Spatial objects and their relationships
- Temporal relations
- Person profiles/user models
- Events
- Device profiles
- Digital document models
- Security and privacy policies

The spatial and temporal domains have been the primary focus of GI Science. A temporal ontology generally describes time and temporal relations using "time instants" and "time intervals." Similarily Zipf and Krüger (2001) presented a conceptual model and its application to GML.

But, of course, semantic modeling might not be the miracle cure solving all the open problems in integrating data from heterogeneous sources. Ontologies are necessary to model and represent semantic knowledge. But the question of which are the stable basic concepts from which to build the whole domain such that reasoning about those concepts will be powerful enough to support future applications is still under debate. The early attempts, made some 15 to 20 years ago, under names like "expert systems" were only moderately successful and typically limited to a very narrow domain. This time, the real challenge is to model spatial semantics (and all the others) such that their integration will yield a useful formal model of what is sometimes referred to as "common knowledge," something projects like Cyc have strived to accomplish for years (Lenat, Guha, Pittman, Pratt, and Shepherd 1990).

Applications (Dimension 2)

UbiGIS: Ubiquitous GI Services

Ubiquitous Computing (UbiComp, UC) is regarded as one of the coming long-term trends in information technology. The term describes the pervasive use of computer services – the possibility of the use of computer-aided services as a ubiquity. The underlying system will contain billions of tiny, wireless, connected computing devices (mostly not computers as we know them), which are integrated into a multitude of objects of everyday life (Weiser 1991). Among others, the following items characterize Ubiquitous Computing:

- Spontaneous networks, service description, service discovery
- Wireless and mobile communication
- New man-machine interfaces and interaction paradigms
- Adaptation to context and situation, in particular localization

We want to focus on the last item in the list, which has a more fundamental relationship to GISc. In the light of UbiComp, the relationship between spatial data infrastructures (SDI) and LBS becomes obvious: both concepts support the access to GI Services at any time at any place using different clients based on an infrastructure providing open interfaces. The need for transparent access to computerized services independent of further restrictions is also one of the main objectives of UbiComp. As a broader topic behind all of this we ask the following question: "How can GISc support the ubiquitous access to and use of the wide variety of geographic information and applications in an optimal way?" So the term "ubiquitous" extends the anywhere, anytime, to anyone approach of Location Based Services (LBS) to the paradigm: the right thing at the right time the right way to the right person(s) (See Chapter 32 by Brimicombe in this volume, for more on location-based services and GIS).

Implied by the idea of an interoperable spatial data infrastructure (SDI) as well as a consequence of new developments in mobile computing and Human Computer Interaction (HCI), one can expect GI Services to be available ubiquitously to all users. For this the term Ubiquitous Geographic Information Services = Ubiquitous GIS = UbiGIS has been suggested (Zipf 2004, see also http://www.ubigis.org). This term can be defined as: *pervasive services based on UbiComp technology and devices, supporting context-dependent (that is, adaptive) interaction, realized by information and functions of geographic information services based on interoperable SDI.*

Often LBS are parameterized by coordinates in some spatial frame of reference. In the UbiComp-approach, however, a more general approach is needed, taking the *context* of the overall situation into account. Dey and Abowd (2000) characterize context as "any information that can be used to characterize the situation of an entity. An entity is a person, place, or object that is considered relevant for the interaction between a user and an application, including the user and the application themselves." Therefore any information that is available at the time of an interaction can be considered as context information.

As a simple example, consider a system that allows a user to ask questions such as: "What are the interesting objects in my vicinity?" Depending on where the user is, what he or she wants to achieve and what the general current situation (which might be modeled through a variety of application specific parameters) is, his/her definitions of both "vicinity" and "interesting object" may vary considerably from fashion shops in driving distance to prehistoric artifacts in a museum. The information about the situation and environment (the context) can be (and has been) categorized in many ways, but an agreed-upon formalization is still missing. While the first context-aware systems have presented progress there are still improvements necessary for supporting knowledge sharing and context reasoning.

The adaptability of GI Services to context can be seen as one of the next steps for GISc research in order to achieve more intuitively usable GIS. In this overview we only can present hints as to which adaptive services might be suitable within the context of GI Services. This is derived from first results regarding adaptive mobile GI Services in our projects. The following categories of adaptation have been identified:

- Adaptation of the visual presentation of the contents offered – both of the text and the graphic information (pictures, maps, video, VR models);
- Adaptation of route planning (by individual weighting and restrictions);
- Adaptation of queries (combined location- and interest-based tips); and
- Adaptation of the offered contents (for example, concerning degree of detail, topic).

In GISc, the first work on context-awareness has focused on mobile maps (Meng, Zipf, and Reichenbacher 2004; Zipf 1998, 2002), navigation support (Kray 2002) or wayfinding with landmarks (Winter, Raubal, and Nothegger 2004) as well as space-time accessibility (for example, Raubal, Miller, and Bridwell 2004). Further examples for adaptive GI applications include the computation of routes based on context-related criteria (Joest and Stille 2003) or a user-aware spatial push of information (Zipf and Aras 2002).

But apart from the possibility of using context as a parameter for adapting GI Services, there is one even more important aspect to context that makes it important for the GISc community to work on this issue in more detail: context parameters are related to space. This is, in fact, sort of a corollary to Tobler's First Law of Geography (Tobler 1970). In more detail the following principles apply to context (Schmidt 2002):

- Context has an origin location;
- The relevance of the context reaches its maximum at this origin;
- The relevance of the context decreases with distance from that origin;
- Exceeding a certain distance the context is no longer relevant; and
- If there are multiple identical sensors (for sensing context information) available, the one which is spatially most close has the highest relevance.

This relationship can for example be modeled by fuzzy functions. Generally this is not stationary, but can move (for example, with the user) and may be distorted in a direction, for instance where the user walks to. Furthermore there is usually a large range of context factors present which overlap in space. All of these principles apply not only to space, but also to time, resulting in an interesting spatio-temporal modeling task.

Platforms (Dimension 3)

Spatial data infrastructures (SDI)

The current main effort regarding the integration of GI Services and data is the development of spatial data infrastructures (SDI). From a simplified technical point of view, SDI can be regarded as the provision of distributed GI Services and geo-data by means of web services using open standards. On the other hand, mobile GIS offer GI functionality on handheld computers (for example, for mapping, data acquisition, or infrastructure maintenance, etc.) dependent from their location. This requires an infrastructure of GI Services and wireless access to geodata, which is based on open interfaces – in other words a wireless SDI. We cannot dig deeper into those technical issues, but one can see that National Spatial Data Infrastructures (NSDI) are currently evolving in all parts of the world. This should lead to a new quality of improved access to spatial data in the mid-range future.

Integrating ubiquitous positioning facilities and location models

Positioning is a basic functionality for LBS and ubiquitous services, with indoor positioning being as important as global positioning. Considerable work has been conducted in mobile computing, leading to location models that are not based on the earth-bound coordinate systems we are accustomed to in GIS. So GISc needs to provide mappings between emerging ego-centric or object-centric location models based, for example, on network topology (for example, Beigl, et al. 2001) as depicted in Figure 34.1 and conventional geographic or geodetic spatial reference systems. Applying these to intelligent positioning and navigation support, for example, would combine several approaches that mutually improve or replace themselves in case of partial failure (Kray 2003).

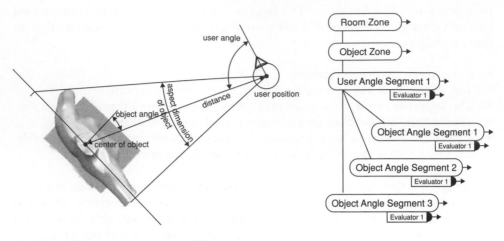

Fig. 34.1   Object-centric location models
Goßmann and Specht 2002

Networks of location-aware sensors

Classically, sensors in the context of GI comprise those built for measuring environmental data (remote sensing being the most prominent example), but recently networks of stationary or mobile sensors for heterogeneous parameters are coming into focus. Therefore researchers are investigating the interoperable integration of sensors through the OGC Sensor Web Enablement initiative. Apart from this classical view, location-aware sensors for whatever parameters constitute important elements for both context awareness and personalization within UbiComp. Through the availability of such new sensors the development of adaptive applications becomes possible. Such a location-aware interoperable sensor infrastructure is also important for non-GI applications, because of the relation between space and context as outlined above. Such sensors are being equipped with basic processing and communication capabilities, for instance in projects like "Smart Dust" (Kahn, Katz, and Pister 1999) or "smart-its" (http://www.smart-its.org), leading to a georeferenced distributed global computing platform. When we envisage such large future networks of sensors delivering masses of data all the time (again remote sensing serves as a primary example) we are faced with the problem of mining these huge amounts of distributed data sources for interesting patterns. This requires new middleware concepts, which leads us to that the idea of Grid computing (see Chapter 19 by Miller in this volume for more detailed discussion of recent work on geographic data mining and knowledge discovery).

The GeoGRID: Towards geospatial GRID computing

GRID Computing (GC) is a new concept for distributed high-performance computing through a coordinated use of geographically distributed large virtual collections of computation resources realized through use of computer clusters. This should not be confused with the term "Grid" in the sense of data structure for raster data, as is frequently used in GISc. GC applications utilize high-speed networks and a new generation of middleware linking networks, computing resources, and traditional geospatial applications. Tasks of this middleware include security and resource management for example. As the concept of GC also supports the aim of making high-performance computer-processing power available ubiquitously, there is a clear relationship to Ubiquitous Computing, which is also highlighted in the Next Generation Grids Report (NGG2 2004). Through the distribution of data, software, and computing resources we find three aspects of GC related to space. In order to realize a Grid the use of standards and open protocols and interfaces is necessary, which builds a bridge to the already mentioned activities of both the OGC and the development of SDI. Again we find our topic of integration, as these worlds need to come together to realize the vision of the Global Ubiquitous Computer that could be used in a range of tasks. A scenario in which all of this can easily be integrated would be the case of support for disaster management, that really brings together the vision of Ubiquitous Computing and Grid computing (NGG2 2004, Zipf 2004) with several clear relationships to GISc and therefore realizes an ideal scenario for UbiGIS. The first actual examples of GI applications using GC technology are the Globus Toolkit middleware for spatial interpolation or watershed modeling on raster data (Wang, Armstrong, and Bennett 2002).

In order for the Grid to achieve its integrative goal, more is needed than just a set of middleware standards. As is pointed out in Gray (2003), seamless integration will also require the nucleus of a "global information schema," which can unify the following categories: (1) units, (2) accuracy, (3) precision, (4) definition of quantities, (5) representation, and (6) semantics. Those unifying means have to be cross-domain, and they have to support automatic schema generation and translation. And again, since many data have a spatial connotation, GIS will have to contribute to building this global information schema.

SUMMARY

The above discussion has focused on various projects and ideas related to the three dimensions of progress we had identified in the beginning. But one could ask, "Assuming that integration is indeed the prevailing direction of innovation, what could be the result of integrating GI services into other platforms?"

That's a hard question; but let us try to come up with an answer anyway. Consider today's mobile phones. They provide you with an infrastructure for communicating from (almost) any location; the contents can be voice, images, and data. It would be extremely useful to augment this technology with location and navigation services in all conceivable contexts and frames of reference. Of course, each of those contexts refers to a different type of application, but there needs to be an enabling platform that supports them all and allows for seamlessly switching from one to the next. Providing the principles and specifications for such a platform and integrating it into the existing framework of mobile computing is definitely an attractive and attainable goal.

Besides technical questions, a range of social issues need to be examined that we could not examine here concerning acceptance, social consequences, data security, and privacy (for example, Dobson and Fisher 2003; but see also Cho, in Chapter 29 of this volume, for an extended discussion of the current ways in which privacy issues are being handled in various countries). Hence, it is often claimed that the acquisition of personal and context information needs to be open to the end user, which is, in fact, seldom realized. This means that the user should always have full knowledge and control over the data that is being collected and that only authorized persons and systems can access these.

ACKNOWLEDGEMENTS

We should like to thank all our colleagues at EML for the fruitful discussions that helped to shape the ideas expressed here.

REFERENCES

Berners-Lee, T., Hendler, J., and Lassila, O. 2001. The semantic web. *Scientific American* 284(5): 34–45.

Beigl, M., Gellersen, H.-W., and Schmidt, A. 2001. Mediacups: Experience with design and use of computer-augmented everyday artifacts. *Computer Networks* 35: 401–9.

Beigl, M., Zimmer, T., and Decker, C. 2002. A location model for communicating and processing of context. *Personal and Ubiquitous Computing* 6: 341–57.

Britt, R. R. 2004. Black Holes Multiply in New Observations. WWW document, http://www.cnn.com/2004/TECH/space/05/31/black.holes/index.html.

Chen, H. 2004. SOUPA: Standard Ontology for Ubiquitous and Pervasive Applications. WWW document, http://pervasive.semanticweb.org/.

Dey, A. K. and Abowd, G. D. 2000. Towards a better understanding of context and context awareness. In *Proceedings of the CHI 2000 Workshop on the What, Who, Where, When, and How of ContextAwareness*, The Hague, Netherlands. New York: Association of Computing Machinery.

Dobson, J. E. and Fisher, P. F. 2003. Geoslavery. *IEEE Technology and Society Magazine* 22: 47–52.

Frank, A. U. 1997. Spatial ontology: A geographical information point of view. In O. Stock (ed.) *Spatial and Temporal Reasoning*. Dordrecht: Kluwer, pp. 135–53.

Fonseca, F., Egenhofer, M., Agouris, P., and Câmara, G. 2002. Using ontologies for integrated geographic information systems. *Transactions in GIS* 6: 231–57.

Goßmann, J. and Specht, M. 2002. Location models for augmented environments. *Personal and Ubiquitous Computing* 6: 334–40.

Gray, J. 2003. Making grid computing ubiquitous. Unpublished talk presented at Globus World Conference, January 15, San Diego, CA, USA (available at http://research.microsoft.com/~Gray/talks/Open%20Issues%20in%20Grid.ppt).

Gray, J. 2004. The next database revolution. In *Proceedings of the ACM SIGMOD 2004 Conference on Management of Data*, Paris, France. New York: Association of Computing Machinery, pp. 1–4.

Joest, M. and Stille, W. 2002. A user-aware tour proposal framework using a hybrid optimization approach. In *Proceedings of the Tenth ACM International Symposium on Advances in Geographic Information Systems*, McLean, VA, USA. New York: Association of Computing Machinery.

Kahn, J. M., Katz, R. H., and Pister, K. S. J. 1999. Next century challenges: Mobile networking for "smart dust." In *Proceedings of the Fifth Annual ACM/IEEE International Conference on Mobile Computing and Networking (MobiCom '99)*, Seattle, WA, USA. New York: Association of Computing Machinery, pp. 271–8.

Kray, C. 2003. *Situated Interaction on Spatial Topics*. Berlin: AKA Verlag DISKI Series No. 274.

Kray, C., Baus, J., and Krüger, A. 2002. Position information and navigation assistance. In J. Strobl and A. Zipf (eds) Mobile Geoinformation. Heidelberg, Hüthig Verlag, pp. 98–108 (in German).

Lenat, D. B., Guha, R. V., Pittman, K., Pratt, D., and Shepherd, M. 1990. Cyc: Toward programs with common sense. *Communications of the ACM* 33: 30–49.

Meng, L., Zipf, A., and Reichenbacher, T. (eds). 2004. *Map-based Mobile Services: Theories, Methods and Implementations*. Heidelberg: Springer-Verlag.

NGG2. 2004. Next Generation Grids Report: Final Report by EU Expert Group on Next Generation Grids. WWW document, http://www.cordis.lu/.

Raubal, M., Miller, H., and Bridwell, S. 2004. User centered time geography for location-based services. *Geografiska Annaler B* 86: 245–65.

Schmidt, A. 2002. Ubiquitous Computing: Computing in Context. Unpublished PhD Dissertation, Department of Computer Science, Lancaster University.

Smith, M. K., Welty, C., and McGuinness, D. 2003. Owl Web Ontology Language Guide. WWW document, http://www.w3.org/TR/owl-guide/.

Tobler, W. R. 1970. A computer movie: Simulation of population change in the Detroit region. *Economic Geography* 46: 234–40.

Wang, S., Armstrong, M. P., and Bennett, D. A. 2002. Conceptual basics of middleware design to support grid computing of geographic information. In *Proceedings of the Second International Conference on Geographic Information Science*, Boulder, CO, USA. Santa Barbara, CA: National Center for Geographic Information and Analysis.

Weiser, M. 1991. The computer for the twenty-first century. *Scientific American* 265(3): 94–104.

Winter, S. (ed.). 2000. *Geographical Domain and Geographical Information Systems*. Vienna, Institute for Geoinformation.

Winter, S., Raubal, M., and Nothegger, C. 2004. Focalizing measures of salience for route directions. In A. Zipf, L. Meng, and T. Reichenbacher (eds). *Map-based Mobile Services: Theories, Methods, and Implementations*. Berlin: Springer, pp. 127–42.

Zipf, A. 1998. DEEP MAP: A prototype context sensitive tourism information system for the city of Heidelberg. In *Proceedings of GIS Planet 98*, Lisbon, Portugal.

Zipf. A. 2002. User-Adaptive Maps for Location-Based Services (LBS) for Tourism. In K. Woeber, A. Frew, M. Hitz (eds) *Information and Communication Technologies in Tourism 2002*. Heidelberg: Springer, pp. 329–38.

Zipf, A. 2004. Mobile Anwendungen auf Basis von Geodateninfrastrukturen: von LBS zu UbiGIS. In L. Bernard, J. Fitzke, and R. Wagner (eds) *Geodateninfrastrukturen*. Heidelberg: Wichmann Verlag, pp. 225–34.

Zipf, A. and Aras, H. 2002. Proactive exploitation of the spatial context in LBS through interoperable integration of GIS services with a multi agent system (MAS). In *Proceedings of the International Conference on Geographic Information Science of the Association of Geographic Information Laboratories in Europe*, Palma, Spain. Wageningen: Association of Geographic Information Laboratories in Europe.

Zipf, A. and Krüger, S. 2001. TGML – Extending GML by temporal constructs: A proposal for a spatiotemporal framework in XML. In *Proceedings of the Ninth ACM International Symposium on Advances in Geographic Information Systems*, Atlanta, GA, USA. New York: Association of Computing Machinery, pp. 94–9.

Index